Lecture Notes in Mathematics

Edited by A. Dold and B. Eckmann

1019

Cabal Seminar 79–81

Proceedings, Caltech-UCLA Logic Seminar 1979–81

Edited by A. S. Kechris, D. A. Martin and Y. N. Moschovakis

Springer-Verlag
Berlin Heidelberg New York Tokyo 1983

Editors

Alexander S. Kechris
Department of Mathematics, California Institute of Technology
Pasadena, California 91125, USA

Donald A. Martin
Yiannis N. Moschovakis
Department of Mathematics, University of California
Los Angeles, California 90024, USA

AMS Subject Classifikations (1980): 03EXX, 03DXX, 04-XX

ISBN 3-540-12688-0 Springer-Verlag Berlin Heidelberg New York Tokyo
ISBN 0-387-12688-0 Springer-Verlag New York Heidelberg Berlin Tokyo

Printed in Germany

Printing and binding: Beltz Offsetdruck, Hemsbach/Bergstr.
2146/3140-543210

الثالثه ثابته

- Arabic proverb

INTRODUCTION

This is the third volume of the proceedings of the Caltech-UCLA Logic Seminar, based essentially on material presented and discussed in the period 1979-1981. The last paper "Introduction to Q-theory" includes some very recent work, but it also gives the first exposition in print of some results going back to 1972.

Papers 5-10 form a unit and deal primarily with the question of the extent of definable scales.

Los Angeles
May 1983

Alexander S. Kechris
Donald A. Martin
Yiannis N. Moschovakis

TABLE OF CONTENTS

MORE SATURATED IDEALS

Matthew Foreman
Department of Mathematics
University of California
Los Angeles, California 90024

In this paper we prove three theorems relating the consistency strengths of huge cardinals with saturated ideals on regular cardinals and with model theoretic transfer properties.

We prove:

Theorem. Con (ZFC + there is a 2-huge cardinal) $\Rightarrow$ Con (ZFC + for all $m, n \in \omega$ with $m > n$, $(\aleph_{m+1}, \aleph_m) \twoheadrightarrow (\aleph_{n+1}, \aleph_n)$).

Theorem. Con (ZFC + there is a huge cardinal) $\Rightarrow$ Con (ZFC + for all $n \in \omega$, there is a normal, $\aleph_n$-complete, $\aleph_{n+1}$-saturated ideal on $\aleph_n$ + there is a normal, $\aleph_{\omega+1}$ - complete, $\aleph_{\omega+2}$-saturated ideal on $\aleph_{\omega+1}$).

The theorem above contains all the new ideas necessary to prove the following theorem:

Theorem. Con (ZFC + there is a huge cardinal) $\Rightarrow$ Con (ZFC + every regular cardinal κ carries a κ^+-saturated ideal).

We now make some definitions: Let $\mathcal{L}$ be a countable language with a unary predicate U. A $\mathcal{L}$-structure $\mathfrak{A}$ is said to have type (κ,λ) iff $|\mathfrak{A}| = \kappa$ and $|U^{\mathfrak{A}}| = \lambda$. If $\kappa \geq \kappa'$, $\lambda \geq \lambda'$ we say that $(\kappa,\lambda) \twoheadrightarrow (\kappa',\lambda')$ iff every structure $\mathfrak{A}$ of type (κ,λ) has an elementary substructure $\mathfrak{B} \prec \mathfrak{A}$ of type (κ',λ').

An ideal $\mathcal{J} \subseteq \mathcal{P}$ is said to be α-complete iff whenever $\{X_\gamma : \gamma < \beta\} \subseteq \mathcal{J}$ and $\beta < \alpha$, $\bigcup_{\gamma<\beta} X_\alpha \in \mathcal{J}$. A set $A \subseteq \mathcal{P}(\kappa)$ is said to be positive if $A \notin \mathcal{J}$. $\mathcal{J}$ is normal iff for every positive set $A \subseteq \kappa$ and every regressive function f defined on A there is a $\beta \in \kappa$ such that $\{\alpha : F(\alpha) = \beta\}$ is positive. $\mathcal{J}$ is said to be λ-saturated iff it is normal and $\mathcal{P}(\kappa)/\mathcal{J}$ has the λ-chain condition. (We will never consider "non-normal" saturated ideals.) There is an extensive literature on saturated ideals. (See [6], [7].)

Let $j : V \to M$ be an elementary embedding from V into a transitive class M. Let κ_0 be the critical point of j (i.e. the first ordinal moved by j), Let $\kappa_{i+1} = j(\kappa_i)$. We will call j an n-huge embedding and κ_0 an n-huge cardinal iff M is closed under κ_n-sequences. (This means that if $\langle x_\alpha : \alpha < \kappa_n \rangle \subseteq M$ then

---

This research was partially supported by NSF Grant No. MCS 78-02989.

$\langle x_\alpha : \alpha < \kappa_n\rangle \in M$.) An almost n-huge embedding is an embedding j as above that is closed under $< \kappa_n$-sequences, i.e. if $\beta < \kappa_n$ and $\langle x_\alpha : \alpha < \beta\rangle \subseteq M$ then $\langle x_\alpha : \alpha < \beta\rangle \in M$. We will use the notation "crit(j)" for the critical point of j.

If $\kappa_0 \leq \lambda < \kappa_1$ and j is at least an almost huge embedding, then j induces a supercompact measure on $\mathcal{P}_\kappa(\lambda)$. In particular j induces a normal measure on κ.

Proposition 1. Let j be an n-huge embedding. Let U be the measure on κ induced by j. Then there is a set A of measure one for U such that for all α and $\beta \in A$, there is an almost n-huge embedding $j_{\alpha,\beta}$ such that the critical point of $j_{\alpha,\beta}$ is α and $j_{\alpha,\beta}(\alpha) = \beta$.

Proof. This is a routine reflection argument.

We will now precisely state the theorems we will prove:

Theorem 1. Con (ZFC + there is a sequence of huge embeddings $\langle j_n : n \in \omega\rangle$ with j_n(critical point of j_n) = critical point of j_{n+1}) $\Rightarrow$ Con (ZFC + for all $m, n \in \omega$, $m > n$ implies $(\aleph_{m+1},\aleph_m) \twoheadrightarrow (\aleph_{n+1},\aleph_n)$).

Theorem 2. Con (ZFC + there is a sequence of almost huge embeddings $\langle j_n : n \in \omega\rangle$ with j_n(critical point of j_n) = critical point of j_{n+1}) $\Rightarrow$ Con (ZFC + for all $n \in \omega$ there is a normal, $\aleph_n$-complete, $\aleph_{n+1}$-saturated ideal on $\aleph_n$).

Theorem 3. Con (ZFC + there is a huge cardinal) $\Rightarrow$ Con (ZFC + $\aleph_{\omega+1}$ carries a normal $\aleph_{\omega+2}$-saturated ideal and for all $n \in \omega$, $\aleph_n$ carries an $\aleph_{n+1}$-saturated ideal.)

We will assume that the reader is familiar with iterated forcing. (See [1] for a very good exposition.) All of our partial orderings $\mathbb{P}$ will have a unique greatest element, $1_{\mathbb{P}}$. Our notion of "support" will be the standard one and if p is a condition in an iteration we will write "supp p" for its support. For the inverse limit of a system $\langle P_i : i \in I\rangle$ we will write $\varprojlim \langle P_i : i \in I\rangle$. We will also use the notion of support for products of partial orderings. If $\langle Q_i : i \in I\rangle$ is a collection of partial orderings, we let $\prod\langle Q_i : i \in I\rangle = \{f \mid f$ is a function with domain I and for all $i \in I$ $f(i) \in Q_i\}$. The product $\prod\langle Q_i : i \in I\rangle$ is ordered coordinatewise. If $p \in \prod\langle Q_i : i \in I\rangle$, then $\text{supp } p = \{i : p(i) \neq 1_{Q_i}\}$. If $\mathcal{K} \subseteq \mathcal{P}(I)$ is an ideal then $\prod_{\text{supp } p \in \mathcal{K}} \langle Q_i : i \in I\rangle = \{f \in \prod\langle Q_i : i \in I\rangle \mid \text{supp } f \in \mathcal{K}\}$.

If $\langle Q_i : i \in \omega\rangle$ is a sequence of terms such that $Q_{i+1} \in V^{Q_0 * Q_1 * \cdots * Q_i}$, we write $\overset{n}{\underset{i=0}{*}} Q_i$ for the finite iteration $Q_0 * Q_1 * \cdots * Q_n$. If S is a partial ordering with a uniform definition, we will use $S^{\mathbb{P}}$ to denote the partial ordering S defined in $V^{\mathbb{P}}$. To simplify notation, we will write $\mathbb{P} * \overline{S}$ to mean $\mathbb{P} * S^{\mathbb{P}}$.

If $\mathbb{P}$ is a partial ordering we will use $\mathcal{B}(\mathbb{P})$ to denote the canonical complete boolean algebra obtained from $\mathbb{P}$. If $\varphi(\dot{t}_1,\ldots,\dot{t}_n)$ is a formula in the forcing language of $\mathbb{P}$ we will use $\|\varphi(\dot{t}_1,\ldots,\dot{t}_n)\|_{\mathbb{P}}$ to denote the boolean value of $\varphi(\dot{t}_1,\ldots,\dot{t}_n)$ in $\mathcal{B}(\mathbb{P})$. For $p, q \in \mathbb{P}$, we will use the notation $p \wedge q$, and $p \vee q$ to be the meet and join of p and q in $\mathcal{B}(\mathbb{P})$. Similarly $\neg p$ will denote the complement of p in $\mathcal{B}(\mathbb{P})$. If $b \in \mathcal{B}(\mathbb{P})$, we will say that p "decides" b (in symbols $p \parallel b$) iff $p \leq b$ or $p \leq \neg b$. We will write $p \Vdash b$ if $p \leq b$.

We define $C(\kappa,\gamma) = \langle\{p \mid p : \kappa \to \gamma,\ |p| < \kappa\},\ \supseteq\rangle$. $C(\kappa,\gamma)$ is the partial ordering appropriate for making γ have cardinality κ. We will call this the Levy collapse. Similarly, we will define the Silver collapse $S(\kappa,\lambda)$ by $p \in S(\kappa,\lambda)$ iff

(a) $p : \lambda \times \kappa \to \lambda$

(b) $|p| \leq \kappa$

(c) there is a $\xi < \kappa$, $\operatorname{dom} p \subseteq \lambda \times \xi$

(d) for all $\alpha < \kappa$, $\gamma < \lambda$, $P(\lambda,\alpha) < \gamma$

$S(\kappa,\lambda)$ is ordered by reverse inclusion: Standard arguments show that for inaccessible λ, $S(\kappa,\lambda)$ makes λ into κ^+.

If $\gamma' < \gamma$, and $\kappa < \gamma'$, $C(\kappa,\gamma')$ is a subset of $C(\kappa,\gamma)$. If $p \in C(\kappa,\gamma)$, we define $p \cap C(\kappa,\gamma')$ to be q, where $\operatorname{dom} q = \{\alpha < \kappa : P(\alpha) < \gamma'\}$ and for each $\alpha \in \operatorname{dom} q$, $q(\alpha) = p(\alpha)$. It is easy to see that $p \cap C(\kappa,\gamma') \in C(\kappa,\gamma')$. If $r \in C(\kappa,\gamma')$ and $p \in C(\kappa,\gamma)$ then r is compatible with p in $C(\kappa,\gamma)$ iff r is compatible with $p \cap C(\kappa,\gamma')$. If $p \in C(\kappa,\gamma)$, we define $\sup p$ to be $\sup\{p(\alpha) + 1 : \alpha \in \operatorname{dom} p\}$.

We will write $\vec{\alpha}$ for a finite sequence of ordinals. If $\alpha_0,\alpha_1,\ldots,\alpha_n$ are mentioned in connection with $\vec{\alpha}$ we will assume that $\alpha = (\alpha_0,\alpha_1,\ldots,\alpha_n)$.

If κ and λ are cardinals, with $\kappa < \lambda$ and $x, y \in \mathcal{P}_\kappa(\lambda) = \{z \subseteq \lambda : |z| < \kappa\}$ then we write $x < y$ iff $x \subseteq y$ and the order type of x is less than the order type of $y \cap \kappa$. If $x \in \mathcal{P}_\kappa(\lambda)$, let $\operatorname{crit}(x)$ = order type of $x \cap \kappa$.

If $\mathcal{B}$ and $\mathcal{C}$ are complete Boolean algebras and $\pi : \mathcal{B} \to \mathcal{C}$ is an order preserving function, π is called a projection map if whenever $G \subseteq \mathcal{B}$ is generic, $\pi''G \subseteq \mathcal{C}$ is generic.

Let $\mathbb{P}$ and $\mathbb{Q}$ be partial orderings. If $i : \mathbb{P} \to \mathbb{Q}$ is a one to one, order and incompatibility preserving function with the property that whenever $A \subseteq \mathbb{P}$ is a maximal antichain, $i''A \subseteq \mathbb{Q}$ is a maximal antichain, then we say that i is a <u>neat embedding</u> from $\mathbb{P}$ into $\mathbb{Q}$, and write $i : \mathbb{P} \hookrightarrow \mathbb{Q}$). Standard theory says that if $i : \mathbb{P} \hookrightarrow \mathbb{Q}$ is a neat embedding then there is a projection map $\pi : \mathcal{B}(\mathbb{Q}) \to \mathcal{B}(\mathbb{P})$ such that for all $p \in \mathbb{P}$, $\pi(i(p)) = P$.

If $e : \mathbb{Q} \hookrightarrow \mathbb{P}$ and $S \in V^{\mathbb{Q}}$ is a partial ordering we define $\mathbb{P} *_e S$ to be the iteration with amalgam $e''\mathbb{Q}$. (See [2] for this definition.)

If $\mathbb{P}$ is a partial ordering and $p \in \mathbb{P}$, then $\mathbb{P}/p$ is defined to be $\{q : q \leq p\}$.

§1. In this section we prove Theorems 1 and 2. We begin by mentioning theorems of Magidor and Kunen that we will use extensively in this paper.

Theorem (Kunen [7]) Let $j : V \to M$ be a huge embedding. Suppose $\mathbb{P}' * Q$ is a forcing notion such that

(a) $\mathbb{P}'$ is κ_0 c. c.

(b) Q is κ_0-closed in $V^{\mathbb{P}'}$

(c) $|\mathbb{P}'| = \kappa_0$, $|\mathbb{P}' * Q| = \kappa_1$

(d) there is a projection map $\pi : \mathcal{B}(j(\mathbb{P}')) \to \mathcal{B}(\mathbb{P}' * Q)$ and a $q \in j(\mathbb{P}') * j(Q)$ such that for all $r \in j(\mathbb{P}') * j(Q)$, $r \leq q$ implies $j(\pi(r)) \geq r$ [If $r = (r_0, r_1)$ with $r_0 \in j(\mathbb{P}')$, we take $\pi(r) = \pi(r_0)$]

then in $V^{\mathbb{P}' * Q}$:

(i) κ_0 carries a normal, κ_0-complete, $j(\kappa_0)$-saturated ideal

(ii) $(j(\kappa_0), \kappa_0) \twoheadrightarrow (\kappa_0, < \kappa_0)$

Magidor and others have commented that to get a normal κ_0-complete, $j(\kappa_0)$-saturated ideal on κ_0, it is enough to have j be almost huge and replace (d) by:

(d') there is a neat embedding $i : \mathcal{B}(\mathbb{P}' * Q) \to \mathcal{B}(j(\mathbb{P}'))$.

We shall need the following preservation lemmas.

Lemma 1. Let λ be a regular cardinal. Suppose $\mathbb{P}$ is λ c. c. and Q is λ-closed. (We do not rule out $\mathbb{P} = \{\emptyset\}$.)

(a) If $\kappa < \lambda$ and κ carries a normal λ-saturated ideal in $V^{\mathbb{P}}$, then κ carries a normal λ-saturated ideal in $V^{\mathbb{P} \times Q}$.

(b) If $\lambda' < \lambda$, $\gamma \leq \kappa < \lambda$ and $(\lambda, \lambda') \twoheadrightarrow (\kappa, \gamma)$ in V, then $(\lambda, \lambda') \twoheadrightarrow (\kappa, \gamma)$ in V^Q.

Proof. Since $\mathbb{P}$ is λ c. c., Q is (λ, ∞)-distributive in $V^{\mathbb{P}}$. Thus, if $\mathcal{J}$ is a normal λ-saturated ideal on κ in $V^{\mathbb{P}}$, $\mathcal{J}$ remains a normal ideal in $V^{\mathbb{P} \times Q}$. We must show that $\mathcal{J}$ remains λ-saturated in $(V^{\mathbb{P}})^Q$. Suppose not. Let $\langle \sigma_\alpha : \alpha < \lambda \rangle$ be terms in $V^{\mathbb{P} \times Q}$ for an antichain. We will build a descending sequence $\langle q_\alpha : \alpha < \lambda \rangle \subseteq Q$ in V with the property that there is a term $\tau_\alpha \in V^{\mathbb{P}}$ such that $\| q_\alpha \Vdash_Q \sigma_\alpha = (\tau_\alpha)^{V^{\mathbb{P}}} \|_{\mathbb{P}} = 1$. We do this by induction on α. Let $q_{-1} = 1_Q$.

Assume we have defined $\{q_\beta : \beta < \alpha\}$. Let $q_\alpha^{-1} \leq \{q_\beta : \beta < \alpha\}$. There is such a q_α^{-1} because $\alpha < \lambda$ and Q is λ-closed. We will simultaneously build $\{q_\alpha^\gamma : \gamma < \gamma_\alpha\} \subseteq Q$ and an antichain $\{p_\gamma : \gamma < \gamma_\alpha\} \subseteq \mathbb{P}$. These will have the property that for each γ there is a term $\tau_\alpha^\gamma \in V^{\mathbb{P}}$ such that $(p_\gamma, q_\alpha) \Vdash_{\mathbb{P} \times Q} \sigma_a = (\tau_\alpha^\gamma)^{V^{\mathbb{P}}}$. Assume we have built $\langle q_\alpha^\gamma : \gamma < \xi \rangle$ and $\langle p_\gamma : \gamma < \xi \rangle$ with the above property. Note that $\xi < \lambda$ since $\mathbb{P}$ has the λ-chain condition. If $\langle p_\gamma : \gamma < \xi \rangle$ is not a maximal antichain, pick p_ξ, q_α^ξ such that $q_\alpha^\xi \leq \langle q_\alpha^\gamma : \gamma < \xi \rangle$ and for some $\tau_\alpha^\xi \in V^{\mathbb{P}}$, $(p_\xi, q_\alpha^\xi) \Vdash_{\mathbb{P} \times Q} (\tau_\alpha^\xi)^{V^{\mathbb{P}}} = \sigma_\alpha$ and p_ξ

is incompatible with each p_γ, $\gamma < \xi$.

Since $\mathbb{P}$ has the λ c. c., for some $\xi < \lambda$, $\langle p_\gamma : \gamma < \xi\rangle$ is a maximal antichain. Let $q_\gamma \leq \langle q_\alpha^\gamma : \gamma < \xi\rangle$ and $\tau_\alpha \in V^{\mathbb{P}}$ be such that $p_\gamma \Vdash \tau_\alpha^\gamma = \tau_\alpha$. Then $\|q_\alpha \Vdash_{Q} (\tau_\alpha)^{V^{\mathbb{P}}} = \sigma_\alpha\|_{\mathbb{P}} = 1$.

In $V^{\mathbb{P}}$, $\langle q_\alpha : \alpha < \gamma\rangle$ identifies a sequence $\langle \tau_\alpha : \alpha < \lambda\rangle \subseteq (\mathcal{P}(\kappa))^{V^{\mathbb{P}}}$. If $\alpha < \beta < \lambda$, then $q_\beta \Vdash_{Q} \tau_\alpha \cap \tau_\beta \in \mathcal{J}$ and $\tau_\alpha, \tau_\beta \notin \mathcal{J}$. But "$\tau_\alpha \cap \tau_\beta \in \mathcal{J}$" and "$\tau_\alpha \notin \mathcal{J}$" is absolute between $V^{\mathbb{P}}$ and $V^{\mathbb{P}\times Q}$. Hence in $V^{\mathbb{P}}$, $\langle \tau_\alpha : \alpha < \lambda\rangle$ is an antichain of size λ in $\mathcal{P}(\kappa)/\mathcal{J}$, a contradiction.

(b) Let $\mathfrak{A} = \langle \lambda; \lambda', f_i\rangle_{i\in\omega}$ be a fully skolemized structure of type (λ, λ') in V^{Q}. Using the λ-closure of Q, in V we can find a sequence $\langle p_\alpha : \alpha < \lambda\rangle \subseteq Q$ and a structure $\mathfrak{A}' = \langle \lambda; \lambda', f_i\rangle_{i\in\omega}$ such that f_i' are defined on all of λ and for each $\vec{\beta} \in \lambda^\omega$ and each i there is an $\alpha < \lambda$ $p_\alpha \Vdash f_i(\vec{\beta}) = f_i'(\vec{\beta})$. Let $\mathfrak{B} < \mathfrak{A}'$ be of type (κ, γ). Since $\kappa < \lambda$ and λ is regular there is an α such that for all $\beta \in \mathfrak{B}$, $p_\alpha \Vdash f_i(\vec{\beta}) = f'(\vec{\beta})$. But then $p_\alpha \Vdash \mathfrak{B}$ is closed under all of the f_i's.

Hence $p_\alpha \Vdash \mathfrak{B} < \mathfrak{A}$.

A partial ordering $\mathbb{P}$ will be called κ-centered iff there is a collection $\{C_\alpha : \alpha < \kappa\}$ of disjoint subsets of $\mathbb{P}$ such that

(a) $\mathbb{P} = \bigcup_{\alpha<\kappa} C_\alpha$

(b) If $p_1, \ldots, p_n \in C_\alpha$ then $\bigwedge_{i=1}^{n} p_i \neq 0$.

In particular, if $|\mathbb{P}| \leq \kappa$, $\mathbb{P}$ is κ-centered.

<u>Lemma 2</u>. Let $\mathcal{J}$ be a κ^{++}-saturated ideal on κ^+. Suppose $\mathbb{P}$ is a κ-centered partial ordering. Then $\mathcal{J}$ generates a normal κ^+-complete, κ^{++}-saturated ideal in $V^{\mathbb{P}}$.

(Laver has proven stronger results unknown to the author at the time this work was done. See [8].)

<u>Proof</u>. $\mathbb{P}$ is manifestly κ^+-c. c. Hence $\mathcal{J}$ remains κ^+-complete and normal. We need to see that $\mathcal{J}$ is κ^{++}-saturated. Let $\langle \dot{X}_\alpha : \alpha < \kappa^{++}\rangle$ be a term for an antichain of size κ^{++}. Let $\dot{Y}_{\alpha,\beta} = \dot{X}_\alpha \cap \dot{X}_\beta$. Then $\dot{Y}_{\alpha,\beta} \in V^{\mathbb{P}}$ and $\|\dot{Y}_{\alpha,\beta} \in \bar{\mathcal{J}}\| = 1$. (Here we are using $\bar{\mathcal{J}}$ to stand for the ideal generated by $\mathcal{J}$ in $V^{\mathbb{P}}$.) Hence, for each pair α, β

$$\| \exists Y \in V\ (Y \in \mathcal{J} \text{ and } Y_{\alpha,\beta} \subseteq Y)\| = 1 .$$

Since $\mathbb{P}$ is κ^+-c. c. and $\mathcal{J}$ is κ^+-complete, there is a sequence in V, $\langle Y_{\alpha,\beta} : \alpha < \beta < \kappa^{++}\rangle \subseteq \mathcal{J}$ and $\|X_\alpha \cap X_\beta \subseteq Y_{\alpha,\beta}\| = 1$.

For each X_α and each $\gamma < \kappa$, let $X_{\alpha,\gamma} = \{\xi \mid$ there is a $p \in C_\gamma$, $p \Vdash \xi \in X_\alpha\}$. Since $\|X_\alpha \subseteq \bigcup_{\gamma<\kappa} X_{\alpha,\gamma}\| = 1$ and $\mathcal{I}$ is κ^+-complete, there must be some γ_α with $X_{\alpha,\gamma_\alpha} \notin \mathcal{I}$. By the pigeon-hole principle, there is a set $S \subseteq \kappa^{++}$, $S \in V$ with $|S| = \kappa^{++}$ such that for all $\alpha, \beta \in S$, $\gamma_\alpha = \gamma_\beta$. Fix such a set S and such a γ. We claim that if $\alpha, \beta \in S$ then $X_{\alpha,\gamma} \cap X_{\beta,\gamma} \subseteq Y_{\alpha,\beta}$.

Let $\xi \in X_{\alpha,\gamma} \cap X_{\beta,\gamma}$. Since C_γ is a collection of compatible elements we can find a $q \in C_\gamma$, $q \Vdash \xi \in X_{\alpha,\gamma} \cap X_{\beta,\gamma}$. But then $q \Vdash \xi \in \check{Y}_{\alpha,\beta}$. Hence $\xi \in Y_{\alpha,\beta}$. We now derive our contradiction: $\langle X_{\alpha,\gamma} : \alpha \in S\rangle$ is a set of positive elements of $\mathcal{P}(\kappa^+)$ with respect to $\mathcal{I}$ and if $\alpha, \beta \in S$, $\alpha \neq \beta$, $X_{\alpha,\gamma} \cap X_{\beta,\gamma} \subseteq Y_{\alpha,\beta} \in \mathcal{I}$. Hence $\langle X_{\alpha,\gamma} : \alpha \in S\rangle$ are an antichain of size κ^{++} in the ground model, a contradiction.

The following lemma is standard (see [6]).

Lemma 3. Suppose $\lambda' < \kappa' \leq \lambda < \kappa$ are regular cardinals and $\mathbb{P}$ has the κ c. c.

(a) If $(\kappa,\lambda) \twoheadrightarrow (\kappa',\lambda')$ then in $V^{\mathbb{P}}$, $(\kappa,\lambda) \twoheadrightarrow (\kappa',\lambda')$

(b) If κ carries a κ^+-saturated ideal $\mathcal{I}$, then in $V^{\mathbb{P}}$, $\mathcal{I}$ is a κ^+-saturated ideal.

Let $\langle j_n : n \in \omega\rangle$ be a sequence of huge embeddings. Let $\kappa_n = \mathrm{crit}(j_n)$ and suppose $j_n(\kappa_n) = \kappa_{n+1}$. Suppose $\langle \mathbb{R}_n : n \in \omega\rangle$ is a sequence of terms for partial orderings with $\mathbb{R}_{n+1} \in V^{\overset{n}{\underset{i=0}{*}} \mathbb{R}_i}$ such that

(a) $\overset{n}{\underset{i=0}{*}} \mathbb{R}_i$ has the κ_n-chain condition and in $V^{\overset{n}{\underset{i=0}{*}} \mathbb{R}_i}$ $\mathbb{R}_{n+1}$ is κ_n-closed

(b) In $V^{\overset{n}{\underset{i=0}{*}} \mathbb{R}_i}$, for all $i \leq n$ $\kappa_i = \aleph_{i+2}$

(c) In $V^{\overset{n}{\underset{i=0}{*}} \mathbb{R}_i}$, $\mathbb{R}_{n+1}$ has cardinality κ_{n+1}.

(d) For each n, there is a projection map $\pi : j_n(\overset{n}{\underset{i=0}{*}} \mathbb{R}_i) \to \overset{n}{\underset{i=0}{*}} \mathbb{R}_i * \mathbb{R}_{n+1}$ and a condition $q \in j_n(\overset{n}{\underset{i=0}{*}} \mathbb{R}_i) * j_n(\mathbb{R}_{n+1})$ such that for all $r \leq q$, $r \in j_n(\overset{n}{\underset{i=0}{*}} \mathbb{R}_i) * j_n(\mathbb{R}_{n+1})$, $j_n(\pi(r)) \geq r$. (i.e. if $\mathbb{P}' = \overset{n}{\underset{i=0}{*}} \mathbb{R}_i$ and $Q = \mathbb{R}_n$, then $\mathbb{P}'$ and Q satisfy the hypothesis (d) in Kunen's Theorem.)

Let $\mathbb{P} = \lim \langle \overset{n}{\underset{i=0}{*}} \mathbb{R}_i : n \in \omega\rangle$. By Kunen's Theorem, in $V^{\overset{n+1}{\underset{i=0}{*}} \mathbb{R}_i}$ there is κ_{n+1}-saturated ideal on κ_n and $(\kappa_{n+1},\kappa_n) \twoheadrightarrow (\kappa_n,\kappa_{n-1})$. (Let $\kappa_{-1} = \aleph_1$.) But $\mathbb{P}/\overset{n+1}{\underset{i=0}{*}} \mathbb{R}_i$ is κ_{n+1}-closed. Hence by Lemma 1 in $V^{\mathbb{P}}$, for all $m, n \in \omega$, $m > n > 0$ $(\aleph_{m+1},\aleph_m) \twoheadrightarrow (\aleph_{n+1},\aleph_n)$ and $\aleph_m$ carries a normal $\aleph_m$-complete

$\aleph_{m+1}$-saturated ideal.

By Lemmas 2 and 3, in $V^{\mathbb{P}*C(\aleph_0,\aleph_1)}$ for all $m, n \in \omega$, $m > n$, $(\aleph_{m+1},\aleph_m) \twoheadrightarrow (\aleph_{n+1},\aleph_n)$ and $\aleph_m$ carries an $\aleph_m$-complete $\aleph_{m+1}$-saturated ideal.

We will now concentrate our attention on building the $\mathbb{R}$'s. The construction given here is considerably simpler than our original one at the expense of introducing somewhat more technicality. The next few definitions and lemmas explore these technicalities.

<u>Lemma 4</u>. Let $\mathbb{P}$ and $\mathbb{Q}$ be partial orderings. Suppose $\pi : \mathbb{P} \to \mathbb{Q}$ has the properties:

(1) $p_1 \leq p_2$ implies $\pi(p_1) \leq \pi(p_2)$

(2) for all $p \in \mathbb{P}$ there is a $q \leq \pi(p)$ such that for all $q' \leq q$ there is a $p' \leq p$ such that $\pi(p') \leq q'$.

$$\begin{array}{ccc}
\underline{\mathbb{P}} & & \underline{\mathbb{Q}} \\
p & \xrightarrow{\ \pi\ } & \pi(p) \\
 & & \vee| \\
 & & q \\
\vee| & & \vee| \\
 & & q' \\
 & & \vee| \\
p' & \xrightarrow{\ \pi\ } & \pi(p')
\end{array}$$

Then π is a projection map.

<u>Proof</u>. This is standard.

<u>Definition</u> (Laver). Let $\mathbb{P}$ be a partial ordering and $\mathbb{Q} \in V^{\mathbb{P}}$ be a term for a partial ordering. We define the <u>termspace</u> partial ordering $\mathbb{Q}^*$ to be the following partial order:

$$\begin{aligned} \operatorname{dom} \mathbb{Q}^* = \{\tau : \tau \in V^{\mathbb{P}} &\text{ and } \|\tau \in \mathbb{Q}\|_{\mathbb{P}} = 1 \text{ and for} \\ &\text{all } \tau' \in V^{\mathbb{P}}, \text{ if } \operatorname{rank}(\tau') < \operatorname{rank}(\tau) \\ &\text{then } \|\tau' = \tau\|_{\mathbb{P}} \neq 1\}\ . \end{aligned}$$

For $\tau, \tau' \in \operatorname{dom} \mathbb{Q}^*$,

$$\tau \leq_{\mathbb{Q}^*} \tau' \quad \text{iff} \quad \|\tau \leq_{\mathbb{Q}} \tau'\|_{\mathbb{P}} = 1\ .$$

(Technically we want to take the universe of $\mathbb{Q}^*$ to be equivalence classes of terms modulo the relation $\tau_1 \sim \tau_2$ iff $\|\tau_1 = \tau_2\|_{\mathbb{P}} = 1$. In practice we ignore this distinction.)

The main use of the termspace partial ordering is summed up by:

Lemma 5. (a) Let $Q \in V^{\mathbb{P}}$ be a term for a partial ordering. There is a projection map $\pi : \mathbb{P} \times Q^* \to \mathbb{P} * Q$

(b) Let $\langle Q_i : i \in \omega \rangle \subseteq V^{\mathbb{P}}$ be terms for partial orderings. Then there is a projection map

$$\pi : \mathbb{P} \times \prod_{i\in\omega} (Q_i)^* \longrightarrow \mathbb{P} * (\prod_{i\in\omega} Q_i)^{V^{\mathbb{P}}}$$

(c) If $\langle Q_i : i \in \omega \rangle$, $\langle \mathbb{P}_i : i \in \omega \rangle$ are two sequences of partial orderings and for each $i \in \omega$ $\varphi_i : Q_i \to \mathbb{P}_i$ is a neat embedding, then there is a $\varphi : \prod_{i\in\omega} Q_i \to \prod_{i\in\omega} \mathbb{P}_i$ extending each φ_i.

Proof. (a) Define π as follows: If $(p,\tau) \in \mathbb{P} \times Q^*$, let $\pi(p,\tau) = (p,\tau) \in \mathbb{P} * Q$. We want to apply Lemma 4. It is enough to see that if $(p,\tau) \in \mathbb{P} \times Q^*$ and $(q,\sigma) \in \mathbb{P} * Q$ with $(q,\sigma) \leq_{\mathbb{P}*Q} (p,\tau)$ there is a $\sigma' \in Q^*$ such that $(q,\sigma') \leq_{\mathbb{P}\times Q^*} (p,\tau)$ and $(q,\sigma') \leq_{\mathbb{P}*Q} (q,\sigma)$. Fix $(p,\tau) \in \mathbb{P} \times Q^*$ and $(q,\sigma) \in \mathbb{P} * Q$ with $(q,\sigma) \leq_{\mathbb{P}*Q} (p,\tau)$. Let σ' be a term of minimal rank such that $\|\sigma' = \sigma\|_{\mathbb{P}} \geq q$ and $\|\sigma' = \tau\|_{\mathbb{P}} \geq \neg q$. Then $q \Vdash \sigma' \leq \sigma$, hence $(q,\sigma') \leq_{\mathbb{P}*Q} (q,\sigma)$. Also $\neg q \leq \|\sigma' = \tau\|$, hence $q \vee (\neg q) \leq \|\sigma' \leq \tau\|$ so $\|\sigma' \leq \tau\| = 1$ and $(q,\sigma') \leq_{\mathbb{P}\times Q^*} (p,\tau)$ as desired.

The proof of (b) is similar and (c) is standard.

If Q is in $V^{\mathbb{P}}$ we will use the notation $A(\mathbb{P};Q)$ for the termspace partial ordering.

We now make a definition we will use to get an easy hold on the chain condition of the termspace partial ordering.

Definition. (a) Let $\mathbb{P}$ and Q be partial orderings. Q is said to be representible in $\mathbb{P}$ iff there is an incompatibility preserving map $i : Q \to \mathbb{P}$. (i.e. if q and q' are incompatible in Q, $i(q)$ and $i(q')$ are incompatible in $\mathbb{P}$.)

(b) Let $\mathfrak{F}(\mathbb{P},S)$ be the partial ordering with domain

$$\{f \mid f : \mathbb{P} \to S \text{ and } f \text{ is order preserving}\}$$

The ordering of $\mathfrak{F}(\mathbb{P},S)$ is $f \leq g$ iff for all $p \in \mathbb{P}$ $f(p) \leq g(p)$.

If Q is representible in $\mathbb{P}$ then the chain condition of $\mathbb{P}$ gives an upper bound on the chain condition for Q.

Example. (a) Let $\mathbb{P}$ and S be partial orderings and $\prod_{p\in\mathbb{P}} S$ be the product of $|\mathbb{P}|$ copies of S, one for each element of $\mathbb{P}$. Then $\mathfrak{F}(\mathbb{P},S)$ is representible in $\prod_{p\in\mathbb{P}} S$, namely map $f \in \mathfrak{F}(\mathbb{P},S)$ to $g \in \prod_{p\in\mathbb{P}} S$ where g has the value $f(p)$ on the pth copy of S.

(b) Let $\mathbb{P}$ be a κ c. c. partial ordering. Let $S^{\mathbb{P}}(\kappa,\lambda)$ be the Silver collapse of λ to κ^+ defined in $V^{\mathbb{P}}$. Let $S(\kappa,\lambda)$ be the Silver collapse of λ to κ^+ defined in V. Then $(S^{\mathbb{P}}(\kappa,\lambda))^*$ is representible in $\mathfrak{F}(\mathbb{P};S(\kappa,\lambda))$: Define a map $i : (S^{\mathbb{P}}(\kappa,\lambda))^* \to \mathfrak{F}(\mathbb{P};S(\kappa,\lambda))$ by $i(\tau) = f_\tau$ where

$$f_\tau(p) = \{(\alpha,\beta,\gamma) : p \Vdash \tau(\alpha,\beta) = \gamma\} .$$

Using the κ c. c. of $\mathbb{P}$ we see that $|f_\tau(p)| \leq \kappa$. Further, $p \Vdash f_\tau(p) \in S^{\mathbb{P}}(\kappa,\lambda)$, so there is a $\xi < \kappa$ such that $\operatorname{dom} f_\tau(p) \subseteq \lambda \times \xi$. Hence $f_\tau(p) \in S(\kappa,\lambda)$. If $p \leq_{\mathbb{P}} p'$ and $p' \Vdash \tau(\alpha,\beta) = \gamma$ then $p \Vdash \tau(\alpha,\beta) = \gamma$. Thus f_τ is order preserving and $f_\tau \in \mathfrak{F}(\mathbb{P},S)$.

We must see that i preserves incompatibility. Let $\tau, \tau' \in (S^{\mathbb{P}}(\kappa,\lambda))^*$, τ and τ' incompatible. Then there is a $p \in \mathbb{P}$,

$$p \Vdash_{\mathbb{P}} \tau \text{ and } \tau' \text{ are incompatible in } S^{\mathbb{P}}(\kappa,\lambda) .$$

Pick $p' \leq p$, $\gamma \neq \gamma'$, $p' \Vdash \tau(\alpha,\beta) = \gamma$ and $p' \Vdash \tau'(\alpha,\beta) = \gamma'$. Then $f_\tau(p')$ is incompatible with $f_{\tau'}(p')$, and hence f_τ is incompatible with $f_{\tau'}$.

(In fact if $|\mathbb{P}| = \kappa$ it is possible to see that $(S^{\mathbb{P}}(\kappa,\lambda))^* \simeq \{f \mid f : \mathbb{P} \to S(\kappa,\lambda),$ f is order preserving and there is a $\xi < \kappa$ $\bigcup_{p\in\mathbb{P}} \operatorname{dom} f(p) \subseteq \lambda \times \xi\}$. We leave this to the reader.)

If $Q \in V^{\mathbb{P}}$ and $\|Q$ is λ-closed$\|_{\mathbb{P}} = 1$ and $\langle \tau_\alpha : \alpha < \beta \rangle$ $(\beta < \lambda)$ is a descending sequence in Q^* then $\|$there is a $\tau \in Q$, $\tau \leq \tau_\alpha$ for all $\alpha\|_{\mathbb{P}} = 1$. Let τ be a term for an element of Q such that for each α $\|\tau \leq \tau_\alpha\|_{\mathbb{P}} = 1$. Then $\tau \leq_{Q^*} \inf_{\alpha<\beta} \tau_\alpha$. Hence Q^* is λ-closed.

<u>Lemma 6</u>. Let $\mathbb{R} \in V^{\mathbb{P}*Q}$ be a term for a partial ordering. Then $A(\mathbb{P};A(Q;\mathbb{R})) \approx A(\mathbb{P} * Q;\mathbb{R})$.

<u>Proof</u>. If $\tau \in A(\mathbb{P};A(Q,\mathbb{R}))$ let $\pi(\tau) \in A(\mathbb{P} * Q;\mathbb{R})$ be a term τ^* such that $\|\tau^* = ((\tau)^{V^{\mathbb{P}}})^{V^{Q}}\|_{\mathbb{P}*Q} = 1$. We verify that π is an isomorphism. π is clearly order preserving. Suppose $\tau, \sigma \in A(\mathbb{P};S(Q;\mathbb{R}))$, $\|\tau = \sigma\|_{\mathbb{P}} \neq 1$. Let $p \in \mathbb{P}$, $p \Vdash \tau \neq \sigma$ in $A(Q;\mathbb{R})$. Pick $q \in V^{\mathbb{P}}$, $\|q \in Q\| \geq p$, $p \Vdash (q \Vdash ((\tau)^{V^{\mathbb{P}}})^{V^{Q}} \neq ((\sigma)^{V^{\mathbb{P}}})^{V^{Q}})$. Then $(p,q) \in \mathbb{P} * Q$ and $(p,q) \Vdash \tau^* \neq \sigma^*$. Let $\tau^* \in V^{\mathbb{P}*Q}$, $\|\tau^* \in \mathbb{R}\|_{\mathbb{P}*Q} = 1$. In $V^{\mathbb{P}}$, there is a term τ for an element of $V^{\mathbb{P}*Q}$ such that $\| \|\tau = \tau^*\|_Q = 1\|_{\mathbb{P}} = 1$. Then $\pi(\tau) = \tau^*$.

The following lemma is standard.

Lemma 7. Let λ be a measurable cardinal. Let $\alpha < \lambda$ and $\langle \mathbb{P}_\gamma : \gamma < \lambda \rangle$ be a sequence of λ-c. c. partial orderings. Then $\prod_{\alpha \text{ supports}} \{\mathbb{P}_\gamma : \gamma < \lambda\}$ has the λ c. c. ("α supports" denotes the ideal of subsets of λ with cardinality $\leq \alpha$.)

We will be interested in constructing partial orderings with nice embedding properties. We make the following definition, which is similar to one that will appear in [3].

Our partial orderings will be in the form $\mathbb{R}(\kappa,\lambda)$ for all regular $\kappa > \omega$ and all measurable $\lambda > \kappa$. $\mathbb{R}$ is the uniform definition given below.

If λ is greater than the first measurable above κ, the partial ordering $\mathbb{R}(\kappa,\lambda)$ will be the product $\prod_{n\in\omega} S^n(\kappa,\lambda)$ for partial orderings $S^n(\kappa,\lambda)$ defined inductively.

Assume that we have defined $\mathbb{R}(\beta,\alpha)$ for all regular β, $\omega < \beta < \alpha$ and all measurable $\alpha < \lambda$. We now define $\mathbb{R}(\beta,\lambda)$.

Case 1. λ is the first inaccessible above β. Let $\mathbb{R}(\beta,\lambda) = S(\beta,\lambda) =$ the Silver collapse of λ to β^+.

Case 2. Otherwise.

By induction on n, we define $S^n(\alpha,\lambda)$ for all regular α, $\beta \leq \alpha < \lambda$. We will have a uniform definition of $S^n(\alpha,\lambda)$ no matter which model we define it in, hence we define it only in the ground model. $(S^n(\alpha,\lambda))^{V^{\mathbb{R}(\beta,\alpha)}}$ is the partial ordering constructed in $V^{\mathbb{R}(\beta,\alpha)}$ using this definition.

If $n = 0$:

$$\text{Let } S^0(\alpha,\lambda) = S(\alpha,\lambda) = \text{the Silver collapse of } \lambda \text{ to } \alpha^+ .$$

Assume that we have defined $S^n(\alpha,\lambda)$ for all regular α.

For $n + 1$:

$$\text{Let } S^{n+1}(\alpha,\lambda) = \prod_{\alpha \text{ supports}} \{A(\mathbb{R}(\alpha,\gamma);S^n(\gamma,\lambda)) : \alpha < \gamma < \lambda \text{ and } \gamma \text{ is measurable}\}.$$

Lemma 8. For all regular $\beta > \omega$ and all measurable λ, $\mathbb{R}(\beta,\lambda)$ is λ-c. c. and has cardinality λ. (In fact λ Mahlo works but we leave this to the reader.)

Proof. By Lemma 7, it is enough to see that each $S^n(\beta,\lambda)$ has the λ-c. c.

We show by induction on λ and n, that for all measurable $\alpha_1,\ldots,\alpha_m$ $A(\mathop{*}_{i=1}^{m-1} \mathbb{R}(\alpha_i,\alpha_{i+1});S^n(\alpha_m,\lambda))$ has cardinality λ and the λ c. c.

Assume that this is true for all $\lambda' < \lambda$. By our example we know that $A(\mathop{*}_{i=1}^{m-1} \mathbb{R}(\alpha_i,\alpha_{i+1});S^0(\alpha_m,\lambda))$ is representible in $\mathfrak{F}(\mathop{*}_{i=1}^{m-1} \mathbb{R}(\alpha_i,\alpha_{i+1});S^0(\alpha_m,\lambda))$. This in turn is representible in the product of α_m copies of $S^0(\alpha_m,\lambda)$. By Lemma 7, this has the λ c. c.

Assume that $A(\overset{m-1}{\underset{i=1}{*}} \mathbb{R}(\alpha_i,\alpha_{i+1});S^n(\alpha_m,\lambda))$ is λ-c. c. for all $\alpha_1,\ldots,\alpha_m$ measurable. We want that for each $\alpha_1,\ldots,\alpha_m$ measurable, $A(\overset{m-1}{\underset{i=1}{*}} \mathbb{R}(\alpha_i,\alpha_{i+1});S^{n+1}(\alpha_m,\lambda))$ is λ-c. c.

Fix such a sequence $\alpha_1,\ldots,\alpha_m$. In $V^{\overset{m-1}{\underset{i=1}{*}} \mathbb{R}(\alpha_i,\alpha_{i+1})}$

$$S^{n+1}(\alpha_m,\lambda) = \prod_{\alpha_{m+1}\text{ support}} \{A(\mathbb{R}(\alpha_m,\beta);S^n(\beta,\lambda)) \quad \alpha_{n+1} < \beta < \lambda \text{ and } \beta \text{ is measurable}\}.$$

Since $\overset{m-1}{\underset{i=1}{*}} \mathbb{R}(\alpha_i,\alpha_{i+1})$ has the α_m-c. c., if $t \in A(\overset{m-1}{\underset{i=1}{*}} \mathbb{R}(\alpha_i,\alpha_{i+1});S^{n+1}(\alpha_m,\lambda))$ there is a set $D \subseteq \lambda$ such that $|D| = \alpha_m$ and $\|\text{supp } t \subseteq D\|_{\overset{m-1}{\underset{i=1}{*}} \mathbb{R}(\alpha_i,\alpha_{i+1})} = 1.$ For each $\beta \in D$, we get a term τ_β such that $\|\tau(\beta) = \tau_\beta\|_{\overset{m-1}{\underset{i=1}{*}} \mathbb{R}(\alpha_i,\alpha_{i+1})} = 1$ and $\tau_\beta \in A(\overset{m-1}{\underset{i=1}{*}} \mathbb{R}(\alpha_i,\alpha_{i+1}); (\mathbb{R}(\alpha_m,\beta);S^n(\beta,\lambda)))$. It is easy to verify that the map $\tau \longmapsto \langle \tau_\beta : \beta \in D\rangle$ is an isomorphism between $A(\overset{m-1}{\underset{i=1}{*}} \mathbb{R}(\alpha_i,\alpha_{i+1});S^{n+1}(\alpha_m,\lambda))$ and $\prod_{\alpha_m \text{ supports}} \{A(\overset{m-1}{\underset{i=1}{*}} \mathbb{R}(\alpha_i,\alpha_{i+1});A(\mathbb{R}(\alpha_m,\beta);S^n(\beta,\lambda))) \quad \alpha_m < \beta < \lambda$ and β is measurable$\}$. By Lemma 6,

$$A(\overset{m-1}{\underset{i=1}{*}} \mathbb{R}(\alpha_i,\alpha_{i+1});A(\mathbb{R}(\alpha_m,\beta);S^n(\beta,\lambda))) \approx A(\overset{m-1}{\underset{i=1}{*}} \mathbb{R}(\alpha_i,\alpha_{i+1}) * \mathbb{R}(\alpha_m,\beta);S^n(\beta,\lambda)).$$

By the induction hypothesis for n, $A(\overset{m-1}{\underset{i=1}{*}} \mathbb{R}(\alpha_i,\alpha_{i+1}) * \mathbb{R}(\alpha_m,\beta);S^n(\beta,\lambda))$ has the λ c. c. and thus by Lemma 7 $A(\overset{m-1}{\underset{i=1}{*}} \mathbb{R}(\alpha_i,\alpha_{i+1});S^{n+1}(\alpha_m,\lambda))$ has the λ-c. c.

We now establish the properties we need to satisfy the conditions of Kunen's Theorem:

<u>Lemma 9</u>. For each $\kappa > \omega$, κ regular and each λ measurable:

(a) $\prod_{\substack{n\in\omega \\ n\geq 1}} S^n(\kappa,\lambda)$ is κ^+-closed

(b) $\prod_{n\in\omega} S^n(\kappa,\lambda)$ is κ-closed

(c) If α is measurable and $id : \mathbb{R}(\kappa,\alpha) \hookrightarrow \mathbb{R}(\kappa,\lambda)$ then there is a map φ extending id such that

$$\varphi : \mathbb{R}(\kappa,\alpha) \times \prod_{n\in\omega} A(\mathbb{R}(\kappa,\alpha);S^n(\alpha,\lambda)) \hookrightarrow \mathbb{R}(\kappa,\lambda)$$

(d) If $id : \mathbb{R}(\kappa,\alpha) \to \mathbb{R}(\kappa,\lambda)$ then there is a map ψ extending id,

$$\psi : \mathbb{B}(\mathbb{R}(\kappa,\alpha) * \mathbb{R}(\alpha,\lambda)) \hookrightarrow \mathbb{B}(\mathbb{R}(\kappa,\lambda))$$

Proof. (a) We show by induction on $n > 0$ that for all regular α, $S^n(\alpha,\lambda)$ is α^+-closed.

n = 1: $S^1(\alpha,\lambda) = \prod_{\alpha \text{ support}} \{A(\mathbb{R}(\alpha,\beta);S^0(\beta,\lambda))\ \alpha<\beta<\lambda \text{ and } \beta \text{ is measurable}\}$

By our remarks preceding Lemma 6, each $A(\mathbb{R}(\alpha,\beta);S^0(\beta,\lambda))$ is β-closed and hence α^+-closed. Thus $\prod_{\alpha \text{ supports}} \{A(\mathbb{R}(\alpha,\beta);S^0(\beta,\lambda))\ \alpha < \beta < \lambda$ and β is measurable$\}$ is α^+-closed.

Assume that for all regular β, $S^n(\beta,\lambda)$ is β^+-closed and $n \geq 1$. Then, $A(\mathbb{R}(\alpha,\beta);S^n(\beta,\lambda))$ is β^+-closed and hence α^+ closed. Again this implies $\prod_{\alpha\text{-supports}} \{A(\mathbb{R}(\alpha,\beta);S^n(\beta,\lambda)) : \alpha < \beta < \lambda$ and β is measurable$\}$ is α^+-closed.

(b) $\mathbb{R}(\kappa,\lambda) = S^0(\kappa,\lambda) \times \prod_{n\geq 1} S^n(\kappa,\lambda)$ where $S^0(\kappa,\lambda)$ is the Silver collapse of λ to κ^+. The Silver collapse is κ-closed and by (a), $\prod_{n\geq 1} S^n(\kappa,\lambda)$ is κ^+-closed. Hence the product is κ-closed.

(c) Suppose $id : \mathbb{R}(\kappa,\alpha) \hookrightarrow \mathbb{R}(\kappa,\lambda)$. By the definition of $S^{n+1}(\kappa,\lambda)$, if α is measurable then $A(\mathbb{R}(\kappa,\alpha);S^n(\alpha,\lambda))$ is a factor of $S^{n+1}(\kappa,\lambda)$. Let φ_n be the map from $A(\mathbb{R}(\kappa,\alpha);S^n(\alpha,\lambda))$ to $S^{n+1}(\kappa,\lambda)$ that sends $p \in A(\mathbb{R}(\kappa,\alpha);S^n(\alpha,\lambda))$ to the element $q \in S^{n+1}(\kappa,\lambda)$ that is 1 on each factor of $S^{n+1}(\kappa,\lambda)$ except $A(\mathbb{R}(\kappa,\alpha);S^n(\alpha,\lambda))$ where it is p. [In essence φ_n is the identity map of $A(\mathbb{R}(\kappa,\alpha);S^n(\alpha,\lambda))$ to its factor in $S^{n+1}(\kappa,\lambda)$.] The product, $\prod_{n\in\omega} A(\mathbb{R}(\kappa,\alpha);S^n(\alpha,\lambda))$ can be embedded in $\prod_{n\in\omega} S^n(\kappa,\lambda)$ by the map

$$\langle p_n : n \in \omega\rangle \stackrel{\varphi_\omega}{\longmapsto} \langle q_n : n \in \omega\rangle$$

where $q_0 = 1$, $q_{n+1} = \varphi_n(p_n)$.

By Lemma 5(c), φ_ω is a neat embedding from $\prod_{n\in\omega} A(\mathbb{R}(\kappa,\alpha);S^n(\alpha,\lambda))$ into $\prod_{n\in\omega} S^n(\kappa,\lambda)$.

Each

$$S^{n+1}(\kappa,\lambda) = \prod_{\kappa \text{ supports}} \{A(\mathbb{R}(\kappa,\beta);S^n(\beta,\lambda)) \mid \kappa<\beta<\alpha \text{ and } \beta \text{ is measurable}\}$$
$$\times \prod_{\kappa \text{ supports}} \{A(\mathbb{R}(\kappa,\beta);S^n(\beta,\lambda)) \mid \alpha\leq\beta<\lambda \text{ and } \beta \text{ is measurable}\} .$$

Let

$$S_0^{n+1}(\kappa,\lambda) = \prod_{\kappa \text{ supports}} \{A(\mathbb{R}(\kappa,\beta);S^n(\beta,\lambda)) \mid \kappa < \beta < \alpha \text{ and } \beta \text{ is measurable}\} ,$$

$$S_1^{n+1}(\kappa,\lambda) = \prod_{\kappa \text{ supports}} \{A(\mathbb{R}(\kappa,\beta);S^n(\beta,\lambda)) \mid \alpha \leq \beta < \lambda \text{ and } \beta \text{ is measurable}\} .$$

By rearranging our product we get

$$\mathbb{R}(\kappa,\lambda) = S^0(\kappa,\lambda) \times \prod_{\substack{n\in\omega \\ n\geq 1}} S_0^n(\kappa,\lambda) \times \prod_{\substack{n\in\omega \\ n\geq 1}} S_1^n(\kappa,\lambda) .$$

For $n \geq 1$, $\mathrm{id}''S^n(\kappa,\alpha) \subseteq S_0^n(\kappa,\lambda)$. Hence, if $\mathrm{id} : \mathbb{R}(\kappa,\alpha) \hookrightarrow \mathbb{R}(\kappa,\lambda)$ then

$$\mathrm{id} : \mathbb{R}(\kappa,\alpha) \hookrightarrow S^0(\kappa,\lambda) \times \prod_{1\leq n\in\omega} S_0^n(\kappa,\lambda) .$$

By the definition of φ_n, $\varphi_n''A(\mathbb{R}(\kappa,\alpha);S^n(\alpha,\lambda)) \subseteq S_1^n(\kappa,\lambda)$. So,

$$\varphi_\omega'' \prod_{n\in\omega} A(\mathbb{R}(\kappa,\alpha);S^n(\alpha,\lambda)) \subseteq \prod_{n\in\omega} S_1^n(\kappa,\lambda) .$$

Thus we have a map $\varphi = \mathrm{id} \times \varphi_\omega$

$$\varphi : \mathbb{R}(\kappa,\alpha) \times \left[\prod_{n\in\omega} A(\mathbb{R}(\kappa,\alpha);S^n(\alpha,\lambda))\right] \hookrightarrow (S^0(\kappa,\lambda) \times \prod_{1\leq n\in\omega} S_0^n(\kappa,\lambda)) \times \left[\prod_{1\leq n\in\omega} S_1^n(\kappa,\lambda)\right]$$

Hence

$$\varphi : \mathbb{R}(\kappa,\alpha) \times (\prod_{n\in\omega} A(\mathbb{R}(\kappa,\alpha);S^n(\alpha,\lambda))) \hookrightarrow \mathbb{R}(\kappa,\lambda)$$

(d) By Lemma 5(b), we have a neat embedding

$$\psi_1 : \mathbb{B}(\mathbb{R}(\kappa,\alpha) * \mathbb{R}(\alpha,\lambda)) \hookrightarrow \mathbb{B}(\mathbb{R}(\kappa,\alpha) \times \prod_{n\in\omega} A(\mathbb{R}(\kappa,\alpha);S^n(\alpha,\lambda)) ,$$

By (c), we have

$$\psi_2 : \mathbb{B}(\mathbb{R}(\kappa,\alpha) \times \prod_{n\in\omega} A(\mathbb{R}(\kappa,\alpha);S^n(\alpha,\lambda)) \hookrightarrow \mathbb{B}(\mathbb{R}(\kappa,\lambda)) .$$

Composing we get:

$$\psi = \psi_2 \circ \psi_1 : \mathbb{B}(\mathbb{R}(\kappa,\alpha) * \mathbb{R}(\alpha,\lambda)) \hookrightarrow \mathbb{B}(\mathbb{R}(\kappa,\lambda)) .$$

This completes the definition and verification of the basic properties of the $\mathbb{R}$'s.

Returning to our proof: Let

$$\mathbb{P} = \varprojlim \langle \mathbb{R}(\aleph_1,\kappa_0) * \overline{\mathbb{R}(\kappa_0,\kappa_1)} * \overline{\mathbb{R}(\kappa_1,\kappa_2)} * \cdots * \overline{\mathbb{R}(\kappa_{n-1},\kappa_n)} : n \in \omega\rangle .$$

To see that in $V^{\mathbb{P}}$:

(a) for all $m > 1$, $\aleph_m$ carries an $\aleph_{m+1}$-saturated ideal

(b) for all $m > n \geq 1$, $(\aleph_{m+1},\aleph_m) \twoheadrightarrow (\aleph_{n+1},\aleph_n)$

It is enough to see that $\mathbb{P}' = \mathbb{R}(\aleph_1,\kappa_0) * \cdots * \overline{\mathbb{R}}(\kappa_{n-1},\kappa_n)$ and $Q = \mathbb{R}^{\mathbb{P}'}(\kappa_n,\kappa_{n+1})$ together with the embedding j_n satisfy the conditions for Kunen's Theorem. Conditions (a), (b) and (c) follow directly from Lemma 8 and the remarks before Lemma 6.

Since $\mathbb{P}'$ is κ_n c. c., $j_n : \mathbb{P}' \to j(\mathbb{P}')$ is a neat embedding. Since $\mathbb{P}' \subseteq V_{\kappa_n}$, $j_n \mid \mathbb{P}' = \text{id}$ and hence $\text{id} : \mathbb{P}' \to j_n(\mathbb{P}')$. Unravelling this we get

$$\text{id} : \mathbb{R}(\aleph_1,\kappa_0) * \overline{\mathbb{R}}(\kappa_0,\kappa_1) * \cdots * \overline{\mathbb{R}}(\kappa_{n-1},\kappa_n) \hookrightarrow \mathbb{R}(\aleph_1,\kappa_0) * \overline{\mathbb{R}}(\kappa_0,\kappa_1) * \cdots * \overline{\mathbb{R}}(\kappa_{n-1},\kappa_{n+1})$$

By Lemma 9 there is a generic object $G * H \subseteq \mathbb{P}' * Q$ in $V^{j_n(\mathbb{P}')}$. In $V^{j(\mathbb{P}')}$, $H \simeq H_0 \times H_{>0}$, where $H_0 \subseteq S^0(\kappa_n,\kappa_{n+1})$ and $H_{>0} \subseteq \prod_{k>0} S^k(\kappa_n,\kappa_{n+1})$. Still in $V^{j(\mathbb{P}')}$: $j''H = j''H_0 \times j''H_{>0} \subseteq S^0(\kappa_{n+1},\kappa_{n+2}) \times \prod_{k>0} S^k(\kappa_{n+1},\kappa_{n+2})$ and $j''H$ has cardinality κ_{n+1}. By Lemma 9, $\prod_{k>0} S^k(\kappa_{n+1},\kappa_{n+2})$ is κ_{n+1}^+-closed and hence there is a $q_{>0} \in \prod_{k>0} S^k(\kappa_{n+1},\kappa_{n+2})$ such that for all $p \in j''H_{>0}$, $q_{>0} \le p$. Let $q_0 = \bigcup_{\tau\in H_0} (j(\tau))^{V^{j(\mathbb{P}')}}$. We show that $q_0 \in S^0(\kappa_{n+1},\kappa_{n+2})$. Clearly q_0 is a function from $\kappa_{n+2} \times \kappa_{n+1}$ satisfying all but the cardinality conditions for being an element of $S^0(\kappa_{n+1},\kappa_{n+2})$. To see that it satisfies these, we note that for each $\tau \in (S^0(\kappa_n,\kappa_{n+1}))^{V^{\mathbb{P}'}}$ there is a $\xi < \kappa_n$ such that $|\text{dom}\ \tau \subseteq \kappa_{n+1} \times \xi|_{\mathbb{P}'} = 1$. Hence, $\text{dom}\ q_0 \subseteq \kappa_{n+2} \times \kappa_n$. Further,

$$|\text{dom}\ q_0| = \Big| \bigcup_{\tau\in H_0} \text{dom}\ j(\tau)^{V^{j(\mathbb{P}')}} \Big| = \kappa_{n+1} \times \kappa_{n+1} = \kappa_{n+1} .$$

Hence $q_0 \in (S^0(\kappa_{n+1},\kappa_{n+2}))^{V^{j(\mathbb{P}')}}$. Let

$$q = (q_0,q_{>0}) \in S^0(\kappa_{n+1},\kappa_{n+2}) \times \prod_{k>0} S^k(\kappa_{n+1},\kappa_{n+2}) .$$

It is easy to check that in $V^{j(\mathbb{P})}$, if $p \in H$, $j(p) \ge q$. From this we conclude that $(1,q) \in j(\mathbb{P}') * \mathbb{R}(\kappa_{n+1},\kappa_{n+2})$ functions as a master condition and satisfies (d) of Kunen's Theorem.

Forcing with $\mathbb{P} \times C(\aleph_0,\aleph_1)$ we get a model as in Theorem 1. This completes the proof of Theorem 1. For Theorem 2 we remark that if our original sequence $\langle j_n : n \in \omega\rangle$ was a sequence of almost huge embeddings then the $\mathbb{R}$'s satisfy the Magidor conditions and our proof would go through to get the saturated ideals. (Although apparently not the strong transfer principles.)

We end this section with some problems. First, not much is known about transfer principles that go "infinite distances". A problem along this line might be:

<u>Problem</u>. Is it consistent to have $(\aleph_{\omega+1},\aleph_\omega) \twoheadrightarrow (\aleph_1,\aleph_0)$?

Magidor has remarked that if the G. C. H. holds then $(\aleph_{\omega+1},\aleph_\omega) \not\twoheadrightarrow (\aleph_{m+1},\aleph_m)$ for $m > 0$. Along these lines, the author has shown:

Con (ZFC + there is a 2-huge cardinal) $\Rightarrow$

$\Rightarrow$ Con (ZFC + there is a supercompact cardinal κ with $(\kappa^+,\kappa) \twoheadrightarrow (\aleph_1,\aleph_0)$) .

This might be relevant to this problem.

In this paper we have finite distance "gap one" transfer principals. In [2] it was shown:

Con (ZFC + there is a 2-huge cardinal) $\Rightarrow$ Con (ZFC + $(\aleph_3,\aleph_1) \twoheadrightarrow (\aleph_2,\aleph_0)$) .

Problem. Is it consistent to have for all $m > n$ $(\aleph_{m+2},\aleph_m) \twoheadrightarrow (\aleph_{n+2},\aleph_n)$?

A general solution to this problem would probably solve many problems in this area.

Another problem is:

Problem. Does "for all $m > n$ $(\aleph_{m+1},\aleph_m) \twoheadrightarrow (\aleph_{n+1},\aleph_n)$" imply "$\aleph_\omega$ is Jonsson"?

Finally, since the method of proof for Theorems 1 and 2 was the same it would be interesting to establish a direct relationship between saturated ideals and there transfer properties. Shelah has shown that "there is an $\aleph_2$-complete ideal $\mathcal{J}$ on $\aleph_2$ such that $\mathcal{P}(\aleph_2)/\mathcal{J}$ has a dense ω-closed subset" implies $(\aleph_2,\aleph_1) \twoheadrightarrow (\aleph_1,\aleph_0)$. The proof however is tied very closely to the cardinals $\aleph_2$, $\aleph_1$ and $\aleph_0$. What is desired is some more general statement for cardinals bigger than $\aleph_2$. For the sake of concreteness we mention a possibility:

Problem. Does "$\aleph_2$ carries an $\aleph_3$-saturated ideal and $\aleph_3$ carries a precipitous ideal" imply $(\aleph_3,\aleph_2) \twoheadrightarrow (\aleph_2,\aleph_1)$?

§2. We now turn our attention to proving Theorem 3. For this section, a "saturated ideal" on κ will be a normal, κ-complete, κ^+-saturated ideal on κ.

Let κ be a huge cardinal. Then there is a set $A \subseteq \kappa$ with measure one with respect to the measure on κ induced by the huge embedding, such that for all $\alpha, \beta \in A$, $\beta > \alpha$ implies: there is an almost huge embedding $j_{\alpha,\beta}$ with $\mathrm{crit}(j_{\alpha,\beta}) = \alpha$ and $j_{\alpha,\beta} = \beta$.

Fix such a set A. If $\alpha \in A$, let α' be the successor of α in A. We need to redefine our $\mathbb{R}(\kappa,\lambda)$'s. Assume by induction that we have defined $\mathbb{R}(\kappa,\lambda)$ for all regular $\kappa > \omega$ and all measurable $\lambda' < \lambda$, and that $\mathbb{R}(\kappa,\lambda')$ is κ-closed, has cardinality λ' and is λ' c. c.

If $X \subseteq \lambda$ is a set of measurable cardinals we define the <u>standard iteration</u> I_X as follows:

(a) I_X is an iteration of length $\leq \lambda$ with Easton supports.

We will define the notion of an ordinal in X being mentioned in $(I_X)_\alpha$ by induction on α. To start with, no ordinals are mentioned in $(I_X)_0$.

(b) At stage $\alpha + 1$: If there are five elements of X not mentioned in $(I_X)_\alpha$, let $\gamma_0 < \gamma_1 < \cdots < \gamma_4$ be the first five elements of X not mentioned in $(I_X)_\alpha$. If $\alpha \geq \gamma_0$, let

$$(I_X)_{\alpha+1} = (I_X)_\alpha * [\overline{\mathbb{R}}(\gamma_0,\gamma_1) * \overline{\mathbb{R}}(\gamma_1,\gamma_2) * \overline{\mathbb{R}}(\gamma_2,\gamma_3) * \overline{\mathbb{R}}(\gamma_3,\gamma_4)] .$$

The elements of X mentioned in $(I_X)_{\alpha+1}$ are defined to be the elements of X mentioned in $(I_X)_\alpha \cup \{\gamma_0,\ldots,\gamma_4\}$. If there are not five elements of X not mentioned in $(I_X)_\alpha$, then $I_X = (I_X)_\alpha$. If $\alpha < \gamma_0$, then $(I_X)_{\alpha+1} = (I_X)_\alpha * 1$. (This iteration is the result of iterating, five at a time, the elements of X strung together by the $\mathbb{R}$'s.)

<u>Proposition 2</u>. For every X, I_X is λ-c. c. and $I_X/(I_X)_\alpha$ is γ_0-closed in $V^{(I_X)_\alpha}$, where γ_0 is the least ordinal mentioned in $(I_X)_{\alpha+1}$ not mentioned in $(I_X)_\alpha$.

<u>Proof</u>. This uses standard facts about iterations with Easton support.

We can now define our new version of $\mathbb{R}(\kappa,\lambda)$ for all regular $\kappa > \omega$. Again $\mathbb{R}(\kappa,\lambda) = \prod_{n\in\omega} S^n(\kappa,\lambda)$ where the $S^n(\alpha,\lambda)$ are defined by induction on n, simultaneously for all regular α, $\omega < \alpha < \lambda$.

Let $S^0(\alpha,\lambda)$ = the Silver collapse of λ to α^+.

$$S^{n+1}(\alpha,\lambda) = \prod_{\alpha \text{ supports}} \{A(\mathbb{R}(\alpha,\beta);C(\beta,\gamma) \times [I_X * \overline{S^n}(\gamma,\lambda)])$$

$$| \alpha<\beta<\gamma<\lambda,\ \beta,\ \gamma \text{ are measurable and } X \subseteq \gamma \sim \beta \text{ is a set of measurable cardinals}\}$$

Let $\mathbb{R}(\alpha,\lambda) = \prod_{n\in\omega} S^n(\alpha,\lambda)$.

Using the techniques of §1 it is now easy to verify that if $\omega < \alpha < \beta < \gamma < \lambda$, α is regular, β, γ and λ are measurable and $X \subseteq \gamma \sim \beta$ is a set of measurable cardinals then there is a map ψ

$$\psi : \prod_{n\in\omega} A(\mathbb{R}(\alpha,\beta);C(\beta,\gamma) \times [I_X * \overline{S}^n(\gamma,\lambda)]) \hookrightarrow \mathbb{R}(\alpha,\lambda) .$$

From this (as before) we conclude that if $id : \mathbb{R}(\alpha,\beta) \hookrightarrow \mathbb{R}(\alpha,\lambda)$ then there is a map φ such that

$$\varphi : \mathbb{R}(\alpha,\beta) \times \prod_{n\in\omega} A(\mathbb{R}(\alpha,\beta);C(\beta,\gamma) \times [I_X * S^n(\gamma,\lambda)]) \hookrightarrow \mathbb{R}(\alpha,\lambda) .$$

Again this implies that: if $id : \mathbb{R}(\alpha,\beta) \hookrightarrow \mathbb{R}(\alpha,\lambda)$ then there is a map φ

$$\varphi : \mathbb{B}(\mathbb{R}(\alpha,\beta) * [\overline{C}(\beta,\gamma) \times (I_X * \overline{\mathbb{R}}(\gamma,\lambda))]) \hookrightarrow \mathbb{B}(\mathbb{R}(\kappa,\lambda))$$

Lemma 10. If $\kappa > \omega$ is regular and $\lambda > \kappa$ is measurable

(a) $\mathbb{R}(\kappa,\lambda)$ has the λ-c. c.

(b) $\mathbb{R}(\kappa,\lambda)$ is κ-closed

(c) In $V^{\mathbb{R}(\kappa,\lambda)}$, $\lambda = \kappa^+$

(d) $|\mathbb{R}(\kappa,\lambda)| = \lambda$

(e) $\prod_{\substack{n\in\omega \\ n\geq 1}} S^n(\kappa,\lambda)$ is κ^+-closed

(f) $\prod_{n\in\omega} S^n(\kappa,\lambda)$ is κ-closed

Proof. Essentially the same as in Section 1.

To build our model of Theorem 3 we start by first doing the following preparatory forcing. (Recall κ is our huge cardinal, j our huge embedding, and the definition of A.)

Let

$$\mathbb{P}_{\leq\kappa} = I_A * \overline{\mathbb{R}}(\kappa,\kappa') * \overline{\mathbb{R}}(\kappa',\kappa'') * \overline{\mathbb{R}}(\kappa'',\kappa''') * \overline{\mathbb{R}}(\kappa''',\kappa^{iv}) ,$$

where I_A is the standard iteration of A and κ', κ'', κ''', κ^{iv} are the first four elements of $j(A)$ above κ.

Proposition 3. In $\mathbb{P}_{<\kappa}$:

(a) $\kappa' = \kappa^+$, $\kappa'' = \kappa^{++}$, $\kappa''' = \kappa^{+3}$, $\kappa^{iv} = \kappa^{+4}$

(b) κ remains κ^{+5} supercompact by a measure μ_5 such that there is a set of μ_5-measure one worth of $x \in P_\kappa(\kappa^{+5})$ such that: if $\alpha = crit(x)$, α^+, α^{+2}, α^{+3} carry saturated ideals

(c) $2^\kappa = \kappa^+$

Proof. I_A is κ-c. c. by earlier comments. In V^{I_A}, $\mathbb{R}(\kappa,\kappa')$ makes κ' into κ^+ and is κ-closed. In $V^{I_A*\overline{\mathbb{R}}(\kappa,\kappa')}$, $\overline{\mathbb{R}}(\kappa',\kappa'') * \overline{\mathbb{R}}(\kappa'',\kappa''') * \overline{\mathbb{R}}(\kappa''',\kappa^{iv})$ is κ' closed. Hence, in $V^{\mathbb{P}_{\leq\kappa}}$, $\kappa' = \kappa^+$. In $V^{I_A*\mathbb{R}(\kappa,\kappa')}$, $\mathbb{R}(\kappa',\kappa'')$ makes κ'' into $(\kappa')^+$. Since $\mathbb{R}(\kappa'',\kappa''') * \overline{\mathbb{R}}(\kappa''',\kappa^{iv})$ is $< \kappa''$-closed in $V^{I_A*\mathbb{R}(\kappa,\kappa')*\mathbb{R}(\kappa',\kappa'')}$, $\kappa'' = (\kappa')^+$ in $V^{\mathbb{P}_{\leq\kappa}}$. Hence $\kappa'' = \kappa^{+2}$ in $V^{\mathbb{P}_{\leq\kappa}}$.

Similarly we see that in $V^{\mathbb{P}_{\leq\kappa}}$, $\kappa''' = \kappa^{+3}$ and $\kappa^{iv} = \kappa^{+4}$.

(b) This is a standard argument using the fact that $\mathbb{P}_{\leq\kappa}$ is an initial segment of $j(\mathbb{P}_{\leq\kappa})$ and $j(\mathbb{P}_{\leq\kappa})/\mathbb{P}_{\leq\kappa}$ is $(\beth_3(\kappa^{+5}))^{V^{\mathbb{P}_{\leq\kappa}}}$-closed in $V^{\mathbb{P}_{\leq\kappa}}$. (See [1].) In fact we can require that the measure μ_5 induces on κ extends the measure j induces on κ. Hence $\{\alpha$: at some stage γ in the iteration I_A, α is the least ordinal not mentioned in $(I_A)_\gamma\}$ has measure one with respect to the measure μ_5 induces on κ.

At some stage γ, if α is the least ordinal not mentioned in $(I_A)_\gamma$, and $\gamma \geq \alpha$,

$$I_A = (I_A)_\gamma * \overline{\mathbb{R}}(\alpha,\alpha') * \overline{\mathbb{R}}(\alpha',\alpha'') * \overline{\mathbb{R}}(\alpha'',\alpha''') * \overline{\mathbb{R}}(\alpha''',\alpha^{iv}) * (I_A/(I_A)_{\gamma+1}) .$$

By Lemma 1, it is enough to see that in $V^{(I_A)_{\gamma+1}}$, α^+, α^{++}, α^{+3} carry saturated ideals.

Since α', $\alpha'' \in A$, there is an almost huge embedding in V, $j_{\alpha',\alpha''}$ such that $j_{\alpha'}(\alpha') = \alpha''$ and $\mathrm{crit}(j_\alpha) = \alpha'$. By Magidor's modification of the Kunen theorem (with $\mathbb{P}' = (I_A)_\alpha * \overline{\mathbb{R}}(\alpha,\alpha')$ and $\varphi = \overline{\mathbb{R}}(\alpha',\alpha'')$ and $j_{\alpha',\alpha''}$ the almost huge embedding) there is a saturated ideal on α' in $V^{(I_A)_\alpha * \overline{\mathbb{R}}(\alpha,\alpha') * \overline{\mathbb{R}}(\alpha',\alpha'')}$.
Since $(I_A)_{\alpha+1}/((I)_\alpha * \mathbb{R}(\alpha,\alpha') * \mathbb{R}(\alpha',\alpha''))$ is α''-closed we can apply Lemma 1 to get that there is a saturated ideal on α^+ in $V^{(I_A)_{\alpha+1}}$. Entirely similar arguments work for α^{+2} and α^{+3}.

(c) is standard.

In our extension κ will become $\aleph_\omega$ using a technique of Magidor [10]. Let $V' = V^{\mathbb{P}_{\leq\kappa}}$.

<u>Lemma 8</u>. Let $\gamma < \beta$ be inaccessible elements of A in V'. In $(V')^{C(\gamma^{+4},\beta)}$, γ^{+4} carries a $(\gamma^{+4})^+$-saturated ideal.

<u>Proof</u>. In $V'^{C(\gamma^{+4},\beta)}$, $(\gamma^{+4})^+ = (\beta^+)^{V'}$. Let α_1 and α_2 be the stages where γ and β are the least elements of A not mentioned in $(I_A)_{\alpha_1}$ and $(I_A)_{\alpha_2}$ respectively.

By Proposition 2, $\mathbb{P}_{\leq\kappa}/(\mathbb{P}_{\alpha_2} * \overline{\mathbb{R}}(\beta,\beta'))$ is β'-closed. Further, $C(\gamma^{+4},\beta)^{V'} = C(\gamma^{+4},\beta)^{V^{\mathbb{P}_{\alpha_1+1}}}$ and has cardinality β. Thus we are in a position to apply Lemma 1 in $V^{\mathbb{P}_{\alpha_2} * \mathbb{P}(\beta,\beta')}$. Note $(V')^{C(\gamma^{+4},\beta)}$ is an extension of $V^{\mathbb{P}_{\alpha_2} * \mathbb{P}(\beta,\beta')}$ of the form

$$(\beta' \text{ c. c.}) \times (\beta'\text{-closed})$$

$$(\text{i.e. } C(\gamma^{+4},\beta) \times \mathbb{P}_{\leq\kappa}/(\mathbb{P}_{\alpha_2} * \mathbb{P}(\beta,\beta')))$$

Thus by Lemma 1, it is enough to see that γ^{+4} carries a β'-saturated ideal in $V^{\mathbb{P}_{\alpha_2} * \overline{\mathbb{R}}(\beta,\beta') * \overline{C}(\gamma^{+4},\beta)}$. Let $Q' = \mathbb{P}_{\alpha_1} * \overline{\mathbb{R}}(\gamma,\gamma') * \overline{\mathbb{R}}(\gamma',\gamma'') * \overline{\mathbb{R}}(\gamma'',\gamma''')$. Then $\mathbb{P}_{\alpha_1+1} = Q' * \mathbb{R}(\gamma''',\gamma^{iv})$. Let $j_{\gamma^{iv},\beta'}$ be the almost huge embedding in V with $\text{crit}(j_{\gamma^{iv},\beta'}) = \gamma^{iv}$ and $j_{\gamma^{iv},\beta'}(\gamma^{iv}) = \beta'$. Then

$$j_{\gamma^{iv},\beta'}(Q' * \overline{\mathbb{R}}(\gamma''',\gamma^{iv})) = Q' * \mathbb{R}(\gamma''',\beta') .$$

Since $Q' * \overline{\mathbb{R}}(\gamma''',\gamma^{iv}) \subseteq V_{\gamma^{iv}}$ and $Q' * \overline{\mathbb{R}}(\gamma''',\gamma^{iv})$ is γ^{iv}-c. c., $j \mid Q' * \mathbb{R}(\gamma''',\gamma^{iv}) = \text{id}$ and hence

$$\text{id} : Q' * \overline{\mathbb{R}}(\gamma''',\gamma^{iv}) \hookrightarrow Q' * \overline{\mathbb{R}}(\gamma''',\beta') .$$

Thus by our construction of $\overline{\mathbb{R}}(\gamma''',\beta')$, there is an embedding

$$i : Q' * \overline{\mathbb{R}}(\gamma''',\gamma^{iv}) * [\overline{C}(\gamma^{iv},\beta) \times \{(\mathbb{P}_{\alpha_2}/\mathbb{P}_{\alpha_1+1}) * \overline{\mathbb{R}}(\beta,\beta')\}] \hookrightarrow Q' * \overline{\mathbb{R}}(\gamma''',\beta') .$$

Since $\overline{C}(\gamma^{iv},\beta) \times \{(\mathbb{P}_{\alpha_2}/\mathbb{P}_{\alpha_1+1}) * \overline{\mathbb{R}}(\beta,\beta')\}$ is γ^{iv}-closed in $V^{Q' * \mathbb{R}(\gamma''',\gamma^{iv})}$, Magidor's remarks about Kunen's theorem apply with $\mathbb{P}' = Q' * \overline{\mathbb{R}}(\gamma''',\gamma^{iv})$ and $Q = \overline{C}(\gamma^{iv},\beta) \times \{(\mathbb{P}_{\alpha_2}/\mathbb{P}_{\alpha_1+1}) * \overline{\mathbb{R}}(\beta,\beta')\}$ (see page 4) and we conclude that in $V^{Q' * \mathbb{R}(\gamma''',\gamma^{iv}) * [\overline{C}(\gamma^{iv},\beta) \times \{\mathbb{P}_{\alpha_2}/\mathbb{P}_{\alpha_1+1} * \mathbb{R}(\beta,\beta')\}]}$, γ^{iv} carries a $(\gamma^{iv})^{+}$-saturated ideal. Pictorially:

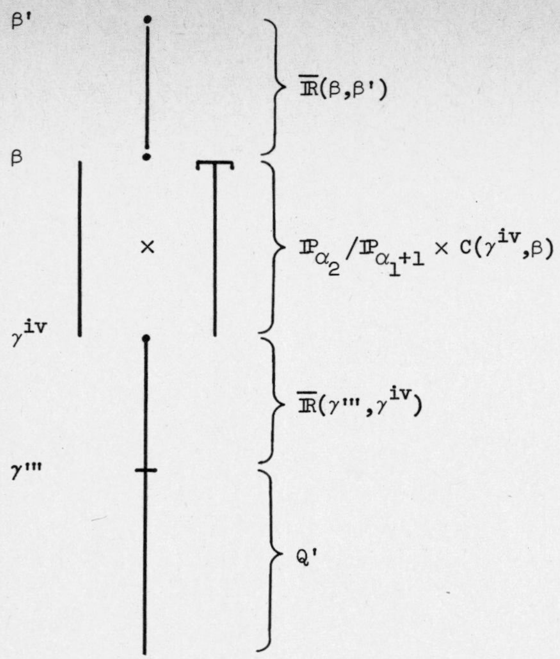

We are now ready to define our forcing conditions for the model in Theorem 3.

This forcing is a direct variant of the Magidor forcing [10]. However, in our forcing we collapse the points on the Prikry sequence. This necessitates a slightly different argument which we show below.

For the moment we will regard V' as our ground model. Note that in V', κ^+ carries a saturated ideal. We will define two forcings, $\mathbb{P}$ and $\mathbb{P}^{\pi}$ with a projection map

$$\pi : \mathbb{P} \to \mathbb{P}^{\pi} .$$

Our final model will be $(V')^{\mathbb{P}^{\pi}}$. We will use $\mathbb{P}$ to show that $\mathbb{P}^{\pi}$ has nice Prikvy-type properties.

Let μ_3 be the κ^{+3}-supercompact measure on $\mathcal{P}_{\kappa}(\kappa^{+3})$ in V' that is induced by μ_5. Let

$$N = \{x : x \in \mathcal{P}_{\kappa}(\kappa^{+3}) \text{ and } x \cap \kappa \text{ is an inaccessible cardinal, } x \cap \kappa \in A\} .$$

Standard arguments show that N has μ_3-measure one and that μ_3 is closed under the following kind of diagonal intersection:

Suppose $\rho : \mathcal{P}_{\kappa}(\kappa^{+3}) \times V_{\kappa} \to \mathcal{P}(\mathcal{P}_{\kappa}(\kappa^{+3}))$ is such that for all (x,y), $\rho(x,y)$ has μ_3-measure one. Define $\Delta\rho = \{z :$ for all $x < z$ and for all $y \in V_{\mathrm{crit}(z)}$

$z \in \rho(x,y)\}$. Then $\Delta\rho$ has μ_3-measure one. For $x \in \mathcal{P}_\kappa(\kappa^{+3})$, let $x \cap \kappa = \kappa_x$.

A condition $p \in \mathbb{P}$ will be of the form

$$p = (x_0\, p_0\, x_1\, p_1 \cdots x_n\, p_n\; f\; B)$$

where

(a) $x_i \in N$, $x_i < x_{i+1}$

(b) $p_i \in C(\kappa_{x_i}^{iv}, \kappa_{x_{i+1}})$

(c) $B \subseteq \mathcal{P}_\kappa(\kappa^{+3}) \cap N$ and B has μ_3-measure one.

(d) $f : B \to V_\kappa$ and for all $x \in B$, $f(x) \in C(\kappa_x^{iv}, \kappa)$

If $q = (x_0\, q_0\, x_1\, q_1 \cdots x_n\, q_n\, y_1\, r_1\, y_2\, r_2 \cdots y_m\, r_m\; g\; C)$ $q \leq_{\mathbb{P}} p$ iff

(a) $y_i \in B$ for all $i \leq m$

(b) $q_i \leq p_i$ for all $i \leq n$

(c) $r_i \leq f(y_i)$ for all $i \leq m$

(d) $C \subseteq B$

(e) for all $x \in C$, $g(x) \leq f(x)$.

$\mathbb{P}$ has the Prikry property in the following sense:

<u>Lemma 9</u>. Let $b \in \mathcal{B}(\mathbb{P})$.

(a) There is a condition (g,D) that decides b. (i.e. $(g,D) \parallel b$.)

(b) If $p = (x_0\, p_0 \cdots x_n\, p_n\; g\; D) \in \mathbb{P}$ there is a condition $q \leq p$ such that $a = (x_0\, q_0 \cdots x_n\, q_n\; g'\; D') \parallel b$ and the length of q is equal to the length of p.

A proof of this appears in Magidor [10], or see [4]. The fact that the conditions have a slightly different form does not change the proof.

If $G \subseteq \mathbb{P}$ is generic and $p = (x_0\, p_0 \cdots x_n\, p_n\; g\; D) \in G$, then G_p is defined to be $\{(x_0\, q_0 \cdots q_{n-1}\, x_n\, q_n)$: there is a $q \in G$, length(q) = length(p), $q = (x_0\, q_0 \cdots q_{n-1}\, x_n\, q_n\; g'\; D')\}$. If $\lambda < \kappa$ then $G \restriction \lambda$ is defined to be $\{(x_0, q_0 \cdots x_n\, q_n)$: there is a $q \in G$ $q = (x_0\, q_0 \cdots x_n\, q_n\; g\; D$ and $\sup q_n < \lambda\}$.

Using Lemma 9 we can easily establish the following corollary:

<u>Corollary 1</u>. (a) Suppose $p \in \mathbb{P}$ and $p \Vdash \dot{\tau} \subseteq \alpha$ and $\alpha < \kappa_x$ for some $x \in P_\kappa(\kappa^{+3})$ occurring in p. Then $p \Vdash \dot{\tau} \in V[G_p]$.

(b) If $p = (x_0\, p_0 \cdots x_n\, p_n\; g\; D)$ and $Q = C(\omega, \kappa_{x_0})$. Let $H \subseteq C(\omega, \kappa_{x_0})$ be generic. If $p \Vdash_{\mathbb{P}} \dot{\tau} \subseteq \alpha$ and $\tau \in V'[H]^{\mathbb{P}}$ and $\alpha < \kappa_{x_n}$, then $p \Vdash \dot{\tau} \in V'[\dot{G}_p \times H]$. ($\dot{G}_p$ is a canonical term for the generic object $G \subseteq \mathbb{P}$ restricted to p.)

We now make the following claim.

<u>Claim</u>. If $G \times H \subseteq \mathbb{P} \times C(\aleph_0, \kappa_{x_0})$ is generic over V', then $V'[G \times H] \models$ for all $n \in \omega$, $\aleph_n$ carries an $\aleph_{n+1}$-saturated ideal.

<u>Proof</u>. Let $\langle \kappa_0 : i \in \omega \rangle$ be the sequence of κ_{x_i} such that x_i occurs in some condition in G. By Magidor's arguments, κ becomes $\aleph_\omega$. By the corollary, it is enough to see that for each i,

$V'[G \restriction \kappa_i \times H] \models$ for all $n \in \omega$ ($\aleph_n < \kappa_i$ implies $\aleph_n$ carries a saturated ideal) .

Our sequence of $\aleph_n$'s in $V[G \times H]$ is

$$\aleph_0 = \aleph_0$$
$$\aleph_1 = \kappa_0'$$
$$\aleph_2 = \kappa_0''$$
$$\aleph_3 = \kappa_0'''$$
$$\aleph_4 = \kappa_0^{iv}$$
$$\aleph_5 = \kappa_1'$$
$$\vdots$$
$$\aleph_{4n+k} = \kappa_n^{+k} \qquad 1 \leq k \leq 4$$

If $p = (x_1 p_1 \cdots x_{n-1} p_{n-1} g D)$, then in V', $G \restriction \kappa_n \times H$ is generic over

$$[C(\kappa_1^{iv}, \kappa_2) \times \cdots \times C(\kappa_{n-1}^{iv}, \kappa_n)] \times C(\omega, \kappa_0) .$$

Hence, if $\aleph_m = \kappa_i^k$, $1 \leq k \leq 3$, we claim $\aleph_m$ carries a saturated ideal in $V'[G \restriction \kappa_n \times H]$ since it does in V' and our forcing is of the form

$$(\text{Cardinality} < \aleph_m) \times (\aleph_{m+1}\text{-closed}) .$$

Any forcing of cardinality $< \aleph_m$ preserves a saturated ideal on $\aleph_m$ by Lemma 2. Thus by Lemma 1 we can conclude that $\aleph_m$ has a saturated ideal in $V'[G \restriction \kappa_n \times H]$. If $\aleph_m = \kappa_i^{iv}$ for some $i < n$, by Lemma 8, $\aleph_m$ carries a saturated ideal in $V'^{C(\kappa_i^{iv}, \kappa_{i+1})}$. To pick up $V'[G \restriction \kappa_n \times H]$ from $(V')^{C(\kappa_i^{iv}, \kappa_{i+1})}$ we need forcing of the form

$$(\text{Cardinality} < \aleph_m) \times ((\aleph_{\kappa_{i+1}}^{iv}, \infty)\text{-distributive})$$

(namely:

$$[C(\omega, \kappa_0') \times C(\kappa_1^{iv}, \kappa_2) \times \cdots \times C(\kappa_{i-1}^{iv}, \kappa_i)] \times [C(\kappa_{i+1}^{iv}, \kappa_{i+2}) \times \cdots] \;) .$$

Thus by Lemma 2, $\aleph_m$ carries a saturated ideal in $V'[G \upharpoonright \kappa_n \times H]$. This establishes the claim.

Unfortunately, in $V'[G \times H]$, $(\kappa^+)^{V'}$ has cardinality κ, since $\bigcup\{x_i : x_i$ occurs in some $p \in G\} = (\kappa^\kappa)^{V'}$. We now define a κ^+-c. c. partial ordering $\mathbb{P}^\pi$, a condition $p_0 \in \mathbb{P}$ and a projection map

$$\pi : \mathbb{P}/p_0 \to \mathbb{P}^\pi .$$

If $G \times H \subseteq \mathbb{P} \times (\omega, \kappa_{x_0})$ then $V'[G \times H]$ and $V'[\pi''G \times H]$ will have the same bounded subsets of κ.

Let U be the measure on κ induced by μ_3. Let Q be the following partial ordering: $p \in Q$ iff $p : X \to V_\kappa$ for a set $X \subseteq \kappa$, $X \in U$ and if $\alpha \in X$, $p(\alpha) \in \mathcal{B}(C(\alpha^{+4}, \kappa))$.

If p and q are elements of Q, $p \leq_Q q$ iff $\{\alpha : p(\alpha) \leq q(\alpha)\} \in U$.

If $B \subseteq \mathcal{P}_\kappa(\kappa^{+3})$ is a set of measure one (for μ_3) and $g : B \to V_\kappa$ is a function such that for all $x \in \mathcal{B}$, $g(x) \in C(\kappa_x^{+4}, \kappa)$, we can associate an element $\bar{g}_B$ of Q with domain $\pi(B) = \{\alpha : \exists\, x \in B,\ \kappa_x = \alpha$ and $\alpha \in A\}$ by

$$\bar{g}_B(\alpha) = \bigvee_{\substack{\kappa_x = \alpha \\ x \in B}} g(x) .$$

Then if $q \in C(\alpha^{+4}, \kappa)$ for some $\alpha \in \pi(B)$ and $q \wedge \bar{g}_B(\alpha) \neq 0$, we can find an $x \in B$, $\kappa_x = \alpha$ and $q \wedge g(x) \neq 0$.

A function g as above naturally gives rise to a filter $\mathfrak{F}_g$, namely

$$\mathfrak{F}_g = \{p \in Q : \text{there is a set } B \in \mu_3,\ g \text{ is defined on } B \text{ and } p \geq \bar{g}_B .\} .$$

If $g : B \to R_\kappa$, $h : C \to R_\kappa$ are functions as above with $\mu_3(B) = \mu_3(C) = 1$ and $g \leq h$ almost everywhere with respect to μ_3, then $\mathfrak{F}_g \supseteq \mathfrak{F}_h$. (If $g \leq h$ almost everywhere we will say $g \leq_{\mu_3} h$.)

<u>Claim</u>. If $\langle g_\alpha : \alpha < \beta\rangle \subseteq \{h :$ there is a $B \subseteq N$, $\mu_3(B) = 1$, $h : B \to R_\kappa$, and for all $x \in B$, $h(x) \in C(\kappa_x^{+4}, \kappa)\}$ is a $\leq_{\mu_3}$-descending sequence of length $\leq \kappa^{+3}$ then there is a g_β and a set B_β, $g_\beta : B_\beta \to R_\kappa$ and for all $x \in B_\beta$, $g(x) \in C(\kappa_x^{+4}, \kappa)$ and for all $\alpha < \beta$, $g_\beta \leq_{\mu_3} g_\alpha$.

<u>Proof</u>. We view each g_α as $[g_\alpha] \in V'^{\mathcal{P}_\kappa(\kappa^{+3})}/\mu_3 = M$. Then $[g_\alpha] \in C(\kappa^{+4}, j(\kappa))^M$. If $\alpha < \alpha' < \beta$ then $[g_\alpha] \geq [g_{\alpha'}]$. $C(\kappa^{+4}, j(\kappa))$ is κ^{+4}-closed in M, hence, using κ^{+3}-supercompactness, there is a $q \in C(\kappa^{+4}, j(\kappa))^M$, such that $q \leq [g_\alpha]$ for all $\alpha < \beta$. Let $g_\beta : \mathcal{P}_\kappa(\kappa^{+3}) \to R_\kappa$ be such that $[g_\beta] = q$. Let $B_\beta = \{x : g\ (x) \in C(\kappa_x^{+4}, \kappa)\}$. Then it is easy to verify that q_β and B_β work in the claim.

We now build a sequence $\langle g_\alpha : \alpha < \kappa^{++}\rangle$ such that if $g \le_{\mu_3} g_\alpha$ for all α and $h \le_{\mu_3} g$, then $\mathfrak{F}_g = \mathfrak{F}_h$.

To do this: At successor stages $\alpha + 1$ pick a $g_{\alpha+1} \le_{\mu_3} g_\alpha$ such that $\mathfrak{F}_{g_{\alpha+1}} \supsetneq \mathfrak{F}_{g_\alpha}$ if possible. At limit stages β, pick a $g_\beta \le_{\mu_3} g_\alpha$ for all $\alpha < \beta$. Let $g \le_{\mu_3} g_\alpha$ for all $\alpha < \kappa^{++}$. Since $2^\kappa = \kappa^+$ in V', we cannot have a strictly increasing chain of filters of order type κ^{++}. Hence for a tail of κ^{++}, $\mathfrak{F}_{g_\alpha}$ is constant. From this we conclude that if $h \le g$, $\mathfrak{F}_g = \mathfrak{F}_h$. (In fact it is possible to show that $\mathfrak{F}_g$ is an ultrafilter.) Fix such a g.

A condition in the projected forcing $\mathbb{P}^\pi$ will be of the form

$$p = (\alpha_0\, p_0\, \alpha_1\, p_1 \cdots \alpha_n\, p_n\; b\; B)$$

where

(a) $\alpha_0 < \alpha_1 < \alpha_2 < \cdots < \alpha_n \in A$
(b) $p_i \in C(\alpha_i^{+4}, \alpha_{i+1})$
(c) $B \subseteq \kappa$, $B \in U$
(d) $b \in Q$, $b \in \mathfrak{F}_g$ and b is defined on B.

$$(\alpha_0\, p_0'\, \alpha_1\, p_1' \cdots \alpha_n\, p_n'\, \beta_1\, q_1 \cdots \beta_m\, q_m\; c\; C) \le (\alpha_0\, p_0\, \alpha_1\, p_1 \cdots \alpha_n\, p_n\; b\; B)$$

iff

(a) $p_i' \le p_i$ for all $i \le n$
(b) for all $j \le m$, $\beta_j \in B$
(c) for all j, $q_j \le b(\beta_j)$
(d) $C \subseteq B$
(e) $c \le b$ in Q.

If $D \in \mu_3$, let $\pi(D) = \{\alpha :$ there is an $x \in D$ $\kappa_x = \alpha\}$. Then $\pi(D) \in U$. Let $p_0 = (g, \mathrm{dom}\, g)$, then $p_0 \in \mathbb{P}$. If $p \in \mathbb{P}$, $p \le p_0$, $p = (x_0\, p_0\, x_1\, p_1 \cdots x_n\, p_n\; h\; B)$, let $\pi(p) = (\kappa_{x_0}\, p_0\, \kappa_{x_1}\, p_1 \cdots \kappa_{x_n}\, p_n\; \bar{h}_B\; \pi(B))$. Then π is clearly order preserving. By Lemma 4, to see that this is a projection map we must show that for all $p \in \mathbb{P}/p_0$ there is a $q \in \mathbb{P}^\pi$, $q \le \pi(p)$ such that for all $q' \le q$ there is a $p' \le p$ with $\pi(p') \le q'$. (It is here that this argument differs slightly from earlier arguments of this type, hence we include it in the manuscript.)

Let $p \in \mathbb{P}/p_0$, $p = (x_0\, p_0 \cdots x_n\, p_n\; h\; B)$, and $\pi(p) = (\kappa_{x_0}\, p_0 \cdots \kappa_{x_n}\, p_n\; \bar{h}_B, \pi(B))$. We want to define $q \subseteq \pi(p)$. We assume that for all $y \in B$, $x_n < y$. We will build a descending sequence $\langle A_i : i \in \omega\rangle \subseteq \mu_3$ with the property that for all $x \in A_{i+1}$ and for all $\alpha < \kappa_x$, if for some p there is a $y \in A_i$, $\alpha = \kappa_y$,

$$p \in C(\alpha^{+4}, \kappa_x)$$

$$p \wedge h_{A_i}(\alpha) \ne 0 \qquad (\text{in } \mathbb{B}(C(\alpha^{+4}, \kappa)))$$

then there is a $y \in A_i$, $\kappa_y = \alpha$ such that

(a) $p \wedge h(y) \neq 0$

(b) $y < x$

(c) $h(y) \in C(\alpha^{+4}, \kappa_x)$

To do this we exploit $C(\alpha^{+4}, \kappa_x) \subseteq C(\alpha^{+4}, \kappa)$.

Let $A_0 = B \cap \{x : \kappa_x > \sup p_n \text{ and } x_n < x\}$. Assume A_i has been defined. For each $\alpha \in \pi(A_i)$ and each regular $\beta > \alpha$, pick a set $Y_\beta \subseteq A_i$ of cardinality β such that

$$\{h(y) \cap C(\alpha^{+4}, \beta) : y \in A_i\} = \{h(y) \cap C(\alpha^{+4}, \beta) : y \in Y_\beta\} .$$

Let $X_\beta = \{x \in A_i : \text{for all } y \in Y_\beta,\ y < x \text{ and } \kappa_x > \sup h(y)\}$. Let $Z_\alpha = \underset{\beta<\kappa}{\Delta} X_\beta = \{x : \text{for all } \beta < \kappa_x,\ x \in X_\beta\}$. Then, Z_α has μ_3-measure one and if $x \in Z_\alpha$, $p \in C(\alpha^{+4}, \kappa_x)$ and p is compatible with $\bigvee_{\substack{y \in A_i \\ \kappa_y = \alpha}} h(y)$, there is a $y \in A_i$, $\kappa_y = \alpha$ and p is compatible with $h(y)$. But, $h(y) \cap (\alpha^{+4}, \kappa_x)$ is bounded in κ_x, hence there is a $\beta < \kappa_x$, $\beta > \sup p$, $h(y) \cap (\alpha^{+4}, \kappa_x) = h(y) \cap (\alpha^{+4}, \beta)$. Pick $y' \in Y_\beta$ such that

$$h(y) \cap (\alpha^{+4}, \beta) = h(y') \cap (\alpha^{+4}, \beta) .$$

Then $y' < x$, $\sup h(y') < x$ and $h(y')$ is compatible with p.

Let $A_{i+1} = \underset{\alpha<\kappa}{\Delta} Z_\alpha = \{x : \text{for all } \alpha < \kappa_x,\ x \in Z_\alpha\}$. Then A_{i+1} has the desired property.

Let $q = (\kappa_{x_0}\, p_0\, \kappa_{x_1}\, p_1 \cdots \kappa_{x_n}\, p_n\, h_{\cap A_i}\ \pi(\cap A_i))$. Let $q' \leq q$, $q' = (\kappa_{x_0}\, q_0 \cdots \kappa_{x_n}\, q_n\, \alpha_1\, q_{n+1} \cdots \alpha_m\, q_{n+m}\ c\ C)$. We must define $p' \in \mathbb{P}$ with $\pi(p') \leq q'$. Pick $y_m \in A_m$, $h(y_m) \wedge q_{n+m} \neq 0$ and $\kappa_{y_m} = \alpha_m$. Pick $p'_{n+m} \leq h(y_m) \wedge q_{n+m}$. Since $y_m \in A_m$ there is a $y_{m-1} \in A_{m-1}$, $\kappa_{y_{m-1}} = \alpha_{m-1}$, $\sup h(y_{m-1}) < \kappa_{y_m}$ and $h(y_{m-1}) \wedge q_{n+m-1} \neq 0$. Pick such a y_{m-1} and $p'_{n+m-1} \leq q_{n+m-1} \wedge h(y_{m-1})$. Continue in this way for $0 \leq k < m$ to pick $y_{m-(k+1)} \in A_{m-(k+1)}$ with $y_{m-(k+1)} < y_{m-k}$, $\kappa_{y_{m-(k+1)}} = \alpha_{m-(k+1)}$, $\sup h(y_{m-(k+1)}) < \kappa_{y_{m-k}}$ and $q_{n+m-(k+1)}$ is compatible with $h(y_{m-(k+1)})$. Pick $p'_{m-(k+1)} \leq q_{n+m-(k+1)} \wedge h(y_{m-(k+1)})$. Let $p' = (x_1\, q_1\, x_2\, q_2 \cdots x_n\, q_n\, y_1\, p'_{n+1}\, y_2\, p'_{n+2} \cdots y_m\, p'_{n+m}\ g\ D)$ where $g \leq h$, $D \subseteq \cap A_i$ and $\bar{g}_D \leq c$ and $\pi(D) \subseteq C$.

It is clear that $p' \leq p$ and $\pi(p') \leq q'$. Thus $\pi : \mathbb{P}/p_0 \to \mathbb{P}^\pi$ is a projection map.

Let $G \times H \subseteq \mathbb{P} \times C(\aleph_0, \kappa_0)$ be V'-generic. Let G^π be $\pi''G$. Then $G^\pi \times H$ is V'-generic for $\mathbb{P}^\pi \times C(\aleph_0, \kappa_0)$. If $\tau \subseteq \alpha < \kappa$, $\tau \in V'[G \times H]$ then $\tau \in V'[G_p \times H]$ for some $p \in G$ and from this we conclude $\tau \in V'[G^\pi \times H]$.

Hence $V'[G^{\pi} \times H] \models$ for all $n \in \omega$, $\aleph_n$ carries a saturated ideal.

To see that $\aleph_{\omega+1}$ carries a saturated ideal it suffices to show that $P^{\pi} \times C(\aleph_0, \kappa_0)$ is κ-centered.

To see this we define an equivalence relation on $\mathbb{P}^{\pi} \times C(\aleph_0, \kappa_0)$ with κ equivalence classes and show that if $p_1, \ldots, p_n$ lie in the same equivalence class then $\bigwedge_{i=1}^{n} p_i \neq 0$.

If $p = ((\alpha_1, p_1 \cdots \alpha_n\, p_n\ c\ C), p') \in \mathbb{P}^{\pi} \times C(\aleph_0, \kappa_0)$ and $q = ((\beta_1\, q_1 \cdots \beta_m\, q_m\ b\ B), q') \in \mathbb{P}^{\pi} \times C(\aleph_0, \kappa_0)$ set $p \sim q$ iff

(a) $n = m$ and for all $k \leq n$ $\alpha_k = \beta_k$, $p_k = q_k$

(b) $p' = q'$.

There are only $|\kappa^{\omega}| = \kappa$ equivalence classes. If we have equivalent elements $p_1, \ldots, p_n$ then to find a $p \leq \bigwedge_{i=1}^{n} p_i$ we merely take the intersection of the boolean values in Q (this is non-zero since they all lie in the filter $\mathfrak{F}_g$) and the intersection of the sets of measure one. This establishes Theorem 3.

With a little more work, starting from a 2-huge cardinal, we can also get $(\aleph_{n+1}, \aleph_n) \twoheadrightarrow (\aleph_{m+1}, \aleph_m)$ for all $n > m$.

To prove Con (ZFC + there is a huge cardinal) $\Rightarrow$ Con (ZFC + all regular cardinals κ carry a saturated ideal) we do the same preparatory forcing $\mathbb{P}_{\leq\kappa}$. We then do the modified Radin forcing that appears in [4].

The auxilliary function again takes its values in the appropriate collapse algebra. There are no essential new ideas needed to do this that do not appear in this paper. Thus we omit repeating the arguments in [4].

References

[1] J. Baumgartner, Iterated forcing. Manuscript.

[2] M. Foreman, Large cardinals and strong model theoretic transfer properties. To appear in TAMS.

[3] M. Foreman and R. Laver, A reflection property of graphs. In preparation.

[4] M. Foreman and H. Woodin, The G. C. H. can fail everywhere. In preparation.

[5] T. Jech, Set Theory, Academic Press, New York, San Francisco, London, 1978.

[6] A. Kanamori and M. Magidor, The evolution of large cardinal axioms in set theory, Lecture Notes in Mathematics, Springer-Verlag, Vol. 669, 1978.

[7] K. Kunen, Saturated ideals, J. Symbolic Logic 43 (1978), 65-76.

[8] R. Laver, Saturated ideals and non-regular ultrafilters, Proc. 1980 ASL meeting in Patras, Greece.

[9] R. Laver, An $(\aleph_2, \aleph_2, \aleph_0)$-saturated ideal on ω_1. Manuscript.

[10] M. Magidor, On the singular cardinals problem I, Israel Journal of Math. 28 (1977), 1-31.

[11] R. Solovay, W. Reinhardt and A. Kanamori, Strong axioms of infinity and elementary embeddings, Annals of Math. Logic 13 (1978), 73-116.

SOME RESULTS IN THE WADGE HIERARCHY OF BOREL SETS

A. Louveau

C. N. R. S.
Equipe d'Analyse
Université Paris VI, France

Department of Mathematics
University of California
Los Angeles, California 90024

This paper has two goals: First, to provide construction principles, by means of boolean operations, of the Wadge classes of Borel sets, and in a second step, to use these construction principles to define lightface versions of the Wadge classes, and prove that the notion of Wadge class, roughly speaking, is a Δ^1_1 notion: If a Δ^1_1 set is in some (boldface) Wadge class Γ, it belongs to the corresponding Δ^1_1-recursive lightface class.

The necessary background concerning Wadge's hierarchy can be found in Van Wesep [1976]. Let us recall that a family $\Gamma \subseteq (\omega^\omega)$ is a <u>class</u> if it is closed under inverse images by continuous functions from ω^ω into itself. If Γ is of the form $[A] = \{f^{-1}(A) \mid f : \omega^\omega \to \omega^\omega, \text{ continuous}\}$ for some set A, it is a <u>Wadge class</u> (the Wadge class of A). For a set A, $\check{A}$ denotes its complement, $\check{A} = \omega^\omega - A$, and $\check{\Gamma} = \{\check{A} \mid A \in \Gamma\}$ is the <u>dual</u> class of the class Γ. $\Delta = \Delta(\Gamma)$ is the <u>ambiguous class</u> associated with Γ, and is defined by $\Delta = \Gamma \cap \check{\Gamma}$. A class Γ is <u>self dual</u> if $\Gamma = \check{\Gamma}$.

The Wadge hierarchy is obtained by (partially) ordering the Wadge classes by strict inclusion: $\Gamma < \Gamma'$ if $\Gamma \subseteq \Gamma'$ and $\Gamma \neq \Gamma'$. We similarly define $\Gamma \leq \Gamma'$ if $\Gamma \subseteq \Gamma'$. This ordering admits a game theoretical analysis: If A, B are two subsets of ω^ω, let $G_W(A, B)$ be the game where Players I and II play alternatively integers, Player I constructing $\alpha \in \omega^\omega$, and Player II having the possibility of passing, as long as he constructs $\beta \in \omega^\omega$. Player II wins this game if $\alpha \in A \leftrightarrow \beta \in B$. It is easy to check that II has a winning strategy in $G_W(A, B) \Longleftrightarrow [A] \leq [B]$.

Let us now consider only Wadge classes of Borel sets. Then we can use Borel Determinacy (Martin [1975]). This gives that $<$ is almost a linear ordering (Wadge [1976]): The only class not comparable to Γ is the class $\check{\Gamma}$ in case Γ is non self-dual. So by identifying twin pairs $(\Gamma, \check{\Gamma})$ of non self-dual classes, we obtain a linear ordering. Moreover by Martin [1973], the ordering $<$ is well-founded, so that we can associate with each Wadge class of Borel sets Γ an ordinal $o(\Gamma)$. The pattern of self-dual and non self-dual classes is as follows (see Van Wesep [1976]): Self-dual and non self-dual twin pairs alternate at successor stages. The hierarchy begins with the twin pair $\{\emptyset\}$, $\{\omega^\omega\}$; at limit stages of cofinality ω stands a self-dual class, and at limit stages of cofinality ω_1 a non self-dual twin pair.

Of course, the Wadge hierarchy is a refinement of the classical hierarchies, the Borel hierarchy $(\underset{\sim}{\Sigma}^0_\xi, \underset{\sim}{\Pi}^0_\xi)_{\xi<\omega_1}$, and the hierarchy of differences of Hausdorff and Kuratowski. But it contains also a lot of more "exotic" classes. The picture presented in part 1 closely follows unpublished results of Wadge [1976], except that Wadge's description exhibits, for each non self-dual class Γ, a set A for which $\Gamma = [A]$, whereas we define a boolean operation which enables to construct Γ in terms of the preceding classes.

I would like to thank John Steel for giving me access to Wadge's papers, and also for enlightening discussions about Wadge classes.

§1. A description of Wadge classes of Borel sets. First, let us make some heuristic comments on what follows. The usual Borel hierarchy of sets, $\{\underset{\sim}{\Sigma}^0_\xi \mid \xi < \omega_1\}$, is our starting point in analysing the Wadge classes. This obviously does not give a complete description, and we certainly must refine it by adding, between $\underset{\sim}{\Sigma}^0_\xi$ and $\underset{\sim}{\Sigma}^0_{\xi+1}$, the hierarchy of differences of $\underset{\sim}{\Sigma}^0_\xi$ sets. This gives a first refinement, which again happens to be insufficient, at least for $\xi \geq 2$. The first expectation is that by refining again (may be a couple of times), one should reach the complete picture of Wadge degrees. It is indeed what happens, and we shall define a set of levels by using successive refinements; but in order to obtain the picture between $\underset{\sim}{\Sigma}^0_\xi$ and $\underset{\sim}{\Sigma}^0_{\xi+1}$, ξ refinements are necessary. And because the Wadge ordering is a well-ordering, these refinements are not well-ordered by inclusion. In fact, it turns out that the reverse ordering is well-founded, and so the ordinal we shall associate with the refinements will measure the "degree of simplicity" of each level, rather than its degree of complexity: the hierarchy of differences is given level ξ, whereas the most complicated classes are those of level 1. At this last level occur all successor classes and all limit classes of cofinality ω. The "simplicity" of each class may be measured, in mathematical terms, by its closure properties. We shall see (lemma 1.4 below) that the closure properties do increase with the level of the class.

For constructing the classes, we have selected a small set of operations, which is certainly not the least possible one, but is convenient for a nice description of the classes. Starting from the classes $\underset{\sim}{\Sigma}^0_\xi$, and applying successively these operations will give the desired description. The main point here is that each operation increases the Wadge ordinal, but decreases the level: Intuitively, it means that the resulting class is more complicated than the original one. A final word of comment: By our general knowledge of the Wadge hierarchy, it is not necessary to give a construction principle from below for the self-dual classes, which are exactly the Δ-parts of successor non self-dual classes, and among the non self-dual classes, we can choose to describe only one of the twin dual classes. One mathematically nice way for doing this choice is given by a result of Steel [1980]:

Among each twin pair, exactly one class has the separation property. For technical reasons, we have chosen to describe the other ones, i.e. the non self-dual classes which do not possess the separation property. In particular, we begin the construction with the $\underset{\sim}{\Sigma}^0_\xi$ classes, not the $\underset{\sim}{\Pi}^0_\xi$ ones.

Let us now begin with the definition of the operations.

1.1. Definition. (a) Differences (Hausdorff, Kuratowski). Let $\xi \geq 1$ be a countable ordinal. If $\langle C_\eta \mid \eta < \xi \rangle$ is an increasing sequence of sets, the set $A = D_\xi(\langle C_\eta \mid \eta < \xi \rangle)$ is defined by

$$A = \cup \{C_\eta - \bigcup_{\eta' < \eta} C_{\eta'} \mid \eta \text{ odd}, \eta < \xi\} \quad \text{for } \xi \text{ even}$$

$$\cup \{C_\eta - \bigcup_{\eta' < \eta} C_{\eta'} \mid \eta \text{ even}, \eta < \xi\} \quad \text{for } \xi \text{ odd}$$

[So $D_1(\langle C_0 \rangle) = C_0$, $D_2(\langle C_0, C_1 \rangle) = C_1 - C_0$, ... This definition is not the usual one, as given in Kuratowski [1966], which deals with decreasing sequences -- and is applied to $\underset{\sim}{\Pi}^0_\xi$.]

If Γ is some class, $D_\xi(\Gamma)$ is the class of all $D_\xi(\langle C_\eta \mid \eta < \xi \rangle)$ for some increasing sequence of sets in Γ. $D_1(\Gamma)$ is simply written Γ.

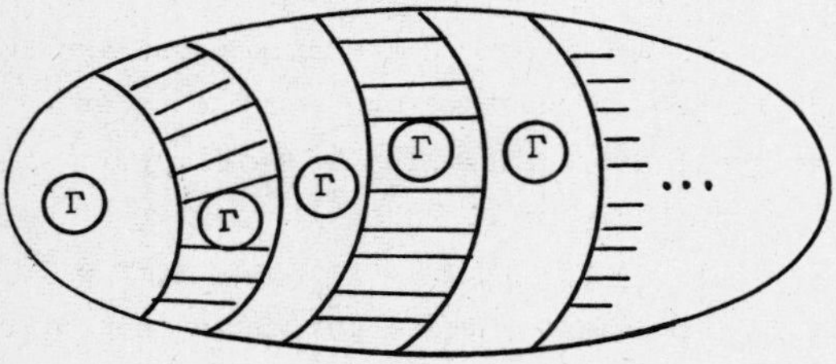

a. $D_\xi(\Gamma)$, for ξ even.

(b) Separated unions (Wadge). We define $A = SU(\langle C_n \mid n \in \omega \rangle, \langle A_n \mid n \in \omega \rangle)$, in case the sets C_n are pairwise disjoint, by $A = \bigcup_n (A_n \cap C_n)$. The set $C = \bigcup_n C_n$ is the corresponding envelop of A.

For classes Γ, Γ', we let $SU(\Gamma, \Gamma')$ be the class of all $SU(\langle C_n \rangle, \langle A_n \rangle)$ with the C_n's in Γ and the A_n in Γ'. The set $\langle \Gamma, SU(\Gamma, \Gamma') \rangle$ is the set of pairs $\langle C, A \rangle$ where $A = SU(\langle C_n \rangle, \langle A_n \rangle)$ is in $SU(\Gamma, \Gamma')$ and $C = \bigcup_n C_n$ is the corresponding envelop.

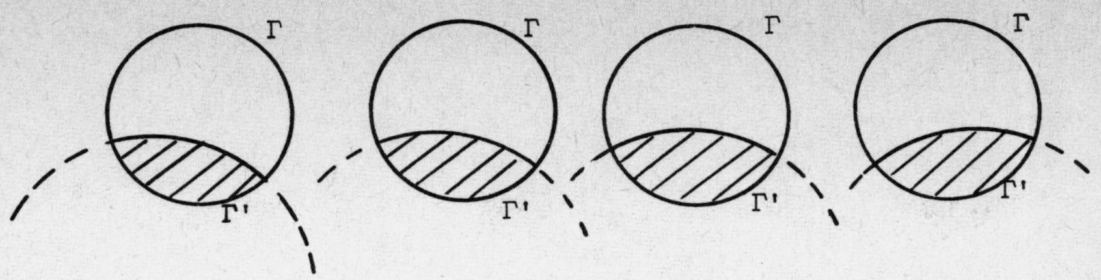

b. $S\,U(\Gamma, \Gamma')$

(c) <u>One-sided separated unions</u> (Myers, Wadge). We say that $A = \text{Sep}\,(C, B_1, B_2)$ if $A = (C \cap B_1) \cup (B_2 - C)$ [This is of course a particular case of (b)].

If Γ, Γ' are two classes, $\text{Sep}\,(\Gamma, \Gamma')$ is the class of all $\text{Sep}\,(C, B_1, B_2)$ where $C \in \Gamma$, $B_1 \in \check{\Gamma}'$ and $B_2 \in \Gamma'$ [This is not symmetric in Γ' and $\check{\Gamma}'$].

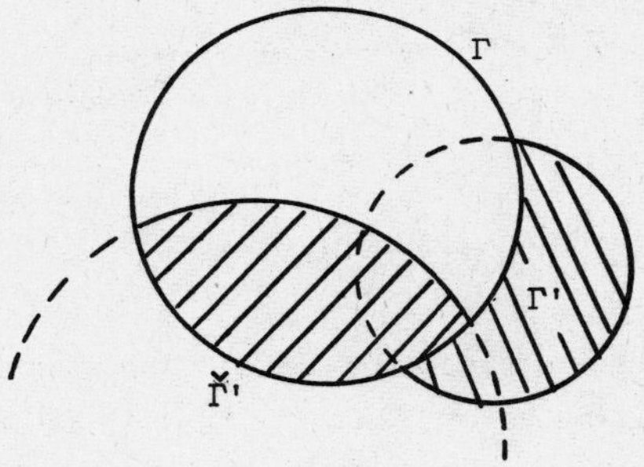

c. $\text{Sep}(\Gamma,\Gamma')$

(d) <u>Two-sided separated unions</u> [This is again a particular case of (b)]. We let $A = \text{Bisep}\,(C_1, C_2, A_1, A_2, B)$, in case C_1 and C_2 are disjoint, be the set $A = (C_1 \cap A_1) \cup (C_2 \cap A_2) \cup (B - (C_1 \cup C_2))$.

If $\Gamma, \Gamma', \Gamma''$ are classes, $\text{Bisep}\,(\Gamma, \Gamma', \Gamma'')$ is the set of all $\text{Bisep}\,(C_1, C_2, A_1, A_2, B)$ with C_1, C_2 in Γ, A_1 in $\check{\Gamma}'$, A_2 in Γ' and B in Γ''. Moreover, if $\Gamma'' = \{\emptyset\}$, we just write $\text{Bisep}\,(\Gamma, \Gamma')$ [The definition is symmetric in Γ' and $\check{\Gamma}'$].

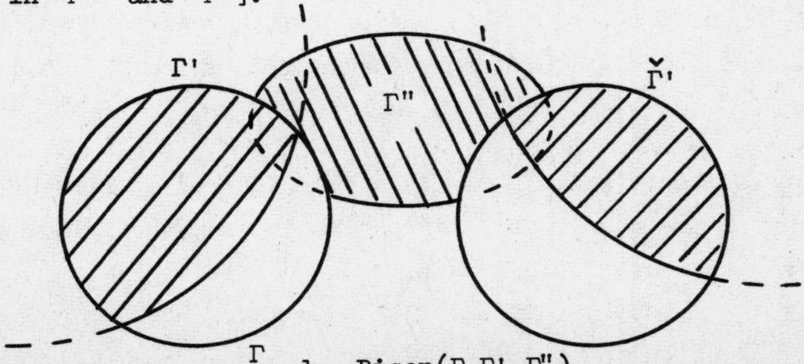

d. $\text{Bisep}(\Gamma,\Gamma',\Gamma'')$

(e) Separated differences. For $\xi \geq 2$ a countable ordinal, we define $A = SD_\xi(\langle C_\eta \mid \eta < \xi\rangle, \langle A_\eta \mid \eta < \xi\rangle, B)$ in case the C_η's and the A_η's are increasing, with $A_\eta \subseteq C_\eta \subseteq A_{\eta+1}$, by

$$A = \bigcup_{\eta<\xi} (A_\eta - \bigcup_{\eta'<\eta} C_{\eta'}) \cup (B - \bigcup_{\eta<\xi} C_\eta) .$$

If $\Lambda \subseteq \Gamma \times \Gamma'$, and Γ'' is a class, $SD_\xi(\Lambda, \Gamma'')$ is the set of all $SD_\xi(\langle C_\eta \mid \eta < \xi\rangle, \langle A_\eta \mid \eta < \xi\rangle, B)$, for $B \in \Gamma''$ and, for each $\eta < \xi$, the pair (C_η, A_η) is in Λ. Again, for $\Gamma'' = \{\emptyset\}$ we write $SD(\Lambda)$ [This operation will be used only for $\Lambda = \langle\Gamma, SU(\Gamma, \Gamma')\rangle$, as defined in (b)].

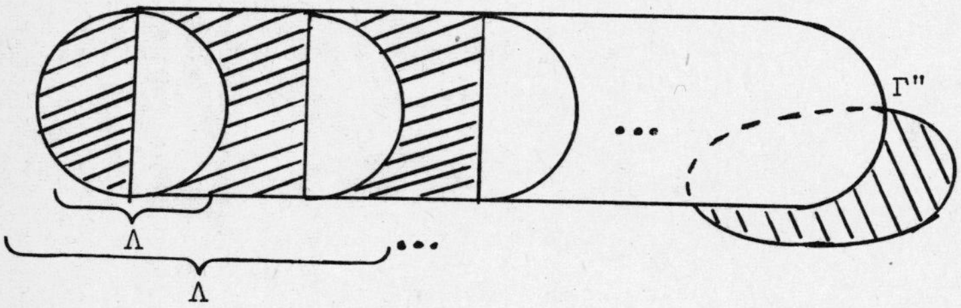

e. $SD_\xi(\Lambda,\Gamma'')$

We now proceed to apply these operations. Particular ways of combining these operations will be encoded by elements of ${}^\omega\omega_1$. Of course, we put restrictions on the allowed combinations. This leads to the following inductive definition of a "description" and a corresponding "described class".

Let u be some element of ${}^\omega\omega_1$. We let $u = \langle u_0, u_1\rangle$, where $u_0(n) = u(2n)$ and $u_1(n) = u(2n + 1)$. Similarly we let $u = \langle (u)_n \mid n \in \omega\rangle$ where $(u)_n(m) = u(\langle n, m\rangle)$. $\underline{0}$ is the element of ${}^\omega\omega_1$ defined by $\underline{0}(n) = 0$. We now define inductively the relations "u is a description" and "u describes the class Γ" (written $\Gamma_u = \Gamma$). Descriptions are elements of ${}^\omega\omega_1$. The first ordinal $u(0)$ will give the level of Γ_u (as informally discussed in the introduction), $u(1)$ and sometimes $u(2)$ the operation used to obtain Γ_u, and the remaining of u the description of the classes from which Γ_u has been obtained.

1.2. Definition. The relations "u is a description" (written $u \in D$), and "u describes Γ", are the least relations satisfying the following conditions:

(a) If $u(0) = 0$, u is a description, and $\Gamma_u = \{\emptyset\}$.

(b) If $u(0) = \xi \geq 1$, $u(1) = 1$ and $u(2) = \eta \geq 1$, then $u \in D$ and $\Gamma_u = D_\eta(\Sigma^0_\xi)$.

(c) If $u = \xi^\frown 2^\frown \eta^\frown u^*$, where $\xi \geq 1$, $\eta \geq 1$, $u^* \in D$ and $u^*(0) > \xi$, then $u \in D$ and $\Gamma_u = \text{Sep}\,(D_\eta(\underset{\sim}{\Sigma}^0_\xi), \Gamma_{u^*})$.

(d) If $u = \xi^\frown 3^\frown \eta^\frown \langle u_0, u_1 \rangle$, where $\xi \geq 1$, $\eta \geq 1$, u_0 and u_1 are in D, $u_0(0) > \xi$, $u_1(0) \geq \xi$ or $u_1(0) = 0$, and $\Gamma_{u_1} < \Gamma_{u_0}$, then $u \in D$ and $\Gamma_u = \text{Bisep}\,(D_\eta(\underset{\sim}{\Sigma}^0_\xi), \Gamma_{u_0}, \Gamma_{u_1})$.

(e) If $u = \xi^\frown 4^\frown \langle u_n \mid n \in \omega \rangle$, where $\xi \geq 1$ and each u_n is in D, and either for all n $u_n(0) = \xi_1 > \xi$, and the Γ_{u_n} are strictly increasing, or $u_n(0) = \xi_n$ and the ξ_n are strictly increasing with $\xi < \sup_n \xi_n$, then $u \in D$ and $\Gamma_u = \text{SU}\,(\underset{\sim}{\Sigma}^0_\xi, \bigcup_n \Gamma_{u_n})$.

(f) If $u = \xi^\frown 5^\frown \eta^\frown \langle u_0, u_1 \rangle$, where $\xi \geq 1$, $\eta \geq 2$, u_0 and u_1 are in D, and $u_0(0) = \xi$, $u_0(1) = 4$ [so that $\langle \underset{\sim}{\Sigma}^0_\xi, \Gamma_{u_0} \rangle$ is defined], and $u_1(0) \geq \xi$ or $u_1(0) = 0$, and $\Gamma_{u_1} < \Gamma_{u_0}$, then $u \in D$ and $\Gamma_u = \text{SD}_\eta\,(\langle \underset{\sim}{\Sigma}^0_\xi, \Gamma_{u_0} \rangle, \Gamma_{u_1})$.

Our aim is to prove that the preceding descriptions give the complete picture of the Wadge classes of Borel sets. We begin with a simple fact.

1.3. <u>Proposition</u>. The described classes are non self-dual Borel Wadge classes.

<u>Proof</u>. The "Borel" part is clear. To prove that these classes are non self-dual, it is enough to exhibit a universal set, and this is easy by induction. The only fact to note here is that by using the reduction property, one can find a sequence of $D_\eta(\underset{\sim}{\Sigma}^0_\xi)$ sets, in ${}^\omega\omega \times {}^\omega\omega$, which is universal for sequences of pairwise disjoint $D_\eta(\underset{\sim}{\Sigma}^0_\xi)$ sets of ${}^\omega\omega$.

It is clear that for each $u \in D$, there corresponds exactly one described class Γ_u. We now show that the <u>level</u> $u(0)$ gives the closure properties of Γ_u.

1.4. <u>Lemma</u>. Let u be a description, with $u(0) = \xi \geq 1$. Then

(a) Γ_u is closed under union with a $\underset{\sim}{\Delta}^0_\xi$ set

(b) $\text{SU}\,(\underset{\sim}{\Sigma}^0_\xi, \Gamma_u) = \Gamma_u$ (written Γ_u is closed under $\underset{\sim}{\Sigma}^0_\xi$ - SU).

<u>Proof</u>. By induction:

<u>Case 1</u>. $u(1) = 1$, so $\Gamma_u = D_\eta(\underset{\sim}{\Sigma}^0_\xi)$.

(b) Let C_n be the separating $\underset{\sim}{\Sigma}^0_\xi$ sets, and $A^n = D_\eta(\langle A^n_\zeta \mid \zeta < \eta \rangle)$ be the $D_\eta(\underset{\sim}{\Sigma}^0_\xi)$ sets. Consider $A_\zeta = \bigcup_n (A^n_\zeta \cap C_n)$. The A_ζ are clearly $\underset{\sim}{\Sigma}^0_\xi$ and increasing, and moreover $\text{SU}\,(\langle C_n \mid n \in \omega \rangle, \langle A^n \mid n \in \omega \rangle) = D_\eta(\langle A_\zeta \mid \zeta < \eta \rangle)$. This proves (b).

(a) Let $A = D_\eta(\langle A_\zeta \mid \zeta < \eta \rangle)$, with $A_\zeta \in \underset{\sim}{\Sigma}^0_\xi$, and let $B \in \underset{\sim}{\Delta}^0_\xi$. If η is odd, then $A \cup B = E_\eta(\langle A_\zeta \cup B, \ \zeta < \eta \rangle)$, and the $\langle A_\zeta \cup B \mid \zeta < \eta \rangle$ are an increasing sequence of $\underset{\sim}{\Sigma}^0_\xi$ sets. If η is even, let $A'_0 = A_0 - B$, and let $A'_\xi = A_\zeta \cup B$, for $\zeta \geq 1$. Then again $A \cup B = D_\eta(\langle A'_\xi \mid \zeta < \eta \rangle)$, and the A'_ξ are $\underset{\sim}{\Sigma}^0_\xi$ and increasing.

Case 2. $u(1) = 2$, so $\Gamma_u = \mathrm{Sep}\,(D_\eta(\boldsymbol{\Sigma}^0_\xi), \Gamma_{u^*})$, with $u^*(0) > \xi$. By the induction hypothesis, Γ_{u^*} is closed under union with a $\Delta^0_{\xi+1}$ set, and under $\boldsymbol{\Sigma}^0_{\xi+1}$- SU. Now intersection with a $\boldsymbol{\Sigma}^0_{\xi+1}$ set is a particular case of $\boldsymbol{\Sigma}^0_{\xi+1}$ - SU, so Γ_{u^*} and $\check{\Gamma}_{u^*}$ are closed under intersection and union with $\boldsymbol{\Delta}^0_{\xi+1}$ sets. This clearly implies that $\check{\Gamma}_{u^*}$ is closed under $\boldsymbol{\Sigma}^0_\xi$ - SU.

(a) Let $A_n = \mathrm{Sep}\,(C_n, A^n_1, A^n_2)$, with $A^n_1 \in \check{\Gamma}_{u^*}$, $A^n_2 \in \Gamma_{u^*}$, $C_n \in D_\eta(\boldsymbol{\Sigma}^0_\xi)$, and let $A = \mathrm{SU}(\langle C'_n \mid n \in \omega\rangle, \langle A_n, n \in \omega\rangle)$, where the C'_n are pairwise disjoint $\boldsymbol{\Sigma}^0_\xi$ sets. Then clearly $A = \mathrm{Sep}\,(\bigcup_n (C_n \cap C'_n), \bigcup_n (A^n_1 \cap C'_n), \bigcup_n (A^n_2 \cap C'_n))$, with $\bigcup_n (C_n \cap C'_n) \in D_\eta(\boldsymbol{\Sigma}^0_\xi)$, $\bigcup_n (A^n_1 \cap C'_n) \in \mathrm{SU}(\langle C'_n \mid n \in \omega\rangle, \langle A^n_1, n \in \omega\rangle)$ is in $\check{\Gamma}_{u^*}$ and $\bigcup_n (A^n_2 \cap C'_n) = \mathrm{SU}(\langle C'_n \mid n \in \omega\rangle, \langle A^n_2 \mid n \in \omega\rangle)$ is in Γ_{u^*}. This shows (a).

(b) Let $A = \mathrm{Sep}\,(C, A_1, A_2)$, $C \in D_\eta(\boldsymbol{\Sigma}^0_\xi)$, $A_1 \in \check{\Gamma}_{u^*}$, $A_2 \in \Gamma_{u^*}$, and let $B \in \boldsymbol{\Delta}^0_\xi$. Then $A \cup B = \mathrm{Sep}\,(C, A_1 \cup B, A_2 \cup B)$, and the induction hypothesis gives (b).

Case 3. $u(1) = 3$.

(a) If $A_n = \mathrm{Bisep}\,(C^n_1, C^n_2, A^n_1, A^n_2, B^n)$, where $C^n_i \in D_\eta(\boldsymbol{\Sigma}^0_\xi)$, $A^n_1 \in \check{\Gamma}_{u_0}$, $A^n_2 \in \Gamma_{u_0}$ and $B^n \in \Gamma_{u_1}$, with $u_0(0) > \xi$ and $u_1(0) \geq \xi$ or $u_1(0) = 0$, and $A = \mathrm{SU}(\langle C_n, n \in \omega\rangle, \langle A_n, n \in \omega\rangle)$, where the C_n are pairwise disjoint $\boldsymbol{\Sigma}^0_\xi$ sets, then $A = \mathrm{Bisep}\,(\bigcup_n (C^n_1 \cap C_n), \bigcup_n (C^n_2 \cap C_n), \bigcup_n (A^n_1 \cap C_n), \bigcup_n (A^n_2 \cap C_n), \bigcup_n (B^n \cap C_n))$ which, together with the induction hypothesis, proves (a).

(b) Let again $A = \mathrm{Bisep}\,(C_1, C_2, A_1, A_2, B)$ with the sets in the same classes as before, and let $D \in \boldsymbol{\Delta}^0_\xi$. If $u_1(0) \geq 1$, take $B' = B \cup D$, $A'_1 = A_1 \cup B$, $A'_2 = A_2 \cup B$. Then $A \cup B = \mathrm{Bisep}\,(C_1, C_2, A'_1, A'_2, B')$, which proves (b) in this case. If $\Gamma_{u_1} = \{\emptyset\}$, so $B = \emptyset$, consider the sets $C_1 \cup D$ and $C_2 \cup D$. These are $D_\eta(\boldsymbol{\Sigma}^0_\xi)$ sets, by case 1. Let C^*_1, C^*_2 reduce them. Then $A = \mathrm{Bisep}\,(C^*_1, C^*_2, A_1 \cup D, A_2 \cup D)$, which proves (b) in that case.

Case 4. $u(1) = 4$. In this case $\Gamma_u = \mathrm{SU}(\boldsymbol{\Sigma}^0_\xi, \bigcup_n \Gamma_{u_n})$, and the Γ_{u_n} are of level $> \xi$ (at least for $n \geq n_0$). (a) is almost trivial. For (b), let $A = \mathrm{SU}(\langle C_n, n \in \omega\rangle, \langle A_n, n \in \omega\rangle)$, with $A_n \in \bigcup_p \Gamma_{u_p}$, and $C_n \in \boldsymbol{\Sigma}^0_\xi$, and let $B \in \boldsymbol{\Delta}^0_\xi$. Let B^*, $\langle C^*_n \mid n \in \omega\rangle$ be $\boldsymbol{\Sigma}^0_\xi$ sets reducing the sets B, $\langle C_n, n \in \omega\rangle$. Then $A \cup B = \mathrm{SU}(\langle C'_n \mid n \in \omega\rangle, \langle A'_n \mid n \in \omega\rangle)$ where $C'_0 = B^*$, $C'_n = C^*_{n-1}$, $n \geq 1$, and $A'_0 = B$, $A'_n = A_{n-1}$, $n \geq 1$.

Case 5. $u(1) = 5$.

(a) Suppose $A_n = \mathrm{SD}_\eta(\langle C^n_\zeta \mid \zeta < \eta\rangle, \langle A^n_\zeta \mid \zeta < \eta\rangle, B^n)$, where the pairs (C^n_ζ, A^n_ζ) are in $\langle \boldsymbol{\Sigma}^0_\xi, \Gamma_{u_0}\rangle$, with $u_0(0) = \xi$ and $u_0(1) = 4$, and $B^n \in \Gamma_{u_1}$, and $A = \mathrm{SU}(\langle C_n \mid n \in \omega\rangle, \langle A_n \mid n \in \omega\rangle)$. Then $A = \mathrm{SD}_\eta(\langle \bigcup_n (C_n \cap C^n_\zeta), \zeta < \eta\rangle, \langle \bigcup_n (C_n \cap A^n_\zeta), \zeta < \eta\rangle, \langle \bigcup_n (C_n \cap B^n)\rangle)$, and the induction hypothesis immediately yields (a).

(b) Let $A = SD_\eta(\langle C_\zeta, \zeta < \eta\rangle, \langle A_\zeta, \zeta < \eta\rangle, B)$, and let $D \in \underset{\sim}{\Delta}^0_\xi$. Then clearly $A \cup D = SD_\eta(\langle C_\zeta \cup D \mid \zeta < \eta\rangle, \langle A_\zeta \cup D, \zeta < \eta\rangle, B)$, and by the proof of case 4, (b), the $C_\zeta \cup D$ are envelops of the $A_\zeta \cup D$. This shows (b) in this case.

1.5. <u>Lemma</u>. If A, B are two disjoint $D_\eta(\underset{\sim}{\Sigma}^0_\xi)$ sets, there are sets A^*, B^* in $D_\eta(\underset{\sim}{\Sigma}^0_\xi)$ such that

(a) A^* and B^* are disjoint

(b) $A \subset A^*$, $B \subset B^*$

(c) $A^* \cup B^* \in \underset{\sim}{\Sigma}^0_\xi$

<u>Proof</u>. Let C (resp. D) be the union of the $\underset{\sim}{\Sigma}^0_\xi$ sets in a $D_\eta(\underset{\sim}{\Sigma}^0_\xi)$ definition of A (resp. B), and let C^*, D^* reduce C, D. It is easily checked that $A^* = (A \cap C^*) \cup (D^* - B)$ and $B^* = (C^* - A) \cup (D^* \cap B)$ satisfy the desired properties.

1.6. <u>Corollary</u> (Normal form for the Bisep operation). Suppose $A \in \text{Bisep}(D_\eta(\underset{\sim}{\Sigma}^0_\xi), \Gamma_{u_0}, \Gamma_{u_1})$, where Γ_{u_0} is of level at least $\xi + 1$, and $\Gamma_{u_1} < \Gamma_{u_0}$. Then for some set C in $\underset{\sim}{\Sigma}^0_\xi$, $A = (A \cap C) \cup (B - C)$ where $A \cap C \in \text{Bisep}(D_\eta(\underset{\sim}{\Sigma}^0_\xi), \Gamma_{u_0})$, $C - A \in \text{Bisep}(D_\eta(\underset{\sim}{\Sigma}^0_\xi), \Gamma_{u_0})$, and $B \in \Gamma_{u_1}$.

<u>Proof</u>. Extend the two $D_\eta(\underset{\sim}{\Sigma}^0_\xi)$ separating sets, using 1.5. The union C of the extended $D_\eta(\underset{\sim}{\Sigma}^0_\xi)$ sets clearly works.

Let us now define a notion of type for each description u. This type intuitively corresponds to the character (successor, limit of cofinality ω or limit of cofinality ω_1) of the class Γ_u, among the Wadge classes of level at least u(0).

1.7. <u>Definition</u>. The type t(u) is defined by the following conditions:

(a) If $u(0) = 0$, $t(u) = 0$ ($\{\emptyset\}$ is the first class)

(b) If $u(0) = \xi \geq 1$, and $u(1) = 1$,
u is of type 1 if u(2) is successor, and
u is of type 2 if u(2) is limit

(c) If $u(0) \geq 1$ and $u(1) = 2$, u is of type 3

(d) If $u(0) \geq 1$ and $u(1) = 3$, then

$$u \text{ is of type } \begin{cases} 1 & \text{if } u_1 \text{ is of type 0 and } u(2) \text{ is successor} \\ 2 & \text{if } u_1 \text{ is of type 0 and } u(2) \text{ is limit} \\ t(u_1) & \text{if } u_1(0) = u(0) \\ 3 & \text{if } u_1(0) > u(0) \end{cases}$$

(e) If $u(0) \geq 1$ and $u(1) = 4$, u is of type 2

(f) If $u(0) \geq 1$ and $u(1) = 5$,

$$u \text{ is of type } \begin{cases} 2 & \text{if } u_1 \text{ is of type } 0 \\ t(u_1) & \text{if } u_1(0) = u(0) \\ 3 & \text{if } u_1(0) > u(0) \end{cases}$$

1.8. Definition. We define $D^0 = \{u \mid t(u) = 0\} = \{u \mid u(0) = 0\}$, $D^+ = \{u \mid u(0) = 1 \text{ and } t(u) = 1\}$, $D^\omega = \{u \mid u(0) = 1 \text{ and } t(u) = 2\}$ and $D^{\omega_1} = D - (D^0 \cup D^+ \cup D^\omega) = \{u \mid u(0) = 1 \text{ and } t(u) = 3\} \cup \{u \mid u(0) > 1\}$.

We are now able to restate Wadge's result in terms of our notion of description:

1.9. Theorem. Assume Borel Determinacy. Let $\mathfrak{w} = \{\Gamma_u, u \in D\} \cup \{\check{\Gamma}_u, u \in D\} \cup \{\Delta(\Gamma_u), u \in D^+ \cup D^\omega\}$. Then $\mathfrak{w}$ is exactly the set of all Borel Wadge classes.

The (long) proof of this theorem goes roughly as follows. We shall define, for each description $u \notin D^0$, a set Q_u of descriptions, satisfying the following properties:

(A) If $u \in D^+$, then $Q_u = \{\bar{u}\}$, $\Gamma_{\bar{u}} < \Gamma_u$ and the only Wadge class Γ such that $\Gamma_{\bar{u}} < \Gamma < \Gamma_u$ is $\Delta(\Gamma_u)$.

(B) If $u \in D^\omega$, then $Q_u = \{u_n, n \in \omega\}$, for each n $\Gamma_{u_n} < \Gamma_u$, and the only Wadge class Γ such that $\forall n\ \Gamma_{u_n} < \Gamma < \Gamma_u$ is $\Delta(\Gamma_u)$.

(C) If $u \in D^{\omega_1}$, then Q_u is a set of descriptions of cardinality ω_1, and $\Delta(\Gamma_u) = \bigcup \{\Gamma_{u'}, u' \in Q_u\}$.

This will finish the proof of the theorem. For suppose there is a Borel Wadge class not in $\mathfrak{w}$, let Γ be the $<$-least counterexample (or one of the two $<$-least counterexamples, in a non self-dual case, for both must be outside $\mathfrak{w}$ by definition of it). As the sequence $(\underset{\sim}{\Sigma}^0_\xi, \xi < \omega_1)$ is cofinal in the Borel Wadge classes, there is a $<$-least described class Γ_u such that $\Gamma < \Gamma_u$, and clearly $u \notin D^0$. Now each of the remaining cases, $u \in D^+$, $u \in D^\omega$ and $u \in D^{\omega_1}$ gives immediately, using (A), (B) and (C), that Γ is in $\mathfrak{w}$, a contradiction.

In the following proof, we shall only indicate the main steps. Some arguments are just sketched, and others are missing (they can be found either in Kuratowski [1966] or in Wadge [1976]).

We first consider the case of a description u of type 1. This will take care of (A), but as we shall see also of a part of (C).

1.10. Definition. For each description u of type 1, we define a description $\bar{u}$ by the following conditions [and we indicate the corresponding classes]:

(a) $u(1) = 1$, $u(2) = \eta + 1$

-- if $\eta = 0$, $\bar{u} = \underline{0}$ $\quad [\Gamma_u = \underset{\sim}{\Sigma}^0_\xi, \Gamma_{\bar{u}} = \{0\}]$

-- if $\eta > 0$, $\bar{u} = u(0)^\frown 1^\frown \eta^\frown \underline{0}$ $\quad [\Gamma_u = D_{\eta+1}(\underset{\sim}{\Sigma}^0_\xi), \Gamma_{\bar{u}} = D_\eta(\underset{\sim}{\Sigma}^0_\xi)]$

(b) $u(1) = 3$, $t(u_1) = 0$, and $u(2) = \eta + 1$
-- if $\eta = 0$, $\bar{u} = u_0$ $[\Gamma_u = \text{Bisep}\ (\underset{\sim}{\Sigma}^0_\xi, \Gamma_{u_0}), \Gamma_{\bar{u}} = \Gamma_{u_0}]$
-- if $\eta > 0$, $\bar{u} = u(0)^\frown 2^\frown \eta^\frown u_0$ $[\Gamma_u = \text{Bisep}(D_{\eta+1}(\underset{\sim}{\Sigma}^0_\xi), \Gamma_{u_0}), \Gamma_{\bar{u}} = \text{Sep}(D_\eta(\underset{\sim}{\Sigma}^0_\xi), \Gamma_{u_0})]$

(c) $u(1) = 3$ or 5, $t(u_1) = 1$ and $u_1(0) = u(0)$
$\bar{u} = u(0)^\frown u(1)^\frown u(2)^\frown \langle u_0, \bar{u}_1\rangle$

1.11. Lemma. For each u of type 1
(a) $\Gamma_u = \text{Bisep}\ (\underset{\sim}{\Sigma}^0_{u(0)}, \Gamma_{\bar{u}})$
(b) $\Delta(\Gamma_u) = \text{Bisep}\ (\Delta(\underset{\sim}{\Sigma}^0_{u(0)}), \Gamma_{\bar{u}})$

Proof.

(a) Case 1. $u(1) = 1$, $u(2) = 0$. The equality $\underset{\sim}{\Sigma}^0_\xi = \text{Bisep}\ (\Sigma^0_\xi, \{\emptyset\})$ is trivial.

Case 2. $u(1) = 1$, $u(2) = \eta + 1$ for $\eta \geq 1$. First $\check{\Gamma}_{\bar{u}}$ and $\Gamma_{\bar{u}}$ are contained in $\Gamma_u = D_{\eta+1}(\underset{\sim}{\Sigma}^0_\xi)$, so by the closure properties of Γ_u, $\text{Bisep}\ (\underset{\sim}{\Sigma}^0_\xi, \Gamma_{\bar{u}}) \subset \Gamma_u$. On the other hand, if $A \in D_{\eta+1}(\underset{\sim}{\Sigma}^0_\xi)$, then for some C in $\underset{\sim}{\Sigma}^0_\xi$ and some B in $D_\eta(\underset{\sim}{\Sigma}^0_\xi) = \Gamma_{\bar{u}}$, $A = C - B$. So A is in $\text{Bisep}\ (\underset{\sim}{\Sigma}^0_\xi, \Gamma_{\bar{u}})$.

Case 3. $u(1) = 3$, $t(u_1) = 0$ and $u(2) = 1$. Then $\Gamma_u = \text{Bisep}\ (\underset{\sim}{\Sigma}^0_\xi, \Gamma_{u_0})$ and $\Gamma_{\bar{u}} = \Gamma_{u_0}$, so the equality is trivial.

Case 4. $u(1) = 3$, $t(u_1) = 0$ and $u(2) = \eta + 1$, with $\eta \geq 1$. Then $\Gamma_u = \text{Bisep}\ (D_{\eta+1}(\underset{\sim}{\Sigma}^0_\xi), \Gamma_{\bar{u}_0})$, and $\Gamma_{\bar{u}} = \text{Bisep}\ (D_\eta(\underset{\sim}{\Sigma}^0_{\xi+1}), \Gamma_{u_0})$. The inclusion $\text{Bisep}\ (\underset{\sim}{\Sigma}^0_\xi, \Gamma_{\bar{u}}) \subset \Gamma_u$ is again easy. For the converse, assume $A \in \Gamma_u$. By the normal form for the Bisep operation, let C_1, C_2 in $\underset{\sim}{\Sigma}^0_\xi$, D_1, D_2 in $D_\eta(\underset{\sim}{\Sigma}^0_\xi)$ be such that $C_1 - D_1 \cap C_2 - D_2 = \emptyset$, $D_2 \subset C_1$, $C_2 \subset D_1$, and $A \subset C_1 \cup C_2$, $A \cap C_1 - D_1 \in \Gamma_{u_0}$, $A \cap C_2 - D_2 \in \Gamma_{u_0}$. Let C^*_1, C^*_2 be $\underset{\sim}{\Sigma}^0_\xi$ sets reducing C_1, C_2, and remark that $A \cap C^*_1 = (A \cap D_1 \cap C^*_1) \cup (A \cap (C_1 - D_1) \cap C^*_1)$ is in $\check{\Gamma}_{\bar{u}}$, and $A \cap C^*_2 = (A \cap D_2 \cap C^*_2) \cup (A \cap (A \cap (C_2 - D_2) \cap C^*_2)$ is in $\Gamma_{\bar{u}}$, so that, as $A \subset C^*_1 \cup C^*_2$, we obtain $A \in \text{Bisep}\ (\underset{\sim}{\Sigma}^0_\xi, \Gamma_{\bar{u}})$.

Case 5 (induction step). Suppose $u(1) = 3$, $t(u_1) = 1$ and $u_1(0) = u(0)$. So $\Gamma_u = \text{Bisep}\ (D_\eta(\Sigma^0_\xi), \Gamma_{u_0}, \Gamma_{u_1})$ and $\Gamma_{\bar{u}} = \text{Bisep}\ (D_\eta(\underset{\sim}{\Sigma}^0_\xi), \Gamma_{u_0}, \Gamma_{\bar{u}_1})$, and by the induction hypothesis, we can assume $\Gamma_{u_1} = \text{Bisep}\ (\underset{\sim}{\Sigma}^0_\xi, \Gamma_{\bar{u}_1})$ (for $u_1(0) = \xi$). The inclusion $\text{Bisep}\ (\underset{\sim}{\Sigma}^0_\xi, \Gamma_{\bar{u}}) \subset \Gamma_u$ is easy. Suppose $A \in \Gamma_u$. Then by the normal form for Bisep, we have $A = (A \cap C) \cup (B - C)$, where $B \in \Gamma_{u_1}$ and $C \in \underset{\sim}{\Sigma}^0_\xi$, $A \cap C$ and $C - A$ are in $\text{Bisep}\ (D_\eta(\underset{\sim}{\Sigma}^0_{\xi+1}), \Gamma_{u_0})$. Now $B = (B \cap C_1) \cup (B \cap C_2)$, where C_1, C_2 are two disjoint $\underset{\sim}{\Sigma}^0_\xi$ sets, and $B \cap C_1 \in \check{\Gamma}_{\bar{u}_1}$, $B \cap C_2 \in \Gamma_{\bar{u}_1}$. Let C^*_1, C^*_2 in $\underset{\sim}{\Sigma}^0_\xi$ reduce the pair $C \cup C_1$, $C \cup C_2$. Then $A = (A \cap C^*_1) \cup (A \cap C^*_2)$, and it is clear that $A \cap C^*_1 \in \text{Bisep}\ (D_\eta(\underset{\sim}{\Sigma}^0_\xi), \Gamma_{u_0}, \check{\Gamma}_{\bar{u}_1}) = \check{\Gamma}_{\bar{u}}$, and

$A \cap C_2^* \in \text{Bisep}\ (D_\eta(\underset{\sim}{\Sigma}_\xi^0),\ \Gamma_{u_0},\ \check{\Gamma}_{\bar{u}_1}) = \check{\Gamma}_{\bar{u}}$, so that $A \in \text{Bisep}\ (\underset{\sim}{\Sigma}_\xi^0,\ \Gamma_{\bar{u}})$.

The second case of the induction case $(u(1) = 5,\ t(u_1) = 1$ and $u_1(0) = u(0))$ is entirely analogous, and we omit it.

(b) is an easy consequence of (a). The inclusion $\text{Bisep}\ (\Delta(\underset{\sim}{\Sigma}^0_{u(0)}),\ \Gamma_{\bar{u}}) \subset \Delta(\Gamma_u)$ is obvious. Suppose now $A \in \Delta(\Gamma_u)$. Then by (a), $A = (A \cap C_1) \cup (A \cap C_2)$ where C_1, C_2 are disjoint $\underset{\sim}{\Sigma}_\xi^0$ sets, and $A \cap C_1 \in \check{\Gamma}_{\bar{u}}$, $A \cap C_2 \in \Gamma_{\bar{u}}$. Similarly $\check{A} = (\check{A} \cap D_1) \cup (\check{A} \cap D_2)$, where D_1, D_2 are disjoint $\underset{\sim}{\Sigma}_\xi^0$ sets and $\check{A} \cap D_1 \in \check{\Gamma}_{\bar{u}}$, $\check{A} \cap D_2 \in \Gamma_{\bar{u}}$. Let C^*, D^* be $\underset{\sim}{\Sigma}_\xi^0$ sets reducing the pair $C_1 \cup C_2$, $D_1 \cup D_2$. As $C^* \cup D^* = \omega^\omega$, C^*, D^* and the sets $C_1^* = C_1 \cap C^*$, $C_2^* = C_2 \cap C^*$, $D_1^* = D_1 \cap D^*$, $D_2^* = D_2 \cap D^*$ are $\Delta(\underset{\sim}{\Sigma}_\xi^0)$. But clearly $A = [(C_1^* \cap A) \cup (D_1^* - \check{A})] \cup [(D_2^* \cap A) \cup (C_2^* - \check{A})]$, which shows $A \in \text{Bisep}\ (\Delta(\underset{\sim}{\Sigma}_\xi^0),\ \Gamma_{\bar{u}})$.

1.12. <u>Corollary</u> (Statement (A)). Let u be some description in D^+, and let $Qu = \{\bar{u}\}$. Then the only Wadge class Γ such that $\Gamma_{\bar{u}} < \Gamma < \Gamma_u$ is $\Delta(\Gamma_u)$.

<u>Proof</u>. Using lemma 1.11 for $u(0) = 1$, we see that $\Delta(\Gamma_u) = \text{Bisep}\ (\underset{\sim}{\Delta}_1^0,\ \Gamma_{\bar{u}})$. Using the game-theoretical characterization of the Wadge ordering this immediately implies $\Delta(\Gamma_u) = \Gamma_{\bar{u}}^+$.

1.13. <u>Definition</u>. Let u be some description in D^{ω_1}, with $u(0) = \xi + 1$, for $\xi \geq 1$, and $t(u) = 1$. We define

$$Qu = \{\xi^\frown 3^\frown \eta^\frown \langle \bar{u},\ \underline{0}\rangle \mid 1 \leq \eta < \omega_1\}.$$

1.14. <u>Corollary</u> (Statement (C) for $u(0) = \xi + 1$, $t(u) = 1$). Let u be a description with $u(0) = \xi + 1$, $\xi \geq 1$ and $t(u) = 1$. Then Qu is a set of descriptions (of level ξ), and $\Delta(\Gamma_u) = \bigcup \{\Gamma_{u'} \mid u' \in Qu\}$.

<u>Proof</u>. (a) The case $u(0) = \xi + 1$, $u(1) = 1$, $u(2) = 1$, i.e. $\Gamma_u = \Sigma^0_{\xi+1}$ is solved by the Hausdorff-Kuratowski theorem: $\underset{\sim}{\Delta}^0_{\xi+1} = \bigcup \{D_\eta(\underset{\sim}{\Sigma}_\xi^0) \mid 1 \leq \eta < \omega_1\}$.

(b) In the general case, we have that $\Delta(\Gamma_u) = \text{Bisep}\ (\Delta(\underset{\sim}{\Sigma}^0_{\xi+1}),\ \Gamma_{\bar{u}})$ by lemma 1.11. Using the Hausdorff-Kuratowski theorem, we obtain

$$\begin{aligned}
\Delta(\Gamma_u) &= \text{Bisep}\ \Big(\bigcup_{\eta<\omega_1} D_\eta(\underset{\sim}{\Sigma}_\xi^0),\ \Gamma_{\bar{u}}\Big) \\
&= \bigcup \{\text{Bisep}\ (D_\eta(\underset{\sim}{\Sigma}_\xi^0),\ \Gamma_{\bar{u}}) \mid 1 \leq \eta < \omega_1\} \\
&= \bigcup \{\Gamma_{u'} \mid u' \in Qu\} \quad \text{by the definition of } Qu.
\end{aligned}$$

We now turn to the case of $u(0)$ a limit ordinal. What we need here is the analysis, obtained by Wadge, of Δ_λ^0 for limit λ. This analysis is done by iterating the SU operation.

1.15. Definition. Let λ be a countable limit ordinal, and $(\lambda_n)_{n\in\omega}$ be an increasing sequence of ordinals, cofinal in λ.

(a) Let Γ be some class. We define, for each (n, η) in $\omega \times \omega_1$ a class $SU_{n,\eta}(\Gamma)$, by the following induction:

(i) $SU_{n,0}(\Gamma) = SU(\underset{\sim}{\Sigma}^0_{\lambda_n}, \Gamma)$.

(ii) $SU_{n,\eta}(\Gamma) = SU(\underset{\sim}{\Sigma}^0_{\lambda_n}, \bigcup \{SU_{p,\eta'}(\Gamma) : p \in \omega, \ \eta' < \eta\})$ for $\eta > 0$.

(b) Similarly, if $s = \langle u_n, n \in \omega\rangle$ codes a sequence of descriptions, with $\sup_n u_n(0) = \lambda$, we define a family $u_{n,\eta}(s)$ by:

(i) $u_{n,0}(s) = \lambda_n {}^\frown 4 {}^\frown s$

(ii) $u_{n,\eta}(s) = \lambda_n {}^\frown 4 {}^\frown \langle u_{n,\eta'}(s) : n \in \omega, \ \eta' < \eta\rangle$ for $\eta > 0$.

(c) If $u = \lambda {}^\frown 1 {}^\frown 1 {}^\frown u'$, we set $Q_u = \{u_{n,\eta}(s) : n \in \omega, \ \eta \in \omega_1\}$ where $s = \langle \lambda_n {}^\frown 1 {}^\frown 1 {}^\frown \underline{0} : n \in \omega\rangle$.

1.16. Lemma. If $s = \langle u_n, n \in \omega\rangle$ codes a sequence of descriptions with either $u_n(0) = \lambda$ and the Γ_{u_n} increasing, or $u_n(0) = \lambda_n$, then each $u_{n,\eta}(s)$ is a description, and $\Gamma_{u_{n,\eta}(s)} = SU_{n,\eta}(\Gamma_s)$, where $\Gamma_s = \bigcup_n \Gamma_{u_n}$.

Proof. The only thing to check is that the levels of the classes on which SU is performed are acceptable, and we omit it.

The next result is a theorem of Wadge [1976], and gives the analysis of $\underset{\sim}{\Delta}^0_\lambda$, for limit λ, in a way very similar to the Hausdorff-Kuratowski theorem.

1.17. Theorem (Wadge). Let λ be limit, and $(\lambda_n)_{n\in\omega}$ be cofinal in λ. Let $\Gamma = \bigcup_{\eta<\lambda} \underset{\sim}{\Sigma}^0_\eta$. Then

$$\underset{\sim}{\Delta}^0_\lambda = \bigcup \{SU_{n,\eta}(\Gamma) : n \in \omega, \ \eta < \omega_1\} .$$

Hence, with our notations, if u is a description with $u(0) = \lambda$, $u(1) = u(2) = 1$, then $\Delta(\Gamma_u) = \bigcup \{\Gamma_{u'} : u' \in Q_u\}$.

1.18. Definition. Let u be a description of type 1, with $u(0) = \lambda$ limit, and (λ_n) a cofinal sequence in λ. Define a sequence $s_u = \langle u_n, n \in \omega\rangle$ by $u_n = \lambda_n {}^\frown 3 {}^\frown 1 {}^\frown \langle \bar{u}, \underline{0}\rangle$. We set $Q_u = \{u_{n,\eta}(s_u) : n \in \omega, \ \eta < \omega_1\}$.

1.19. Corollary (Statement (C) for $t(u) = 1$ and $u(0)$ limit). Let u be a description of type 1 with $u(0) = \lambda$ limit. Then

$$\Delta(\Gamma_u) = \bigcup \{\Gamma_{u'} : u' \in Q_u\} .$$

Proof. By lemma 1.11, we know that

$$\begin{aligned}\Delta(\Gamma_u) &= \text{Bisep}\ (\Delta(\underset{\sim}{\Sigma}^0_\lambda),\ \Gamma_{\bar{u}}) \\ &= \text{Bisep}\ (\ \cup\ \{S\,U_{n,\eta}(\Gamma) : n \in \omega,\ \ \eta < \omega_1\},\ \ \Gamma_{\bar{u}})\end{aligned}$$

with $\Gamma = \bigcup_n \underset{\sim}{\Sigma}^0_{\lambda_n}$, by using Wadge's theorem. So the only thing to prove is the equality

$$\text{Bisep}\ (S\,U_{n,\eta}(\Gamma),\ \Gamma_{\bar{u}}) = S\,U_{n,\eta}(\ \bigcup_n \text{Bisep}\ (\underset{\sim}{\Sigma}^0_{\lambda_n},\ \Gamma_{\bar{u}}))\ .$$

(a) Suppose first that $\eta = 0$. The left side of the equality is $\Gamma_\ell = \text{Bisep}\ (S\,U\,(\underset{\sim}{\Sigma}^0_{\lambda_n},\ \Gamma),\ \Gamma_{\bar{u}}\ )$ and the right side is $\Gamma_r = S\,U\,(\underset{\sim}{\Sigma}^0_{\lambda_n},\ \text{Bisep}\ (\Gamma,\ \Gamma_{\bar{u}}))$. The inclusion from right to left is obvious. Let A be in Γ_ℓ. For some disjoint sets C_1, C_2 in $S\,U\,(\underset{\sim}{\Sigma}^0_{\lambda_n},\ \Gamma)$, some A_1, A_2 in $\check{\Gamma}_{\bar{u}}$, $\Gamma_{\bar{u}}$ respectively, we have $A = (A_1 \cap C_1) \cup (A_2 \cap C_2)$. Now, $C_1 = \bigcup_p (H^1_p \cap C^1_p)$ with the H^1_p disjoint in $\underset{\sim}{\Sigma}^0_{\lambda_n}$, and the C^1_p in Γ; and similarly we can find corresponding sets H^2_p, C^2_p for C_2. Let K^1_p, K^2_p, $p \in \omega$, be $\underset{\sim}{\Sigma}^0_{\lambda_n}$ sets reducing the sets H^1_p and H^2_p. Then $A \subset (\ \bigcup_p K^1_p) \cup (\ \bigcup_p K^2_p)$, and moreover $A \cap K^1_p = (A \cap K^1_p \cap C^1_p) \cup (A \cap K^1_p \cap \check{C}^1_p)$ is in $\text{Bisep}\ (\Gamma,\ \Gamma_{\bar{u}})$, and similarly for $A \cap K^2_p$. This shows that A is in Γ_r.

(b) Suppose now $\eta > 0$. The left side is now

$$\begin{aligned}\Gamma_\ell &= \text{Bisep}\ (S\,U\,(\underset{\sim}{\Sigma}^0_{\lambda_n},\ \cup\ \{S\,U_{p,\eta'}(\Gamma) : p \in \omega,\ \ \eta' < \eta\}),\ \Gamma_{\bar{u}}) \\ &= S\,U\,(\underset{\sim}{\Sigma}^0_{\lambda_n},\ \cup\ \{\text{Bisep}\ (S\,U_{p,\eta'}(\Gamma) : p \in \omega,\ \ \eta' < \eta\})\end{aligned}$$

by the same proof as in (a), and the right side is

$$\Gamma_r = S\,U\,(\underset{\sim}{\Sigma}^0_{\lambda_n},\ \cup\ \{S\,U_{p,\eta'}(\text{Bisep}\,(\Gamma,\ \Gamma_{\bar{u}})) : p \in \omega,\ \ \eta' < \eta\})\ .$$

The induction hypothesis then immediately gives the result.

We now turn to the case of descriptions of type 2.

1.20. Definition. For each description u of type 2 we define a sequence s_u by the following conditions:

(a) If $u(1) = 1$ and $u(2) = \lambda$ is limit, with cofinal sequence (λ_n), let $s_u = \langle \xi^\frown 1^\frown \lambda_n{}^\frown \underline{0},\ \ n \in \omega\rangle$.

(b) If $u(1) = 3$ and $t(u_1) = 0$, $u(2) = \lambda$ is limit with cofinal sequence (λ_n), let $s_u = \langle \xi^\frown 2^\frown \lambda_n{}^\frown u_0 : n \in \omega\rangle$.

(c) If $u(1) = 4$, so $u = \xi^\frown 4^\frown u'$, let $s_u = u'$.

(d) If $u(1) = 5$ and $t(u_1) = 0$, then
-- if $u(2) = \eta + 1$, with $\eta > 0$, so $u_0 = \xi^\frown 4^\frown \langle u_n : n \in \omega\rangle$, let
$s_u = \langle \xi^\frown 5^\frown \eta^\frown \langle u_0, u_n\rangle : n \in \omega\rangle$.
-- if $u(2) = \lambda$ is limit with cofinal sequence (λ_n), let
$s_u = \langle \xi^\frown 5^\frown \lambda_n^\frown \langle u_0, u_1\rangle : n \in \omega\rangle$.
(e) (induction step) If $u(1) = 3$ or 5, and $t(u_1) = 2$ and $u_1(0) = u(0)$, then writing $s_{u_1} = \langle u_n^1 : n \in \omega\rangle$, set $s_u = \langle u(0)^\frown u(1)^\frown u(2)^\frown \langle u_0, u_n^1\rangle : n \in \omega\rangle$.

1.21. Definition. For each description u of type 2, we define a set Q_u of descriptions by the following:

(a) If $u(0) = 1$, $Q_u = \{(s_u)_n : n \in \omega\}$.
(b) If $u(0) = \xi + 1$, with $\xi > 0$, $Q_u = \{\xi^\frown 5^\frown \eta^\frown \langle \xi^\frown 4^\frown s_u, \underline{0}\rangle : 1 < \eta < \omega_1\}$.
(c) If $u(0) = \lambda$ is limit, with cofinal sequence (λ_n), $Q_u = \{u_{p,\eta}(s_u) : p \in \omega,\ \eta < \omega_1\}$.

1.22. Definition. Partitioned Unions (Wadge). We say that $A = PU(\langle C_n \mid n \in \omega\rangle, \langle A_n \mid n \in \omega\rangle)$ if $A = SU(\langle C_n\rangle, \langle A_n\rangle)$, and moreover the envelop $C = \bigcup_n C_n$ is ω^ω (so that $\langle C_n \mid n \in \omega\rangle$ is a partition of ω^ω).
$PU(\Gamma, \Gamma')$ is the class of all $PU(\langle C_n\rangle, \langle A_n\rangle)$, with $C_n \in \Gamma$ and $A_n \in \Gamma'$, for each n.

1.23. Lemma. Let u be a description of type 2 and level $\xi \geq 1$. Then

(a) $\Gamma_u = SU(\underset{\sim}{\Sigma}^0_\xi, \bigcup \{\Gamma_{u'} \mid u' \in Qu\})$
(b) $\Delta(\Gamma_u) = PU(\underset{\sim}{\Sigma}^0_\xi, \bigcup \{\Gamma_{u'} \mid u' \in Qu\})$
(c) In particular, if $\xi = 1$, the only Wadge class Γ such that $\forall u' \in Qu$ $\Gamma_{u'} < \Gamma < \Gamma_u$ is $\Delta(\Gamma_u)$. (Assertion (B))

Proof. (a) is by induction.

Case 1. $u(1) = 1$ and $u(2) = \lambda$. Then $\Gamma_u = D_\lambda(\underset{\sim}{\Sigma}^0_\xi)$, and we want to prove $\Gamma_u = SU(\underset{\sim}{\Sigma}^0_\xi, \bigcup_{\eta<\lambda} D_\eta(\underset{\sim}{\Sigma}^0_\xi))$. From right to left the inclusion is obvious. If $A \in \Gamma_u$, let $(A_\eta, \eta < \lambda)$ be an increasing sequence of $\underset{\sim}{\Sigma}^0_\xi$ sets with $A = D_\lambda(\langle A_\eta\rangle)$, and let $(A'_\eta, \eta < \lambda)$ reduce $(A_\eta, \eta < \lambda)$. Then $A = \bigcup_{\eta<\lambda} (A \cap A'_\eta)$, and $A \cap A'_\eta = A_\eta \cap A'_\eta \cap A$ is clearly in $D_\eta(\underset{\sim}{\Sigma}^0_\xi)$.

Case 2. $u(1) = 3$, $t(u_1) = 0$, $u(2) = \lambda$. Then $\Gamma_u = \text{Bisep}(D_\lambda(\underset{\sim}{\Sigma}^0_\xi), \Gamma_{u_0})$, and we want to prove $\Gamma_u = SU(\underset{\sim}{\Sigma}^0_\xi, \bigcup_{\eta<\lambda} \text{Sep}(D_\eta(\underset{\sim}{\Sigma}^0_\xi), \Gamma_{u_0}))$. Again the inclusion from right to left is trivial. So let $A \in \Gamma_u$, and let $C_0, C_1 \in D_\lambda(\underset{\sim}{\Sigma}^0_\xi)$ be the biseparating sets. Using case 1, C_0 and C_1 are in $SU(\underset{\sim}{\Sigma}^0_\xi, \bigcup_{\eta<\lambda} D_\eta(\underset{\sim}{\Sigma}^0_\xi))$, and then it is obvious, using the closure properties of Γ_{u_0}, that $A \in SU(\Sigma^0_\xi, \bigcup_{\eta<\lambda} \text{Sep}(D_\eta(\Sigma^0_\xi), \Gamma_{u_0}))$.

Case 3. $u(1) = 4$, so $\Gamma_u = SU(\underset{\sim}{\Sigma}^0_\xi, \bigcup_n \Gamma_{u_n})$, which is the equality we want.

Case 4. $u(1) = 5$, $u(2) = \eta + 1$ $\eta \geq 1$, $t(u_1) = 0$. Then $\Gamma_u = SD_{\eta+1}(\underset{\sim}{\Sigma}^0_\xi, \Gamma_{u_0})$ where $\Gamma_{u_0} = SU(\underset{\sim}{\Sigma}^0_\xi, \bigcup_n \Gamma_n)$ and we want to prove

$$\Gamma_u = SU(\underset{\sim}{\Sigma}^0_\xi, \bigcup SD_\eta(\underset{\sim}{\Sigma}^0_\xi, \Gamma_{u_0}, \Gamma_{u_n})).$$

It is clear that $SD_\eta(\underset{\sim}{\Sigma}^0_\xi, \Gamma_{u_0}, \Gamma_{u_n})$ is in $\Delta(SD_{\eta+1}(\underset{\sim}{\Sigma}^0_\xi, \Gamma_{u_0}))$, so the inclusion from right to left is obvious. Suppose now $A \in \Gamma_u$. Then for some increasing pairs $\langle A_\zeta, C_\zeta, \zeta \leq \eta\rangle$, with $A_\zeta = \bigcup_n (A^n_\zeta \cap C^n_\zeta)$, $\bigcup_n C^n_\zeta = C_\zeta$, where $A^n_\zeta \in \Gamma_n$ and $C^n_\zeta \in \underset{\sim}{\Sigma}^0_\xi$, we have $A = \bigcup_\zeta (A_\zeta - \bigcup_{\zeta'<\zeta} C_{\zeta'})$. Let $A_{<\eta} \subset C_{<\eta}$. Now let $\langle C^*_n \mid n \in \omega\rangle$ reduce the sequence $(C_{<\eta} \cup C^n_\eta)$, and consider $A \cap C^*_n = (A_{<\eta} \cap C^*_n) \cup (A_\eta - C_{<\eta} \cap C^*_n)$. It is clearly in $SD_\eta(\underset{\sim}{\Sigma}^0_\xi, \Gamma_{u_0}, \Gamma_{u_n})$. Moreover $A \subset \bigcup_n C^*_n = C_\eta$, so that $A \in SU(\underset{\sim}{\Sigma}^0_\xi, \bigcup SD_\eta(\underset{\sim}{\Sigma}^0_\xi, \Gamma_{u_0}, \Gamma_{u_n}))$.

Case 5. $u(1) = 5$, $u(2) = \lambda$ is limit and $t(u_1) = 0$. The proof is entirely analogous to case 4, and we omit it.

Case 6 (induction step). Suppose that $u(1) = 3$ (the case $u(1) = 5$ being analogous) and $t(u_1) = 2$, $u_1(0) = 1$. We have $\Gamma_u = \text{Bisep}(D_\eta(\underset{\sim}{\Sigma}^0_\xi), \Gamma_{u_0}, \Gamma_{u_1})$ with $\Gamma_{u_1} = SU(\underset{\sim}{\Sigma}^0_\xi, \bigcup\{\Gamma_{u'} \mid u' \in Qu_1\})$ and we want to prove that

$$\Gamma_u = SU(\underset{\sim}{\Sigma}^0_\xi, \bigcup\{\text{Bisep}(D_\eta(\underset{\sim}{\Sigma}^0_\xi), \Gamma_{u_0}, \Gamma_{u'}) \mid u' \in Qu_1\}).$$

Let $\Gamma^* = \bigcup\{\Gamma_{u'} \mid u' \in Qu_1\}$, so that the left side is $\Gamma_u = \text{Bisep}(D_\eta(\underset{\sim}{\Sigma}^0_\xi), \Gamma_{u_0}, SU(\underset{\sim}{\Sigma}^0_\xi, \Gamma^*))$, and the right side $\Gamma_r = SU(\underset{\sim}{\Sigma}^0_\xi, \text{Bisep}(D_\eta(\underset{\sim}{\Sigma}^0_\xi), \Gamma_{u_0}, \Gamma^*))$. The proof that $\Gamma_r = \Gamma_u$ is entirely analogous to the one in Corollary 1.19(a). We omit it.

(b) The proof of part (b) follows easily from part (a). For if $A, \check{A} \in SU(\underset{\sim}{\Sigma}^0_\xi, \bigcup_n \Gamma_n)$, so that $A = \bigcup_n (A_n \cap C_n)$, $A_n \in \Gamma_n$, $C_n \in \underset{\sim}{\Sigma}^0_\xi$, (pairwise disjoint sets) and $\check{A} = \bigcup_n (A'_n \cap C'_n)$, $A'_n \in \Gamma_n$, $C'_n \in \Sigma^0_\xi$, then choosing a sequence C^*_n, C'^*_n of $\underset{\sim}{\Sigma}^0_\xi$ sets reducing the sequence $\langle C_n, n \in \omega\rangle$, $\langle C'_n, n \in \omega\rangle$, we obtain $A = \bigcup_n (A_n \cap C^*_n) \cup \bigcup_n (A_n \cap C'^*_n)$, which, because $\bigcup_n C^*_n \cup \bigcup_n C'^*_n = \omega^\omega$, shows that $A \in PU(\underset{\sim}{\Sigma}^0_\xi, \bigcup_n \Gamma_n)$.

(c) Assertion (B) then follows easily from the game theoretical characterization of Wadge's ordering.

The following next two lemmas take care of assertion (C) for descriptions u of type 2 and of level, respectively, a successor and a limit ordinal.

1.24. Lemma. Assume u is a description of type 2 and level $u(0) = \xi + 1$, with $\xi \geq 1$. Then $\Delta(\Gamma_u) = \bigcup\{\Gamma_{u'} \mid u' \in Qu\}$.

Proof. By lemma 1.23, we know that

$$\Delta(\Gamma_u) = PU(\underset{\sim}{\Sigma}^0_{\xi+1}, \bigcup_n \Gamma_{s_u(n)})$$

and we want to prove that

$$\Delta(\Gamma_u) = \bigcup \{S D_\eta (\underset{\sim}{\Sigma}^0_\xi, S U (\underset{\sim}{\Sigma}^0_\xi, \bigcup_n \Gamma_{s_u(n)})) \mid 2 \leq \eta < \omega_1\}$$

Let $\Gamma^* = \bigcup_n \Gamma_{s_u(n)}$. By the definition of s_u, each $\Gamma_{s_u(n)}$ is of level $\geq \xi + 1$. From right to left, the inclusion is easy: If $A \in S D_{\eta_0} (\underset{\sim}{\Sigma}^0_\xi, S U(\underset{\sim}{\Sigma}^0_\xi, \Gamma^*))$ for some $\eta_0 < \omega_1$, let A_η, C_η, $\eta < \eta_0$, be pairs witnessing this fact, $A_\eta \in S U (\underset{\sim}{\Sigma}^0_\xi, \Gamma^*)$, $C_\eta \in \underset{\sim}{\Sigma}^0_\xi$ and $A = \bigcup_{\eta < \eta_0} (A_\eta - \bigcup_{\eta' < \eta} C_{\eta'})$.

Each $A^* = A_\eta - \bigcup_{\eta' < \eta} C_{\eta'}$ is clearly in $S U (D_{\eta+2}(\underset{\sim}{\Sigma}^0_\xi, \Gamma^*))$, so is in $P U (\underset{\sim}{\Sigma}^0_{\xi+1}, \Gamma^*)$. But the sets A^* are disjoint, and separated by the $C^*_\eta = C_\eta - \bigcup_{\eta' < \eta} C_{\eta'}$, which are disjoint and in $D_\eta(\underset{\sim}{\Sigma}^0_\xi)$. This clearly proves that $A \in \Delta(\Gamma_u) = P U (\underset{\sim}{\Sigma}^0_{\xi+1}, \Gamma^*)$.

For the other inclusion, let us look at the case $\xi = 1$. Suppose $A \in \Delta(\Gamma_u) = P U (\underset{\sim}{\Sigma}^0_2, \Gamma^*)$. Because each $\underset{\sim}{\Sigma}^0_2$ set is the disjoint union of $\underset{\sim}{\Pi}^0_1$ sets, we also have $A \in P U (\underset{\sim}{\Pi}^0_1, \Gamma^*)$, say $A = \bigcup_n A_n$, where $A_n \in \Gamma^*$ and for some sequence F_n of closed sets, $A_n \subset F_n$, and the F_n are a partition of ω^ω.

Define a transfinite sequence O_η, $\eta < \omega_1$, by $O_0 = \bigcup \{O \in \underset{\sim}{\Sigma}^0_1 \; A \cap O \in \Gamma^*\}$, and more generally $O_\eta = \bigcup \{O \in \underset{\sim}{\Sigma}^0_1 \; O \cap (A - \bigcup_{\eta' < \eta} O_{\eta'}) \in \Gamma^*\}$. The sequence O_η is increasing, hence is stationary after $\eta_0 < \omega_1$. We claim that the sequence $\langle A \cap O_\eta \cup \bigcup_{\eta' < \eta} O_{\eta'}, O_\eta \rangle$, $\eta \leq \eta_0$ witnesses that $A \in S D_{\eta_0+1} (\underset{\sim}{\Sigma}^0_1, S U (\underset{\sim}{\Sigma}^0_1, \Gamma^*)$.
We have to prove two things: First, that $A \cap (O_\eta - \bigcup_{\eta' < \eta} O_{\eta'})$ is in $S U (\underset{\sim}{\Sigma}^0_1, \Gamma^*)$ which will imply that $(A \cap O_\eta) \cup \bigcup_{\eta' < \eta} O_{\eta'}$ is also in it, and secondly that $A \subset O_{\eta_0}$. The first fact comes from the definition of O_η: O_η is the union of a disjoint family of open sets O^n_η with $O^n_\eta \cap (A - \bigcup_{\eta' < \eta} O_{\eta'}) \in \Gamma^*$. This is what we wanted. The second fact is an obvious use of the Baire category theorem applied to the partition of $(\bigcup_{\eta' < \eta} O_{\eta'})$ induced by the sets F_n.

The case of an arbitrary ξ can be reduced to the preceding case using Kuratowski's technique of generalized homeomorphisms. An alternative proof would use the effective topologies we introduce in section 2.

We shall also omit the proof of the next result, which again uses an argument of transfer, as in the proof of Wadge's lemma, and in the preceding lemma.

1.25. Lemma. Assume u is a description of type 2 and level $u(0) = \lambda$ a limit ordinal. Then $\Delta(\Gamma_u) = \bigcup \{\Gamma_{u'} : u' \in Q_u\}$.

We now turn to the last step of the proof of theorem 1.9: the case of a description of type 3.

1.26. __Lemma__. Let u be a description of type 3, with $u(0) = \xi$. Then Γ_u is closed under intersection with a $\underset{\sim}{\Pi}^0_\xi$ set. Moreover $\Delta(\Gamma_u)$ has the same closure property.

__Proof__. The second assertion is an immediate corollary of the first, and the fact that Γ_u is closed under union with a $\underset{\sim}{\Sigma}^0_\xi$ set, which can easily be seen from the proof of lemma 1.4 (b). [The only classes of level ξ not closed under union with a $\underset{\sim}{\Sigma}^0_\xi$ set are the $D_n \underset{\sim}{\Sigma}^0_\xi$, for $n \in \omega$, n even.] For if $A \in \Delta(\Gamma_u)$ and $B \in \underset{\sim}{\Pi}^0_\xi$, then $A \cap B \in \Gamma_u$ by the first assertion of the lemma, and $(A \check{\cap} B) = \check{A} \cup \check{B} \in \Gamma_u$ by the preceding remark.

To prove that Γ_u is closed under intersection with a $\underset{\sim}{\Pi}^0_\xi$ set, argue by induction. If $u(1) = 3$ or 5, then either we have $u_1(0) > u(0)$, and then Γ_{u_1} is clearly closed under intersection with a $\underset{\sim}{\Pi}^0_\xi$ set (in fact with a $\underset{\sim}{\Sigma}^0_{\xi+1}$ set, by lemma 1.4), or u_1 is itself of type 3, and we can apply the induction hypothesis. The only other case is if $u(1) = 2$, so $\Gamma_u = \mathrm{Sep}\ (D_\eta(\underset{\sim}{\Sigma}^0_\xi), \Gamma^*_u)$. But then $u^*(0) > \xi$, so Γ_{u^*} is closed under intersection with a $\underset{\sim}{\Pi}^0_\xi$, and the conclusion follows immediately.

1.27. __Definition__. For each u of type 3, we define a set Qu of descriptions by the following conditions:

__Case 1__. $u(1) = 2$, so $\Gamma_u = \mathrm{Sep}\ (D_\eta(\underset{\sim}{\Sigma}^0_\xi), \Gamma_{u^*})$. Define $Qu = \{\xi^\frown 3^\frown \eta^\frown \langle u^*, u'\rangle \mid u' \in D,\ \Gamma_{u'} < \Gamma_{u^*}$ and $u'(0) \geq \xi\}$.

__Case 2__. $u(1) = 3$ or 5, and $t(u_1) = 3$ (inductive step). Define $Qu = \{u(0)^\frown u(1)^\frown u(2)^\frown \langle u_0, u'\rangle \mid u' \in Qu_1\}$.

__Case 3__. $u(1) = 3$ or 5, and $t(u_1) = 1$ or 2 (so $u_1(0) > u(0)$). Then Qu_1 has been previously defined, and define $Qu = \{u(0)^\frown u(1)^\frown u(2)^\frown \langle u_0, u'\rangle \mid u' \in Qu_1$ and $u'(0) \geq \xi\}$.

1.28. __Lemma__ (Assertion (c) for u of type 3). Let u be a description of type 3. Then $\Delta(\Gamma_u) = \{\Gamma_{u'} \mid u' \in Qu\}$.

__Proof__.

__Case 1__. $u(1) = 2$, so $\Gamma_u = \mathrm{Sep}\ (D_\eta(\underset{\sim}{\Sigma}^0_\xi), \Gamma_{u^*})$, with $u^*(0) > \xi$. We want to prove that

$$\Delta(\Gamma_u) = \bigcup \{\mathrm{Bisep}\ (D_\eta(\underset{\sim}{\Sigma}^0_\xi), \Gamma_{u^*}, \Gamma_{u'}) \mid \Gamma_{u'} < \Gamma_{u^*},\ u'(0) \geq \xi\}$$

Suppose first A is in the right hand side class. Then, by the closure properties of Γ_{u^*}, there are disjoint $D_\eta \underset{\sim}{\Sigma}^0_\xi$ sets C_1 and C_2 such that $A_1 = A \cap C_1 \in \check{\Gamma}_{u^*}$, $A_2 = A \cap C_2 \in \Gamma_{u^*}$, and $B = A - (C_1 \cup C_2) \in \Delta(\Gamma_{u^*})$. Now A and B are two Γ_{u^*} sets separated by the $\underset{\sim}{\Sigma}^0_{\xi+1}$ sets C_2 and $\check{C}_2$, so $A_2 \cup B \in \check{\Gamma}_{u*}$, and $A \in \mathrm{Sep}\,(D_\eta(\underset{\sim}{\Sigma}^0_\xi), \Gamma_{u*})$. Similarly A_1 and B are two $\check{\Gamma}_{u*}$ sets separated by the $\underset{\sim}{\Sigma}^0_{\xi+1}$ sets C_1 and $\check{C}_1$, so $A_1 \cup B \in \check{\Gamma}_{u*}$ and $A \in \mathrm{Sep}\,(D_\eta(\underset{\sim}{\Sigma}^0_\xi), \Gamma_{u*})$. This gives $A \in \Delta(\Gamma_u)$.

For the converse, we suppose $A \in \Delta(\Gamma_u)$, and we want to find disjoint $D_\eta(\underset{\sim}{\Sigma}^0_\xi)$ sets C_1 and C_2 such that $A \cap C_1 \in \check{\Gamma}_{u*}$, $A \cap C_2 \in \Gamma_{u*}$ and $A - (C_1 \cup C_2) \in \Delta(\Gamma_{u*})$, for then the inductive hypothesis will give the result. Let C, C' be two $D_\eta(\underset{\sim}{\Sigma}^0_\xi)$ sets such that $A \cap C \in \check{\Gamma}_{u*}$, $A - C \in \Gamma_{u*}$, and $A \cap C' \in \check{\Gamma}_{u*}$, $\check{A} - C' \in \Gamma_{u*}$. Let C_1 and C_2 be $D_\eta(\underset{\sim}{\Sigma}^0_\xi)$ sets reducing the pair (C, C'). Then $(A \cap C_1) = A \cap C \cap C_1 \in \check{\Gamma}_{u*}$, $A \cap C_2 = C_2 - \check{A} = C_2 - (C' \cap \check{A})$ is in Γ_{u*}, and finally $B = A - (C_1 \cup C_2) = A - (C \cup C') = (A - C) - C'$ is in Γ_{u*} and $\check{B} = (C_1 \cup C_2) \cup \check{A} = (C \cup C') \cup \check{A} = C \cup C' \cup \check{A} - C'$ is in Γ_{u*}. This proves case 1.

Case 2. $u(1) = 3$, $t(u_1)$ arbitrary. We have $\Gamma_u = \mathrm{Bisep}\,(D_\eta(\underset{\sim}{\Sigma}^0_\xi), \Gamma_{u_0}, \Gamma_{u_1})$, and we know that $u_0(0) > \xi = u(0)$, and either $u_1(0) > u(0)$, or $u_1(0) = u(0)$ and u_1 is of type 3. We may assume that $\cup\{\Gamma_{u'} \mid u' \in Qu_1 \text{ and } u'(0) \geq \xi\} = \Delta(\Gamma_{u_1})$: This is the induction hypothesis if u_1 is of type 3, and if u_1 is of type 1 or 2, a look at the definition of Qu_1 shows that $\cup\{\Gamma_{u'} \mid u' \in Qu_1 \text{ and } u'(0) \geq \xi\} = \cup\{\Gamma_{u'} \mid u' \in Qu_1\}$, if $u_1(0) > \xi$. But we have already proved that this last class is $\Delta(\Gamma_{u_1})$. So we want to prove that $\Delta(\Gamma_u) = \mathrm{Bisep}\,(D_\eta(\underset{\sim}{\Sigma}^0_\xi), \Gamma_{u_0}, \Delta(\Gamma_{u_1}))$. The inclusion from right to left is obvious. Suppose now $A \in \Delta(\Gamma_u)$. Using the normal form for the Bisep operation, plus the fact that $\Delta(\Gamma_{u_1})$ is closed under intersection with a $\underset{\sim}{\Pi}^0_\xi$ set (lemma 1.26), we have $A = (C_0 \cap A) \cup (C_1 \cap A) \cup B$, where C_0, C_1 are disjoint $D_\eta(\underset{\sim}{\Sigma}^0_\xi)$ sets, with $C_0 \cup C_1 \in \underset{\sim}{\Sigma}^0_\xi$, and $C_0 \cap A \in \check{\Gamma}_{u_0}$, $C_1 \cap A \in \Gamma_{u_0}$ and $B \in \Gamma_{u_1}$, $B \cap (C_0 \cup C_1) = \emptyset$. Similarly we have $\check{A} = (C'_0 \cap \check{A}) \cup (C'_1 \cap \check{A}) \cup B'$, with similar properties. Let C^*_0, C^*_1 be two $\underset{\sim}{\Sigma}^0_\xi$ sets reducing the pair $C_0 \cup C_1$, $C_0 \cup C_1$. Then $A \cap C^*_0$ is in $\mathrm{Bisep}\,(D_\eta(\underset{\sim}{\Sigma}^0_\xi), \Gamma_{u_0})$, and $\check{A} \cap C^*_1$, so $A \cap C^*_1$, is in $\mathrm{Bisep}\,(D_\eta(\underset{\sim}{\Sigma}^0_\xi), \Gamma_{u_0})$. By the closure properties of Γ_{u_0}, it follows that $A \cap (C^* \cup C^*_1)$ is also in $\mathrm{Bisep}\,(D_\eta(\underset{\sim}{\Sigma}^0_\xi), \Gamma_{u_0})$. So we just have to show that $A - (C^*_0 \cup C^*_1) \in \Delta(\Gamma_{u_1})$. But $A - C^*_0 \cup C^*_1 = A - (C_0 \cup C_1 \cup C'_0 \cup C'_1) = B - (C'_0 \cup C'_1) = B' - (C_0 \cup C_1)$. We clearly have $A - (C^*_0 \cup C^*_1) \in \Delta(\Gamma_{u_1})$.

Case 3, for $u(1) = 5$, is entirely similar, and we omit it.

Lemmas 1.12, 1.14, 1.16, 1.17, 1.19, 1.23, 1.24, 1.25 and 1.28, put together, give a proof of the assertions (A), (B) and (C) of page and hence prove theorem 1.9.

§2. Effective results in the Borel Wadge hierarchy. The Wadge classes considered in the first part are boldface classes. We now are interested in their lightface counterparts, and in order to define them, we need a coding system, both for classes and for sets in each class.

For the classes, there is no problem: it is enough to code by reals sequences of countable ordinals, and this is obvious: we say that α is a D-code, written $\alpha \in \ulcorner D \urcorner$, if for every n $(\alpha)_n \in WO$, and the coded sequence $u_\alpha = \langle |(\alpha)_n|, n \in \omega \rangle$ is a description. Going back to the definition of descriptions shows immediately that $\ulcorner D \urcorner$ is a Π^1_1 set. We shall denote by Γ_α the class Γ_{u_α} (although it is a bit ambiguous, as some descriptions may be reals).

Encoding the sets in each Γ_α is also easy, but technical. Fix some recursive real $\bar{1}$ in WO, with $|\bar{1}| = 1$. Start from a pair (W, C) in Π^1_1 which is universal for $\mathbf{\Sigma}^0_\xi$ sets in the following sense:

1. $W \subseteq \omega^\omega \times \omega^\omega$, and $\exists \gamma\ (\alpha, \gamma) \in W \leftrightarrow \alpha \in WO$

2. $C \subseteq \omega^\omega \times \omega^\omega \times \omega^\omega$, and $\forall \alpha \in WO$ $C_\alpha = \{(\gamma, \delta) \mid (\alpha, \gamma, \delta) \in C\}$ is universal for $\mathbf{\Sigma}^0_{|\alpha|}$ subsets of ω^ω

3. C is Δ^1_1 on W, i.e., the relation $(\alpha, \gamma) \in W \wedge (\alpha, \gamma, \delta) \notin C$ is Π^1_1.

It is then easy to construct a Π^1_1 pair (W^{is}, C^{is}) such that

1. $\exists \gamma\ (\alpha, \beta, \gamma) \in W^{is} \leftrightarrow \alpha \in WO \wedge \beta \in WO$

2. $C^{is}_{\alpha,\beta}$ is universal for $<_\beta$-increasing sequences of $\mathbf{\Sigma}^0_{|\alpha|}$ sets $(C^{is} \subset \omega^\omega \times \omega^\omega \times \omega^\omega \times \omega \times \omega^\omega)$

3. C^{is} is Δ^1_1 on W^{is}.

[Define $(\alpha, \beta, \gamma) \in W^{is} \leftrightarrow \beta \in WO \wedge \forall n\ (\alpha, (\gamma)_n) \in W \wedge \forall n\ \forall m\ \forall \delta$ $((\beta(n, m) = 0 \wedge (\alpha, (\gamma)_n, \delta) \in C \rightarrow (\alpha, (\gamma)_m, \delta) \in C)$ and $(\alpha, \beta, \gamma, n, \delta) \in C^{is} \Longleftrightarrow (\alpha, \beta, \gamma) \in W^{is} \wedge (\alpha, (\gamma)_n, \delta) \in C.$]

Similarly, one can define a Π^1_1 pair (W^{ds}, C^{ds}) such that

(a) $\gamma(\alpha, \gamma) \in W^{ds} \leftrightarrow \alpha \in WO$

(b) For $\alpha \in WO$ C^{ds}_α is universal for pairwise disjoint sequences of $\mathbf{\Sigma}^0_{|\alpha|}$ sets.

(c) C^{ds} is Δ^1_1 on W^{ds}

[Let $(\alpha, \gamma) \in W^{ds} \leftrightarrow \forall n\ (\alpha, (\gamma)_n) \in W$, and if $C_1 = \{(\alpha, \gamma, n, \delta) \mid (\alpha, \gamma) \in W^{is} \wedge (\alpha, (\gamma)_n, \delta) \in C$, let C^{ds} reduce C_1 (with respect to n), in such a way that for each α $(C)^{ds}_{\alpha,\gamma,n}$ is $\mathbf{\Sigma}^0_{|\alpha|}$.]

With these preliminary constructions, we can define a relation β is a Γ_α-code, together with "β codes the set $\Gamma_{\alpha,\beta}$".

2.1. Definition. For $\alpha \in \ulcorner D \urcorner$, the relations "$\beta$ is a Γ_α-code", and "The set in Γ_α coded by β is A" (written $\Gamma_{\alpha,\beta} = A$) are defined inductively by the following conditions:

(a) If $|(\alpha)_0| = 0$ and $\beta(0) = 0$, β is a Γ_α-code and $\Gamma_{\alpha,\beta} = \emptyset$

(b) Suppose $|(\alpha)_0| \geq 1$ and $|(\alpha)_1| = 1$ and $\beta = 1^\frown\beta^*$. Then if $((\alpha)_0, (\alpha)_2, \beta^*) \in W^{is}$, then β is a Γ_α-code, and

$$\Gamma_{\alpha,\beta} = D_{|(\alpha)_2|}(\langle A_\zeta \mid \zeta < |(\alpha)_2|\rangle) ,$$

where $A_\zeta = C^{is}_{(\alpha)_0,(\alpha)_2,\beta^*,n}$ for the unique n which has order type ζ in $<_{(\alpha)_2}$.

(c) Suppose $|(\alpha)_0| \geq 1$ and $|(\alpha)_1| = 2$. Let $\alpha_0 = (\alpha)_0{}^\frown 1^\frown(\alpha)_2{}^\frown \underline{0}^\frown\underline{0} \ldots$ and let $\alpha_1 = \langle(\alpha)_{n+3} \mid n \in \omega\rangle$. Then β is a Γ_α code if $\beta = 2^\frown\langle\beta_0, \beta_1, \beta_2\rangle$, where β_0 is a Γ_{α_0}-code, β_1 and β_2 are Γ_{α_1}-codes. Moreover $\Gamma_{\alpha,\beta} = \text{Sep}\,(\Gamma_{\alpha_0,\beta_0}, \check{\Gamma}_{\alpha_1,\beta_1}, \Gamma_{\alpha_2,\beta_2})$.

(d) Suppose now $|(\alpha)_0| \geq 1$ and $|(\alpha)_1| = 3$. Let $\alpha_0 = (\alpha)_0{}^\frown\overline{1}^\frown(\alpha)_2{}^\frown\underline{0}^\frown\underline{0} \ldots$, and let α_1, α_2 be such that $u_\alpha = \xi^\frown 3^\frown\eta^\frown\langle u_{\alpha_1}, u_{\alpha_2}\rangle$. (Such α_1, α_2 could be defined precisely, and are supposed to be recursive in α.) Then β is a Γ_α-code $\beta = 3^\frown\langle\beta_0, \beta_1, \beta_2, \beta_3, \beta_4\rangle$, where β_0, β_1 are Γ_{α_0}-codes, β_1, β_2 are Γ_{α_1}-codes, β_4 is a Γ_{α_2}-code, and $\Gamma_{\alpha_0,\beta_0} \cap \Gamma_{\alpha_0,\beta_1} = \emptyset$; and then $\Gamma_{\alpha,\beta} = \text{Bisep}\,(\Gamma_{\alpha_0,\beta_0}, \Gamma_{\alpha_0,\beta_1}, \check{\Gamma}_{\alpha_1,\beta_2}, \Gamma_{\alpha_1,\beta_3}, \Gamma_{\alpha_2,\beta_4})$.

(e) If $|(\alpha)_1| = 4$, then let α_n be (recursively in α) a sequence such that $u_\alpha = \xi^\frown 4^\frown\langle u_{\alpha_n} \mid n \in \omega\rangle$; then β is a Γ_α-code if $\beta = 4^\frown\langle\beta^*, \beta^{**}, \beta_n \mid n \in \omega\rangle$, where $((\alpha)_0, \beta^*) \in W^{ds}$, $((\alpha)_0, \beta^{**}) \in W$ and codes the union of the disjoint sequence coded by β^*, and for each n β_n is a Γ_{α_n}-code. Then $\Gamma_{\alpha,\beta} = S\,U(\langle C^{ds}_{(\alpha)_0,\beta^*,n} \mid n \in \omega\rangle, \langle\Gamma_{\alpha_n,\beta_n} \mid n \in \omega\rangle)$.

(f) Finally if $|(\alpha)_1| = 5$, let α_0, α_1 (recursively in α) be such that $u_\alpha = \xi^\frown 5^\frown\eta^\frown\langle u_{\alpha_0}, u_{\alpha_1}\rangle$. Then β is a Γ_α-code if $\beta = 5^\frown\langle\beta_1, \gamma_n \in n \in \omega\rangle$, where β_1 is a Γ_{α_1}-code, for each n γ_n is a Γ_{α_0} code, say $\gamma_n = 4^\frown\langle\gamma^*_n, \gamma^{**}_n, \gamma^p_n \mid p \in \omega\rangle$ and the sequence of pairs (A_ζ, C_ζ), $\zeta < |(\alpha)_2|$ defined by $A_\zeta = \Gamma_{(\alpha)_0,\gamma_n}$, $C_\zeta = C_{(\alpha)_0,\gamma^{**}_n}$ for the only n of order type ζ in $<_{(\alpha)_2}$, is an increasing sequence with $C_\zeta \subset A_{\zeta+1}$. And then

$$\Gamma_{\alpha,\beta} = S\,D_{|(\alpha)_2|}(\langle C_\zeta \mid \zeta < |(\alpha)_2|\rangle, \langle A_\zeta \mid \zeta < |(\alpha)_2|\rangle, \Gamma_{\alpha_1,\beta_1})$$

It is clear, from the preceding definition, that the coding relations

$$\alpha \in \ulcorner D \urcorner$$

$$\alpha \in \ulcorner D \urcorner \wedge \beta \text{ is a } \Gamma_\alpha\text{-code}$$

$$\alpha \in \ulcorner D \urcorner \wedge \beta \text{ is a } \Gamma_\alpha\text{-code} \wedge \gamma \in \Gamma_{\alpha,\beta}$$

and

$$\alpha \in \ulcorner D \urcorner \wedge \beta \text{ is a } \Gamma_\alpha\text{-code} \wedge \gamma \notin \Gamma_{\alpha,\beta}$$

are all Π^1_1.

From the proof of the main theorem of part 1, it is also clear that some variant of the preceding coding would enable to prove "recursive" analogs of the Hausdorff-Kuratowski-type results we have quoted. Such a variant would involve coding by partial recursive functions, in the spirit of what is done for the classes Σ^0_ξ.

Anyway, we are more interested here in "Δ^1_1-recursive" results, for which the coding we defined above is good enough. From now on, Wadge classes will be written boldface, to distinguish from their lightface counterparts.

2.2. Definition. A described Wadge class $\boldsymbol{\Gamma}$ is a Δ^1_1-class if it admits a Δ^1_1 code (i.e., $\boldsymbol{\Gamma} = \boldsymbol{\Gamma}_\alpha$ for some Δ^1_1 real α in D). So in particular the Δ^1_1 classes, among the $\boldsymbol{\Sigma}^0_\xi$'s, are the $\boldsymbol{\Sigma}^0_\xi$ for $\xi < \omega_1^{ck}$.

2.3. Definition. Let $\boldsymbol{\Gamma}_\alpha$, $\alpha \in \Delta^1_1$, be a Δ^1_1-class. We define the lightface classes Γ_α, $\Gamma_\alpha(\beta)$, Γ^1_α by

$$\Gamma_\alpha = \{\boldsymbol{\Gamma}_{\alpha,\beta} \mid \beta \text{ a recursive } \boldsymbol{\Gamma}_\alpha\text{-code}\}$$

$$\Gamma_\alpha(\beta) = \{\boldsymbol{\Gamma}_{\alpha,\gamma} \mid \gamma \text{ a recursive-in-}\beta \ \boldsymbol{\Gamma}_\alpha\text{-code}\}$$

and

$$\Gamma^1_\alpha = \cup \{\Gamma_\alpha(\beta) \mid \beta \in \Delta^1_1\} = \{\boldsymbol{\Gamma}_{\alpha,\beta} \mid \beta \in \Delta^1_1\}$$

Because of the coding we chose, it is not clear that the lightface class Γ_α is really well defined, i.e., does not depend on the particular code α for $\boldsymbol{\Gamma}_\alpha$, even in case α is recursive. But it can be seen that Γ^1_α does not depend on the particular choice of $\alpha \in \Delta^1_1$, but only on the class $\boldsymbol{\Gamma}_\alpha$. [This can be seen directly, but is also an immediate corollary of the main result below.]

In Louveau [1980], we studied $(\boldsymbol{\Sigma}^0_\xi)^1 = \bigcup_{\alpha\in\Delta^1_1} \Sigma^0_\xi(\alpha)$, for $\xi < \omega_1^{ck}$, and we proved that $(\boldsymbol{\Sigma}^0_\xi)^1 = \boldsymbol{\Sigma}^0_\xi \cap \Delta^1_1$, i.e., that every $\boldsymbol{\Sigma}^0_\xi$ set in Δ^1_1 admits a Δ^1_1 $\boldsymbol{\Sigma}^0_\xi$-code.

The main theorem in this section is the extension of this result to all Δ^1_1 (non self-dual) Borel Wadge classes.

2.4. Theorem. Let $\boldsymbol{\Gamma}_\alpha$ be a described Wadge class, with $\alpha \in \Delta^1_1$. Then each Δ^1_1 set in $\boldsymbol{\Gamma}_\alpha$ admits a $\boldsymbol{\Gamma}_\alpha$ code which is Δ^1_1, i.e., $\Gamma^1_\alpha = \boldsymbol{\Gamma}_\alpha \cap \Delta^1_1$.

In order to prove this theorem, we need some tools from Louveau [1980]. For $\xi < \omega_1^{ck}$, we define T_ξ to be the topology on ω^ω generated by the Σ^1_1 sets which are in $\bigcup_{\eta<\xi} \boldsymbol{\Pi}^0_\eta$. T_∞, the Harrington topology on ω^ω, is the topology generated by all Σ^1_1 subsets of ω^ω. For this topology, ω^ω is a Baire space (i.e., no non-empty T_∞-open set is T_∞-meager). Say that a property of reals is true ∞-a.e. if it is false for a set of reals which is T_∞-meager. In the induction used to prove the result for $\boldsymbol{\Sigma}^0_\xi$ sets, we proved the following two results:

2.5. <u>Proposition</u>. Let A be a $\underset{\sim}{\Sigma}^0_\xi$ set. Then there is a $\underset{\sim}{\Sigma}^0_\xi$ set A' which is T_ξ-open and satisfies $A = A'$ ∞-a.e.

2.6. <u>Proposition</u> (Separation result). Let A, B be two Σ^1_1 sets, and suppose there is a $\underset{\sim}{\Sigma}^0_\xi$ set C such that $A \subseteq C \subseteq B$ ∞-a.e. Then there is a $(\underset{\sim}{\Sigma}^0_\xi)^1$ set D such that $A \subseteq D \subseteq B$.

Proposition 2.6 clearly implies the result for $\underset{\sim}{\Sigma}^0_\xi$ classes; working analogously, we shall prove Theorem 2.4 by proving first a separation result for $\underset{\sim}{\Gamma}_\alpha$, by induction on α.

2.7. <u>Theorem</u>. Let $\underset{\sim}{\Gamma}_\alpha$ be a described Wadge class, with $\alpha \in \Delta^1_1$. If A, B are two Σ^1_1 sets, and there is a $\underset{\sim}{\Gamma}_\alpha$ set C such that $A \subseteq C \subseteq B$ ∞-a.e., then there is a Γ^1_α set D with $A \subseteq D \subseteq B$.

<u>Proof</u>. It is clear that if α is Δ^1_1, the class $\underset{\sim}{\Gamma}_\alpha$ has been constructed from previous classes which are also Δ^1_1, so that we can prove 1.7 by induction. There are five cases (we forget $\underset{\sim}{\Gamma} = \{\emptyset\}$!).

<u>Case 1</u>. $(\alpha)_1 = 1$, so $\underset{\sim}{\Gamma} = D_\eta(\underset{\sim}{\Sigma}^0_\xi)$, with $\eta = |(\alpha)_2|$ and $\xi = |(\alpha)_0|$ (so ξ and η are recursive ordinals). We assume η is an even ordinal, the other case being similar. We first define a sequence $\langle A_\zeta \mid \zeta < \eta\rangle$ by the following: If is even, let A_ζ be the largest T_ξ-open set disjoint from $A - \bigcup_{\zeta' < \zeta} A_{\zeta'}$. If ζ is odd, let A_ζ be the largest T_ξ-open set disjoint from $B - \bigcup_{\zeta' < \zeta} A_{\zeta'}$.

Clearly the sequence $\langle A_\zeta \mid \zeta < \eta\rangle$ is an increasing sequence of $\underset{\sim}{\Sigma}^0_\xi$ sets. Moreover we have:

(a) The relation $\beta \in A_{\zeta_n}$, where ζ_n is the order type of the predecessors of n in $<_{(\alpha)_2}$, is Π^1_1 (in β and n).

(b) If $\langle C_\zeta \mid \zeta < \eta\rangle$ is any sequence of $\underset{\sim}{\Sigma}^0_\xi$ sets with $A \subseteq D_\eta(\langle C_\zeta \mid \zeta < \eta\rangle) \subseteq B$ ∞-a.e., then for each $\zeta < \eta$ $C_\zeta \subseteq A_\zeta$ ∞-a.e., and $A \subseteq D_\eta(\langle A_\zeta \mid \zeta < \eta\rangle) \subseteq B$.

To prove (a). Suppose H is a Σ^1_1 set, and $\xi < \omega_1^{ck}$. Then the largest T_∞-open set O disjoint from H is a Π^1_1 set. In fact

$$\begin{aligned} x \in O &\Leftrightarrow \exists G \in \Sigma^0_\xi \cap \Sigma^1_1 \quad (x \in G \quad G \cap H = \emptyset) \\ &\Leftrightarrow \exists G \in (\Sigma^0_\xi)^1 \quad (x \in G \quad G \cap H = \emptyset) \end{aligned}$$

the second equivalence being obtained by using proposition 2.6. This clearly implies that the relation $\beta \in A_{\zeta_n}$ is Π^1_1.

To prove (b). Using Proposition 2.5, we may assume that the C_ζ are T_∞-open. Then by induction on ζ, we prove that $C_\zeta \subseteq A_\zeta$. Suppose ζ is even. Then

$C_\zeta - \bigcup_{\zeta'<\zeta} C_{\zeta'}$ is disjoint from $D_\eta(\langle C_\zeta \mid \zeta < \eta\rangle)$, so is disjoint from A, ∞-a.e. This implies that C_ζ is ∞-a.e. disjoint from $A - \bigcup_{\zeta'<\zeta} C_{\zeta'}$, and using the ind. hyp., also from $A - \bigcup_{\zeta'<\zeta} A_{\zeta'}$. But this implies, because this last set is Σ^1_1 by part (a), that $C_\zeta \cap (A - \bigcup_{\zeta'<\zeta} A_\zeta) = \emptyset$, using the Baire category theorem for T_∞. So by the definition of A_ζ, $C_\zeta \subseteq A_\zeta$. The odd case is similar, and we omit it. To prove the last assertion, i.e., that $A \subseteq D_\eta(\langle A_\zeta \mid \zeta < \eta\rangle) \subseteq B$, it is enough to prove that $A \subseteq \bigcup_{\zeta<\eta} A_\zeta$. But $A \subseteq \bigcup_{\zeta<\eta} C_\zeta$ ∞-a.e., so using the preceding result, $A \subseteq \bigcup_{\zeta<\eta} A_\zeta$ ∞-a.e.

Finally $\bigcup_{\zeta<\eta} A_\zeta$ is Π^1_1 and A is Σ^1_1, so using again the Baire category theorem for T_∞ gives $A \subseteq \bigcup_{\zeta<\eta} A_\zeta$.

The last part of the proof consists in replacing the Π^1_1 sequence $\langle A_\zeta \mid \zeta < \eta\rangle$ by a Δ^1_1 sequence with the same properties. For each recursive ordinal θ, say that a Δ^1_1 increasing sequence $\langle D_\zeta \mid \zeta < \theta\rangle$ of $\underset{\sim}{\Sigma}^0_\xi$ sets is a θ-system for (A, B) if D_ζ is disjoint from $A - \bigcup_{\zeta'<\zeta} D_\zeta$ if ζ is even, from $B - \bigcup_{\zeta'<\zeta} D_\zeta$ if ζ is odd, and say that this θ-system covers H if $H \subseteq \bigcup_{\zeta<\theta} D_\zeta$. What we want to construct is an η-system covering the set A. So it is enough to prove the following claim:

<u>Claim</u>. Let θ be some even ordinal, $\theta \subseteq \eta$, and let H be a Σ^1_1 set, $H \subseteq \bigcup_{\zeta<\theta} A_\zeta$. Then there exists a θ-system covering H.

The proof of the claim is by induction on θ. Suppose first θ is successor, so that $\theta = \theta' + 2$ with θ' even. By the hypothesis $H \subseteq A_{\theta'+1}$, so we can find a set $D^0 \in (\underset{\sim}{\Sigma}^0_\xi)^1$ with $H \subseteq D^0 \subseteq A_{\theta'+1}$. But as $A_{\theta'+1}$ is disjoint from $B - A_{\theta'}$, $D^0 \cap B \subseteq A_{\theta'}$, and so we can find a $(\underset{\sim}{\Sigma}^0_\xi)^1$ set $D^1 \subseteq D^0$ with $D^0 \cap B \subseteq D^1 \subseteq A_{\theta'}$. By the same reasoning, we must have $A \cap D^1 \subseteq \bigcup_{\zeta<\theta'} A_\zeta$, so by the induction hypothesis, we can find a θ'-system $\langle D_\zeta \mid \zeta < \theta'\rangle$ covering $A \cap D^1$. We then extend this θ' system by setting $D_{\theta'} = \bigcup_{\zeta<\theta'} D_\zeta \cup D^1$, and $D_{\theta'+1} = \bigcup_{\zeta<\theta'} D_\zeta \cup D^0$. This clearly gives the desired θ-system.

Suppose now θ is limit. As $H \subseteq \bigcup_{\zeta<\theta} A_\zeta = \bigcup_{\substack{\zeta \text{ even} \\ \zeta<\theta}} A_\zeta$, we can first choose a Δ^1_1 sequence $\langle H_\zeta, \zeta < \theta, \zeta \text{ even}\rangle$ with $H \subseteq \bigcup H_\zeta$, and for ζ even, $\zeta < \theta$, $H_\zeta \subseteq A_\zeta$. By the induction hypothesis, and Δ^1_1 selection, there is, for each even $\zeta < \theta$, a ζ-system $\langle D^\zeta_{\zeta'} \mid \zeta' < \zeta\rangle$, covering H_ζ, and such that the double sequence is Δ^1_1. Define then a sequence $\langle D_\zeta, \zeta < \theta\rangle$ by

$$D_\zeta = \bigcup_{\zeta'<\zeta} D_{\zeta'} \cup \bigcup_{\zeta<\zeta'<\theta} D^{\zeta'}_\zeta .$$

We claim that the Δ^1_1 and increasing sequence $\langle D_\zeta, \zeta < \theta\rangle$ is the θ-system we wanted. It certainly covers H, so the only thing we have to prove is that it is a θ-system. Suppose $\zeta < \theta$ is even. Then

$$D_\zeta \cap (B - \bigcup_{\zeta'' < \zeta} D_{\zeta''}) \subseteq \bigcup_{\substack{\zeta < \zeta' < \theta \\ \zeta' \text{ even}}} (D^{\zeta'}_\zeta \cap (B - \bigcup_{\zeta'' < \zeta} D_{\zeta''}))$$

so fix ζ' even, with $\zeta < \zeta' < \theta$. As $D^{\zeta'}_{\zeta''} \subseteq D_{\zeta''}$ for any $\zeta'' < \zeta$

$$D^{\zeta'}_\zeta \cap (B - \bigcup_{\zeta'' < \zeta} D_{\zeta''}) \subseteq D^{\zeta'}_\zeta \cap (B - \bigcup_{\zeta'' < \zeta} D^{\zeta}_{\zeta''}) = \emptyset .$$

Similarly if ζ is odd

$$D_\zeta \cap (A - \bigcup_{\zeta'' < \zeta} D_{\zeta''}) \subseteq \bigcup_{\substack{\zeta < \zeta' < \theta \\ \zeta' \text{ even}}} (D^{\zeta'}_\zeta \cap (B - \bigcup_{\zeta'' < \zeta} D_{\zeta''}))$$

and for each ζ' even, $\zeta < \zeta' < \theta$, as $D^{\zeta'}_{\zeta''} \subseteq D_{\zeta''}$ for any $\zeta'' < \zeta$

$$D^{\zeta'}_\zeta \cap (A - \bigcup_{\zeta'' < \zeta} D_{\zeta''}) \subseteq D^{\zeta'}_\zeta \cap (A - \bigcup_{\zeta'' < \zeta} D^{\zeta'}_{\zeta''}) = \emptyset .$$

This proves the claim, and so finishes the proof of case 1.

<u>Case 2</u>. We now suppose α codes a description u with $u(1) = 2$, so that $\underset{\sim}{\Gamma}_\alpha = \text{Sep}\,(D_\eta(\underset{\sim}{\Sigma}^0_\xi), \underset{\sim}{\Gamma}_{\alpha_1})$, where $\xi = |(\alpha)_0|$ and $\eta = |(\alpha)_2|$ are recursive ordinals, and $\alpha_1 \in \Delta^1_1$, so that by the induction hypothesis we can assume the theorem is true for $\underset{\sim}{\Gamma}_{\alpha_1}$. Let us suppose that η is even, the other case being similar. Given A and B, we define a sequence $\langle A_\zeta, \zeta < \eta\rangle$ of $\underset{\sim}{\Sigma}^0_\xi$ sets by the following: If ζ is even, we let A_ζ be the union of all T_ξ open sets O such that for some set C in $\underset{\sim}{\Gamma}_{\alpha_1}$, $(A - \bigcup_{\zeta' < \zeta} A_{\zeta'}) \cap O \subseteq C \subseteq B$ ∞-a.e., and similarly if ζ is odd, A_ζ is the union of all T_ξ open sets O such that for some set C in $\underset{\sim}{\Gamma}_{\alpha_1}$ $(A - \bigcup_{\zeta' < \zeta} A_{\zeta'}) \cap O \subseteq C \subseteq B$.

We claim that the sequence $\langle A_\zeta \mid \zeta < \eta\rangle$, which is clearly an increasing sequence of $\underset{\sim}{\Sigma}^0_\xi$ sets, has the following properties:

(a) The relation $\beta \in A_{\zeta_n}$ is Π^1_1 (in β and n)

(b) If $\langle C_\zeta \mid \zeta < \eta\rangle$ is an increasing sequence of $\underset{\sim}{\Sigma}^0_\xi$ sets such that $C = D_\eta(\langle C_\zeta \mid \zeta < \eta\rangle)$ satisfies that $A \cap C \subseteq H \subseteq B$ ∞-a.e. for some H in $\check{\underset{\sim}{\Gamma}}_{\alpha_1}$, and $A - C \subseteq H' \subseteq B$ ∞-a.e. for some H' in $\underset{\sim}{\Gamma}_{\alpha_1}$, then for each $\zeta < \eta$ $C_\zeta \subseteq A_\zeta$ ∞-a.e.

To prove (a), let (C, D) be the union of all T_ξ-open sets O such that for some set H in $\underset{\sim}{\Gamma}_{\alpha_1}$, $C \cap O \subseteq H \subseteq D$ ∞-a.e. We want to prove that if C, D are

Σ^1_1, (C, D) is Π^1_1. But it is Π^1_1, for

$$x \in (C, D) \Longleftrightarrow \exists G \in \Sigma^1_1 \cap \underset{\sim}{\Sigma}^0_\xi \ \exists H\, (x \in G \wedge H \in \underset{\sim}{\Gamma}_{\alpha_1} \wedge G \cap C \subseteq H \subseteq \check{D} \ \infty\text{-a.e.})$$

$$\Longleftrightarrow \exists G' \in (\underset{\sim}{\Sigma}^0_\xi)^1 \ \exists H' \in \Gamma^1_{\alpha_1} \ (x \in G' \wedge G' \cap C' \subseteq H' \subseteq \check{D})$$

The second equivalence is justified by the following fact: If G is Σ^1_1 and $x \in G$, and for some $H \in \underset{\sim}{\Gamma}_{\alpha_1}$ $G \cap C \subseteq H \subseteq D$ ∞-a.e., then using the induction hypothesis $\exists H' \in \Gamma^1_{\alpha_1}$ with $G \cap C \subseteq H' \subseteq \check{D}$. But then G is disjoint from $C - H'$, which is Σ^1_1, so we can find B' in $(\underset{\sim}{\Sigma}^0_\xi)^1$ with $G \subset G'$ and $G' \cap C - H' = \emptyset$, so that $G' \cap C \subseteq H' \subseteq \check{D}$, and $x \in G'$.

To prove (b), use proposition 2.5 to replace the $\underset{\sim}{\Sigma}^0_\xi$-sets C_ζ by T_ξ-open sets. Then it is immediate, by induction on ζ, that $C_\zeta \subseteq A_\zeta$.

In order to finish the proof of case 2, we argue as follows. Consider $A - \bigcup_{\zeta<\eta} A_\zeta$. It is a Σ^1_1 set, by (a), and by (b) it is a subset, ∞-a.e., of $(\bigcup_{\zeta<\eta} \check{C}_\zeta)$, where $\langle C_\zeta \mid \zeta < \eta\rangle$ is a sequence as in (b) (we know that such a sequence exists). Now it follows that there is an $H \in \underset{\sim}{\Gamma}_{\alpha_1}$ with $A - \bigcup_{\zeta<\eta} A_\zeta \subseteq H \subseteq B$ ∞-a.e., so by the induction hyp., we can find such an H_η in $\Gamma^1_{\alpha_1}$. Now $A \cap \check{H}_\eta \subseteq \bigcup_{\zeta<\eta} A_\zeta$, and so by an argument very similar to the one we used for case 1, we can replace the sequence $\langle A_\zeta \mid \zeta < \eta\rangle$ by a Δ^1_1 sequence $\langle C_\zeta \mid \zeta < \eta\rangle$ having the same properties, namely

-- it is a Δ^1_1 sequence of $\underset{\sim}{\Sigma}^0_\xi$ sets, increasing, and

-- if ζ is even, C_ζ is a union of T_ξ-open sets O such that for some H in $\underset{\sim}{\Gamma}_{\alpha_1}$ $(A - \bigcup_{\zeta'<\zeta} C_{\zeta'}) \cap O \subseteq H \subseteq \check{B}$ ∞-a.e.

-- if ζ is odd, C_ζ is a union of $\check{T}_\xi$-open sets O such that for some H in $\underset{\sim}{\Gamma}_{\alpha_1}$, $(A - \bigcup_{\zeta'<\zeta} C_{\zeta'}) \cap O \subseteq H \subseteq \check{B}$ ∞-a.e.

-- and finally the sequence covers $A \cap \check{H}_\eta$.

We now define a Δ^1_1-sequence $\langle H_\zeta,\ \zeta < \eta\rangle$ of sets: Suppose ζ is even. Then as C_ζ is a union of T_ξ-open sets O such that for some H in $\underset{\sim}{\Gamma}_{\alpha_1}$ $A - \bigcup_{\zeta'<\zeta} C_{\zeta'} \cap O \subseteq H \subseteq \check{B}$ ∞-a.e., we can, by the induction hypothesis and using Δ^1_1 selection, find two Δ^1_1 sequences $\langle C^n \mid n \in \omega\rangle$ and $\langle H^n \mid n \in \omega\rangle$ such that the C^n_ζ are pairwise disjoint $\underset{\sim}{\Sigma}^0_\xi$ sets, of union C_ζ, and H^n_ζ are in $\underset{\sim}{\Gamma}^1_{\alpha_1}$ with $A - \bigcup_{\zeta'<\zeta} C_{\zeta'} \cap C^n_\zeta \subseteq H^n_\zeta \subset B$. Let $H_\zeta = S\,U(\langle C^n_\zeta,\ n \in \omega\rangle, \langle H^n \mid n \in \omega\rangle)$. $H_\zeta \in \underset{\sim}{\Gamma}^1_{\alpha_1}$, so that C_ζ itself satisfies $\exists H \in \underset{\sim}{\Gamma}^1_{\alpha_1}\ A - \bigcup_{\zeta'<\zeta} C_{\zeta'} \cap C_\zeta \subseteq H \subseteq \check{B}$. Similarly for ζ odd, there is an $H \in \check{\underset{\sim}{\Gamma}}^1_{\alpha_1}$ with $A - \bigcup_{\zeta'<\zeta} C_{\zeta'} \cap C_\zeta \subseteq H \subseteq \check{B}$. Using Δ^1_1-selection, we can find such a sequence $\langle H_\zeta \mid \zeta < \eta\rangle$ in a Δ^1_1 way. We now put

$$H_0 = \bigcup_{\substack{\zeta \text{ odd} \\ \zeta < \eta}} (H_\zeta \cap (C_\zeta - \bigcup_{\zeta' < \zeta} C_{\zeta'}))$$

and

$$H_1 = \bigcup_{\substack{\zeta \text{ even} \\ \zeta < \eta}} (H_\zeta \cap (C_\zeta - \bigcup_{\zeta' < \zeta} C_{\zeta'})) \cup (H_\eta - \bigcup_{\zeta < \eta} C_\zeta)$$

It is clear that $H_0 \in \check{\underset{\sim}{\Gamma}}^1_{\alpha_1}$, $H_1 \in \Gamma^1_{\alpha_1}$, and that the set $D = \text{Sep}\,(D_\eta(\langle C_\zeta \mid \zeta < \eta\rangle), H_0, H_1)$ is in Γ^1_α and satisfies $A \subseteq D \subseteq \check{B}$.

Case 3. Suppose α codes a description u with $u(1) = 3$, so that $\underset{\sim}{\Gamma}_\alpha = \text{Bisep}\,(D_\eta(\underset{\sim}{\Sigma}^0_\xi), \underset{\sim}{\Gamma}_{\alpha_1}, \underset{\sim}{\Gamma}_{\alpha_2})$, with η, ξ as before, and α_1, α_2 are Δ^1_1 codes for descriptions. We assume again η is even. Starting with A, B in Σ^1_1, we first construct the sequence $\langle A_\zeta \mid \zeta < \eta\rangle$ as in case 2, and a similar sequence $\langle B_\zeta \mid \zeta < \eta\rangle$, defined by exchanging the role of $\underset{\sim}{\Gamma}_{\alpha_1}$ and $\check{\underset{\sim}{\Gamma}}_{\alpha_1}$ in the definition. Let $A^0 = \bigcup_{\zeta<\eta} A_\zeta$, $A^1 = \bigcup_{\zeta<\eta} B_\zeta$. These sets are Π^1_1 and $\underset{\sim}{\Sigma}^0_\xi$. Now using the normal form for the Bisep operation (lemma 1.5), we know that there are disjoint sets C^0, C^1 in $\underset{\sim}{\Sigma}^0_\xi$ sets, with $\bigcup_{\zeta<\eta} C^0_\zeta = C^0$, $\bigcup_{\zeta<\eta} C^1_\zeta = C^1$, and sets $H^0_0, H^0_1, H^1_0, H^1_1, H_2$ with H^0_0, H^1_0 in $\check{\underset{\sim}{\Gamma}}_{\alpha_1}$, H^0_1, H^1_1 in $\underset{\sim}{\Gamma}_{\alpha_1}$ and H_2 in $\underset{\sim}{\Gamma}_{\alpha_2}$ such that

$$\begin{aligned} A \subseteq\ & (D_\eta(\langle C^0_\zeta \mid \zeta < \eta\rangle) \cap H^0_0) \cup (C^0 - D_\eta(\langle C^0_\zeta \mid \zeta < \eta\rangle) \cap H^0_1) \cup \\ & \cup (D_\eta(\langle C^1_\zeta \mid \zeta < \eta\rangle) \cap H^1_1) \cup (C^1 - D_\eta(\langle C^1_\zeta \mid \zeta < \eta\rangle) \cap H^1_0) \cup \\ & \cup H_2 - (C^0 \cup C^1) \subseteq \check{B} \quad \infty\text{-a.e.} \end{aligned}$$

It easily follows that for each $\zeta < \eta$ $C^0_\zeta \subseteq A_\zeta$, $C^1_\zeta \subseteq B_\zeta$ and $C^0 \subseteq A^0$, $C^1 \subseteq A^1$ ∞-a.e. From this, it follows that $A - (A^0 \cup A^1) \subseteq H_2 \subseteq \check{B}$ ∞-a.e., so using the inductive hypothesis, we can find a set $D \in \Gamma^1_{\alpha_2}$ with $A - (A^0 \cup A^1) \subseteq D \subseteq \check{B}$. Now by the argument previously used, one can shrink the Π^1_1 sequences $\langle A_\zeta \mid \zeta < \eta\rangle$ and $\langle B_\zeta \mid \zeta < \eta\rangle$ into Δ^1_1 sequences $\langle D^0_\zeta \mid \zeta < \eta\rangle$ and $\langle D^1_\zeta \mid \zeta < \eta\rangle$ with the same properties, in such a way that setting $D^0 = \bigcup_{\zeta<\eta} D^0_\zeta$ and $D^1 = \bigcup_{\zeta<\eta} D^1_\zeta$, D^0, D^1 are disjoint and $D^0 \cup D^1$ covers the Σ^1_1 set $A \cap \check{D}$. Again imitating the proof of case 2, we can find sets H^0_0 and H^1_0 in $\check{\Gamma}^1_{\alpha_1}$ and H^0_1, H^1_1 in $\Gamma^1_{\alpha_1}$ such that

$$A \cap D_\eta(\langle D^0_\zeta \mid \zeta < \eta\rangle) \subseteq H^0_0 \subseteq \check{B}$$

$$A \cap D^0 - D_\eta(\langle D^0_\zeta \mid \zeta < \eta\rangle) \subseteq H^0_1 \subseteq \check{B}$$

$$A \cap D_\eta(\langle D_\zeta \mid \zeta < \eta\rangle) \subseteq H^1_1 \subseteq \check{B}$$

$$A \cap D^1 - D_\eta(\langle D^1_\zeta \mid \zeta < \eta\rangle) \subseteq H^1_0 \subseteq \check{B}$$

so that the Γ^1_α set

$$H = [H^0_0 \cap D_\eta(\langle D^0_\zeta \mid \zeta < \eta\rangle)] \cup [H^0_1 \cap (D^0 - D_\eta(\langle D^0_\zeta \mid \zeta < \eta\rangle))] \cup$$

$$\cup [H^1_0 \cap (D^1 - D_\eta(\langle D^1_\zeta \mid \zeta < \eta\rangle))] \cup [H^1_1 \cap D_\eta(\langle D^1_\zeta \mid \zeta < \eta\rangle)] \cup$$

$$\cup [D - (D^0 \cup D^1)]$$

separates A from B.

Case 4. Suppose α code a description u with $u(1) = 4$, so that $\Gamma_\alpha = SU(\underset{\sim}{\Sigma}^0_\xi, \bigcup_n \underset{\sim}{\Gamma}_{\alpha_n})$, where the sequence $\langle \alpha_n \mid n \in \omega\rangle$ is a Δ^1_1 sequence of codes of descriptions. Starting from A, B in Σ^1_1, we let A_n be the union of all T_ξ-open sets O such that for some H in $\underset{\sim}{\Gamma}_{\alpha_n}$ $A \cap O \subseteq H \subseteq \check{B}$ ∞-a.e. As before, it is not hard to see that A_n is a Π^1_1 increasing sequence of $\underset{\sim}{\Sigma}^0_\xi$ sets, and that $A \subseteq \bigcup_n A_n$. Let $\langle C_n \mid n \in \omega\rangle$ be a Δ^1_1 sequence of pairwise disjoint $\underset{\sim}{\Sigma}^0_\xi$ sets with $A \subseteq \bigcup_n A_n$ and $C_n \subseteq A_n$, and let $\langle H_n \mid n \in \omega\rangle$ be a Δ^1_1 sequence, with $H_n \in \Gamma^1_{\alpha_n}$, and $A \cap C_n \subseteq H_n \subseteq \check{B}$. This is easy to find using the induction hypothesis. Then the set $H = SU(\langle C_n \mid n \in \omega\rangle, \langle H_n \mid n \in \omega\rangle)$ separates A from B and is in $\underset{\sim}{\Gamma}^1_\alpha$.

Case 5. Suppose $u_\alpha(1) = 5$, so that $\underset{\sim}{\Gamma}_\alpha = SD_\eta(\underset{\sim}{\Sigma}^0_\xi, \underset{\sim}{\Gamma}_{\alpha_1}, \underset{\sim}{\Gamma}_{\alpha_2})$, with $\alpha_1, \alpha_2 \in \Delta^1_1$, and $\eta, \xi < \omega^{ck}_1$. We again define inductively a sequence $\langle A_\zeta \mid \zeta < \eta\rangle$: A_ζ is the union of all T-open sets O such that for some set H in $\underset{\sim}{\Gamma}_{\alpha_1}$ with envelop O, $(A - \bigcup_{\zeta' < \zeta} A_{\zeta'}) \cap O \subseteq H \subseteq \check{B}$ ∞-a.e. Again it can be seen that the sequence $\langle A_\zeta \mid \zeta < \eta\rangle$ is Π^1_1 and increasing, and moreover that $A - \bigcup_{\zeta<\eta} A_\zeta \subseteq H \subseteq B$ ∞-a.e., for some H in $\underset{\sim}{\Gamma}_{\alpha_2}$. The rest of the proof is analogous to case 2, and we leave the details to the reader.

Remarks.

1. All the preceding results are of effective type. But as usual, they can be translated into non-effective, uniform results concerning analytic and Borel sets in the plane, using Δ^1_1-selection.

2. The results in §2 do not use Borel determinacy, but without it we are unable to show that theorem 2.7 covers all non self-dual Wadge classes of Borel sets.

In a recent paper, Thomas John has proved Wadge Det $(\underset{\sim}{\Pi}^0_4)$ (i.e., that all Wadge games $G_W(A, B)$, with A, B in $\underset{\sim}{\Pi}^0_4$ are determined), and the stronger statement that every $\underset{\sim}{\Pi}^0_n - \underset{\sim}{\Delta}^0_n$ set is $\underset{\sim}{\Pi}^0_n$-complete, for n in ω, in second order arithmetic. His proof uses as main tool the characterization $\underset{\sim}{\Pi}^0_n \cap \Delta^1_1 = (\Pi^0_n)^1$. This fact, together with threorem 2.7, is a bit of evidence in support of the conjecture that unlike Borel determinacy, Wadge Det $(\underset{\sim}{\Delta}^1_1)$ could be proved in second order arithmetic.

References

K. Kuratowski [1966], Topology, Vol. 1, Academic Press, New York.

A. Louveau [1980], A separation theorem for Σ_1^1 sets, T. A. M. S. 260, 2 (1980), 363-378.

D. A. Martin [1975], Borel determinacy, Annals of Math., 102 (1975), 363.

D. A. Martin [1973], The Wadge degrees are well-ordered (handwritten notes).

J. Steel [1976], Ph.D. Thesis, Berkeley.

J. Steel [1980], Determinateness and the separation property, J. S. L. 45 (1980), 143-146.

Van Wesep [1976], Wadge degrees and descriptive set theory, Cabal Seminar 76-77, Lecture Notes in Mathematics, Vol. 689, Springer-Verlag 1978.

W. Wadge [1976], Ph.D. Thesis, Berkeley (handwritten notes).

POINTCLASSES AND WELL-ORDERED UNIONS

Steve Jackson
Department of Mathematics
University of California
Los Angeles, California 90024

Donald A. Martin*
Department of Mathematics
University of California
Los Angeles, California 90024

§0. Introduction. Our basic notation and terminology is that of [4]. In particular, recall that a boldface pointclass is a pointclass containing all clopen sets and closed under continuous preimages.

A.S. Kechris [4] proved the following theorem, assuming the Axiom of Determinacy (AD) and the Axiom of Dependent Choices (DC):

Let Γ be a boldface pointclass closed under countable unions, countable intersections, and $\exists^{\mathbb{R}}$ but not under $\forall^{\mathbb{R}}$. If Reduction(Γ), then Γ is closed under well-ordered unions.

Kechris' theorem raises two sorts of questions:

Q1. Does Kechris' theorem remain true when we replace "but not under $\forall^{\mathbb{R}}$" by "and under $\forall^{\mathbb{R}}$ but not under complements"?

It seems strange that a failure of closure is needed to prove another kind of closure.

The second question is, roughly: Are there indeed any interesting well-ordered unions for Γ to be closed under? If Γ is closed under countable unions, countable intersections, and $\exists^{\mathbb{R}}$ but not under complements, and Γ has the prewellordering property, then it follows easily from AD + DC and the Moschovakis Coding Lemma (7D.5 of [5]) that Γ is closed under well-ordered unions of length $\leq \kappa$, where κ is the length of any Γ prewellordering.

Q2. Can one show (assuming AD + DC) that, whenever $\langle A_\beta : \beta < \gamma \rangle$ is a strictly increasing sequence of Γ sets and Γ is as above, then there is a Γ prewellordering of length γ?

The second question is especially important when Γ is the class of κ-Souslin sets, for then it bears on the problem of reliable cardinals. A set A is κ-Souslin if there is a scale on A of length $\leq \kappa$. κ is a Souslin cardinal if some set is κ-Souslin but not λ-Souslin for any $\lambda < \kappa$. A cardinal κ is reliable if there is a scale on some set A of length exactly λ.

To see how Q2 relates to the problem of whether all reliable cardinals are Souslin, assume DC and suppose that there are no non-trivial unions of κ-Souslin

*The second author was supported in part by NSF Grant # MCS 78-02989.

sets of length κ^+, i.e., suppose that there is no strictly increasing sequence $\langle A_\beta : \beta < \kappa^+\rangle$ of κ-Souslin sets. Let γ be the least reliable cardinal greater than κ. We show that γ is Souslin. If not, let $\langle \varphi_i : i \in \omega\rangle$ be a scale of length γ on a set A. By DC, κ^+ does not have cofinality ω. Hence some φ_i has length $\geq \kappa^+$. We may then assume that $\kappa^+ \subseteq \text{range } \varphi_i$. Let $A_\beta = \{x : x \in A \,\&\, \varphi_i(x) \leq \beta\}$ for $\beta < \kappa^+$. Since each A_β is γ-Souslin, we may suppose that each A_β is κ-Souslin. But the sequence $\langle A_\beta : \beta < \kappa^+\rangle$ is strictly increasing, contradicting our hypothesis.

Following Kechris [3], we are led to the following special case of Q2:

Q2'. If the class of κ-Souslin sets has the scale property, does it follow from AD + DC that there is no strictly increasing sequence of κ-Souslin sets of length κ^+?

The genesis of this paper was as follows:

Jackson answered Q1 positively (with the minor change of strengthening the assumption Reduction(Γ) to Prewellordering(Γ)). Jackson's method was completely different from Kechris'.

By a minor variant of his proof, Jackson also answered Q2' positively when the κ-Souslin sets are closed under $\forall^{\mathbb{R}}$ as well as $\exists^{\mathbb{R}}$. Martin next mixed Jackson's methods with other ideas to settle Q2' positively for a large class of successor κ. J. Steel remarked that Martin's proof could be modified to deal with many limit cardinals, assuming a certain technical lemma. Kechris then proved this technical lemma. These results, together with Steel's analysis [6] of the scale property in $L[\mathbb{R}]$, were sufficient to answer Q2' in $L[\mathbb{R}]$ and also to show that in $L[\mathbb{R}]$ all reliable cardinals are Souslin. (See [6].) Chuang [1] used Jackson's methods to produce a full positive answer to Q1. He proved several other results about well-ordered unions and Souslin cardinals. We used these further results to give a full positive answer to Q2' for κ of cofinality greater than ω. (The case $\kappa = \omega$ probably does not arise, but this is not yet proved.)

In §1 we present the results on classes closed under both $\exists^{\mathbb{R}}$ and $\forall^{\mathbb{R}}$. In §2 we deal with the case κ a successor cardinal. In §3, we indicate how to modify the proof of §2 when κ is a limit cardinal. (We state, but do not prove, Kechris' lemma. See [6] for a proof.)

In §2 and §3, we make use of the results of [1], [3], and [4]. This is purely to make our results general. The reader not familiar with these papers may think of a concrete case (e.g., $\Gamma = \Sigma^1_2$ in §2), whereupon the lemmas using results from these papers will become obvious from older standard results.

§1. Inductive-like pointclasses.

Theorem 1.1 (AD + DC). Let Γ be a boldface pointclass closed under $\exists^{\mathbb{R}}$ and $\forall^{\mathbb{R}}$ but not under complements. Then either Γ or $\check{\Gamma}$ is closed under well-ordered unions.

Proof. We let Γ be as in the hypothesis of the theorem and assume that Γ and $\check{\Gamma}$ are not closed under well-ordered unions. We let θ_1 denote the least ordinal such that, for some θ_1-sequence of sets in Γ, A_ξ, $\xi < \theta_1$, we have $A = \bigcup_{\xi<\theta_1} A_\xi \notin \Gamma$. We similarly let θ_2 be the least ordinal such that $\check{\Gamma}$ is not closed under well-ordered unions of length θ_2. We let $\bigcup\Gamma$ denote the pointclass obtained by taking θ_1-unions of sets in Γ, that is $B \in \bigcup\Gamma$ if there exists a sequence B_ξ, $\xi < \theta_1$, such that $B_\xi \in \Gamma$ for $\xi < \theta_1$ and $B = \bigcup_{\xi<\theta_1} B_\xi$. We are then assuming that $\Gamma \subsetneqq \bigcup\Gamma$. Hence it follows (from Wadge's lemma) that $\check{\Gamma} \subseteq \bigcup\Gamma$. Hence we may write a $\check{\Gamma}$ complete set, which we also denote by A as $A = \bigcup_{\xi<\theta_1} A_\xi$, where $A_\xi \in \Gamma$ for $\xi < \theta_1$.

Since Γ is not closed under complements, we get a coding of Γ sets by reals using a universal Γ set. We let $B = \{x : x$ codes a Γ subset of $A\}$. It is easy to see that B is $\check{\Gamma}$ using the closure hypotheses (which imply that Γ is closed under countable unions and intersections). Hence we may write $B = \bigcup_{\xi<\theta_1} B_\xi$ where $B_\xi \in \Gamma$. We now let $A_\xi^{(2)} = \{x : \exists y\,(y \in B_\xi$ and x belongs to the Γ set coded by $y)\}$. We then have that $A_\xi^{(2)} \in \Gamma$ and $A = \bigcup_{\xi<\theta_1} A_\xi^{(2)}$ since each $A_\xi^{(2)} \subseteq A$ and since $\bigcup_{\xi<\theta_1} A_\xi^{(2)} \supseteq A_\xi$ for each $\xi < \theta_1$ since each A_ξ is a Γ subset of A. We may further assume that the sequence $A_\xi^{(2)}$ is strictly increasing in the sense that $A_\xi^{(2)} \supsetneqq \bigcup_{\xi'<\xi} A_{\xi'}^{(2)}$ for all $\xi < \theta_1$.

Now the union $\bigcup_{\xi<\theta_1} A_\xi^{(2)}$ has the property that it is Γ-bounded, that is, if $B \subseteq \bigcup_{\xi<\theta_1} A_\xi^{(2)}$ and $B \in \Gamma$, then for some $\eta < \theta_1$ we have $B \subseteq A_\eta^{(2)}$. This follows from the definition of the $A_\xi^{(2)}$.

We now play the game where I plays a real x and II plays reals y, z. We say that II wins provided that if x codes a Γ subset of $A = \bigcup_{\xi<\theta} A_\xi^{(2)}$ then y is a Γ-code for some $A_\eta^{(2)}$, $\eta < \theta_1$, where η is larger than the least ordinal ξ such that $A_\xi^{(2)} \supseteq$ the set coded by x and $z \in A_\eta^{(2)} - \bigcup_{\xi<\eta} A_\xi^{(2)}$. We then have that II has a winning strategy for the game above, for, if not, then there would be a Σ_1^1 set C of codes of Γ subsets of A with the property that for any $\eta < \theta_1$ there is an $x \in C$ such that some member of the set coded by x does not belong to $A_\eta^{(2)}$. Since $\Sigma_1^1 \subseteq \Gamma$ we would then have a Γ subset of A unbounded in $A = \bigcup_{\xi<\theta_1} A_\xi^{(2)}$. This contradicts the fact that $A = \bigcup_{\xi<\theta_1}$ is Γ-bounded.

Hence we may assume that II has a winning strategy s. We let $B = \{x : x$ codes a Γ subset of $A\}$, and so $B \in \check{\Gamma}$ and is $\check{\Gamma}$ complete (otherwise A would be in Γ). We define an ordering on B by $x_1 < x_2 \iff x_1 \in B$ & $x_2 \in B$ & $s(x_2)_2 \notin s(x_1)_1$, where $s(x)_1$ and $s(x)_2$ are the y and z respectively played against x according to s. The ordering is in $\check{\Gamma}$ and is easily seen to be a prewellordering. ($x_1 < x_2 \Longleftrightarrow \xi_1 < \xi_2$, where $s(x_1)_1$ codes $A^{(2)}_{\xi_1}$ and $s(x_2)_1$ codes $A^{(2)}_{\xi_2}$.) It follows from the regularity of θ_1 that the rank of the prewellordering $<$ is θ_1.

It now follows from the Coding Lemma and the fact that $\check{\Gamma}$ is closed under $\exists^{\mathbb{R}}$ that $\check{\Gamma}$ is closed under well-ordered unions of length θ_1. Hence $\theta_1 < \theta_2$.

The argument above, however, is symmetric in Γ and $\check{\Gamma}$, and hence we also get $\theta_2 < \theta_1$. This contradiction establishes that either Γ or $\check{\Gamma}$ is closed under well-ordered unions.

Corollary 1.1.1 (AD + DC). If Γ is a boldface pointclass closed under $\exists^{\mathbb{R}}$ and $\forall^{\mathbb{R}}$ but not under complements and has the Prewellordering property, then Γ is closed under well-ordered unions.

Proof. This follows from the fact that, since Γ has the Prewellordering property, $\check{\Gamma}$ cannot be closed under well-ordered unions.

Corollary 1.1.2 (AD + DC). The class of inductive sets is closed under well-ordered unions.

The class of inductive sets also has the scale property. We now turn to the question Q2 for such pointclasses.

Theorem 1.2 (AD + DC). Let Γ be a boldface pointclass closed under $\exists^{\mathbb{R}}$ and $\forall^{\mathbb{R}}$ but not under complements. Let κ be a cardinal and suppose that some complete Γ set admits a Γ norm of length κ. Suppose also that every set in Γ is κ-Souslin. There is no strictly increasing sequence of Γ sets of length κ^+.

Our last two hypotheses hold if Γ has the scale property via a scale of length κ. They also hold by Chuang [1] if the other hypotheses hold and κ is a Souslin cardinal and $\Gamma = S(\kappa)$.

Proof. We let Γ satisfy the hypotheses of the theorem and assume that A_ξ for $\xi < \kappa^+$ is a strictly increasing sequence of Γ sets. We may assume $A_\xi \supsetneq \bigcup_{\xi'<\xi} A_{\xi'}$ for all $\xi < \kappa^+$. We let $A = \bigcup_{\xi<\kappa^+} A_\xi$. We have that $A \in \Gamma$ from Corollary 1.2.1. Since there is a Γ norm of length κ on a complete Γ set, we have a Γ coding for ordinals $<\kappa$, which we denote by $|x|$ for $x \in P$, the complete Γ set. By the Coding Lemma, using the closure properties of Γ, we have that every subset X of κ is Γ in the codes, that is $\{x : |x| \in X\} \in \Gamma$. Thus there is a Γ prewellordering of length η for each $\eta < \kappa^+$.

We choose a Γ coding of Γ relations and let $C = \{x : x$ codes a well-founded relation$\}$. We have that $C \in \check{\Gamma}$. We then play the game where I plays x and II plays y, z and II wins provided $x \in C \Rightarrow y$ is a Γ code for some A_η, $\eta < \kappa^+$ where $\eta >$ the rank of the well-founded relation coded by x, and

$z \in A_\eta - \bigcup_{\eta' < \eta} A_{\eta'}$. We have that II has a winning strategy, for, if I had a winning strategy, then there would be a Σ^1_1 set of codes for well-founded relations unbounded in κ^+, and, since $\Sigma^1_1 \subseteq \Gamma$, there would be a Γ well-founded relation of height κ^+. This, however, contradicts the fact that Γ is κ-Souslin, since κ-Souslin well-founded relations have height $< \kappa^+$ by 2G.2 of [5].

Hence II has a winning strategy s. We then consider the relation defined by

$$x_1 < x_2 \iff x_1 \in B \;\&\; x_2 \in B \;\&\; s(x_2)_2 \notin \text{the set coded by } s(x_1)_1 .$$

We have, as before, that $<$ is a $\check{\Gamma}$ prewellordering. It follows from the regularity of κ^+ (which follows from the Coding Lemma and the fact that $\kappa^+ = \sup\{\beta : \beta$ is the height of a Γ well-founded relation$\}$) that $<$ has length κ^+. It then follows from the Coding Lemma that $\check{\Gamma}$ is closed under well-ordered unions of length κ^+, and hence κ, which contradicts the existence of a Γ norm of length κ on a complete Γ set.

§2. κ-Souslin sets for κ a successor cardinal. For each cardinal κ, let $S(\kappa)$ be the class of all κ-Souslin sets.

Theorem 2. (AD + DC). Let κ be a Souslin cardinal such that $\kappa = \lambda^+$ for some cardinal λ. There is no strictly increasing sequence $\langle A_\beta : \beta < \kappa^+ \rangle$ such that each $A_\beta \in S(\kappa)$.

Proof. We begin by cataloguing some useful facts about κ and $S(\kappa)$.

Lemma 2.1. Let γ be the supremum of the Souslin cardinals $< \kappa$. γ is a Souslin cardinal.

Proof. Since κ is a successor cardinal, κ has cofinality greater than ω. By Kechris [3], $S(\kappa)$ is not closed under complements.

Let $\langle \varphi_i : i \in \omega \rangle$ be a scale of length κ on a set $A \in S(\kappa) - S(\check{\kappa})$. Since κ has cofinality $> \omega$, one of the φ_i, say φ_n, has length κ. For $\beta < \alpha < \kappa$ let $A_{\alpha,\beta} = \{x : \sup_i \varphi_i(x) + 1 < \alpha$ or $(\sup_i \varphi_i(x) + 1 = \alpha \;\&\; \varphi_n(x) \leq \beta)\}$. It is easily seen that $A_{\alpha,\beta}$ is $|\alpha|$-Souslin. Since κ has cofinality $> \omega$, $A = \bigcup_{\beta<\alpha<\kappa} A_{\alpha,\beta}$. If $\varphi_n(x) = \beta$, there is an α, $\beta < \alpha < \kappa$, such that $x \in A_{\alpha,\beta}$ and $x \notin A_{\alpha,\beta'}$ for any $\beta' < \beta$ and $x \notin A_{\alpha',\beta'}$ for any $\alpha' < \alpha$ and any β'. Thus there are κ distinct sets $A_{\alpha,\beta}$. If we order $\{(\alpha,\beta) : \kappa > \alpha > \beta\}$ lexicographically, we arrange the $A_{\alpha,\beta}$ in an increasing sequence. Thus we have shown that there is a strictly increasing sequence $\langle B_\alpha : \alpha < \kappa \rangle$ of sets such that each B_α is δ-Souslin for $\delta < \kappa$.

Fix, for the moment, a Souslin cardinal $\delta < \kappa$, By [4], either $\forall^{\mathbb{R}} S(\delta)$ or $\forall^{\mathbb{R}} \exists^{\mathbb{R}} S(\check{\delta})$ contains $S(\delta)$, is closed under $\forall^{\mathbb{R}}$ and countable unions and intersections, and has the prewellordering property. Let Γ be whichever of these classes satisfies these conditions. If δ_1 and δ_2 are the next two Souslin cardinals after δ, $\exists^{\mathbb{R}} \check{S}(\delta) \subseteq S(\delta_1)$ and $\exists^{\mathbb{R}} \forall^{\mathbb{R}} S(\delta) \subseteq S(\delta_2)$. Thus every $\check{\Gamma}$ prewellordering has length $< \delta_2^+$, by 2G.1 of [5]. By Chuang [1], there is no strictly increasing sequence of Γ sets of length δ_2^{++}. Hence either there is a greatest Souslin cardinal $< \gamma$, and we are done, or there is no strictly increasing sequence of length γ of sets in $S(\delta)$.

We may then assume that, for each Souslin cardinal $\delta < \kappa$, there are fewer than γ of the B_α which belong to $S(\delta)$. Since there are at most γ Souslin cardinals $< \kappa$, we get the contradiction that there are no more than γ of the B_α.

Lemma 2.2. λ is a Souslin cardinal. $S(\check{\lambda})$ has the prewellordering property via an $S(\check{\lambda})$ norm of length κ. $S(\kappa) = \exists^{\mathbb{R}} S(\check{\lambda})$. $S(\kappa)$ has the scale property.

Proof. Since κ has uncountable cofinality and γ is the greatest Souslin cardinal $< \kappa$, the lemma follows from Theorem 5 of [1] and its proof.

Lemma 2.3. $S(\lambda) \cap S(\check{\lambda}) \supseteq \mathcal{B}_\kappa$, the closure of the open sets under well-ordered unions of length κ, well-ordered intersections of length κ, and complements.

Proof. The proof of 7D.9 of [5] essentially proves the lemma.

Lemma 2.4. Let μ be the closed unbounded filter on κ. μ is a normal, κ-complete ultrafilter. If j is the embedding associated with the ultrapower by μ, $j(\kappa) = \kappa^+$.

Proof. The proofs are like those of Theorems 11.2 and 14.3 of [2].

Our next aim is to introduce the Kunen trees associated with $S(\kappa)$ well-founded relations. These trees provide us with our method for representing ordinals $< \kappa^+$.

Let $R \subseteq (\omega^\omega)^3$ belong to $S(\kappa)$ and be universal for $S(\kappa)$ subsets of $(\omega^\omega)^2$. Let T be a tree on $\omega \times \omega \times \omega \times \kappa$ witnessing that R is κ-Souslin.

Let $(i,j) \mapsto n_{ij}$ be a one-one surjection of $\omega \times \omega$ onto ω. If $\tau \in \omega^\omega$ let $\tau_i(j) = \tau(n_{ij})$ if $n_{ij} < \ell h\, \tau$. Similarly define $\rho_i(j)$ for $\rho \in \kappa^{<\omega}$.

We define our Kunen tree $\mathcal{J}$ by

$$\mathcal{J} = \{(\sigma,\tau,\rho) : \ell h\, \sigma = \ell h\, \tau = \ell h\, \rho \ \&\ \forall\ i,k \le \ell h\, \sigma \ (\text{if } \tau_i \restriction k,\ \tau_{i+1} \restriction k, \text{ and } \rho_i \restriction k \text{ are all defined, then } (\sigma \restriction k,\ \tau_{i+1} \restriction k,\ \tau_i \restriction k,\ \rho_i \restriction k) \in T\}.$$

For $x \in \omega^\omega$ let $\mathcal{J}_x = \{(\tau,\rho) : (x \restriction \ell h\, \tau, \tau, \rho) \in \mathcal{J}\}$. For $\alpha < \kappa$, let $\mathcal{J} \restriction \alpha = \{(\sigma,\tau,\rho) : (\sigma,\tau,\rho) \in \mathcal{J}\ \&\ \rho \in \alpha^{<\omega}\}$. Similarly define $\mathcal{J}_x \restriction \alpha$ $(= (\mathcal{J} \restriction \alpha)_x)$.

Lemma 2.6 (Kunen). R_x is a well-founded relation $\iff \mathcal{J}_x$ is well-founded. If R_x is well-founded, $|R_x| \leq |\mathcal{J}_x|$, where $|\ |$ is the height function for well-founded relations.

Proof. A branch through $\mathcal{J}_x$ is essentially an infinite descending chain in R_x with witnesses. For the second assertion, let $R_x^* = \{((y,f),(z,g)) : \forall i\ (x\restriction i,\ y\restriction i,\ z\restriction i,\ f\restriction i) \in T\}$. Clearly the tree of all finite descending chains in R_x^* can be embedded in $\mathcal{J}_x$ Hence $|R_x^*| \leq |\mathcal{J}_x|$. It is not hard to see that $|R_x| = |R_x^*|$.

Choose a universal $S(\kappa)$ set and so a coding of $S(\kappa)$ sets by elements of ω^ω.

Assume, contrary to the theorem, that $\langle A_\beta : \beta < \kappa^+\rangle$ is a sequence of $S(\kappa)$ sets with $\bigcup_{\beta'<\beta} A_{\beta'} \subsetneq A_\beta$ for each $\beta < \kappa^+$.

We play a game as follows:

I chooses $x \in \omega^\omega$.

II chooses $y \in \omega^\omega$ and $z \in \omega^\omega$.

If R_x and R_y are both well-founded, II wins just in case $|\mathcal{J}_x| < |\mathcal{J}_y|$ and z is a code for $A_{|\mathcal{J}_y|}$.

If R_x and R_y are both well-founded, let $\alpha < \kappa$ be the least ordinal such that $\mathcal{J}_x\restriction\alpha$ or $\mathcal{J}_y\restriction\alpha$ is not well-founded. I wins $\iff \mathcal{J}_x\restriction\alpha$ is well-founded.

Lemma 2.7. I has no winning strategy.

Proof. Suppose s is a winning strategy for I. Let $\alpha < \kappa$. Let $B_\alpha = \{y : \forall\beta < \alpha\ (\mathcal{J}_y\restriction\beta$ is well-founded and $|\mathcal{J}_y\restriction\beta| < \alpha\}$. $B_\alpha \in \mathcal{B}_\kappa$ and so $B_\alpha \in S(\lambda) \cap S(\check{\lambda})$. Let S be a tree on $\omega \times \lambda$ witnessing that B_α is λ-Souslin. Let $\mathcal{J}^*$ be the tree defined as follows:

$$\mathcal{J}^* = \{(\sigma_0,\tau_0,\rho_0,\sigma,\tau,\rho) : (\sigma_0,\rho_0) \in S \ \&\ (\sigma,\tau,\rho) \in \mathcal{J}\restriction\alpha \ \&\ \sigma \text{ agrees with the reply to } (\sigma_0,\tau_0) \text{ according to } s\}.$$

Since s is a winning strategy $\mathcal{J}_x\restriction\alpha$ is well-founded whenever $x = s(y,z)$ and $y \in B_\alpha$. Thus $\mathcal{J}^*$ is well-founded. Since $\mathcal{J}^*$ is a tree on $\max(\alpha,\lambda)$, $|\mathcal{J}^*| < \kappa$. It is easy to see that if $y \in B_\alpha$ and $x = s(y,z)$, then $|\mathcal{J}_x\restriction\alpha| \leq |\mathcal{J}^*|$.

We have thus shown that there is an $f(\alpha) < \kappa$ (namely, $|\mathcal{J}^*|$) such that if II plays a $y \in B_\alpha$ and I plays according to s, then $|\mathcal{J}_x\restriction\alpha| < f(\alpha)$. Let $C \subseteq \kappa$ be closed, unbounded and satisfy $(\eta \in C \ \&\ \alpha < \eta) \Rightarrow f(\alpha) < \eta$.

Now let $\beta \in j(C)$, $\beta > \kappa$. Let $h : \kappa \leftrightarrow \beta$ be a bijection. Let φ be a $S(\kappa)$ norm of length κ. Let $(z,w) \in E \iff h(\varphi(z)) < h(\varphi(w))$. $E \in S(\kappa)$. Let $E = R_y$. R_y is well-founded and $|R_y| = \beta$. Thus $|\mathcal{J}_y| \geq \beta$. Let I play y and z coding $A_{|\mathcal{J}_y|}$. Let II play x according to x. There is a closed, unbounded set of $\alpha < \kappa$ such that $y \in B_\alpha$. For each α in this set, $|\mathcal{J}_x\restriction\alpha|$ is less than the next member of C after α. By the normality of μ, $|\mathcal{J}_\beta|$ is less than the next element of $j(C)$ after κ. Thus $|\mathcal{J}_x| < |\mathcal{J}_y|$. It follows that the play is a win for II, a contradiction.

Let t be a winning strategy for II.

Lemma 2.8. There is a closed, unbounded $C^* \subseteq \kappa^+$ such that, if $\beta \in C^*$, I plays x with $|\mathfrak{J}_x| < \beta$, and II plays according to t, then $|\mathfrak{J}_y| < \beta$.

Proof. If $\alpha,\gamma < \kappa$, let $B^\gamma_\alpha = \{x : \mathfrak{J}_x \restriction \alpha$ is well-founded and $|\mathfrak{J}_x \restriction \alpha| < \gamma\}$. $B^\gamma_\alpha \in \mathcal{B}_\kappa$. As before we get an $f(\alpha,\gamma) < \kappa$ such that if $x \in B^\gamma_\alpha$ and II plays y according to t, then $|\mathfrak{J}_y \restriction \alpha| < f(\alpha,\gamma)$. Let $C \subseteq \kappa$ be closed and unbounded such that

$$(\eta \in C \;\&\; \alpha < \eta \;\&\; \gamma < \eta) \Rightarrow f(\alpha,\gamma) < \eta .$$

Now let $C^* = j(C) - (\kappa + 1)$.

Lemma 2.9. For each $\beta \in C^*$, $\{x : \mathfrak{J}_x$ is well-founded and $|\mathfrak{J}_x| < \beta\} \in S(\check{\kappa})$.

Proof. For each x, let $(y(x),z(x))$ be II's play against x according to t. Let $\beta \in C^*$ and $w \in A_\beta - \bigcup_{\beta'<\beta} A_{\beta'}$. Suppose $\mathfrak{J}_x$ is well-founded. Then

$$|\mathfrak{J}_x| < \beta \iff |\mathfrak{J}_{y(x)}| < \beta \iff w \text{ does not belong to the set coded by } z(x) .$$

This last condition is $S(\check{\kappa})$. Since $S(\check{\kappa})$ is closed under $\forall^{\mathbb{R}}$, $\{x : \mathfrak{J}_x$ is well-founded$\} = \{x : R_x$ is well-founded$\} \in S(\check{\kappa})$. Since $S(\check{\kappa})$ is closed under intersections, the lemma is proved.

Lemma 2.10. For every sufficiently large $\beta < \kappa^+$, every $S(\kappa)$ set is Wadge reducible to $\{x : \mathfrak{J}_x$ is well-founded and $|\mathfrak{J}_x| < \beta\}$.

Proof. Let φ be a $S(\kappa)$ norm of length κ on a set $H \in S(\kappa) - S(\check{\kappa})$. Let

$$R^* = \{(x,y,z) : y \in H \;\&\; z \in H \;\&\; \exists y' \,\exists z' \,(\varphi(y') = \varphi(y) \;\&\; \varphi(z') = \varphi(z) \;\&$$
$$(x',y',z') \in R\} .$$

Define R^*_x in the obvious way. R^*_x is thus just R_x fixed up to be well defined on ordinals as coded by φ.

Let f be continuous such that, for each (x,y),

$$R_{f(x,y)} = \{(z,w) : z \in H \;\&\; w \in H \;\&\; (x \notin H \vee (\varphi(x) > \varphi(z) \;\&\; \varphi(x) > \varphi(w)) \;\&$$
$$(z,w) \in R^*_y\} .$$

Let

$$V = \{(x,y) : x \in H \;\&\; R^*_y \cap \{(z,w) : \varphi(z) < \varphi(x) \;\&\; \varphi(w) < \varphi(x)\}$$
$$\text{is well-founded}\} .$$

To see that $V \in S(\kappa)$, note that

$$(x,y) \in V \iff \exists \alpha < \kappa\ (\varphi(x) = \alpha \ \&\ R_y^* \cap \{(z,w) : \varphi(z) < \alpha \ \&\ \varphi(w) < \alpha\} \text{ is well-founded}\}).$$

Thus V is a union of κ sets, so it suffices by Lemma 2.2 and the Coding Lemma to show that each of these sets belongs to $S(\kappa)$. $\{x : \varphi(x) = \alpha\} \in S(\kappa) \cap \check{S}(\kappa)$, so it is enough to show that the second conjunct defines a set in $S(\kappa)$. $\{(x,y) : R_y^* \cap \{(z,w) : \varphi(z) < \alpha \ \&\ \varphi(w) < \alpha\}\}$ is induced by a relation on α. If it is well-founded, it has height $< \kappa$. Thus the second conjunct defines a union of κ sets of the form

$$\{(x,y) : R_y^* \cap \{(z,w) : \varphi(z) < \alpha \ \&\ \varphi(w) < \alpha\} \text{ has height } < \gamma\} .$$

These sets all belong to $\mathcal{B}_\kappa$ and so to $S(\lambda) \cap \check{S}(\lambda)$.

For each $(x,y) \in V$, $R_{f(x,y)}$ is well-founded. If $\{|\mathcal{J}_{f(x,y)}| : (x,y) \in V\}$ were unbounded in κ^+, we could, using the fact that V is κ-Souslin, put together all the $\mathcal{J}_{f(x,y)}$ for $(x,y) \in V$ to get a well-founded tree of height κ^+.

Thus let $\kappa^+ > \beta > |\mathcal{J}_{f(x,y)}|$ for all $(x,y) \in V$. As in the proof of Lemma 2.7, let y be such that $R_y = R_y^*$ and is a well-ordering of order type $\geq \beta$.

$$x \in H \Rightarrow (x,y) \in V \Rightarrow \mathcal{J}_{f(x,y)} \text{ well-founded } \& \ |\mathcal{J}_{f(x,y)}| < \beta .$$

Also

$$x \notin B \Rightarrow R_{f(x,y)} = R_y^* \Rightarrow |R_{f(x,y)}| \geq \beta \Rightarrow |\mathcal{J}_{f(x,y)}| \geq \beta .$$

We have thus shown that $H \in S(\kappa) - \check{S}(\kappa)$ is Wadge reducible to the required set.

Lemmas 2.9 and 2.10 give the contradiction which proves the theorem.

§3. κ-Souslin sets for κ a limit cardinal of uncountable cofinality.

Theorem 3. Let κ be a Souslin cardinal, and a limit cardinal of uncountable cofinality. Assume that $S(\kappa)$ has the reduction property. There is no strictly increasing sequence $\langle A_\beta : \beta < \kappa^+\rangle$ such that $A_\beta \in S(\kappa)$.

Note: The hypothesis that $S(\kappa)$ has the reduction property can be eliminated.

Proof. We prove the theorem from a lemma of Kechris proved in [6].

Lemma 3.1. κ is a limit of Souslin cardinals.

Proof. The proof of Lemma 2.1 shows that either there is a greatest Souslin cardinal $\gamma < \kappa$ or else κ is a limit of Souslin cardinals. In the former case, Theorem 5 of Chuang [1] implies that $\kappa = \gamma^+$.

Lemma 3.2. $S(\kappa)$ has the scale property.

Proof. This follows from [1] and [4].

Lemma 3.3. There is an ω-closed, unbounded set of $\alpha < \kappa$ such that $\mathcal{B}_{\alpha^+} \subseteq S(\alpha) \cap S(\check{\alpha})$.

Proof. If δ is a Souslin cardinal $< \kappa$, there is a $\delta' < \kappa$ such that every set in $S(\delta)$ admits a scale of length δ which belongs to $S(\delta')$. To see this, let $A \in S(\delta)$ and let $\langle \varphi_i : i \in \omega \rangle$ be a scale on A of length δ. Since $\{x : x \in A \ \& \ \varphi_i(x) < \alpha\}$ and $\{x : x \in A \ \& \ \varphi_i(x) \leq \alpha\}$ are both δ-Souslin for each $i \in \omega$ and $\alpha < \delta$, $\{x : x \in A \ \& \ \varphi_i(x) = \alpha\}$ is a difference of δ-Souslin sets. If δ_1 is the next Souslin cardinal after δ, $\{x : x \in A \ \& \ \varphi_i(x) = \alpha\} \in S(\delta_1)$ for each $i \in \omega$ and $\alpha < \delta$. Either $S(\delta_1)$ or $\exists^{\mathbb{R}} S(\check{\delta}_1)$ is closed under $\exists^{\mathbb{R}}$, under countable unions and intersections, and has the prewellordering property, by [4]. Thus one of these classes is closed under well-ordered unions, by Theorem 1.1 and the Kechris theorem quoted in §0. Hence we may take δ' as the least Souslin cardinal $> \delta_1$.

There is an ω-closed unbounded set of $\delta < \kappa$ such that δ is a limit of Souslin cardinals and, if $\delta_1 < \delta$, there is a $\delta' < \delta$ such that every set in $S(\delta_1)$ admits a scale in $S(\delta')$. For δ in this ω-cube set, let $\Lambda_\delta = \bigcup_{\delta' < \delta} S(\delta')$. Let Γ_δ be the collection of countable unions of members of Λ_δ. It is easily seen that every set in Γ_δ admits a Γ_δ scale of length $\leq \delta$. By first periodicity, $\forall^{\mathbb{R}} \Gamma_\delta$ has the prewellordering property. Thus, as in Lemma 2.3, $\mathcal{B}_{\delta+} \subseteq \forall^{\mathbb{R}} \Gamma_\delta \cap \exists^{\mathbb{R}} \check{\Gamma}_\delta$. But $\exists^{\mathbb{R}} \check{\Gamma}_\delta \subseteq S(\delta)$. By Theorem 2E.2 of [5], $S(\delta) \cap S(\check{\delta}) \subseteq \mathcal{B}_{\delta+}$, so $\exists^{\mathbb{R}} \check{\Gamma}_\delta = S(\delta)$, and the lemma is proved.

Lemma 3.4 (Kechris). There is a measure μ on κ such that μ extends the ω-closed, unbounded filter and, in the ultrapower by μ, the function $f(\alpha) = \alpha^+$ represents κ^+.

For a proof, see [6].

We define the Kunen tree $\mathcal{J}$ exactly as in §2.

The rest of the proof is like that in §2. Lemma 2.6 goes through as before. Our game is as before. The analogue of Lemma 2.7 is proved as before, except that $f(\alpha) < \alpha^+$ and we choose β such that $\beta < \kappa^+$ and $\beta >$ the ordinal represented by f. The analogue of Lemma 2.8 is proved as before except that $f(\alpha,\gamma)$ is defined for α in the set given by Lemma 3.3. and $\gamma < \alpha^+$, and $f(\alpha,\gamma) < \alpha^+$. We let $C_\alpha \subseteq \alpha^+$ be closed and unbounded such that

$$(\eta \in C_\alpha \ \& \ \gamma < \eta) \Rightarrow f(\alpha,\gamma) < \eta .$$

We let C^* be the subset of κ represented by $g(\alpha) = C_\alpha$. The analogue of Lemma 2.9 is proved as before, as is the analogue of Lemma 2.10.

References

[1] Chen-Lian Chuang, The Propagation of Scales by Game Quantifiers, Ph.D. Thesis, UCLA, 1982.

[2] A.S. Kechris, AD and projective ordinals, Cabal Seminar 76-77, A.S. Kechris and Y.N. Moschovakis, eds., Springer-Verlag, 1978, 91-132.

[3] A.S. Kechris, Souslin cardinals, κ-Souslin sets, and the scale property in the hyperprojective hierarchy, Cabal Seminar 77-79, A.S. Kechris, D.A. Martin, and Y.N. Moschovakis, eds., Springer-Verlag, 1981, 127-146.

[4] The Axiom of Determinacy and the prewellordering property, Cabal Seminar 77-79, A.S. Kechris, D.A. Martin, and Y.N. Moschovakis, eds., Springer-Verlag, 1981, 101-126.

[5] Y.N. Moschovakis, Descriptive Set Theory, North Holland, 1980.

[6] J.R. Steel, The scale property in $L[\mathbb{R}]$, this volume.

AD AND THE UNIQUENESS OF THE SUPERCOMPACT MEASURES ON $P\omega_1(\lambda)$

W. Hugh Woodin[1]
Department of Mathematics
California Institute of Technology
Pasadena, California 91125

An early consequence of determinacy was the supercompactness of ω_1. Specifically Solovay [4] showed that assuming the determinacy of real games $(AD_{\mathbb{R}})$ then for each $\lambda < \Theta$ there is a supercompact measure on $P\omega_1(\lambda)$ where $\Theta = \sup\{\zeta \mid \text{there is an onto map } f : \mathbb{R} \to \zeta\}$.

Recall that $P\omega_1(\lambda) = \{\sigma \subseteq \lambda \mid \sigma \text{ is countable}\}$ and that a measure μ on $P\omega_1(\lambda)$ is a supercompact measure if it is normal and fine which is to say that μ satisfies the following two conditions:

(1) (normality) Suppose $f : P\omega_1(\lambda) \to \lambda$ is a function such that $\{\sigma \mid f(\sigma) \in \sigma\}$ is of (μ) measure one. Then for some $\alpha < \lambda$, $\{\sigma \mid f(\sigma) = \alpha\}$ is of measure one.

(2) (fineness) For each $\alpha < \lambda$, $\{\sigma \mid \alpha \in \sigma\}$ is of measure one.

Harrington and Kechris [2] improved Solovay's result by weakening the assumption to the determinacy of integer games (AD). However the price paid is a weaker result. They proved that if λ lies below a Souslin cardinal then ω_1 is λ supercompact. Becker [1] then showed (assuming AD) that if λ is a Souslin cardinal then the supercompact measure on $P\omega_1(\lambda)$ is in fact unique.

This left open the general question of uniqueness of the supercompact measure on $P\omega_1(\lambda)$ for arbitrary λ. A typical case is $\lambda = \omega_2$. Assuming AD ω_2 admits exactly two normal measures. Each of these may be used to construct a supercompact measure on $P\omega_1(\omega_2)$ using the normal measure that AD provides on ω_1. It is not immediately clear that these two supercompact measures should be equal.

We show that the supercompact measure on $P\omega_1(\omega_2)$ is unique assuming AD. Our arguments are general enough to show uniqueness for the case of an arbitrary λ below a Souslin cardinal assuming AD or for any $\lambda < \Theta$ assuming $AD_{\mathbb{R}}$.

We assume familiarity with the fundamentals of set theory in the context of determinacy, for a reference see [3]. We work in ZF + DC throughout this paper.

---

[1]Research partially supported by NSF Grant MCS80-21468.

§1. Suppose $\lambda < \Theta$. The use of determinacy to construct a supercompact measure on $P\omega_1(\lambda)$ is through a simulation of the following ordinal game. Fix $A \subseteq P\omega_1(\lambda)$. Associated to A is the game G_A:

I	II	
α_0	β_0	$\alpha_i < \lambda, \ \beta_i < \lambda$
α_1	β_1	I wins $\leftrightarrow \{\alpha_0, \beta_0, \ldots\} \in A$
$\vdots$	$\vdots$	

The players alternately play ordinals less than λ. Player I wins provided $\{\alpha_0, \beta_0, \ldots\} \in A$.

Fix a map $\pi : \mathbb{R} \to \lambda$ that is onto.

We say that a real r codes the ordinal α if $\pi(r) = \alpha$. We can now simulate the ordinal game G_A with the obvious real game. Let G_A^r denote this real game, where there is an obvious implicit dependence on π.

We define the notion of a quasi strategy for Player I in G_A^r.

Definition. A quasi strategy (for Player I) is a set, τ, of finite sequences of reals such that if $p \in \tau$ is of even length then $p^\frown x \in \tau$ for each $x \in \mathbb{R}$, otherwise $p^\frown x \in \tau$ for some $x \in \mathbb{R}$.

Given a quasi strategy τ let

$$p_\pi(\tau) = \{\sigma \in P\omega_1(\lambda) \mid \text{for some } f \in \mathbb{R}^\omega, \ f \restriction n \in \tau \text{ for each } n \text{ and } \sigma = \{\pi(f(n)) \mid n \in \omega\}\}.$$

Thus $p_\pi(\tau)$ denotes the set of elements of $P_1(\lambda)$ that can be enumerated by a play against τ. Note that $p_\pi(\tau)$ depends upon the notion of coding, i.e. on the map $\pi : \mathbb{R} \to \lambda$.

For $A \subseteq P\omega_1(\lambda)$ we say that G_A^r is determined provided there is a quasi strategy τ such that $p_\pi(\tau) \subseteq A$ or $p_\pi(\tau) \subseteq P\omega_1(\lambda) \setminus A$.

Let

$$\mathfrak{F} = \{p_\pi(\tau) \mid \tau \text{ is a quasi strategy and } \pi : \mathbb{R} \to \lambda \text{ is onto}\}.$$

$\mathfrak{F}$ is easily seen to be a fine filter over $P\omega_1(\lambda)$.

If one assumes the determinacy of real games ($AD_{\mathbb{R}}$) then for each $A \subseteq P\omega_1(\lambda)$, G_A^r is determined. The point is that if Player II wins the real game G_A^r then Player I must win G_B^r, $B = P\omega_1(\lambda) \setminus A$. Hence $\mathfrak{F}$ defines an ultrafilter over $P\omega_1(\lambda)$. A standard dovetailing of strategies show that the ultrafilter is normal (see [4]).

It follows from the results of Harrington and Kechris [2] that assuming AD, for each $A \subseteq P\omega_1(\lambda)$, G^r_A is determined given that λ lies below a Souslin cardinal. Hence again $\bar{\mathfrak{F}}$ defines a normal measure.

We show that any normal fine measure on $P\omega_1(\lambda)$ must extend $\bar{\mathfrak{F}}$. The desired uniqueness results will then be immediate.

Theorem 1. (ZF + DC). Suppose λ is an ordinal less than Θ and that τ is a (real) quasi strategy on λ. Let μ be a supercompact measure on $P\omega_1(\lambda)$. Then $\mu(p_\pi(\tau)) = 1$.

Proof. Fix τ, μ and λ. Let $\pi : \mathbb{R} \to \lambda$ be the underlying notion of coding.

Assume toward a contradiction that $\mu(p(\tau)) = 0$. Let $B = P\omega_1(\lambda)\backslash p_\pi(\tau)$. Hence $\mu(B) = 1$.

For each $\sigma \in B$ consider the following game g_σ:

I	II	
α_0	s_0	$\alpha_i \in \sigma$, $s_i \in \tau$, s_{i+1} extends
α_1	s_1	s_i and $\alpha_i \in \{\pi(r) \mid r \in s_i\} \subseteq \sigma$.
$\vdots$	$\vdots$	

Player I plays ordinals in σ, Player II plays elements of τ. Player II must play such that s_{i+1} extends s_i and $\alpha_i \in \{\pi(r) \mid r \in s_i\} \subseteq \sigma$.

Player I wins if at some stage Player II cannot play.

The game g_σ is open for Player I, therefore it is determined. More precisely, either I or II has a winning "many valued" strategy. If such is the case for II then using DC one can easily show that $\sigma \in p_\pi(\tau)$. Hence I has the winning "many valued" strategy. Since I plays ordinals he must therefore have a winning strategy. Further this strategy may be chosen canonically in terms of decreasing rank. Let ρ_σ denote this strategy. View ρ_σ as a set of (legal) positions in g_σ.

Suppose $q = \langle \alpha_0 s_0 \cdots \alpha_i s_i \rangle$ is such that for μ-almost all σ, $q \in \rho_\sigma$. Hence for each such σ there is an $\alpha \in \sigma$ such that $q^\frown\alpha \in \rho_\sigma$. This defines a function $f : P\omega_1(\lambda) \to \lambda$ with $f(\sigma) \in \sigma$ on a set of μ-measure one. Thus by the normality of μ there is an $\alpha^* < \lambda$ so that for μ-almost all σ, $q^\frown\alpha^* \in \rho_\sigma$.

Using DC and the fineness of μ construct a sequence $\langle \alpha_0 s_0 \cdots \alpha_i s_i \cdots \rangle$ such that for all i and μ-almost all σ, $\langle \alpha_0 s_0 \cdots \alpha_i s_i \rangle$ is a position in g_σ with $\langle \alpha_0 s_0 \cdots \alpha_i s_i \rangle \in \rho_\sigma$.

Thus by the countable completeness of μ there is a fixed σ^* such that for all i, $\langle \alpha_0 s_0 \cdots \alpha_i s_i \rangle$ is a position in g_{σ^*}, $\langle \alpha_0 s_0 \cdots \alpha_i s_i \rangle \in \rho_{\sigma^*}$. But therefore $\langle \alpha_0 s_0 \cdots \alpha_i s_i \cdots \rangle$ is an infinite play against ρ_{σ^*}, a contradiction. Thus $\mu(p_\pi(\tau)) = 1$. ⊣

By our previous remarks the following are immediate corollaries of Theorem 1.

Theorem 2 (AD + DC). Assume $\lambda \leq \kappa$ and that κ is a Souslin cardinal. Then the supercompact measure on $P\omega_1(\lambda)$ is unique.

Theorem 3 ($AD_{\mathbb{R}}$). Assume $\lambda < \Theta$. Then the supercompact measure on $P\omega_1(\lambda)$ is unique.

One can actually prove a slightly more general version of Theorem 3. Suppose S is a set. Let $P\omega_1(S) = \{\sigma \subseteq S \mid \sigma$ is countable$\}$. A supercompact measure on $P\omega_1(S)$ is a fine measure μ satisfying the following modified normality condition. Suppose $f : P\omega_1(S) \to P\omega_1(S)$ is such that for μ-almost all σ, $f(\sigma) \subseteq \sigma$. Then for some $a \in S$, $a \in f(\sigma)$ on a set of μ measure one.

If $S = \lambda$ this is just the usual notion of a supercompact measure. The modified normality condition is important in avoiding trivialities in the case of more general S. For instance if $S = \mathbb{R}$ then the usual notion of normality reduces to countable completeness.

Theorem 4 ($AD_{\mathbb{R}}$). Let S be an arbitrary set such that there is an onto map $g : \mathbb{R} \to S$. Then there is a supercompact measure on $P\omega_1(S)$ and it is unique.

Proof. For existence simply project the supercompact measure on $P\omega_1(\mathbb{R})$ constructed by Solovay [4] or play the obvious real games.

The situation for uniqueness is identical to case of $S = \lambda$ except that in the proof of Theorem 1 one must work with "many valued" strategies. The modified normality condition is sufficient for the requirements of the proof. ⊣

§2. Further problems. There are many questions left open regarding supercompact measures in the context of AD. We state two typical of these.

(1) Assume $V = L(\mathbb{R})$ and AD. How supercompact is ω_1? In particular is there a supercompact measure on $P\omega_1(\lambda)$ for each $\lambda < \Theta$?

(2) Assume AD (or $AD_{\mathbb{R}}$). Is $L(\mathbb{R},\mu)$ unique given $L(\mathbb{R},\mu) \models \mu$ is a supercompact measure on $P\omega_1(\mathbb{R})$?

Assume $AD_{\mathbb{R}}$ ($AD + V \neq L(\mathbb{R})$ will suffice). Suppose $\lambda < \Theta^{L(\mathbb{R})}$. Let μ be the supercompact measure on $P\omega_1(\lambda)$. Let j denote the embedding corresponding to the ultrapower of the ordinals by μ. One can show that $j(\omega_1) \leq \delta$ for δ the first measurable cardinal past λ. In particular $j(\omega_1) < \Theta^{L(\mathbb{R})}$. This suggests that μ should lie in $L(\mathbb{R})$.

The following conjecture settles (1) positively. Fix, $\leq$, a prewellordering of $\mathbb{R}$ of length λ. Let $g : \mathbb{R} \to \lambda$ be the corresponding map. A λ-splitting tree, T, is a set of finite sequences of reals such that if $q \in T$ and $\alpha < \lambda$ then $q^\frown x \in T$ for some $x \in \mathbb{R}$, $g(x) = \alpha$.

<u>Conjecture</u>: (a strong coding lemma). Assume $V = L(\mathbb{R})$ and AD. Suppose $\leq$ is a prewellordering of $\mathbb{R}$ of length λ. Let T be a λ-splitting tree. Then there is a λ-splitting tree $T^* \subseteq T$ such that T^* is projective in $\leq$.

References

[1] H. Becker, AD and the supercompactness of $\aleph_1$, Journal of Symbolic Logic 46 (1981), 822-841.

[2] L. A. Harrington and A. S. Kechris, On the determinacy of games on ordinals, Ann. Math. Logic, 20 (1981), 109-154.

[3] Y. N. Moschovakis, Descriptive set theory, North-Holland, Amsterdam, 1980.

[4] R. M. Solovay, The independence of DC from AD, Cabal Seminar 76-77, Lecture Notes in Mathematics, Springer-Verlag, Vol. 689, (1978), 171-184.

SCALES ON Σ^1_1 SETS

John R. Steel
Department of Mathematics
University of California
Los Angeles, California 90024

Let Λ be the class of sets which are $\omega n - \Pi^1_1$ for some $n < \omega$. We shall show that every set in Λ admits a very good scale all of whose norms are in Λ. Even for Σ^1_1 sets, this is the best possible definability bound on very good scales. Busch shows in [1] that every Σ^1_1 set admits a not very good scale all of whose norms are $\omega + 3 - \Pi^1_1$.

Some terminology: if L is an ordered set, then "$\leq_{lex}$" denotes the lexicographic ordering on $L^{<\omega} \cup {}^{\omega}L$. A tree on L is a subset of $L^{<\omega}$ closed under initial segment. $[T]$ is the set of infinite branches of the tree T. If T is a tree on $L \times M$ and $x \in {}^{\omega}L$, then $T(x) = \{\tau \in M^{<\omega} \mid (x \upharpoonright \ell h\tau, \tau) \in T\}$, and $p[T] = \{x \in {}^{\omega}L \mid [T(x)] \neq \emptyset\}$. If M is wellordered and $x \in p[T]$, then f^T_x is the leftmost branch of $T(x)$. That is, $f^T_x \in [T(x)]$ and $\forall g \in [T(x)]\ (f^T_x \leq_{lex} g)$. Finally, if $\langle A_\alpha \mid \alpha < \beta\rangle$ is a sequence of sets, then

$$\mathrm{Diff}_{\alpha<\beta} A_\alpha = \{x \mid \exists \alpha < \beta\ (\alpha \text{ is odd} \wedge x \in A_\alpha \wedge \forall \gamma < \alpha\ (x \notin A_\gamma))\}.$$

A set $A \subseteq {}^{\omega}\omega$ is $\beta - \underset{\sim}{\Pi}^1_1$ iff $A = \mathrm{Diff}_{\alpha<\beta} A_\alpha$ for some sequence $\langle A_\alpha \mid \alpha < \beta\rangle$ of $\underset{\sim}{\Sigma}^1_1$ sets. If the sequence of Σ^1_1 indices for the A_α's can be taken to be recursive, then A is $\beta - \Pi^1_1$.

<u>Lemma</u>. If A and B are $\omega n - \Pi^1_1$, where $n < \omega$, then $A \cap B$ is $\omega(n^2) - \Pi^1_1$.

<u>Proof</u>. We claim first that there is a wellorder $\leq^*$ of $\omega^2 \times \omega^2$ of order type ω^2 which extends the product partial order. (That is, if $\alpha \leq \beta$ and $\gamma \leq \delta$, then $(\alpha,\gamma) \leq^* (\beta,\delta)$.) For simply let

$$(\omega i_0 + j_0, \omega k_0 + \ell_0) \leq^* (\omega i_1 + j_1, \omega k_1 + \ell_1) \text{ iff}$$
$$(2^{i_0}3^{k_0} < 2^{i_1}3^{k_1} \text{ or}$$
$$(2^{i_0}3^{k_0} = 2^{i_1}3^{k_1} \text{ and } 2^{j_0}3^{\ell_0} \leq 2^{j_1}3^{\ell_1})).$$

Notice that $\leq^* \upharpoonright (\omega n \times \omega n)$ has order type $\omega(n^2)$.

Now let $A = \mathrm{Diff}_{\alpha<\omega n} A_\alpha$ and $B = \mathrm{Diff}_{\alpha<\omega n} B_\alpha$, where the A_α's and B_α's are Σ^1_1. Fix an $\alpha < \omega(n^2)$, and let (β,γ) be the αth element of $\omega n \times \omega n$ under $\leq^*$. If β and γ are both odd, set $C_{2\alpha} = \emptyset$ and $C_{2\alpha+1} = A_\beta \cap B_\gamma$; otherwise set $C_{2\alpha} = A_\beta \cap B_\gamma$ and $C_{2\alpha+1} = \emptyset$. We claim that $A \cap B = \mathrm{Diff}_{\alpha<\omega(n^2)} C_\alpha$. For example, let $x \in A \cap B$. Let β be least so that $x \in A_\beta$ and γ be least so that $x \in B_\gamma$. Then β and γ are odd, so $x \in C_{2\alpha+1}$, where (β,γ) is the αth element of $\omega n \times \omega n$ under $\leq^*$. If $x \in C_\delta$ for some $\delta < 2\alpha + 1$, then $x \in A_{\beta'} \cap B_{\gamma'}$ for some $(\beta',\gamma') <^* (\beta,\gamma)$. But $\beta \leq \beta'$ and $\gamma \leq \gamma'$ by the definitions of β and γ, so $(\beta,\gamma) \leq^* (\beta',\gamma')$, a contradiction. Thus $x \notin C_\delta$ for $\delta < 2\alpha + 1$, so that $x \in \mathrm{Diff}_{\alpha<\omega(n^2)} C_\alpha$. It is equally easy to check that $\mathrm{Diff}_{\alpha<\omega(n^2)} C_\alpha \subseteq A \cap B$. Thus $A \cap B$ is $\omega(n^2) - \Pi^1_1$. □

With a little care, one can show that the intersection of two $\omega n - \Pi^1_1$ sets is $\omega(2n - 1) - \Pi^1_1$.

Since the complement of an $\alpha - \Pi^1_1$ set is $\alpha + 2 - \Pi^1_1$, the class $\Lambda = \bigcup_{n<\omega} (\omega n - \Pi^1_1)$ is closed under $\sim$, $\cap$, and $\cup$. This means that "interweaving" Λ-norms produces a Λ-norm. In particular, interweaving the norms of the Busch scale on a Σ^1_1 set produces a very good scale on that set with norms in Λ. We give now a direct construction of such a scale.

<u>Theorem 1</u>. Every Σ^1_1 set A admits a very good scale $\{\varphi_n\}$ such that for all n,

$$\varphi_n : A \longrightarrow \omega(n + 1)$$

and

$$\varphi_n \text{ is an } \omega(n + 1) - \Pi^1_1 \text{ norm} .$$

<u>Proof</u>. Let $A = p[T]$, where T is a recursive tree on $\omega \times \omega$. We may assume that if $(\sigma,\tau) \in T$, then $\forall i < \ell h\sigma\ \exists k\ (\tau(i) = 2^{\sigma(i)}3^k)$, so that any branch of $T(x)$ codes x. Let $\langle \sigma_i \mid i < \omega \rangle$ be a recursive enumeration of $\omega^{<\omega}$ such that for all n, $\{\sigma_i \mid i \leq n\}$ is a tree. For $n < \omega$, let

$$E_n = \{\tau \in \omega^{<\omega} \mid \tau \notin \{\sigma_i \mid i \leq n\} \ \wedge\ \tau \restriction (\ell h\tau - 1) \in \{\sigma_i \mid i \leq n\}\} .$$

For $x \in A$ and $n < \omega$, let

$$\tau^x_n = \text{unique } \tau \in E_n\ (\tau \subseteq f^T_x)$$

and

$$\varphi_n(x) = \text{the ordinal rank of } \tau^x_n \text{ in } \leq_{\text{lex}} \restriction E_n .$$

Notice that $\leq_{lex} \upharpoonright E_n$ has order type $\omega(n+1)$, so that $\varphi_n : A \longrightarrow \omega(n+1)$. We now check that $\{\varphi_n\}$ is a very good scale on A. Clearly, if $x \in A$ and $n \leq m$, then $\tau_n^x \subseteq \tau_m^x$. So if $x,y \in A$ and $\tau_n^x <_{lex} \tau_n^y$, then $\tau_m^x <_{lex} \tau_m^y$ for $m \geq n$, as required by the refinement property of very goodness. Now let $\langle x_i \mid i < \omega \rangle$ be a sequence of elements of A such that for all $n < \omega$, the sequence $\langle \varphi_n(x_i) \mid i < \omega \rangle$ is eventually constant, say with constant value (the ordinal rank in $\leq_{lex} \upharpoonright E_n$ of) τ_n. Then for any n, $\tau_n = \tau_n^{x_i}$ and $\tau_{n+1} = \tau_{n+1}^{x_i}$ for sufficiently large i, so $\tau_n \subseteq \tau_{n+1}$. But $\tau_n \notin \{\sigma_i \mid i \leq n\}$, so $\ell h \tau_n \to \infty$ as $n \to \infty$. Let $f = \bigcup_n \tau_n$. Now for any n, $(x_i \upharpoonright n, f \upharpoonright n) \in T$ for sufficiently large i, so by our assumption on T, $\lim_{i\to\infty} x_i$ exists. Let $x = \lim_{i\to\infty} x_i$. Then $(x,f) \in [T]$, so $x \in A$; moreover $f_x^T \leq_{lex} f$, so $\tau_n^x \leq_{lex} \tau_n$ for all n. Thus $\{\varphi_n\}$ is a very good scale on A.

It is easy to compute that φ_n is an $\omega(n+1) - \Pi_1^1$ norm. □

The universal Σ_1^1 set admits no very good scale whose norms are all $\omega n - \mathbf{\Pi}_1^1$ for some fixed n, at least if $\forall x\ (x^\#$ exists). For if it did, the Third Periodicity Theorem ([3], p. 335) would give an $x \in {}^\omega\omega$ so that every $\Sigma_1^1(x)$ game has a winning strategy which is $\mathfrak{D}(\omega n - \Pi_1^1(x))$. But every $\mathfrak{D}(\omega n - \Pi_1^1(x))$ real is recursive in the type of the first $n+1$ indiscernibles for $L[x]$, and hence a member of $L[x]$, by Martin's analysis of $\omega n - \Pi_1^1$ games. This is impossible since there are $\Sigma_1^1(x)$ games without winning strategies in $L[x]$.

<u>Theorem 2</u>. Let Λ be the class of sets which are $\omega n - \Pi_1^1$ for some $n < \omega$. Then every set in Λ admits a very good scale whose norms are in Λ.

<u>Proof</u>. Let $A = \mathrm{Diff}_{\alpha<\omega n} A_\alpha$, where the A_α's are Σ_1^1. For $\alpha < \omega n$, let $\{\varphi_i^\alpha\}$ be a scale on A_α as given in Theorem 1 (or in Busch [1]), and let $\{\theta_i^\alpha\}$ be a very good Π_1^1 scale on $\{x \mid \forall\beta < \alpha\ (x \notin A_\beta)\}$. Let $\langle F_i \mid i < \omega \rangle$ be an increasing sequence of finite sets whose union is $\{\alpha < \omega n \mid \alpha$ is odd$\}$. For $x \in A$, let

$$\alpha_x = \text{least } \beta\ (x \in A_\beta)\ ,$$

and let

$$\psi_i(x) = \langle \alpha_x \rangle \qquad \text{if } \alpha_x \notin F_i\ ,$$

$$\psi_i(x) = \langle \alpha_x, \varphi_0^{\alpha_x}(x), \theta_0^{\alpha_x}(x), \ldots, \varphi_i^{\alpha_x}(x), \theta_i^{\alpha_x}(x) \rangle \qquad \text{if } \alpha_x \in F_i\ .$$

Here, of course, $\psi_i(x)$ is identified with an ordinal using $\leq_{lex} \upharpoonright \omega_1^{\leq 2i+1}$.

It is routine to check that $\{\psi_i\}$ is a very good scale on A. We shall check that ψ_i is a Λ norm; for notational simplicity, we consider only $i = 0$.

First, notice that there is a relation $R \in \Lambda$ so that for $x,y \in A$, $R(x,y)$ iff $\alpha_x < \alpha_y$. For simply let

$$B_{2\alpha} = \{(x,y) \mid y \in A_\alpha\}$$

and

$$B_{2\alpha+1} = \{(x,y) \mid x \in A_\alpha\}$$

for $\alpha < \omega n$, and then set $R = \mathrm{Diff}_{\alpha<\omega n} B_\alpha$. Let $E(x,y)$ iff $\sim(R(x,y) \vee R(y,x))$, so that $E \in \Lambda$ and for $x,y \in A$, $E(x,y)$ iff $\alpha_x = \alpha_y$. Now let

$$x \leq_0 y \quad \text{iff} \quad x \in A \wedge (y \notin A \vee \psi_0(x) \leq \psi_0(y)),$$

$$x <_0 y \quad \text{iff} \quad x \in A \wedge (y \notin A \vee \psi_0(x) < \psi_0(y)).$$

Then

$$x \leq_0 y \text{ iff } [x \in A \wedge (y \notin A \vee [y \in A \wedge (R(x,y) \vee [E(x,y) \wedge$$

$$\bigwedge_{\beta\in F_0} (x \in A_\beta - \bigcup_{\gamma<\beta} A_\gamma \Rightarrow \langle\varphi_0^\beta(x),\theta_0^\beta(x)\rangle \leq_{lex} \langle\varphi_0^\beta(y),\theta_0^\beta(y)\rangle)])])]$$

Our lemma implies that $\leq_0 \in \Lambda$. Similarly $<_0 \in \Lambda$, so that ψ_0 is a Λ norm. □

According to the Third Periodicity Theorem, if every Γ set has a very good scale with norms in Γ, then every $\Game\Gamma$ set has a (very good) scale with norms in Γ. So we have

<u>Corollary</u>. Let $\Lambda = \bigcup_n \omega n - \Pi^1_1)$. Then for all $k < \omega$, every $\Game^k\Lambda$ set admits a very good scale with norms in $\Game^k\Lambda$.

In particular, every Π^1_2 $(= \Game\Sigma^1_1)$ set admits a scale with $\Game\Lambda$ norms. The Martin-Solovay scale ([2], Theorem 6.3) also has this property. In fact, if one propagates the scale of Theorem 1 to Π^1_2 via the second periodicity construction, one obtains the Martin-Solovay scale. (Martin and the author proved this independently.) One can see this by looking at the auxilliary closed games associated to the $\omega n - \Pi^1_1$ games involved in propagating the scale of Theorem 1.

Theorem 2 generalizes easily to the class $\Lambda_\alpha = \bigcup_n (\omega^2\alpha + \omega n - \underset{\sim}{\Pi}^1_1)$, where $\alpha < \omega_1$.

References

[1] Douglas R. Busch, λ-Scales, κ-Souslin sets and a new definition of analytic sets, J. Symbolic Logic, vol. 41, June 1976, p. 373.

[2] Alexander S. Kechris, AD and projective ordinals, Cabal Seminar 76-77, Lecture Notes in Mathematics, vol. 689, Springer-Verlag.

[3] Yiannis N. Moschovakis, Descriptive set theory, Studies in Logic, vol. 100, North Holland.

SCALES ON COINDUCTIVE SETS

Yiannis N. Moschovakis*
Department of Mathematics
University of California
Los Angeles, California 90024

The main purpose of this note is to prove (under appropriate determinacy hypotheses) that coinductive pointsets admit definable scales. This solves the second Victoria Delfino problem [1] and was announced in [3], together with many stronger related results of Martin and Steel; see also the papers following this one in this volume.

§1. Lemmas on preservation of scales. When this question was first posed in [4], there was a feeling (certainly on my part) that its answer might require methods for constructing scales that would be radically different from the simple and well-understood ideas of the periodicity theorems. As it happens, this is not true: one can define a scale on a given coinductive set quite easily, by a judicious mix of the constructions in the second periodicity theorems 6C.1 and 6C.3 of [5]. In fact, the simplest way to explain the proof of this theorem is to reexamine the proofs of these two results and extract from them a bit more than was stated explicitly in [5].

Suppose P is a pointset, i.e. $P \subseteq \mathcal{X} = X_1 \times \cdots \times X_n$ with each $X_i = \omega$ or $X_i = \mathcal{N} = {}^{\omega}\omega$. A putative scale on P is a sequence

$$\overline{\varphi} = \{\varphi_i\}_{i\in\omega}$$

of norms on P, each norm

$$\varphi_i : P \longrightarrow \text{Ordinals}$$

mapping P into the ordinals.

A putative scale $\overline{\varphi}$ on P defines in a natural way a notion of convergence for sequences of points in P: we say that x_n converges to x modulo $\overline{\varphi}$ and write

$$x_n \longrightarrow x \ [\text{mod } \overline{\varphi}]$$

if $\lim_{n\to\infty} x_n = x$ in the usual topology on $\mathcal{X}$ and if in addition, for each fixed i,

---

*During the preparation of this paper the author was partially supported by NSF Grant MCS78-02989.

the sequence of ordinals $\varphi_i(x_0), \varphi_i(x_1), \varphi_i(x_2), \ldots$ is ultimately constant.

We say that $\overline{\varphi}$ is a semiscale on P if for every sequence $\{x_n\}$ in P,

(1) $$x_n \longrightarrow x \,[\text{mod } \overline{\varphi}] \Rightarrow x \in P\,.$$

A norm $\psi : P \to$ Ordinals is lower semicontinuous relative to the putative scale $\overline{\psi}$, if for each sequence $\{x_n\}$ in P,

(2) $$x_n \longrightarrow x \,[\text{mod } \overline{\varphi}] \Rightarrow [x \in P \,\&\, \psi(x) \leq \lim_{n\to\infty} \psi(x_n)]\ ;$$

we call $\overline{\varphi}$ a scale if each φ_i is lower semicontinuous relative to $\overline{\varphi}$.

To give a topological interpretation of these definitions (for those familiar with uniform spaces), let $\mathfrak{T}(\overline{\varphi})$ be the topology on P generated by the neighborhoods

$$N(x_0,k) = \{x : x(0) = x_0(0) \,\&\, \cdots \,\&\, x(k) = x_0(k),$$
$$\varphi_0(x) = \varphi_0(x_0) \,\&\, \cdots \,\&\, \varphi_k(x) = \varphi_k(x_0)\}$$

relative to a putative scale $\overline{\varphi}$. It is easy to check that $\mathfrak{T}(\overline{\varphi})$ is a uniform topology in the sense of Kelley [2] and that

$$\overline{\varphi} \text{ is a semiscale} \Longleftrightarrow P \text{ is complete}\ ,$$

$$\overline{\varphi} \text{ is a scale} \Longleftrightarrow \text{for each } i, \lambda, \text{ the "closed ball" } \{x : \varphi_i(x) \leq \lambda\} \text{ is complete}\ .$$

Of course the theory of uniform spaces is tailored to the examples where the topology is determined by a sequence of pseudo-metrics with real values and is not useful in the present context of ordinal-valued norms.

If $\overline{\varphi}$ and $\overline{\varphi}'$ are both putative scales on P, we say that $\overline{\varphi}$ is finer than $\overline{\varphi}'$ if for every $x_0, x_1, \ldots$ in P,

$$x_n \longrightarrow x \,[\text{mod } \overline{\varphi}] \Rightarrow x_n \longrightarrow x \,[\text{mod } \overline{\varphi}']\ .$$

For example one can refine a putative scale $\overline{\varphi}$ by adding some norms or interweaving, i.e. setting

$$\psi_n(x) = \langle \varphi_0(x), \varphi_1(x), \ldots, \varphi_n(x) \rangle$$

where

$$(\xi_0, \ldots, \xi_n) \longmapsto \langle \xi_0, \ldots, \xi_n \rangle$$

is some order-preserving map (depending on n) of the (n + 1)-tuples of ordinals (ordered lexicographically) into the ordinals.

Suppose that

$$Q(x) \iff (\exists\alpha)P(x,\alpha)$$

and that we are given norms $\varphi_0,\ldots,\varphi_n$ on P. We define the norm

$$\psi_n = \text{"inf"}\ \{\varphi_0,\ldots,\varphi_n\}$$

on Q by

$$\psi_n(x) = \text{infimum}\{\langle\varphi_0(x,\alpha),\alpha(0),\varphi_1(x,\alpha),\alpha(1),\ldots,\varphi_n(x,\alpha),\alpha(n)\rangle : P(x,\alpha)\}$$

where $\langle\ldots\rangle$ again stands for an order-preserving map of 2n-tuples into the ordinals. If $\overline{\varphi} = \{\varphi_n\}$ is a putative scale on P, then this definition associates with $\overline{\varphi}$ the infimum putative scale

$$\overline{\psi} = \text{"inf"}\ \overline{\varphi} = \{\psi_n\}$$

on Q -- but <u>it is important that the definition of each</u> ψ_n <u>in</u> "inf" $\overline{\varphi}$ <u>depends only on</u> $\varphi_0,\ldots,\varphi_n$ <u>and not on the whole sequence</u> $\overline{\varphi}$.

Now the proof of 6C.1 in [5] (stated there for "very good scales") actually establishes the following result.

<u>The Infimum Lemma</u>. Suppose $\overline{\varphi}$ is a putative scale on $P \subseteq \mathcal{X} \times \mathcal{N}$,

$$Q(x) \iff (\exists\alpha)P(x,\alpha)$$

and $\overline{\psi} = \text{"inf"}\ \overline{\varphi}$.

(a) If $\overline{\psi}'$ is any putative scale finer than $\overline{\psi}$ and if $\{x_n\}$ is a sequence of points in Q such that

$$x_n \to x\ [\text{mod}\ \overline{\psi}'] ,$$

then we can find a subsequence

$$x_k^* = x_{n_k} \qquad (n_0 < n_1 < \cdots)$$

and irrationals $\alpha,\alpha_0,\alpha_1,\ldots,$ such that $P(x_k,\alpha_k)$ holds for each k and

$$(x_k^*,\alpha_k) \to (x,\alpha)\ [\text{mod}\ \overline{\varphi}] .$$

(b) If $\overline{\varphi}$ is a semiscale on P and for $i = 0,\ldots,m$ each φ_i is lower semicontinuous relative to $\overline{\varphi}$, then ψ_m is lower semicontinuous relative to $\overline{\psi}$.

In particular, if $\overline{\varphi}$ is a scale on P, then $\overline{\psi} = \text{"inf"}\ \overline{\varphi}$ is a scale on Q.

<u>Proof</u>. (a) Notice that it is enough to consider the case $\overline{\psi} = \text{"inf"}\ \overline{\varphi}$ and choose

$$n_k = \text{least } n \text{ such that for } i = 0,\ldots,k, \text{ all } m \geq n$$

$$\psi_i(x_m) = \psi_i(x_n) .$$

By definition,

$$\psi_i(x_k^*) = \langle \varphi_0(x_k^*,\alpha_i^k),\alpha_i^k(0),\varphi_1(x_k^*,\alpha_i^k),\alpha_i^k(1),\ldots,\varphi_i(x_k^*,\alpha_i^k),\alpha_i^k(i)\rangle$$

for some α_i^k and for $k \geq i$, $\ell \geq i$

$$\alpha_i^k(0) = \alpha_i(0),\ \ldots,\ \alpha_i^k(i) = \alpha_i^\ell(i)$$

so we can take

$$\alpha_k = \alpha_k^k$$

and the result follows immediately.

(b) is not hard to prove directly from the definition of the infimum norms and we will skip it. ⊣

We will also need the analogous modification of the second periodicity theorem 6C.3 of [5] which applies to arbitrary putative scales that need not be semiscales or scales.

Suppose then that

$$Q(x) \Longleftrightarrow (\forall\alpha)P(x,\alpha)$$

and $\overline{\varphi}$ is a given putative scale on Q and let $u(0),u(1),\ldots$ be a canonical enumeration of all finite sequences of integers such that if $u(i)$ is an initial segment of $u(j)$, then $i < j$. As in 6C.3 of [5] (<u>but interweaving the norms in $\overline{\varphi}$</u>), we define for each $x,y \in \mathcal{X}$ and each n a game $G_n(x,y)$ where I plays α', II plays β'

$$G_n(x,y) \quad \begin{array}{lllllll} u(n) & \text{I} & a_0 & & a_1 & & \alpha';\ \alpha = u(n)^\frown\alpha' \\ & & \searrow & \nearrow & \searrow & \cdots & \\ u(n) & \text{II} & & b_0 & & b_1 & \beta';\ \beta = u(n)^\frown\beta' \end{array}$$

and

$$\text{II wins} \Longleftrightarrow \langle\varphi_0(x,\alpha),\varphi_1(x,\alpha),\ldots,\varphi_n(x,\alpha)\rangle \leq \langle\varphi_0(y,\beta),\varphi_1(y,\beta),\ldots,\varphi_n(y,\beta)\rangle .$$

The relation

$$x \leq_n y \Longleftrightarrow x,y \in Q \ \&\ \text{II wins } G_n(x,y)$$

is a prewellordering (granting the determinacy of these games), so we can define on Q for each n the norm

$$\psi_n = \text{"sup"}\ \{\varphi_0,\ldots,\varphi_n\}$$

by

$$\psi_n(x) = \text{rank of } x \text{ in the prewellordering } \leq_n .$$

The putative scale $\overline{\psi} = \{\psi_n\}$ on Q is the (fake) supremum putative scale associated with $\overline{\varphi}$ on P, in symbols

$$\overline{\psi} = \text{"sup"}\ \overline{\varphi} .$$

Notice again that the definition of each ψ_n in "sup" $\overline{\varphi}$ depends only on $\varphi_0,\ldots,\varphi_n$.

The Fake Supremum Lemma. Suppose $\overline{\varphi}$ is a putative scale on $P \subseteq \mathcal{X} \times \mathcal{N}$,

$$Q(x) \Longleftrightarrow (\forall\alpha)P(x,\alpha) ,$$

assume that each φ_i is a Γ-norm where Γ is an adequate pointclass satisfying $\mathrm{Det}(\underset{\sim}{\Delta})$ and let $\overline{\psi} = \text{"sup"}\ \overline{\varphi}$.

(a) If $\overline{\psi}'$ is any putative scale on Q which is finer than $\overline{\psi}$ and if $\{x_n\}$ is a sequence of points in Q such that

$$x_n \longrightarrow x\ [\mathrm{mod}\ \overline{\psi}'] ,$$

then for each α, there exists a subsequence

$$x_k^* = x_{n_k} \qquad (n_0 < n_1 < \cdots)$$

and irrationals $\alpha_0,\alpha_1,\ldots,$ such that

$$(x_k^*,\alpha_k) \rightarrow (x,\alpha)\ [\mathrm{mod}\ \overline{\varphi}] .$$

(b) If $\overline{\varphi}$ is a semiscale on P and for $i = 0,\ldots,m$ each φ_i is lower semicontinuous relative to $\overline{\varphi}$, then ψ_m is lower semicontinuous relative to $\overline{\psi}$.

In particular, if $\overline{\varphi}$ is a scale on P, then $\overline{\psi} = \text{"sup"}\ \overline{\varphi}$ is a scale on Q. $\dashv$

The proof of this is exactly the proof of 6C.3 in [5] which was stated with additional hypotheses there.

§2. The main result. A pointset $Q \subseteq \mathcal{X}$ is inductive if there is a projective $P \subseteq \mathcal{X} \times \mathcal{N}$ such that

(3) $$Q(x) \Longleftrightarrow \{(\forall\alpha_0)(\exists\,\alpha_1)(\forall\alpha_2)\cdots\}(\exists t)P(x,\langle\alpha_0,\ldots,\alpha_{t-1}\rangle) .$$

This can be taken as the definition of inductive sets or it can be proved easily from the more natural definition in terms of positive elementary induction on $\mathcal{N}$, see section 7C of [5]. (A quick summary of some of the basic properties of inductive sets can be found in [4].)

For the proof in this paper, we will only need the following very simple normal form for inductive sets.

Lemma. A set $Q \subseteq \mathcal{X}$ is inductive if and only if there is a number-theoretic relation R such that

$$(4) \qquad Q(x) \Longleftrightarrow \{(\forall\alpha_0)(\exists\alpha_1)(\forall\alpha_2)\cdots\}(\exists t)R(\bar{x}(t),\langle\bar{\alpha}_0(t),\ldots,\bar{\alpha}_{t-1}(t)\rangle) \;;$$

consequently, Q is coinductive if for some R,

$$Q(x) \Longleftrightarrow \{(\exists\alpha_0)(\forall\alpha_1)(\exists\alpha_2)\cdots\}(\forall t)R(\bar{x}(t),\langle\bar{\alpha}_0(t),\ldots,\bar{\alpha}_{t-1}(t)\rangle) \;.$$

Proof. Suppose, for example, that (3) above holds with some $P(x,\alpha)$ in $\underset{\sim}{\Sigma}^1_2$, so that

$$P(x,\alpha) \Longleftrightarrow (\exists\beta)(\forall\gamma)(\exists s)R(\bar{x}(s),\bar{\alpha}(x),\bar{\beta}(s),\bar{\gamma}(s))$$

with a number-theoretic R. An easy game-argument shows that

$$\begin{aligned}(5) \quad Q(x) \Longleftrightarrow \{(\forall\alpha_0)(\exists\alpha_1)(\exists u_1)(\forall\alpha_2)(\exists\alpha_3)(\exists u_3)\cdots\}(\exists t)\{&u_t = 1 \;\& \\ \& (\forall i < t,\; i \text{ odd})\,&[u_i = 0] \\ \& [\text{if } s = \alpha_{t+2}(0), \text{ then } R(\bar{x}(s),\langle\bar{\alpha}(s),\ldots,&\alpha_{t-1}(s)\rangle,\bar{\alpha}_t(s),\bar{\alpha}_{t+1}(s))]\} \;.\end{aligned}$$

(From a strategy that establishes $Q(x)$ by (3), get one that establishes $Q(x)$ by (5) and vice versa; notice that both (3) and (5) are interpreted via open games.) Now (5) gives a normal form for $Q(x)$ as in (4) (with a different R) by trivial recursive manipulations of the infinite quantifier string and the matrix. ⊣

Let

$$\underset{\sim}{\Sigma}^*_0 = \text{all Boolean combinations of inductive and coinductive pointsets} \;,$$

and for $n \geq 1$, by induction,

$$Q \in \underset{\sim}{\Sigma}^*_n \Longleftrightarrow \text{for some } P \text{ in } \underset{\sim}{\Sigma}^*_{n-1}, \text{ and all } x,\; Q(x) \Longleftrightarrow (\exists\alpha)\neg P(x,\alpha) \;.$$

Theorem. If every game with payoff in $\bigcup_n \underset{\sim}{\Sigma}^*_n$ is determined, then every coinductive pointset Q admits a scale $\bar{\psi} = \{\psi_i\}$, such that each ψ_i is a $\underset{\sim}{\Sigma}^*_i$-norm.

Proof. Suppose by the lemma that

$$Q(x) \Longleftrightarrow \{(\exists\alpha_0)(\forall\alpha_1)\cdots\}(\forall t)R(\bar{x}(t),\langle\bar{\alpha}_0(t),\ldots,\bar{\alpha}_{t-1}(t)\rangle)$$

and for each even n define

$$(6)\qquad Q_n(x,\alpha_0,\ldots,\alpha_{n-1}) \Longleftrightarrow \{(\exists\alpha_n)(\forall\alpha_{n+1})\cdots\}(\forall t)R(\bar{x}(t),\langle\bar{\alpha}_0(t),\ldots,\bar{\alpha}_{t-1}(t)\rangle) .$$

For odd n let

$$Q_n(x,\alpha_0,\ldots,\alpha_{n-1}) \Longleftrightarrow (\forall\alpha_n)Q_{n+1}(x,\alpha_0,\ldots,\alpha_{n-1},\alpha_n) .$$

We will define a putative scale $\bar{\psi}_n = \{\psi_i^n\}$ on each Q_n and then show that $\bar{\psi}^0$ on $Q = Q_0$ is actually a scale.

The definition of ψ_i^n is by induction on i, simultaneously for all n. We will use "vector notation for tuples,

$$\tilde{\alpha} = \alpha_0,\ldots,\alpha_{n-1}, \qquad \tilde{\beta} = \beta_0,\ldots,\beta_{n-1} .$$

Case 1. n is even and $i = 0$.

If $Q_n(x,\tilde{\alpha})$ holds, put

$$\psi_0^n(x,\tilde{\alpha}) = 0 .$$

This norm just records (by being defined) that $Q_n(x,\tilde{\alpha})$ holds. It is a Σ_0^*-norm since all Q_n are clearly coinductive and (in the notation of 4B of [5]),

$$(x,\tilde{\alpha}) \le^*_{\psi_0^n} (y,\tilde{\beta}) \Longleftrightarrow Q_n(x,\tilde{\alpha}) ,$$

$$(x,\tilde{\alpha}) <^*_{\psi_0^n} (y,\tilde{\beta}) \Longleftrightarrow Q_n(x,\tilde{\alpha})\ \&\ \neg\, Q_n(y,\tilde{\beta}) .$$

It is not hard to check that in general we cannot expect ψ_0^n to be either inductive or coinductive.

Case 2. n is odd.

Now

$$Q_n(x,\tilde{\alpha}) \Longleftrightarrow (\forall\alpha_{n+1})Q_{n+1}(x,\tilde{\alpha},\alpha_{n+1})$$

and assuming that ψ_j^{n+1} is defined for $j \le i$, we put

$$\psi_i^n = \text{"sup"}\{\psi_0^{n+1},\ldots,\psi_i^{n+1}\} .$$

Case 3. n is even and $i > 0$.

Now

$$Q_n(x,\tilde{\alpha}) \Longleftrightarrow (\exists\alpha_{n+1})Q_{n+1}(x,\tilde{\alpha},\alpha_{n+1})$$

and assuming that ψ_j^{n+1} is defined for all $j < i$, we put

$$\psi_i^n = \text{"inf"}\{\psi_0^{n+1},\ldots,\psi_{i-1}^{n+1}\} .$$

This completes the definition of a putative scale $\overline{\psi}_n^n$ on each Q_n and the definability assertion in the theorem is immediate from the proofs of 6C.1 and 6C.3 in [5], by induction on i.

Notice that by the construction, immediately, for each even n, $\overline{\psi}^n$ is finer than $\text{"inf"}\overline{\psi}^{n+1}$ and for each odd n, $\overline{\psi}^n = \text{"sup"}\overline{\psi}^{n+1}$, so the basic lemmas apply.

To check first that $\overline{\psi}^0$ is a semiscale on $Q = Q_0$, suppose

$$x_m \longrightarrow x \ [\text{mod } \overline{\psi}^0] ;$$

to prove that $Q(x)$ holds, we must describe a strategy for $\exists$ to win the closed game which interprets

$$\{(\exists\alpha_0)(\forall\alpha_1)(\exists\alpha_1)\cdots\}(\forall t)R(\overline{x}(t),\langle\overline{\alpha}_0(t),\ldots,\overline{\alpha}_{t-1}(t)\rangle) .$$

By the infimum lemma, choose a subsequence $\{x_m^0\}$ of $\{x_m\}$ and irrationals $\alpha_0,\alpha_{0,1}^0,\alpha_{0,2}^0,\ldots$ such that

$$(x_m^0,\alpha_{0,m}^0) \longrightarrow (x,\alpha_0) \ [\text{mod } \overline{\psi}^1]$$

and play this α_0. If your opponent responds with some α_1, choose by the supremum lemma a subsequence $\{x_m^1,\alpha_{0,m}^1\}$ of $\{x_m^0,\alpha_{0,m}^0\}$ and irrationals $\alpha_{1,m}^1$ such that

$$(x_m^1,\alpha_{0,m}^1,\alpha_{1,m}^1) \longrightarrow (x,\alpha_0,\alpha_1) \ [\text{mod } \overline{\psi}^2]$$

and then apply the infimum lemma again to $\{x_m^1,\alpha_{0,m}^1,\alpha_{1,m}^1\}$ to get a subsequence $\{x_m^2,\alpha_{0,m}^2,\alpha_{1,m}^2\}$ and irrationals α_2 and $\{\alpha_{2,m}^2\}$ such that

$$(x_m^2,\alpha_{0,m}^2,\alpha_{1,m}^2,\alpha_{2,m}^2) \longrightarrow (x,\alpha_0,\alpha_1,\alpha_2) \ [\text{mod } \overline{\psi}^3]$$

and play this α_2.

It is clear that $\exists$ can continue to play in this manner indefinitely, so we have defined a strategy for him. To see that he wins, notice that by the construction, for each n we know that for each m,

$$Q_n(x_m^{n-1},\alpha_{0,m}^{n-1},\alpha_{1,m}^{n-1},\ldots,\alpha_{n-1,m}^{n-1})$$

so that by the definition of Q_n, taking $t = n$ in the matrix, we have for each m

$$R(\overline{x}_m^{n-1}(n),\langle\overline{\alpha}_{0,m}^{n-1}(n),\ldots,\overline{\alpha}_{n-1,m}^{n-1}(m)\rangle) ;$$

letting $m \to \infty$, since

$$(x_m^{n-1}, \alpha_{0,m}^{n-1}, \ldots, \alpha_{n-1,m}^{n-1}) \to (x, \alpha_0, \ldots, \alpha_{n-1}) ,$$

we have for each n

$$R(\bar{x}(n), \langle \bar{\alpha}_0(n), \ldots, \bar{\alpha}_{n-2}(n) \rangle)$$

which is precisely the condition for to win. Thus $\vec{\psi}^0$ is a semiscale on Q_0.

Now the same argument shows easily that each $\vec{\psi}^n$ is a semiscale on Q_n, with just some added notation. To show that these are actually scales, notice first that for n even, ψ_0^n is surely semicontinuous since it is constant; the Infimum and Fake Supremum Lemmas then imply immediately that all ψ_i^n are lower semicontinuous relative to $\vec{\psi}^n$, by induction on i. $\dashv$

References

[1] Cabal Seminar 76-77, edited by A. S. Kechris and Y. N. Moschovakis, Springer Lecture Notes in Mathematics, Vol. 689, 1978.

[2] J. Kelley, General Topology, The University series in higher mathematics, Van Nostrand, Princeton, N. J., 1955.

[3] D. A. Martin, Y. N. Moschovakis and J. R. Steel, The extent of definable scales, Bulletin Amer. Math. Soc.

[4] Y. N. Moschovakis, Inductive scales on inductive sets, Cabal Seminar 76-77, see above.

[5] Y. N. Moschovakis, Descriptive Set Theory, Studies in Logic, Vol. 100, North Holland Publishing Co., Amsterdam, 1980.

THE EXTENT OF SCALES IN $L(\mathbb{R})$

Donald A. Martin
Department of Mathematics
University of California
Los Angeles, California 90024

John R. Steel
Department of Mathematics
University of California
Los Angeles, California 90024

We shall show that every set of reals which is Σ_1 definable over $L(\mathbb{R})$ from parameters in $\mathbb{R} \cup \{\mathbb{R}\}$ admits a scale in $L(\mathbb{R})$, and in fact that the class of sets so definable has the scale property. Kechris and Solovay observed some time ago that no other sets admit scales in $L(\mathbb{R})$, so that our result determines precisely the extent of scales in $L(\mathbb{R})$.

Some preliminaries: We work in ZF + DC, and state our additional determinacy hypotheses as we need them. We let $\mathbb{R} = {}^{\omega}\omega$, the Baire space, and call its elements reals. Variables $z,y,x,\ldots$ range over $\mathbb{R}$, while $\alpha,\beta,\gamma,\ldots$ range over the class OR of ordinals. Let $*$ be a recursive homeomorphism from ${}^{\omega}\mathbb{R}$ to $\mathbb{R}$ such that $\langle x_n \mid n < \omega\rangle^* \restriction i$ depends only on $\langle x_n \mid n < i\rangle$. We use $\langle x_n \mid n < i\rangle^*$ to denote the common value of $\langle x_n \mid n < \omega\rangle^* \restriction i$ for all extensions $\langle x_n \mid n < \omega\rangle$ of $\langle x_n \mid n < i\rangle$.

A pointclass is a class of relations on $\mathbb{R}$ closed under recursive substitutions; a boldface pointclass is a pointclass closed under continuous substitutions. Let $\Sigma_n(M,X)$ be the class of relations on $\mathbb{R}$ which are Σ_n definable over M from parameters in X. We are mainly interested in the pointclass $\Sigma_1(L(\mathbb{R}), \{\mathbb{R}\})$, and its boldface counterpart $\Sigma_1(L(\mathbb{R}), \mathbb{R} \cup \{\mathbb{R}\})$. It is easy to show that $\Sigma_1(L(\mathbb{R}),\{\mathbb{R}\}) = (\Sigma^2_1)^{L(\mathbb{R})}$ and $\Sigma_1(L(\mathbb{R}), \mathbb{R} \cup \{\mathbb{R}\}) = (\utilde{\Sigma}^2_1)^{L(\mathbb{R})}$ (cf. [8], Lemma 1.12). We shall have no use for this fact, however.

For any X, $X^{<\omega}$ is the set of finite sequences from X. A tree on X is a subset of $X^{<\omega}$ closed under initial segment. If T is a tree on X, $[T] = \{f \in {}^{\omega}X \mid \forall n(f \restriction n \in T)\}$; T is well founded just in case $[T] = \emptyset$. We sometimes regard a sequence of tuples as a tuple of sequences (of the same length), so that a tree on $X_1 \times \cdots \times X_k$ becomes a subset of $X_1^{<\omega} \times \cdots \times X_k^{<\omega}$. If T is a tree on $X \times Y$ and $f \in {}^{\omega}X$, then $T(f) = \{u \in Y^{<\omega} \mid (f \restriction \ell h(u),u) \in T\}$. Notice that $T(f)$ is a tree on Y depending continuously on f.

A quasi-strategy for I in a game on X is a nonempty tree S on X such that if $u \in S$ and $\ell h(u)$ is even, then $\exists a \in X \forall b \in X(u^\frown\langle a\rangle^\frown\langle b\rangle \in S)$. We say S is a winning quasi-strategy (wqs) for I in the game with payoff A just in case $[S] \subseteq A$. If G is the game on $\mathbb{R}$ with payoff A, and Γ is a pointclass, we call G a Γ game iff $\{\langle x_n \mid n < \omega\rangle^* \mid \langle x_n \mid n < \omega\rangle \in A\}$ is in Γ. $\Gamma\text{-}AD_{\mathbb{R}}$ is the assertion that all Γ games on $\mathbb{R}$ are determined. Similarly, if $B \subseteq \mathbb{R} \times \mathbb{R}$, let

---

*The preparation of this paper was partially supported by NSF Grant Number MCS78-02989.

$$\partial^{R}B = \{x \mid \text{I has a wqs in the game with payoff } \{\langle y_n \mid n<\omega\rangle \mid B(x,\langle y_n \mid n<\omega\rangle^*)\}\},$$
$$= \{x \mid \exists y_0 \forall y_1 \exists y_2 \cdots B(x,\langle y_n \mid n<\omega\rangle^*)\} ,$$

and if Γ is a pointclass, let $\partial^{R}\Gamma = \{\partial^{R}B \mid B \in \Gamma\}$.

For whatever else we use from Descriptive Set Theory, and in particular for the notions of a scale and of the scale property, we refer the reader to [4].

Our main theorem builds directly on a scale construction due to Moschovakis ([6]). We shall describe briefly the slight generalization of this construction we need. Suppose that $\alpha \in OR$, and that for each $x \in R$ we have a game G_x in which I's moves come from $R \times \alpha$ while II's moves come from R. Thus a typical run of G_x has the form

I	x_0, β_0		x_2, β_1		...
II		x_1		x_3	

where $x_i \in R$ and $\beta_i < \alpha$ for all i. Suppose that G_x is closed and continuously associated to x in the strong sense that for some tree T on $\omega \times \omega \times \alpha$

$$\text{I wins } G_x \text{ iff } (x,\langle x_n \mid n<\omega\rangle^*, \langle\beta_n \mid n<\omega\rangle) \in [T] .$$

Let

$$\mathbf{P}_k(x,u) \text{ iff } u \text{ is a position of length } k \text{ from which I has a wqs in } G_x ,$$

and

$$\mathbf{P}(x) \text{ iff } P_0(x,\emptyset) .$$

If sufficiently many games are determined, then Moschovakis' construction yields a scale on $\mathbf{P}$. [As in [5], we extend the concept of a scale to relations with arguments from OR as well as R by giving OR the discrete topology. Following [6] then, we define putative scales $\vec{\varphi}^k$ on P_k for all k simultaneously. If $P_k(x,u)$, then $\varphi_0^k(x,u)=0$. (Otherwise $\varphi_0^k(x,u)$ is undefined.) Now either $(P_k(x,u) \Longleftrightarrow \forall y\, P_{k+1}(x,u^\frown\langle y\rangle))$ or $(P_k(x,u) \Longleftrightarrow \exists y\, P_{k+1}(x,u^\frown\langle y\rangle))$ or $(P_k(x,u) \Longleftrightarrow \exists\beta\, P_{k+1}(x,u^\frown\langle\beta\rangle))$, depending on whose turn it is to play what at move k. In the first two cases, we define φ_{i+1}^k from $\langle\varphi_0^{k+1},\ldots,\varphi_i^{k+1}\rangle$ as in [6]. In the third case, if $P_k(x,u)$ then

$$\varphi_{i+1}^k(x,u) = \langle\beta,\varphi_0^{k+1}(x,u^\frown\langle\beta\rangle),\ldots,\varphi_i^{k+1}(x,u^\frown\langle\beta\rangle)\rangle ,$$

where β is least so that $P_{k+1}(x,u^\frown\langle\beta\rangle)$. (More precisely, $\varphi_{i+1}^k(x,u)$ is the ordinal of this tuple in the lexicographic order.) As in [6], one can show that each $\vec{\varphi}^k$ is in fact a scale on P_k, and thus $\vec{\varphi}^0$ is a scale on P, as desired.] The prewellordering $\leq_i$ induced by the ith norm of this scale on P is definable from the $\mathbf{P}_k$'s for $k \leq i$ by means of recursive substitution, the Boolean operations, and quantification over $R\times\alpha$. In particular, if the map $k \mapsto \mathbf{P}_k$ is

$\Sigma_1(L(\mathbb{R}),\{\alpha,\mathbb{R}\})$, then so is the map $i \mapsto \leq_i$. The determinacy required to construct this scale on P is closely related to the definability of the scale constructed. In particular, if the map $k \mapsto P_k$ is in $L(\mathbb{R})$, then the determinacy of all games on ω in $L(\mathbb{R})$ suffices.

In the circumstances just described we call the map $x \mapsto G_x$ a <u>closed game representation</u> of P. One can show directly that every $\Sigma_1(L(\mathbb{R}),\mathbb{R} \cup \{\mathbb{R}\})$ set admits a closed game representation, and hence a scale, in $L(\mathbb{R})$. This is done in [8]. We shall take a slightly more circuitous route here, our reward being some additional information concerning $\underset{\sim}{\Pi}^1_1$ games on $\mathbb{R}$.

<u>Lemma 1.</u> Every $\Game^{\mathbb{R}}\underset{\sim}{\Pi}^1_1$ set admits a closed game representation in $L(\mathbb{R})$.

<u>Proof</u>. Let P be $\Game^{\mathbb{R}}\underset{\sim}{\Pi}^1_1$, and let T be a tree on $\omega \times \omega \times \omega$ so that

$$P(x) \quad \text{iff I has a wqs in } G_x \quad ,$$

where G_x is the game on $\mathbb{R}$ whose payoff for I is $\{\langle x_n | n < \omega\rangle \mid T(x,\langle x_n | n < \omega\rangle^*)$ is wellfounded$\}$. We shall associate to each G_x auxilliary closed games $G^*_{x,\alpha}$ for $\alpha \in OR$ as in Martin's proof of $\underset{\sim}{\Pi}^1_1$ determinacy. A typical run of $G^*_{x,\alpha}$ has the form

I	x_0,β_0		x_2,β_1		...
II		x_1		x_3	

where $x_i \in \mathbb{R}$ and $\beta_i < \alpha$ for all i. Let $\langle \sigma_i \mid i < \omega\rangle$ be an enumeration of $\omega^{<\omega}$ such that if $\sigma_i \subseteq \sigma_j$ then $i \leq j$. We say I wins the run of $G^*_{x,\alpha}$ just displayed just in case whenever $\sigma_i,\sigma_j \in T(x,\langle x_n \mid n < \omega\rangle^*)$ and $\sigma_i \subsetneq \sigma_j$, we have $\beta_i > \beta_j$.

<u>Claim</u>. I has a wqs in G_x iff $\exists\alpha$ (I has a wqs in $G^*_{x,\alpha}$) .

<u>Proof</u>. If I has a wqs in $G^*_{x,\alpha}$, then I has a wqs in G_x, since G_x requires less of him. So suppose S is a wqs for I in G_x. Let

$$U = \{(u,v) \in \mathbb{R}^{<\omega} \times \omega^{<\omega} \mid \ell h(u) = \ell h(v) \text{ and } u \in S \text{ and } (x \restriction \ell h(u), u^*, v) \in T\} \quad .$$

Then U is a tree on $\mathbb{R} \times \omega$, and since S is a wqs for I, U is wellfounded. Let $\|(u,v)\|_U$ be the rank of (u,v) in U if $(u,v) \in U$, and 0 otherwise. Let $\alpha = \|(\emptyset,\emptyset)\|_U + 1$. Let

$$S^*_0 = \{\langle x_0,\beta_0,x_1,\ldots,x_{2k},\beta_k,x_{2k+1}\rangle \mid \langle x_0,x_1,\ldots,x_{2k+1}\rangle \mid \langle x_0,x_1,\ldots,x_{2k+1}\rangle \in S$$
$$\text{and } \forall i \leq k(\beta_i = \|(\langle x_0,\ldots,x_\ell\rangle,\sigma_i\|_U, \text{ where } \ell = \ell h(\sigma_i) - 1)\} \quad ,$$

and let S^* be the closure of S^*_0 under initial segment. It is easy to verify that S^* is a wqs for I in $G^*_{x,\alpha}$. This proves the claim.

If I has a wqs in G_x, let α_x be the least α such that I has a wqs in $G^*_{x,\alpha}$. Let $\lambda = \sup\{\alpha_x \mid \text{I has a wqs in } G_x\}$. Since any wqs for I in $G^*_{x,\alpha}$ is a wqs for I in all $G^*_{x,\beta}$ with $\beta \geq \alpha$, the map $x \mapsto G^*_{x,\lambda}$ is a closed game representation of $\mathbf{P}$. The map $x \mapsto G^*_{x,\lambda}$ is clearly definable over $L(\mathbb{R})$ (from T and λ). ⊠

The proof of Lemma 1 easily gives the following corollary.

<u>Corollary 1</u>. If G is a $\underset{\sim}{\Pi}^1_1$ game on $\mathbb{R}$ and I has a wqs for G, then I has a wqs for G in $L(\mathbb{R})$. Thus $\mathfrak{D}^{\mathbb{R}}\Pi^1_1 = (\mathfrak{D}^{\mathbb{R}}\Pi^1_1)^{L(\mathbb{R})}$.

Corollary 1 is the best absoluteness result for $L(\mathbb{R})$ of its kind, at least if Π^1_1 - $AD_{\mathbb{R}}$ holds. For in that case, Corollary 3 below implies that there is a Σ^1_1 game on $\mathbb{R}$ for which Player I has a wqs in V, but no wqs in $L(\mathbb{R})$.

In proofs of determinacy from large cardinal hypotheses one associates to a complicated game G a closed game G^* in which one or both players make additional ordinal moves. One then uses measures on the possible additional moves to "integrate" strategies for G^*, and thereby show that whoever wins G^* wins G (and vice-versa, since G^* is determined). We could have proved Lemma 1 this way, but would have required the existence of $\mathbb{R}^{\#}$ in order to do so. The proof of Lemma 1 we did give shows that whoever wins G wins G* (and vice-versa, <u>if G</u> is determined). In the large cardinals/determinacy problem this simpler method for proving the equivalence of G with G* may give some guidance in discovering G*, although integration is unavoidable in actually proving the determinacy of G.

<u>Lemma 2</u>. $\mathfrak{D}^{\mathbb{R}}\Pi^1_1 = \Sigma_1(L(\mathbb{R}),\{\mathbb{R}\})$.

<u>Proof</u>. Since quantification over $\mathbb{R}$ is bounded, $(\mathfrak{D}^{\mathbb{R}}\Pi^1_1)^{L(\mathbb{R})} \subseteq \Sigma_1(L(\mathbb{R}),\{\mathbb{R}\})$. By Corollary 1 then, $\mathfrak{D}^{\mathbb{R}}\Pi^1_1 \subseteq \Sigma_1(L(\mathbb{R}),\{\mathbb{R}\})$. For the other inclusion, let $\varphi(v_0,v_1)$ be a Σ_1 formula. We must associate recursively to each real x a $\Pi^1_1(x)$ game G_x on $\mathbb{R}$ so that $L(\mathbb{R}) \models \varphi[x,\mathbb{R}]$ iff I has a wqs in G_x. Our plan is to force Player I in G_x to describe a countable, wellfounded model of $ZF^- + V = L(\mathbb{R}) + \varphi[x,\mathbb{R}]$ which contains all the reals played in the course of G_x, and minimally so. If $L(\mathbb{R}) \models \varphi[x,\mathbb{R}]$, then I will be able to win G_x by describing an elementary submodel of $L_\alpha(\mathbb{R})$, where α is least so that $L_\alpha(\mathbb{R}) \models ZF^- + \varphi[x,\mathbb{R}]$. On the other hand, if I has a wqs in G_x, we shall be able to piece together the models he describes in different runs of G_x according to his strategy, and thereby produce a model of the form $L_\alpha(\mathbb{R})$ satisfying $\varphi[x,\mathbb{R}]$.

By "$V = L(\mathbb{R})$" we mean some (easily constructed) sentence θ in the language of set theory such that if M is a transitive model of ZF^-, then $M \models \theta$ iff $M = L_\alpha(\mathbb{R}^M)$ for some α.

Player I describes his model in the language $\mathcal{L}$ which has, in addition to ε and $=$, constant symbols $\underline{x}_i$ for $i < \omega$. He uses $\underline{x}_i$ to denote the $i^{\underline{th}}$ real played in the course of G_x. Let us fix recursive maps

$$m,n : \{\theta \mid \theta \text{ is an } \mathcal{L} \text{ formula}\} \to \{2n \mid 1 \leq n < \omega\} \quad .$$

which are one-one, have disjoint recursive ranges, and are such that whenever $\underline{x}_i$ occurs in θ, $i < \min(m(\theta), n(\theta))$. These maps give stages sufficiently late in G_x for I to decide certain statements about his model.

Player I's Description must extend the following $\mathcal{L}$ theory T. The axioms of T include

(1) $\quad ZF^- + DC + V = L(R)$

$(2)_i \quad \underline{x}_i \in R$

(3) $\quad \varphi(\underline{x}_0, R) \wedge \forall\beta\,(L_\beta(R) \not\models ZF^- + \varphi[\underline{x}_0, R])$.

Finally, T has axioms which guarantee that for any model $\mathfrak{A}$ of T, the definable closure of $\{\underline{x}_i^{\mathfrak{A}} \mid i < \omega\}$ is an elementary submodel of $\mathfrak{A}$. Let $\theta_0(v_0, v_1, v_2)$ be a Σ_1 formula such that whenever M is transitive and $M \models ZF^- + V = L(R)$, θ_0^M defines the graph of a map from $OR^M \times R^M$ onto M. It is easy to construct such a formula; cf. Lemma 1.4 of [8]. Now, for any $\mathcal{L}$ formula $\varphi(v)$ of one free variable, T has axioms

$(4)_\varphi \quad \exists v\,\varphi(v) \Rightarrow \exists v\,\exists\alpha\,(\varphi(v) \wedge \theta_0(\alpha, \underline{x}_{m(\varphi)}, v))$,

$(5)_\varphi \quad \exists v\,(\varphi(v) \wedge v \in R) \Rightarrow \varphi(\underline{x}_{n(\varphi)})$.

This completes the list of axioms of T.

A typical run of G_x has the form

I	i_0, x		i_1, x_2		....
II		x_1		x_3	

where for all k, $i_k \in \{0,1\}$ and $x_k \in R$. If $u = \langle\langle i_k, x_{2k}, x_{2k+1}\rangle \mid k<n\rangle$ is a position in G_x, then

$$T^*(u) = \{\theta \mid \theta \text{ is a sentence of } \mathcal{L} \wedge n(\theta) < n \wedge i_{n(\theta)} = 0\} \quad ,$$

and if p is a full run of G_x,

$$T^*(p) = \bigcup_{n<\omega} T^*(p \restriction n) \quad .$$

Now let $p = \langle\langle i_k, x_{2k}, x_{2k+1}\rangle \mid k < \omega\rangle$ be a run of G_x; we say that p is a winning run for I iff

(a) $x_0 = x$,

(b) $T^*(p)$ is a complete, consistent extension T such that for all i,m,n "$\underline{x}_i\ (\underline{n}) = \underline{m}$" $\in T^*(p)$ iff $x_i(n) = m$, and

and

(c) There is no sequence $\langle \theta_i \mid i < \omega \rangle$ of $\mathcal{L}$-formulae of one free variable such that for all i, "$\iota v\theta_{i+1}(v)\ \theta\ \iota v\theta_i(v)$" $\in T^*(p)$.

In condition (c) we have used the "unique v" operator as an abbreviation: $\psi(\iota v\theta(v))$ abbreviates "$\exists v(\psi(v) \wedge \forall u(\theta(u) \Longleftrightarrow u = v))$".

It is clear that G_x is a $\Pi^1_1(x)$ game on $\mathbb{R}$, uniformly in x; its complexity comes from the wellfoundedness condition (c). So we need only prove the following claim.

Claim. $L(\mathbb{R}) \models \varphi[x,\mathbb{R}]$ iff I has a wqs in G_x.

Proof. ($\Rightarrow$) Let α be least so that $L_\alpha(\mathbb{R}) \models ZF^- + \varphi[x,\mathbb{R}]$. We call a position $u = \langle\langle i_k, x_{2k}, x_{2k+1}\rangle \mid k < n\rangle$ honest iff

(i) $n > 0 \Rightarrow x_0 = x$,

(ii) if we let $I_u(\underline{x}_i) = x_i$ for $i < 2n$, then all axioms of $T \cup T^*(u)$ thereby interpreted in $(L_\alpha(\mathbb{R}), I_u)$ are true in the structure.

It is easy to check that if u is an honest position of length n, then $\exists i,x\ \forall y\,(u ^\frown \langle i,x,y\rangle)$ is honest. [If $n = n(\theta)$ for some sentence θ of $\mathcal{L}$, put $i = 0$ iff $(L_\alpha(\mathbb{R}), I_u) \models \theta$. Otherwise let i be random. If $n = m(\varphi)$ for φ on $\mathcal{L}$-formula of one free variable, choose x so that $(L_\alpha(\mathbb{R}), I_{u^\frown\langle i,x,y\rangle}) \models$ axiom $(4)_\varphi$ of T. If $n = n(\varphi)$, choose x for the sake of axiom $(5)_\varphi$, and otherwise let x be random.] Now if p is a run of G_x such that $p \restriction n$ is honest for all n, and $I_p = \bigcup_n I_{p\restriction n}$, then $(L_\alpha(\mathbb{R}), I_p) \models T \cup T^*(p)$. It follows at once that p is a winning run for I. Thus $\{u \mid u$ is honest$\}$ determines a wqs for I in G_x.

($\Leftarrow$) Let S be a wqs for I in G_x. If p is a run of G_x, let $\mathbb{R}_p$ be the set of reals played during p, that is, let $\mathbb{R}_p = \mathrm{range}(I_p)$. If $p \in [S]$ then $T^*(p)$ is consistent, and by axioms 4 of T has, up to isomorphism, a unique model $\mathfrak{m}_p$ such that every element of $\mathfrak{m}_p$ is $\mathcal{L}$-definable over $\mathfrak{m}_p$. By I's payoff condition (c), $\mathfrak{m}_p$ is wellfounded, so we may assume that $\mathfrak{m}_p$ is transitive. By axioms 5 of T and I's payoff condition (b), $\mathbb{R}^{\mathfrak{m}_p} = \mathbb{R}_p$. Thus $\mathfrak{m}_p = (L_{\alpha_p}(\mathbb{R}), I_p)$ for some α_p such that no reals beyond those in $\mathbb{R}_p$ appear in $L_{\alpha_p}(\mathbb{R}_p)$. Notice then that by axiom 3 of T, if $q \in [S]$ and $\mathbb{R}_q = \mathbb{R}_p$, then $\alpha_q = \alpha_p$.

Subclaim. If $p,q \in [S]$ and $I_p(\underline{x}_i) = I_q(\underline{x}_{n_i})$ for $i \leq k$, then for any formula $\theta(v_0,\ldots,v_k)$ of the language of set theory, $\theta(\underline{x}_0,\ldots,\underline{x}_k) \in T^*(p)$ iff $\theta(\underline{x}_{n_0},\ldots,\underline{x}_{n_k})\ \theta\ T^*(q)$.

Proof. Suppose not, and let $u = p \upharpoonright \ell$ and $v = q \upharpoonright \ell$ where ℓ is so large that (say) $\theta(\underline{x}_0,\ldots,\underline{x}_k) \in T^*(p \upharpoonright \ell)$ and $-\theta(\underline{x}_{n_0},\ldots,\underline{x}_{n_k}) \in T^*(q \upharpoonright \ell)$. Since II is free in G_x to play whatever reals he pleases, we can find $s,t \in [S]$ so that $u \subseteq s$, $v \subseteq t$, and $R_s = R_t$. But then $\alpha_s = \alpha_t$. Since $\theta(v_0,\ldots,v_k)$ involves no constant symbols $\underline{x}_i$, and since $I_s(\underline{x}_i) = I_t(\underline{x}_{n_i})$ for $i \leq k$, we have $\mathfrak{M}_s \models \theta(\underline{x}_0,\ldots,\underline{x}_k)$ iff $\mathfrak{M}_t \models \theta(\underline{x}_{n_0},\ldots,\underline{x}_{n_k})$. But $\mathfrak{M}_s \models T^*(u)$ and $\mathfrak{M}_t \models T^*(v)$, a contradiction.

By the subclaim, if $p,q \in [S]$ and $R_p \subseteq R_q$, then there is a unique elementary embedding $j_{p,q} : L_{\alpha_p}(R_p) \to L_{\alpha_q}(R_q)$. The uniqueness of the embeddings implies that if $p,q,s \in [S]$ and $R_p \subseteq R_q \subseteq R_s$, then $j_{p,s} = j_{q,s} \circ j_{p,q}$. Thus we can form the direct limit M of the $L_{\alpha_p}(R_p)$'s under these embeddings. Now whenever $\{p_i \mid i < \omega\} \subseteq [S]$ we can find a $q \in [S]$ so that $R_{p_i} \subseteq R_q$ for all i, and therefore M is wellfounded. So we may assume M is transitive. Since $R \subseteq \bigcup\{R_p \mid p \in [S]\}$, $R \subseteq M$, and so $M = L_\alpha(R)$ for some α. But $M \models \varphi[x,R]$.

This completes the proof of the claim, and thereby Lemma 2. ⊠

Our proof of Lemma 2 sharpens some earlier arguments of Solovay [7] which implied, assuming AD_R, that every set of reals in $L(R)$ is $\mathfrak{D}^R \underset{\sim}{\Delta}^1_3$.

Theorem 1. Assume all games in $L(R)$ are determined. Then the pointclasses $\Sigma_1(L(R),\{R\})$ and $\Sigma_1(L(R),R \cup \{R\})$ have the scale property.

Proof. It is enough to show $\Sigma_1(L(R),\{R\})$ has the scale property. Let P be $\Sigma_1(L(R),\{R\})$. Lemmas 1 and 2 yield a map $(x,\alpha) \mapsto G^*_{x;\alpha}$ which is $\Delta_1(L(R), \{R\})$ such that

$$P(x) \text{ iff } \exists\alpha\,(\text{I has a wqs in } G^*_{x,\alpha}) .$$

Let $P^\alpha(x)$ iff I has a wqs in $G^*_{x,\alpha}$. Our slight generalization of [6], and the uniformity in its proof, yield a $\Delta_1(L(R),\{R\})$ map $\alpha \mapsto \vec{\psi}^\alpha$ such that $\vec{\psi}^\alpha$ is a scale on P^α for all α. Now for $x \in P$ let

$$\varphi_0(x) = \text{least } \alpha \text{ such that } P^\alpha(x) ,$$

and

$$\varphi_{i+1}(x) = \langle \varphi_0(x), \psi_i^{\varphi_0(x)}(x) \rangle ,$$

where of course we use the lexicographic order to identify $\varphi_{i+1}(x)$ with an ordinal. It is easy to check that $\vec{\varphi}$ is a $\Sigma_1(L(R),\{R\})$ scale on P. ⊠

Combining Theorem 1 with the result of Kechris and Solovay we mentioned earlier, we have

Corollary 2. Assume all games in $L(R)$ are determined. Then P admits a scale in $L(R)$ iff P is $\Sigma_1(L(R),\{R\})$.

<u>Proof</u>. One direction is Theorem 1; we prove the other for the sake of completeness. For $x,y \in \mathbb{R}$, let

$$C(x,y) \text{ iff } L(\mathbb{R}) \models y \text{ is ordinal definable from } x \ .$$

By the reflection theorem, C is $\Sigma_1(L(\mathbb{R}),\{\mathbb{R}\})$. Now $\sim C$ has no uniformization, hence no scale, in $L(\mathbb{R})$. [If D uniformizes $\sim C$ and $D \in L(\mathbb{R})$, then we can find $x_0 \in \mathbb{R}$ so that $L(\mathbb{R}) \models D$ is ordinal definable from x_0. Then $D(x_0,y_0) \Rightarrow C(x_0,y_0)$, a contradiction since $\exists y \sim C(x_0,y)$.] Now the class Γ of sets admitting scales in $L(\mathbb{R})$ is a boldface pointclass. Since $\sim C \notin \Gamma$ and C is $\Sigma_1(L(\mathbb{R}),\mathbb{R} \cup \{\mathbb{R}\})$, Wadge's lemma implies that $\Gamma \subseteq \Sigma_1(L(\mathbb{R}),\mathbb{R} \cup \{\mathbb{R}\})$. ⊠

As promised, our proof of Theorem 1 gives some information about Π^1_1 games on $\mathbb{R}$.

<u>Corollary 3</u>. If Π^1_1-$AD_{\mathbb{R}}$ holds, then $\Game^{\mathbb{R}}\Sigma^1_1 \neq (\Game^{\mathbb{R}}\Sigma^1_1)^{L(\mathbb{R})}$. Thus Π^1_1-$AD_{\mathbb{R}}$ is false in $L(\mathbb{R})$.

<u>Proof</u>. If Π^1_1-$AD_{\mathbb{R}}$ holds, then $\Game^{\mathbb{R}}\Sigma^1_1$ is the dual of $\Game^{\mathbb{R}}\Pi^1_1$, so that $\Game^{\mathbb{R}}\Sigma^1_1 = \Pi_1(L(\mathbb{R}),\{\mathbb{R}\})$. But $(\Game^{\mathbb{R}}\Sigma^1_1)^{L(\mathbb{R})} \subseteq \Sigma_1(L(\mathbb{R}),\{\mathbb{R}\})$ by a direct computation. ⊠

<u>Corollary 4</u>. Assume AD. Then the following are equivalent:

(a) Π^1_1-$AD_{\mathbb{R}}$,

(b) $\mathbb{R}^{\#}$ exists.

<u>Proof</u>. If $\mathbb{R}^{\#}$ exists, then Π^1_1-$AD_{\mathbb{R}}$ follows by the argument of Martin [2]. (AD is not necessary here.) If Π^1_1-$AD_{\mathbb{R}}$ holds, then $\exists A \subseteq \mathbb{R}(A \notin L(\mathbb{R}))$ by Corollary 3. But in the presence of AD this last statement implies that $\mathbb{R}^{\#}$ exists; cf. [9], ⊠

It seems likely that Π^1_1-$AD_{\mathbb{R}}$ and the existence of $\mathbb{R}^{\#}$ are provably equivalent in ZF + DC. We have not tried to show this.

We have seen that the complete $\Pi_1(L(\mathbb{R}),\{\mathbb{R}\})$ set admits no scale in $L(\mathbb{R})$. On the other hand, this set admits a scale just beyond $L(\mathbb{R})$: Solovay has shown that if $\mathbb{R}^{\#}$ exists and $L(\mathbb{R}) \models AD$, then every set of reals in $L(\mathbb{R})$ admits a scale each of whose norms is in $L(\mathbb{R})$. (By Theorem 1.3.3 of [9], one needs $\mathbb{R}^{\#}$ in order to construct definable scales beyond those in $L(\mathbb{R})$.) This is a special case of a more general result of Martin [3]: if Γ is a reasonably closed pointclass with the scale property, Γ-$AD_{\mathbb{R}}$ holds, and $\Game^{\mathbb{R}}\Gamma$ games on ω are determined, then $\Game^{\mathbb{R}}\Gamma$ has the scale property. (Solovay's result follows by taking Γ to be the class of sets which are ωn-$\underset{\sim}{\Pi}^1_1$ for some $n < \omega$.) We shall conclude by proving part of Martin's theorem in a way which avoids the determinacy of games on $\mathbb{R}$. Our proof generalizes the proof of Lemma 1.

<u>Theorem 2</u>. Assume AD. Then whenever A admits a scale, $\mathfrak{D}^{R}A$ admits a scale.

<u>Proof</u>. Let $\overline{\varphi}$ be a scale on A whose norms map into κ, where $\kappa < \Theta$. Let R be the tree of $\overline{\varphi}$, that is

$$R = \{\langle s,t,u\rangle \in \omega^{<\omega} \times \omega^{<\omega} \times \kappa^{<\omega} \mid$$

$$\ell h(s) = \ell h(t) = \ell h(u) \wedge \exists x,y(A(x,y) \wedge \forall i < \ell h(u)\,(u(i) = \varphi_i(x,y)))\} \quad .$$

Thus $A = \{(x,y) \mid [R(x,y)] \neq \emptyset\}$.

We now obtain a "homogeneous" tree representation of $-A$ in the standard way (cf. [1]). For any $\langle s,t\rangle \in \omega^{<\omega} \times \omega^{<\omega}$ such that $\ell(s) = \ell(t)$, let $W(s,t)$ be the Brouwer-Kleene order of $\kappa^{<\omega}$ restricted to

$$\mathrm{fld}(W(s,t)) = \{u \in \kappa^{<\omega} \mid \ell(u) \leq \ell(s) \wedge (s \restriction \ell(u), t \restriction \ell(u), u) \in R\} \quad .$$

$W(s,t)$ is a wellorder because the se uences in its field have length $\leq \ell(s)$. By AD we have a $\lambda > \kappa$ such that $\lambda \rightarrow (\lambda)^{\lambda}$. For any $C \subseteq \lambda$, let $[C]^{W(s,t)}$ be the set of order-preserving maps from $\mathrm{fld}(W(s,t))$ into C. Now let T be the tree on $\omega \times \omega \times \bigcup_{\langle s,t\rangle} [\lambda]^{W(s,t)}$ given by

$$T = \{\langle s,t,\langle f_1,\ldots,f_\ell\rangle\rangle \mid \ell(s) = \ell(t) = \ell \wedge \forall i \leq \ell(f_i \in [\lambda]^{W(s \restriction i, t \restriction i)})$$

$$\wedge \forall i,j \leq \ell(i \leq j \Rightarrow f_i = f_j \restriction W(s \restriction i, t \restriction i))\} \quad .$$

A path through T is equivalent to a pair (x,y) of reals together with a map of $R(x,y)$ into λ preserving the Brouwer-Kleene order on $R(x,y)$. Thus

$$A = \{(x,y) \mid [T(x,y)] = \emptyset\} \quad .$$

For any $(s,t) \in \omega^{<\omega} \times \omega^{<\omega}$ such that $\ell(s) = \ell(t)$, let

$$T(s,t) = \{u \mid (s,t,u) \in T\} \quad .$$

Each $T(s,t)$ carries a canonical measure $\mu_{s,t}$ given by the strong partition property of λ. [Fix $\alpha \leq \lambda$. For $C \leq \lambda$, let

$$C^{\natural} = \{h \in [\lambda]^{\alpha} \mid \exists g \in [C]^{\omega\alpha}\, \forall \beta < \alpha\ (h(\beta) = \sup_n g(\omega\beta + n))\}$$

Then since $\lambda \rightarrow (\lambda)^{\lambda}$, the sets of the form $C^{\natural}$ for C closed and unbounded in λ are the base for an ultrafilter μ_α on $[\lambda]^{\alpha}$. If α is the order type of $W(s,t)$, then μ_α is isomorphic in an obvious way to an ultrafilter $\mu_{s,t}$ on $[\lambda]^{W(s,t)}$. But an inspection of the definition of T shows that we may identify $T(s,t)$ with $[\lambda]^{W(s,t)}$. (See [1] for further details on this construction.)]. The measures $\mu_{s,t}$ are compatible, in the sense that if $\mu_{s,t}(X) = 1$, and $i < \ell h(s)$, then

$\mu_{s \restriction i, t \restriction i}(\{u \restriction i \mid u \in X\}) = 1$. They also have a "homogeneity" property: if $[T(x,y)] \neq \emptyset$ and $\mu_{x \restriction i, y \restriction i}(Y_i) = 1$ for all $i < \omega$, then $\exists f \in [T(x,y)]\ \forall i (f \restriction i \in Y_i)$. We shall now produce a closed game representation of $\mathfrak{D}^{\mathbb{R}}A$. For $x \in \mathbb{R}$ and $\alpha \in OR$, we define a closed game $G^*_{x,\alpha}$. A typical run of $G^*_{x,\alpha}$ has the form

I	x_0, β_0		x_2, β_1		...
II		x_1		x_3	

where $x_i \in \mathbb{R}$ and $\beta_i < \alpha$ for all i. Player I wins this run of $G^*_{x,\alpha}$ just in case, if we set $y = \langle x_n \mid n < \omega \rangle^*$, and for all $i < \omega$ pick maps $F_i : T(x \restriction i, y \restriction i) \to OR$ such that $\beta_i = [F_i]_{\mu_{x \restriction i, y \restriction i}}$, then whenever $i < k$,

$$\mu_{x \restriction k, y \restriction k}(\{u \mid F_k(u) < F_i(u \restriction i)\}) = 1 \ .$$

The compatibility of the measures implies that I's payoff condition does not depend on which map F_i is chosen to represent β_i in the ultrapower by $\mu_{x \restriction i, y \restriction i}$.

<u>Claim</u>. $x \in \mathfrak{D}^{\mathbb{R}}A$ iff $\exists \alpha$ (I has a wqs in $G^*_{x,\alpha}$).

<u>Proof</u>. Let S^* be a wqs for I in $G^*_{x,\alpha}$. Let S result from S^* by omitting the ordinals from I's moves. Let $\langle x_n \mid n < \omega \rangle \in [S]$. Then the homogeneity property of the $\mu_{s,t}$'s guarantees that $[T(x, \langle x_n \mid n < \omega \rangle^*)] = \emptyset$. Thus $(x, \langle x_n \mid n < \omega \rangle^*) \in A$, and S witnesses that $x \in \mathfrak{D}^{\mathbb{R}}A$.

Conversely, let S be a wqs for I in the game whose payoff is $\{\langle x_n \mid n < \omega \mid A(x, \langle x_n \mid n < \omega \rangle^*)\}$. Let U be the tree on $\mathbb{R} \times \bigcup_{\langle s,t \rangle} [\lambda]^{W(s,t)}$ given by

$$U = \{\langle u,v \rangle \mid \ell h(u) = \ell h(v) \wedge u \in S \wedge \langle x \restriction \ell h(u), u^*, v \rangle \in T\} \ .$$

Then U is wellfounded since S is a wqs for I. For $u \in S$, let $F_u : T(x \restriction \ell h(u), u^*) \to OR$ be defined by

$$F_u(v) = \text{rank of } \langle u,v \rangle \text{ in } U$$

We can now define a wqs S^* for I as follows: let

$$S^*_0 = \{\langle x_0, \beta_0, x_1, \ldots, x_{2k}, \beta_k, x_{2k+1} \rangle \mid \langle x_0, \ldots, x_{2k+1} \rangle \in S \wedge \forall i \leq k\ (\beta_i = [F_i]_\mu,$$

$$\text{where } F_i = F_{\langle x_n \mid n < i \rangle} \text{ and } \mu = \mu_{x \restriction i, \langle x_n \mid n < i \rangle^*})\} \ ,$$

and let S^* be the closure of S^*_0 under initial segment. It is easy to check that S^* is a wqs for I in $G^*_{x,\alpha}$ for all sufficiently large α. This completes the proof of the claim.

Since a wqs for I in $G^*_{x,\alpha}$ is also a wqs for I in $G^*_{x,\beta}$ for all $\beta \geq \alpha$, we can find $\lambda \in \mathbf{OR}$ such that $\forall x (x \in \supset^{\mathbb{R}} A$ iff I has a wqs in $G^*_{x,\lambda})$. This $x \to G^*_{x,\lambda}$ is a closed game representation of $\supset^{\mathbb{R}} A$, and $\supset^{\mathbb{R}} A$ admits a scale. □

REFERENCES

[1] A.S. Kechris, Homogeneous trees and projective scales, in Cabal Seminar 77-79, Lecture Notes in Mathematics #839, Springer 1981, 33-74.

[2] D.A. Martin, Measurable cardinals and analytic games, Fund. Math. LXVI (1970), 287-291.

[3] D.A. Martin, The Real game quantifier propagates scales, this volume.

[4] Y.N. Moschovakis, Descriptive Set Theory, North Holland 1980.

[5] Y.N. Moschovakis, Ordinal games and playful models, in Cabal Seminar 77-79, Lecture Notes in Mathematics #839, Springer 1981, 169-202.

[6] Y.N. Moschovakis, Scales on coinductive sets, this volume.

[7] R.M. Solovay, The Independence of DC from AD, in Cabal Seminar 76-77, Lecture Notes in Mathematics #689, Springer 1978, 171-184.

[8] J.R. Steel, Scales in L(R), this volume.

[9] J.R. Steel and R. Van Wesep, Two consequences of determinacy consistent with choice, Trans. Amer. Math. Soc. 272 (1982), 67-85.

THE LARGEST COUNTABLE THIS, THAT, AND THE OTHER

Donald A. Martin*
Department of Mathematics
University of California
Los Angeles, California 90024

§1. Introduction. ${}^{\omega}\omega \cap L$ has, if it is countable, an implicit descriptive set-theoretic characterization as the largest countable Σ^1_2 set [Mo 1980, p. 538]. In this paper we give an explicit descriptive set-theoretic characterization of this and (under determinacy hypotheses) other related sets.

A subset A of ${}^{k}\omega \times ({}^{\omega}\omega)^m$ is α - $\mathbf{\Pi}^1_1$, where α is an ordinal, if there are $\mathbf{\Pi}^1_1$ sets A_β, $\beta < \alpha$ such that

$$x \in A \Longleftrightarrow \text{the least } \beta \text{ such that } \beta = \alpha \text{ or } x \notin A_\beta \text{ is odd} .$$

If α is small, say recursive, the lightface notion, $\alpha - \Pi^1_1$, has an obvious definition.

A is $\Game\Gamma$ if there is a $B \in \Gamma$ such that

$$x \in A \Longleftrightarrow \text{I has a winning strategy for } G(B) ,$$

where $G(B)$ is the game in which the players cooperatively produce a $y \in {}^{\omega}\omega$ and I wins just in case $(x,y) \in B$.

Let C^{2n} be the largest countable Σ^1_{2n} subset of ${}^{\omega}\omega$, for $n \geq 1$. See [Mo 1980, p. 346]. In §2 and §3 we prove the following theorem:

Theorem (Projective determinacy (PD)). $C^{2k} = \bigcup_n \Game^{2k-1}\, \omega \cdot n - \Pi^1_1$.

Here $\Game^{i+1}\Gamma = \Game\Game^{i}\Gamma$.

Corollary ($0^{\#}$ exists). $L = \bigcup_n \Game\, \omega \cdot n - \Pi^1_1$.

We also use our proof to characterize $0^{\#}$ as the recursive join of a sequence of complete $\Game\, \omega \cdot n - \Pi^1_1$ sets.

Let Σ^*_0 = the class of unions of inductive (see [Mo 1980, p. 410]) and coinductive sets. (We take "inductive" in the lightface sense.) Let Π^*_n be the class of complements of Σ^*_n sets and let Σ^*_{n+1} be the class of projections of Π^*_n sets. In §4 we prove

*Research partially supported by NSF grant

<u>Theorem</u> (Determinacy Σ_n^*, all n). The largest countable inductive set is $\bigcup_n \Sigma_n^*$.

Moschovakis [Mo 1983] shows that every coinductive set admits a scale whose complexity is the join of the Σ_n^*. It follows from the theorem that there is a coinductive set which has no Σ_n^* scale for any n.

§2. <u>Every $\mathfrak{D}^{2k-1}\ \omega \cdot n - \Pi_1^1$ set belongs to</u> C^{2k}. We first fix a coding of $\omega \cdot n - \Pi_1^1$ sets. Let each $z \in {}^{\omega}\omega$ code, in some effective manner, a sequence $\langle A_\beta^z : \beta < \omega^2 \rangle$ of $\underset{\sim}{\Pi}_1^1$ sets. For each z let $A^{z,n}$ be the $\omega \cdot n - \underset{\sim}{\Pi}_1^1$ set determined by the sequence $A_\beta^z : \beta < \omega \cdot n$. For each z, let G_z^n be the game with I's winning set $A^{z,n}$. Do this coding so that $\{z : G_z^n$ is a win for I$\}$ is a complete $\mathfrak{D}\ \omega \cdot n - \Pi_1^1$ set, uniformly in n.

<u>Lemma 2.1</u>. For each z and each sequence $\gamma_1,\ldots,\gamma_n$ of ordinals, there is an open game $G^*_{z,\gamma_1,\ldots,\gamma_n}$, played on ordinals $< \sup\{\gamma_i : i = 1,\ldots,n\}$ such that

(1) The winning conditions for $G^*_{z,\gamma_1,\ldots,\gamma_n}$ are definable in L[z] from z and $\gamma_1,\ldots,\gamma_n$ by a definition independent of $z,\gamma_1,\ldots,\gamma_n$.

(2) If $\gamma_1 < \cdots < \gamma_n$ are indiscernibles for L[z], then whomever has a winning strategy for $G^*_{z,\gamma_1,\ldots,\gamma_n}$ has a winning strategy for G_z^n.

(3) There is a Π_1^1 relation R(n,z,y) such that, if $y_1,\ldots,y_n$ are codes for countable ordinals $\gamma_1,\ldots,\gamma_n$ respectively, then

$$\text{I wins } G^*_{z,\gamma_1,\ldots,\gamma_n} \iff R(n,z,\langle y_1,\ldots,y_n\rangle)\ .$$

<u>Proof</u>. For simplicity, we suppress z and often n from the notation. Let A_β, $\beta < \omega \cdot n$, be the $\underset{\sim}{\Pi}_1^1$ sets given by z. We can associate with each $x \in {}^{\omega}\omega$ and each $\beta < \omega \cdot n$ a linear ordering R_x^β of ω in an effective manner, such that $R_x^\beta \restriction k$ depends only on $x \restriction k$, such that 0 is maximum in R_x^β, and such that $x \in A_\beta \iff R_x^\beta$ is a well-ordering.

Let $h : \omega \to \omega \cdot n \times \omega$ be an effective bijection such that

(a) if $h(k) = (\beta,t)$, then β even $\iff$ k even;

(b) if $h(k_1) = (\beta,t_1)$, $h(k_2) = (\beta,t_2)$, and $t_1 < t_2$, then $k_1 \le k_2$;

(c) if $h(k_1) = (\omega \cdot i + j_1,0)$, $h(k_2) = (\omega \cdot i + j_2,0)$, and $j_1 < j_2$, then $k_1 < k_2$.

We play $G^*_{\gamma_1,\ldots,\gamma_n}$ as follows:

I	II
m_0, ρ_0	
	m_1, ρ_1
m_2, ρ_2	
	m_3, ρ_3
	$\vdots$

The rules of the game are:

(i) $m_i \in \omega$.

(ii) If $h(k) = (\omega \cdot i + j, t)$, then $\rho_k < \gamma_{t+1}$.

(iii) Let $F_\beta(t) = \rho_k$, where $h(k) = (\beta, t)$. F_β must embed the ordering R_x^β into the ordinals, where $x = \langle m_0, m_1, \ldots \rangle$.

The first player to violate the rules loses. Otherwise II wins.

The players are thus trying to verify that the R_β^x are well-orderings. I is responsible for even β, II for odd β. (b) ensures that the maximum value of F_β is played before any other values, and that $R_\beta^x \restriction t + 1$ is known before $F_\beta(t)$ is played. (c) ensures that $F_\beta(0)$ is played before $F_\delta(0)$ whenever $\delta > \beta$ and β and δ are in the same ω-block.

Condition (1) in the statement of the lemma is clear. Condition (3) is also easily verified, since given codes for $\gamma_1, \ldots, \gamma_n$ (i.e., well-orderings of those order types), we can construe $G^*_{\gamma_1, \ldots, \gamma_n}$ as an ordinary open game. Since $\Game\Sigma_1^0 = \Pi_1^1$, this yields (3).

By (1), it suffices to verify (2) in the special case $\gamma_i = \omega_i$, where we may assume $z^\#$ exists.

Assume for definiteness that I has a winning strategy σ^* for $G^* = G^*_{\omega_1, \ldots, \omega_n}$.

Let us say that II plays well if II obeys all rules, $F_{\omega \cdot i+2j+1} > \gamma_i$ for all $i \geq 1$ and $j \in \omega$, $F_{\omega \cdot i+2j+1}(t) > F_{\omega \cdot i+j'}(0)$ for all $j' \leq 2j$, and $F_{\omega \cdot i+2j+1}(t)$ is an indiscernible for $L[z]$ for all $i \in \omega$ and $j \in \omega$.

We define a strategy σ for I for G_z^n by assuming II plays well and I plays σ^*.

Note that, in positions in G^* such that II has played well, I's moves m_i and $F_\beta(t)$ given by σ^* are independent of II's moves $F_{\beta'}(t)$ for $\beta' > \beta$.

Let x be any play of G ($= G_z^n$) consistent with σ. Let β_0 be the least β such that $\beta = \alpha$ or $x \notin A_\beta$. By induction we define F_β for $\beta < \beta_0$. Assume $F_{\beta'}$ is defined for $\beta' < \beta$.

Suppose β is even. Let F_β be given by letting x be played, letting II play $F_{\beta'}$ for odd $\beta' < \beta$, letting II play well otherwise, and letting I according to σ^*.

Suppose β is odd. Let $\eta = \sup_{\beta'<\beta} F_{\beta'}(0)$. Let $F_\beta : \omega \to \omega_{i+1}$ embed R_x^β into the indiscernibles for $L[z]$ between η and γ_{i+1}, where $\beta = \omega \cdot i + j$.

Suppose $\beta_0 = \alpha$. Then the F_β extend x to a play of G^* according to σ^* with all rules obeyed. Since this play is a win for II, we have a contradiction.

Suppose β_0 is even. Let F_{β_0} be given by letting x be played, letting II play F_β for odd $\beta < \beta_0$, letting II play well otherwise, and letting I play σ^*. Eventually F_{β_0} must violate rule (iii). Extending to a position where II has played well, we get a contradiction.

Thus β_0 is odd, and so x is a win for I. Hence x is a winning strategy for I for G. ⊣

<u>Lemma 2.2</u> ($0^\#$ exists). If S is a $\Game\,\omega \cdot n - \Pi_1^1$ set of integers, then $S \in L = C^2$.

<u>Proof</u>. There is a recursive $f : \omega \to {}^\omega\omega$ such that

$$m \in S \iff \text{I wins } G^n_{f(m)} .$$

Let $B = \{S' : \exists \gamma_1,\dots,\gamma_n < \omega_1 \, \forall m \, (\text{I wins } G^*_{f(m),\gamma_1,\dots,\gamma_n} \iff m \in S')$.

By part (2) of Lemma 1, $S \in B$.

It is clear that $S' \in B \Rightarrow S' \in L$, but we prove directly that B is a countable Σ_2^1 set. By part (3) of Lemma 2.1,

$$S' \in B \iff \exists y_1,\dots,y_n \, \forall m$$
$$(y_1,\dots,y_m \text{ are codes for countable ordinals } \&$$
$$(R(n,f(m), y_1,\dots,y_n) \iff m \in S')) .$$

Thus $B \in \Sigma_2^1$. Since B has a (Σ_2^1) well-ordering, B is countable.

<u>Lemma 2.3</u> (PD). For $i \in \omega$, every $\Game^{2i+1}\omega \cdot n - \Pi_1^1$ set $S \subseteq \omega$ belongs to C^{2i+2}.

<u>Proof</u>. There is a recursive function f such that $m \in S$ if and only if I wins the following game $G(m)$ of length $\omega \cdot 2i$:

$$x_1,\dots,x_{2i} \text{ is a win for I just in case}$$
$$\text{I has a winning strategy for } G^n_{f(m,x_1,\dots,x_{2i})} .$$

Given degrees of unsolvability $d_1,\dots,d_{2i}$, consider the game $G(m,d_1,\dots,d_{2i})$:

I	II
σ_1	
	x_1
σ_2	
	x_2
$\vdots$	
σ_{2i}	
	x_{2i}

σ_j must be a strategy for I for an ordinary game on ω and must satisfy degree$(\sigma_j) \leq d_j$. x_j must be a play consistent with σ_j such that degree$(x_j) \leq d_j$. I wins $\Longleftrightarrow x_1,\ldots,x_{2i}$ is a win for I in $G^n_{f(m,x_1,\ldots,x_{2i})}$.

Sublemma 2.4. For almost every $d_1,\ldots,d_{2i}$ (with respect to the iterated product of the usual measure on the degrees),

$$\forall m\ (\text{I wins } G(m) \Longleftrightarrow \text{I wins } G(m,d_1,\ldots,d_{2i}))\ .$$

Proof of Sublemma. Suppose I has a winning strategy σ for $G(m)$. Let d_1 be any degree $\geq$ the degree of I's strategy restricted to positions of finite length. Assume $d_1,\ldots,d_j$ are chosen. Let d_{j+1} be any degree such that for all $x_1,\ldots,x_j$ consistent with σ such that degree$(x_j) \leq d_j$, $d_{j+1} \geq$ the degree of $\sigma \restriction$ the next ω moves, in the position $x_1,\ldots,x_j$. Clearly σ gives a winning strategy for $G(m,d_1,\ldots,d_{2i})$. Since the set of $(d_1,\ldots,d_{2i})$ allowed had measure one, we have proved the first half of the sublemma.

Let τ be a winning strategy for II for $G(m)$. Assume $d_1,\ldots,d_j$ are chosen for $0 \leq j < 2i$. Let d_{j+1} be such that, for any $(x_1,\ldots,x_j)$ consistent with τ such that degree$(x_{j'}) \leq d_{j'}$ for all $j' \leq j$, $d_{j+1} \geq$ the degree of τ for the next ω moves starting at $(x_1,\ldots,x_j)$. If degree$(x_{j'}) \leq d_{j'}$ for all $j' \leq j$ and I plays σ_{j+1} with degree $\sigma_{j+1} \leq d_{j+1}$, then the x_{j+1} given by τ has degree $\leq d_{j+1}$. Hence II wins $G(m,d_1,\ldots,d_{2i})$.

For each $d_1,\ldots,d_{2i}$, and each $\gamma_1,\ldots,\gamma_n < \omega_1$, let $G(m,d_1,\ldots,d_{2i},\gamma_1,\ldots,\gamma_n)$ be the game played like $G(m,d_1,\ldots,d_{2i})$ which I wins just in case I wins $G^*_{f(m,x_1,\ldots,x_{2i}),\gamma_1,\ldots,\gamma_n}$.

Let $B = \{S' :$ for almost all $d_1,\ldots,d_{2i}$ there exist $\gamma_1,\ldots,\gamma_n < \omega_1$, such that $(\forall m)(\text{I wins } G(m,d_1,\ldots,d_{2i},\gamma_1,\ldots,\gamma_n) \Longleftrightarrow m \in S')$.

We first note that B is (projectively) well-orderable. For $S' \in B$, let

$$\varphi_{S'}(d_1,\ldots,d_{2i}) = \text{the lexicographically least } (\gamma_1,\ldots,\gamma_n) \text{ such that}$$
$$(\forall m)(\text{I wins } G(m,d_1,\ldots,d_{2i},\gamma_1,\ldots,\gamma_n) \Longleftrightarrow m \in S') .$$

if it is defined. It is defined for almost all $d_1,\ldots,d_{2i}$. If $\varphi_{S_1'}(d_1,\ldots,d_{2i}) = \varphi_{S_2'}(d_1,\ldots,d_{2i})$ almost everywhere, then $S_1' = S_2'$.

Next we show that B is Σ^1_{2i+1}.

$$S' \in B \Longleftrightarrow \exists d_1 \, \forall d_1' \geq d_1 \, \forall d_2 \, \exists d_2' \geq d_2$$
$$\ldots \forall d_{2i} \, \exists d_{2i}' \geq d_{2i} \, \exists y_1,\ldots,y_m$$
$$(y_1,\ldots,y_n \text{ are codes for countable ordinals } \&$$
$$\forall m \, (\exists \sigma_1 \text{ of degree } \leq d_1' \, \forall x_1 \text{ of degree } \leq d_1'$$
$$\ldots \exists \sigma_{2i} \text{ of degree } \leq d_{2i}' \, \forall x_{2i} \text{ of degree } \leq d_{2i}'$$
$$(R(n,f(m,x_1,\ldots,x_{2i}), y_1,\ldots,y_n) \Longleftrightarrow m \in S'))) .$$

Finally we show that $S \in B$. Let $d_1,\ldots,d_{2i}$ be such that $\forall m$ (I wins $G(m)$ I wins $G(m,d_1,\ldots,d_{2i})$). By the sublemma, almost all $(d_1,\ldots,d_{2i})$ have this property. Now let $\gamma_1 < \cdots < \gamma_n$ be countable indiscernibles for $L[w]$ with degree$(w) \geq d_j$, $j = 1,\ldots,2i$. By Lemma 1.1, if $x_1,\ldots,x_{2i}$ is the result of a play of $G(m,d_1,\ldots,d_{2i})$, then

$$x_1,\ldots,x_{2i} \text{ is a win for I in } G(m,d_1,\ldots,d_{2i})$$
$$\Longleftrightarrow x_1,\ldots,x_{2i} \text{ is a win for I in } G(m,d_1,\ldots,d_{2i},\gamma_1,\ldots,\gamma_n) .$$

Now $\forall m \, (m \in S \Longleftrightarrow \text{I wins } G(m) \Longleftrightarrow \text{I wins } G(m,d_1,\ldots,d_{2i},\gamma_1,\ldots,\gamma_n))$.

§3. Characterization of C^{2i} and of $0^\#$

Lemma 3.1 (Moschovakis) (Sufficient determinacy). Suppose every set in Γ admits a very good scale $\langle \varphi_i : i \in \omega \rangle$ such that each φ_i is a Γ_i norm.

(a) Every set in $\partial\,\Gamma$ admits a very good scale $\langle \psi_i : i \in \omega \rangle$ such that each ψ_i is a $\partial\,\Gamma_i$ norm.

(b) Every Γ game won by I has a winning strategy σ such that σ restricted to positions of length i belongs to $\partial\,\Gamma_i$.

An inspection of the proof of Theorems 6E.1 and 6E.15 of [Mo 1980] will reveal that they prove the lemma.

Lemma 3.2 (Steel [St 1983]). Every Σ^1_1 set admits a very good scale $\langle \varphi_i : i \in \omega \rangle$ such that each φ_i is $\omega \cdot (i + 1) - \Pi^1_1$.

Lemma 3.3 (PD). Every Π^1_{2i+1} [Σ^1_{2i+2}] game won by II has a winning strategy τ such that τ restricted to positions of length n is $\partial^{2i+1}\ \omega(n+1) - \Pi^1_1$ [$\partial^{2i+2}\ \omega(n+1) - \Pi^1_1$].

Proof. The preceding two lemmas allow one to prove this lemma by induction.

Lemma 3.4. For each $i \geq 1$, every member of C^{2i} is $\partial^{2i-1}\ \omega \cdot n - \Pi^1_1$ for some n.

Proof. Let A be any countable Σ^1_{2i} subset of ${}^{\omega}2$. Let $A = \{x : \exists y \langle x,y\rangle \in B\}$ for $B \in \Pi^1_{2i-1}$.

Consider the following game:

$$\begin{array}{cc} \text{I} & \text{II} \\ m_i, s_0 & \\ & \varepsilon_0 \\ m_i, s_1 & \\ & \varepsilon_1 \\ & \vdots \end{array}$$

Each m_j must be a natural number. Each s_j must be an element of $2^{<\omega}$ with $s_{j+1} \supseteq s_j{}^\frown \varepsilon_j$. ε_j must be 0 or 1. If all rules are obeyed, I wins just in case $\langle x,y\rangle \in B$, where $x = \bigcup_j s_j$ and $y = \langle m_i : i \in \omega\rangle$.

If I has a winning strategy, then A has a perfect subset, contradicting its countability.

By Lemma 3.3, let τ be a winning strategy for II such that τ restricted to positions of length n is $\partial^{2n-1}\ \omega \cdot (n+1) - \Pi^1_1$.

Suppose $x,y \in B$. Then there is a position p in our game with last move ε_{j-1} (we permit p to be the initial position, i.e., $j = 0$) such that

(a) $m_{j'} = y(j')$ for all $j' < j$;

(b) $s_{j-1}{}^\frown \varepsilon_{j-1} \subseteq x$ (if $j > 0$);

(c) for any s_j such that $s_{j-1}{}^\frown \varepsilon_{j-1} \subseteq s_j \subseteq x$, if I plays $(y(j), s_j)$ at p, then τ calls for II to play $\varepsilon_j = 1 - x(\text{length } s_j)$.

If no such position existed, then there would be a play of the game consistent with τ with result $\langle x,y\rangle$.

For such a position p, x is clearly recursive in τ restricted to positions of length $2j+1$. Thus every $x \in A$ belongs to $\partial^{2i-1}\ \omega \cdot n - \Pi^1_1$ for some n.

Theorem 3.5 (PD). $C^{2i} = \bigcup_n \partial^{2i-1}\ \omega \cdot n - \Pi^1_1$.

Corollary 3.6 ($0^\#$ exists). $L = \bigcup_n \partial\ \omega \cdot n - \Pi^1_1$.

Theorem 3.7 ($0^\#$ exists). Let B_n be a complete $\mathfrak{D}\,\omega\cdot n$ - Π^1_1 subset of ω, uniformly in n. $0^\#$ is recursively isomorphic with $\{\langle n,m\rangle : m \in B_n\} = B$.

Proof. There is a one-one recursive function f such that $\langle n,m\rangle \in B \Leftrightarrow$ I has a winning strategy for $G^*_{f(n,m),\gamma_1,\ldots,\gamma_n}$, for any indiscernibles $\gamma_1 < \cdots < \gamma_n$ for L. This follows from Lemma 2.1, part (2). By Lemma 2.1, part (1), B is one-one reducible to $0^\#$.

For the other direction, let $\varphi(x_1,\ldots,x_n)$ be any formula in the language of set theory. We play a game G_φ as follows: I chooses (dividing up ω) $x_\beta \in {}^\omega\omega$ for each $\beta < \omega\cdot(n+1)$. II chooses $y_\beta \in {}^\omega\omega$ for each $\beta < \omega(n+1)$. If some x_β or y_β is not a code for a countable ordinal, then I wins if, for the least such β, x_β is a code for a countable ordinal, and II wins otherwise. If x_β [y_β] codes an ordinal, let $|x_\beta|$ [$|y_\beta|$] be the ordinal coded. If all $|x_\beta|$ and $|y_\beta|$ are defined, let

$$\gamma_i = \sup_j\{\max(|x_{\omega\cdot i+j}|,|y_{\omega\cdot i+j}|) : j \in \omega\} .$$

I wins just in case $L_{\gamma_n} \models \varphi[\gamma_0,\ldots,\gamma_{n-1}]$. G_φ is $\omega\cdot(n+1)+1$- Π^1_1, so it is enough to prove that I wins $G_\varphi \Leftrightarrow \#\varphi \in 0^\#$.

Assume for definiteness that I has a winning strategy σ for G_φ. By boundedness, there is a closed, unbounded $C \subseteq \omega$, such that, for all $\beta < \omega(n+1)$, all $\alpha \in C$, and all plays consistent with σ,

$$[\forall\beta' < \beta\ (|y_\beta| \text{ is defined})\ \&\ \sup\{|y_{\beta'}| : \beta' < \beta\} < \alpha] \Rightarrow |x_\beta| < \alpha .$$

Let $\bar{\gamma}_0 < \cdots < \bar{\gamma}_n$ be indiscernibles for L which belong to C. Let II play y_β such that each $\langle |y_{\omega\cdot i+j}| : j \in \omega\rangle$ is an increasing sequence with limit $\bar{\gamma}_i$. If I plays σ, then $\gamma_i = \bar{\gamma}_i$ and so $L_{\bar{\gamma}_n} \models \varphi[\bar{\gamma}_0,\ldots,\bar{\gamma}_{n-1}]$ and so $L \models \varphi[\bar{\gamma}_0,\ldots,\bar{\gamma}_{n-1}]$, i.e., $\#\varphi \in 0^\#$.

§4. The largest countable inductive set.

Lemma 4.1 (Inductive determinacy). If $A \subseteq \omega$ is Σ^*_n, then A belongs to the largest countable inductive set.

Proof. Let A be Π^*_n if n is even and Σ^*_n if n is odd. Then

$$m \in A \Leftrightarrow (Q_1x_1)(Q_2x_2)\cdots(\exists x_n)(\langle m,x_1,\ldots,x_n\rangle \in B \cap C)$$

where the quantifiers are alternating and B is inductive and C is coinductive. By [Mo 1980, p. 419], let φ be an inductive norm on B and let ψ be an inductive norm on $\neg C$.

For degrees $d_1,\ldots,d_n$, let

$$E_{d_1,\ldots,d_n} = \{S \subseteq \omega : (Q_1 x_1 \text{ of degree} \le d_1)(Q_2 x_2 \text{ of degree} \le d_2)\cdots(\exists x_n \text{ of degree} \le d_n)$$
$$\exists \gamma\ \forall m\ (m \in S \Longleftrightarrow [\varphi(m,x_1,\ldots,x_n) < \gamma\ \&\ \psi(m,x_1,\ldots,x_n) \not< \gamma])$$

Since the inductive sets are closed under number quantification, $E_{d_1,\ldots,d_n}$ is inductive uniformly in reals $x_1,\ldots,x_n$ of degrees $d_1,\ldots,d_n$ respectively.

Let

$$E = \{S : \text{For almost all } (d_1,\ldots,d_n),\ S \in E_{d_1,\ldots,d_n}\}$$

E is inductive, since $S \in E$ just in case

$$\exists d_1\ \forall d_1' \ge d_1\ \forall d_2\ \exists d_2' \ge d_2 \cdots Q_n^* d_n' \ge d_n\quad S \in E_{d_1',\ldots,d_n'},$$

and the inductive sets are closed under quantification over ${}^{\omega}\omega$.

E is (inductively) well-orderable, since the function $\varphi_S(d_1,\ldots,d_n)$ = the least γ which witnesses $S \in E_{d_1,\ldots,d_n}$ embeds E in the ultrapower of the ordinals by the iterated product measure (where we only need the measure defined on inductive sets).

Let us finally show that $A \in E$. For almost all $(d_1,\ldots,d_n)$,

$$(Q_1 x_1 \text{ of degree} \le d_1)\cdots(\exists_n x_n \text{ of degree} \le d_n)$$
$$(\forall m)((m,x_1,\ldots,x_n) \in B \cap C \Longleftrightarrow m \in A)\ .$$

Let $\gamma(m,d_1,\ldots,d_n) = \sup\{\varphi(m,x_1,\ldots,x_n) : (m,x_1,\ldots,x_n) \in A\ \&\ \text{degree}(x_i) \le d_i$ all $i \le n\}$. Let $\gamma(d_1,\ldots,d_n) = \sup\{\gamma(m,d_1,\ldots,d_n) : m \in \omega\}$ $\gamma(d_1,\ldots,d_n)$ witnesses that $A \in E_{d_1,\ldots,d_n}$.

<u>Lemma 4.2</u> (Determinacy for all Σ_n^* games). Every member of any countable inductive set is Σ_n^* for some n.

<u>Proof</u>. By Moschovakis [Mo 1982] every coinductive set admits a very good scale $\langle \varphi_i : i \in \omega\rangle$ such that φ_i is Σ_i^*. By Lemma 3.3, every inductive game won by II has a winning strategy τ such that τ restricted to positions of length i is Σ_j^* for some j:

Using a game like that in the proof of Lemma 3.4, except that there are no m_i, we see that every member of a countable inductive set is Σ_n^* for some n.

<u>Theorem 4.3</u>. The largest countable inductive set is $\bigcup_n \Sigma_n^*$.

<u>Corollary 4.4</u> (Determinacy of all Σ_n^* games). There is a coinductive set which does not admit a $\utilde{\Sigma}_n^*$ scale for any n.

Proof. For each x, let C_x be the largest countable set inductive in x. Let $C = \{\langle x,y\rangle : y \in C_x\}. \neg C$ is coinductive}. Suppose $\neg C$ admits a $\underset{\sim}{\Sigma}^*_n$ scale. This scale is Σ^*_n in x for some x. Thus $\neg C_x$ has a member Δ^*_n in x. This contradicts the relativization of the theorem.

The results of this section hold for wider classes than the inductive sets. Suppose Γ has the scale property and is closed under trivial operations and integer and real quantification. With the obvious definitions, Lemma 4.1 holds for Γ (by the same proof). Hence Corollary 4.4 holds for Γ. Lemma 4.2 does not hold in general. We must replace Σ^*_n by, roughly speaking, the $n^{\underline{th}}$ norm on the largest countable $\tilde{\Gamma}$ set.

References

[Mo 1980] Y. N. Moschovakis, Descriptive set theory, North Holland, Amsterdam, New York, Oxford.

[Mo 1983] Y. N. Moschovakis, Scales on coinductive sets, this volume.

[St 1983] J. R. Steel, Scales on Σ^1_1 sets, this volume.

SCALES IN L(R)

John R. Steel
Department of Mathematics
University of California
Los Angeles, California 90024

§0. Introduction. We now know the extent of scales in $L(\mathbb{R})$: $(\underset{\sim}{\Sigma}^2_1)^{L(\mathbb{R})}$ sets admit $(\underset{\sim}{\Sigma}^2_1)^{L(\mathbb{R})}$ scales, while properly $(\underset{\sim}{\Pi}^2_1)^{L(\mathbb{R})}$ sets admit no scales whatsoever in $L(\mathbb{R})$. It follows that $(\underset{\sim}{\Delta}^2_1)^{L(\mathbb{R})}$ sets admit $(\underset{\sim}{\Delta}^2_1)^{L(\mathbb{R})}$ scales, but this is by no means a local result, in that the simplest possible scale on a given $(\underset{\sim}{\Delta}^2_1)^{L(\mathbb{R})}$ set may be substantially more complicated than the set itself. Here we shall consider the problem of finding scales of minimal complexity on sets in $L(\mathbb{R})$, and obtain a fairly complete solution. Given a set A in $L(\mathbb{R})$, we shall identify by means of reflection properties of the Levy hierarchy for $L(\mathbb{R})$ the first level $\underset{\sim}{\Sigma}_n(L_\alpha(\mathbb{R}))$ of this hierarchy at which a scale on A is definable. This level occurs very near the least α such that $A \in L_\alpha(\mathbb{R})$ and for some Σ_1 formula $\varphi(v)$ and real x, $L_{\alpha+1}(\mathbb{R}) \models \varphi[x]$ while $L_\alpha(\mathbb{R}) \models \sim\varphi[x]$. That is, in $L(\mathbb{R})$ the construction of new scales is closely tied to the verification of new Σ_1 statements about reals.

Scales are important in Descriptive Set Theory because they provide the only known general method which will take arbitrary definitions in a given logical form of sets of reals, and produce definitions of members of those sets. This is something a descriptive set theorist will often want to do. It is a pleasing consequence of our work and the earlier work upon which it builds that there is no better genenal method in $L(\mathbb{R})$. We shall see that there are no simpler uniformizations of arbitrary $\underset{\sim}{\Sigma}_n(L_\alpha(\mathbb{R}))$ relations on reals than those given by scales.

Our work knits together earlier work of Kechris and Solovay, Martin [8], Martin-Steel [10], and Moschovakis [12]. We shall credit this work in the appropriate places. What is new is our systematic use of the Levy hierarchy for $L(\mathbb{R})$, especially its reflection properties and fine structure.

The paper is organized as follows. In §1 we exposit rather carefully the basic fine structure theory of $L(\mathbb{R})$. Although there is nothing really new here, we have included this section as a service to the scrupulous reader. In §2 we present the heart of our analysis of the complexity of scales in $L(\mathbb{R})$. §3 is devoted to the one case in this analysis not covered by §2; we have isolated this case because it is technically more involved than the others, and the casual reader might want to skip it. Finally, in §4 we refine the results of §§2 and 3 slightly, and use these results to prove some theorems concerning Suslin cardinals and the pointclasses $S(\kappa)$ of κ-Suslin sets.

Some preliminaries and notation: Except in §4, we work in ZF + DC, and state our additional determinacy hypotheses as we need them. (This is done chiefly as a service to the readers and authors of [7], who must keep close watch on the determinacy we assume in Theorems 2.1 and 3.7.) We let $\mathbb{R} = {}^{\omega}\omega$, the Baire space, and call its elements reals. Variables $z, y, x, w, \ldots$ range over $\mathbb{R}$, while $\alpha, \beta, \gamma, \delta, \ldots$ range over the class OR of ordinals. If $0 \leq k \leq \omega$ and $1 \leq \ell \leq \omega$, then $\omega^k \times (\omega^\omega)^\ell$ is recursively homeomorphic to $\mathbb{R}$, and we sometimes tacitly identify the two. A pointclass is a class of subsets of $\mathbb{R}$ closed under recursive substitutions; a boldface pointclass is a pointclass closed under continuous substitutions. If Γ is a pointclass, then

$$\breve{\Gamma} = \{\mathbb{R} - A \mid A \in \Gamma\} = \text{the dual of } \Gamma ,$$

$$\exists^{\mathbb{R}}\Gamma = \{\exists^{\mathbb{R}}A \mid A \in \Gamma\}, \quad \text{where } \exists^{\mathbb{R}}A = \{x \mid \exists y(\langle x, y\rangle \in A)\} ,$$

and

$$\forall^{\mathbb{R}}\Gamma = (\exists^{\mathbb{R}}\Gamma)\breve{}\,.$$

By "Det (Γ)" we mean the assertion that all games whose payoff set is in Γ are determined. For whatever else we use from Descriptive Set Theory, and in particular for the notions of a scale and of the scale property, we refer the reader to [14].

Our general set theoretic notation is standard. $\mu\alpha P(\alpha)$ is the least ordinal satisfying P. For any set X, $X^{<\omega}$ is the set of finite sequences of elements of X, $[S]^{<\omega}$ is the set of finite subsets of X, and P(X) is the set of all subsets of X. V_α is the set of sets of rank less than α. If $X \cup \{V_{\omega+1}\} \subseteq M$, where M is any set, then $\Sigma_n(M, X)$ is the class of relations on M definable over $(M, \in)$ by a Σ_n formula from parameters in $X \cup \{V_{\omega+1}\}$. Thus we are always allowed $V_{\omega+1}$ itself (not necessarily its elements) as a parameter in definitions. $\Sigma_\omega(M, X) = \bigcup_{n<\omega} \Sigma_n(M, X)$. We write "$\Sigma_n(M)$" for "$\Sigma_n(M, \phi)$", and "$\utilde{\Sigma}_n(M)$" for "$\Sigma_n(M, M)$". Similar conventions apply to the Π_n and Δ_n notations. If $X \cup \{V_{\omega+1}\} \subseteq M \subseteq N$, then "$M \prec_n^X N$" means that for all $\vec{a} \in (X \cup \{V_{\omega+1}\})^{<\omega}$ and all Σ_n formulae φ, $M \models \varphi[\vec{a}]$ iff $N \models \varphi[\vec{a}]$. We write "$M \prec_n N$" for "$M \prec_n^M N$".

In the sequel we shall sometimes refer to "the pointclass $\Sigma_n(M, X)$", or assert "$\Sigma_n(M, X)$ has the scale property". In such contexts we are actually referring to $\Sigma_n(M, X) \cap P(\mathbb{R})$; the context should make this clear.

§1. The fine structure of $L(\mathbb{R})$. This section contains a straightforward generalization to $L(\mathbb{R})$ of Jensen's fine structure theory of L. Although the reader could no doubt supply the generalization himself, we have assumed he would rather not, and so have given a fairly complete exposition. ZF + DC will suffice for the results of this section; no determinacy is required.

Because it is convenient, we shall use the Jensen J-hierarchy for $L(\mathbb{R})$. We presume the basic facts about rudimentary functions set forth in [3]. Let $\operatorname{rud}(M)$ be the closure of $M \cup \{M\}$ under the rudimentary functions. By Corollary 1.7 of [3], $\operatorname{rud}(M) \cap P(M) = \underset{\sim}{\Sigma}_\omega(M)$ for transitive sets M. Now let

$$J_1(\mathbb{R}) = V_{\omega+1} = \{a \mid \operatorname{rank}(a) \leq \omega\}$$

$$J_{\alpha+1}(\mathbb{R}) = \operatorname{rud}(J_\alpha(\mathbb{R})) \quad \text{for } \alpha > 0 \ ,$$

and $$J_\lambda(\mathbb{R}) = \bigcup_{\alpha<\lambda} J_\alpha(\mathbb{R}) \quad \text{for } \lambda \text{ limit} \ .$$

Of course, $L(\mathbb{R}) = \bigcup_{\alpha \in OR} J_\alpha(\mathbb{R})$. For all $\alpha \geq 1$, $J_\alpha(\mathbb{R})$ is transitive and $J_{\alpha+1}(\mathbb{R}) \cap P(J_\alpha(\mathbb{R})) = \underset{\sim}{\Sigma}_\omega(J_\alpha(\mathbb{R}))$. Thus $J_\alpha(\mathbb{R}) \in J_{\alpha+1}(\mathbb{R})$ for $\alpha \geq 1$, and $J_\alpha(\mathbb{R}) \subseteq J_\beta(\mathbb{R})$ for $1 \leq \alpha \leq \beta$. If $\alpha > 1$, then $\operatorname{rank}(J_\alpha(\mathbb{R})) = OR \cap J_\alpha(\mathbb{R}) = \omega\alpha$.

It is useful to refine this hierarchy. Recall from [3] the rudimentary functions $F_1, \ldots, F_8$ from which all rudimentary functions can be generated by composition. Let $F_9(a, b) = \langle a, b \rangle$, $F_{10}(a, b) = a''b = \{c \mid \langle b, c \rangle \in a\}$, and $F_{11}(a, b, c) = \langle a, b, c \rangle$. Define

$$S(M) = M \cup \{M\} \cup \bigcup_{i=1}^{11} F_i''(M \cup \{M\}) \ .$$

Then if M is transitive, so is $S(M)$ (which is why we included F_9, F_{10}, and F_{11}); moreover,

$$\operatorname{rud}(M) = \bigcup_{n<\omega} S^n(M) \ .$$

Now let

$$S_\omega(\mathbb{R}) = J_1(\mathbb{R}) \ ,$$

$$S_{\alpha+1}(\mathbb{R}) = S(S_\alpha(\mathbb{R})) \quad \text{for } \alpha \geq \omega \ ,$$

and $$S_\lambda(\mathbb{R}) = \bigcup_{\alpha<\lambda} S_\alpha(\mathbb{R}) \quad \text{for } \lambda \text{ limit} \ .$$

Clearly the $S_\alpha(\mathbb{R})$'s are transitive and cumulative, $S_\alpha(\mathbb{R}) \in S_{\alpha+1}(\mathbb{R})$, and $S_{\omega\alpha}(\mathbb{R}) = J_\alpha(\mathbb{R})$ for all α.

1.1. Lemma. The sequences $\langle S_\gamma(\mathbb{R}) \mid \gamma < \omega\alpha \rangle$ and $\langle J_\beta(\mathbb{R}) \mid \beta < \alpha \rangle$ are uniformly $\Sigma_1(J_\alpha(\mathbb{R}))$ for $\alpha > 1$.

Proof. The reason is that the two sequences are defined by local Σ_0 recursions. Notice that

$$M = S_\gamma(\mathbb{R}) \Longleftrightarrow \exists f(\Phi(f) \wedge f(\gamma) = M) ,$$

where

$$\Phi(f) \Longleftrightarrow [f \text{ is a function} \wedge \operatorname{dom} f \in OR \wedge f(\omega) = V_{\omega+1} \wedge \forall\alpha \in \operatorname{dom} f$$
$$(\alpha + 1 \in \operatorname{dom} f \Rightarrow f(\alpha+1) = S(f(\alpha)))$$
$$\wedge \forall\lambda \in \operatorname{dom} f \quad (\lambda \text{ limit} \Rightarrow f(\lambda) = \bigcup_{\alpha<\lambda} f(\alpha))] .$$

Φ is Σ_0 since rudimentary functions have Σ_0 graphs and we are always allowed $V_{\omega+1}$ as a parameter. So it is enough to show that for $\gamma < \omega\alpha$

$$M = S_\gamma(\mathbb{R}) \Longleftrightarrow \exists f \in J_\alpha(\mathbb{R}) \ (\Phi(f) \wedge f(\gamma) = M) .$$

That is, we must show that $\langle S_\delta(\mathbb{R}) \mid \delta \leq \gamma\rangle \in J_\alpha(\mathbb{R})$ for all γ and α such that $\gamma < \omega\alpha$. This can be proved by induction on γ. For limit γ, say $\gamma = \omega\beta$, we use the fact that $\langle S_\delta(\mathbb{R}) \mid \delta < \gamma\rangle$ is $\Sigma_1(J_\beta(\mathbb{R}))$ by induction hypothesis.

A similar argument shows that $\langle \omega\beta \mid \beta < \alpha\rangle$ is uniformly $\Sigma_1(J_\alpha(\mathbb{R}))$, and therefore $\langle J_\beta(\mathbb{R}) \mid \beta < \alpha\rangle$ is uniformly $\Sigma_1(J_\alpha(\mathbb{R}))$.

In a similar vein,

1.2. <u>Lemma</u>. There is a Π_2 sentence θ such that for all transitive sets M with $V_{\omega+1} \in M$,

$$M \models \theta \quad \text{iff} \quad \exists\alpha > 1 \ (M = J_\alpha(\mathbb{R})) .$$

<u>Proof</u>. Let $\Phi(f, u)$ be the Σ_0 formula resulting from the Σ_0 formula $\Phi(f)$ of 1.1 by replacing "$V_{\omega+1}$" with the free variable u. Let

$$\psi(u) \Longleftrightarrow \forall a \exists f \exists\gamma \geq \omega \ (\Phi(f, u) \wedge a \in f(\gamma)) \wedge$$
$$\forall\gamma \in OR \ \exists\beta \in OR \ (\gamma < \beta) \wedge$$
$$\forall\gamma \in OR \ \exists f \ (\Phi(f, u) \wedge \gamma \in \operatorname{dom} f) .$$

Then ψ is Π_2, and it is easy to check that for transitive M with $V_{\omega+1} \in M$,

$$M \models \psi[V_{\omega+1}] \quad \text{iff} \quad \exists\alpha > 1 \ (M = J_\alpha(\mathbb{R})) .$$

The desired sentence θ is therefore

$$\forall u \ (\forall x \ (x \in u \Longleftrightarrow \operatorname{rank}(x) \leq \omega) \Rightarrow \psi(u)) .$$

We shall call the Π_2 sentence θ provided by Lemma 1.2 "$V = L(\mathbb{R})$".

1.3. <u>Corollary</u>. If $M \prec_1 J_\alpha(\mathbb{R})$ and $V_{\omega+1} \in M$, then $M \cong J_\beta(\mathbb{R})$ for some $\beta \leq \alpha$.

Of course, if there is any point to this paper there is no definable well-order of $J_\alpha(\mathbb{R})$. The next lemma is as much as we can get in the direction of such a wellorder.

1.4. <u>Lemma</u>. There are uniformly $\Sigma_1(J_\alpha(\mathbb{R}))$ maps f_α such that

$$f_\alpha : [\omega\alpha]^{<\omega} \times \mathbb{R} \xrightarrow{\text{onto}} J_\alpha(\mathbb{R}) .$$

<u>Proof</u>. Fix a rudimentary function h such that for all finite $F \subseteq OR$,

$$h''(\{F\} \times \mathbb{R}) = \{\langle G_1, G_2, G_3\rangle \mid G_i \subseteq F \text{ for } 1 \leq i \leq 3\} .$$

It is easy to construct such an h. We now define by induction on $\gamma \geq \omega$ functions

$$g_\gamma : [\gamma]^{<\omega} \times \mathbb{R} \xrightarrow{\text{onto}} S_\gamma(\mathbb{R})$$

such that $g_\gamma \subseteq g_\delta$ if $\gamma \leq \delta$. We leave g_ω to the reader's discretion. If λ is a limit, let

$$g_\lambda = \bigcup_{\beta<\lambda} g_\beta .$$

Finally, suppose g_γ is given. Let F and x be given, and set

$$g_{\gamma+1}(F, x) = g_\gamma(F, x)$$

if $\gamma \notin F$. If $\gamma \in F$, let $h(F - \{\gamma\}, (x)_0) = \langle G_1, G_2, G_3\rangle$, and for $1 \leq i \leq 3$, let

$$a_i = \begin{cases} g_\gamma(G_i, (x)_i) & \text{if } (x)_4(i) = 0 \\ S_\gamma(\mathbb{R}) & \text{if } (x)_4(i) \neq 0 . \end{cases}$$

Let also $j \in \{1, \ldots, 11\}$ be such that $j \equiv (x)_4(4) \pmod{11}$. Then we set

$$g_{\gamma+1}(F, x) = \begin{cases} F_j(a_1, a_2) & \text{if } j \neq 11 \\ F_j(a_1, a_2, a_3) & \text{if } j = 11 . \end{cases}$$

By an easy induction we have $g_\gamma : [\gamma]^{<\omega} \times \mathbb{R} \xrightarrow{\text{onto}} S_\gamma(\mathbb{R})$ for all γ. But now notice that there is a rudimentary function G such that $g_{\gamma+1} = G(g_\gamma, S_\gamma(\mathbb{R}))$ for all γ. By the proof of 1.1, $\langle g_\gamma \mid \gamma < \omega\alpha\rangle$ is uniformly $\Sigma_1(J_\alpha(\mathbb{R}))$. Thus the desired f_α is given by

$$f_\alpha = \bigcup_{\gamma<\omega\alpha} g_\gamma .$$

Lemma 2.10 of [3] implies that for all α there is a $\underset{\sim}{\Sigma}_1(J_\alpha(\mathbb{R}))$ map of α onto $[\omega\alpha]^{<\omega}$, and therefore a $\underset{\sim}{\Sigma}_1(J_\alpha(\mathbb{R}))$ map of $\omega\alpha \times \mathbb{R}$ onto $J_\alpha(\mathbb{R})$. However, the parameters involved in the Σ_1 definitions of these maps, and the consequent lack of uniformity in their definitions, make the maps described in 1.4 more useful.

One can identify finite sets of ordinals with descending sequences of ordinals. The tree of all such descending sequences is wellfounded, and so its Brouwer-Kleene order is a wellorder. This gives us a wellorder of $[\mathrm{OR}]^{<\omega}$. More explicitly, for $F, G \in [\mathrm{OR}]^{<\omega}$ let

$$F \leq_{BK} G \quad \text{iff} \quad (\exists \alpha \in G\ (G = F - \alpha) \vee \max(G \,\Delta\, F) \in G)\ .$$

Then $\leq_{BK}$ is a Σ_0 wellorder of $[\mathrm{OR}]^{<\omega}$.

We begin now to make our way toward the Σ_n selection theorem.

One central feature of fine structure theory is that it allows us to form Skolem hulls in the most effective way possible. This involves Σ_n selection (or uniformization). Now in $L(\mathbb{R})$ we cannot hope to select from arbitrary sets in a definable way; for example, the relation

$$R(\alpha, x) \Longleftrightarrow \alpha < \omega_1 \wedge x \in \mathbb{R} \wedge x \text{ codes } \alpha$$

cannot be uniformized at all in $L(\mathbb{R})$. However, we shall throw all of $\mathbb{R}$ into the hulls we build anyway, so that in view of 1.4 it suffices to select effectively from subsets of $[\mathrm{OR}]^{<\omega}$. Here we have a chance.

1.5. <u>Definition</u>. $J_\alpha(\mathbb{R})$ <u>satisfies</u> Σ_n <u>selection</u> iff whenever $R \subseteq J_\alpha(\mathbb{R}) \times [\omega\alpha]^{<\omega}$ is $\underset{\sim}{\Sigma}_n(J_\alpha(\mathbb{R}))$, there is a $\underset{\sim}{\Sigma}_n(J_\alpha(\mathbb{R}))$ set $S \subseteq R$ such that

$$\forall a(\exists F\, R(a, F) \Rightarrow \exists !\, F\, S(a, F))\ .$$

1.6. <u>Definition</u>. Let $h : J_\alpha(\mathbb{R}) \times \mathbb{R} \to J_\alpha(\mathbb{R})$ be a partial $\underset{\sim}{\Sigma}_n(J_\alpha(\mathbb{R}))$ map. Then h is a Σ_n <u>Skolem function</u> for $J_\alpha(\mathbb{R})$ iff whenever S is $\Sigma_n(J_\alpha(\mathbb{R}), \{a\})$ and $S \neq 0$, then $\exists x \in \mathbb{R}\ \ (h(a, x) \in S)$.

Recall Corollary 1.13 of [3], according to which the satisfaction relation restricted to Σ_n formulae is uniformly $\Sigma_n(M)$ for transitive, rudimentarily closed M.

1.7. <u>Lemma</u>. If $J_\alpha(\mathbb{R})$ satisfies Σ_n selection, then there is a Σ_n Skolem function for $J_\alpha(\mathbb{R})$.

<u>Proof</u>. Let f_α be the map given by 1.4, and let $\langle \varphi_i \mid i < \omega \rangle$ be an enumeration of the Σ_n formulae of two free variables. Let

$$R(a, x, F) \quad \text{iff} \quad J_\alpha(\mathbb{R}) \models \varphi_{x(0)}[a, f_\alpha(F, \lambda i.x(i+1))]\ .$$

Since R is $\Sigma_n(J_\alpha(R))$, we have a partial $\underset{\sim}{\Sigma}_n(J_\alpha(R))$ map g uniformizing R. Let

$$h(a, x) = f_\alpha(g(a, x), \lambda i.x(i + 1)) .$$

It is easy to check that h is the desired Skolem function.

1.8. Definition. Let $X \subseteq M$ and $1 \leq n \leq \omega$. Then

$$\mathrm{Hull}_n^M(X) = \{a \in M \mid \{a\} \text{ is } \Sigma_n(M, X)\} .$$

We write "$\mathrm{Hull}_n^\alpha(X)$" for "$\mathrm{Hull}_n^{J_\alpha(R)}(X)$". The connection between these hulls and Σ_n Skolem functions is explained in the next lemma.

1.9. Lemma. Let $H = \mathrm{Hull}_n^\alpha(X)$ where $n < \omega$ and $R \subseteq X$. Suppose that for some $p \in H$ there is a Σ_n Skolem function for $J_\alpha(R)$ which is in fact $\Sigma_n(J_\alpha(R), \{p\})$. Then

(a) $H \prec_n J_\alpha(R)$,

(b) $H = h''(X^{<\omega})$ for some partial $\underset{\sim}{\Sigma}_n(H)$ map h, and

(c) if $X = Y \cup R$ where Y is finite, then $H = h''R$ for some partial $\underset{\sim}{\Sigma}_n(H)$ map h.

Proof. (a) Notice that if $\{c\}$ is $\Sigma_n(J_\alpha(R), H)$, then $c \in H$. Now let $a_1, \ldots, a_n \in H$ and suppose $J_\alpha(R) \models \varphi[a_1, \ldots, a_n, b]$ where φ is Σ_n. We want to see that $J_\alpha(R) \models \varphi[a_1, \ldots, a_n, c]$ for some $c \in H$. Let $\tilde{h}$ be a Σ_n Skolem function for $J_\alpha(R)$ which is $\Sigma_n(J_\alpha(R), \{p\})$ with $p \in H$. Then for some $x \in R$, if we set $c = \tilde{h}(\langle a_1, \ldots, a_n\rangle, x)$ then $J_\alpha(R) \models \varphi[a_1, \ldots, a_n, c]$. Since $\{c\}$ is $\Sigma_n(J_\alpha(R), \{a_1, \ldots, a_n, x, p\})$, $c \in H$.

(b) Let $\tilde{h}$ be as in (a), and set

$$h(\langle a_1, \ldots, a_n\rangle) = \tilde{h}(\langle a_1, \ldots, a_{n-1}, p\rangle, a_n) .$$

One can easily check that $h''X^{<\omega} = H$.

(c) This follows easily from (b).

Incidentally, we don't need Σ_n selection to show that for $R \subseteq X$,

$$\mathrm{Hull}_{n+1}^\alpha(X) \prec_n J_\alpha(R)$$

for all $n < \omega$, and therefore

$$\mathrm{Hull}_\omega^\alpha(X) \prec_\omega J_\alpha(R) .$$

The crude Skolem functions obtained by selecting the BK-least F in the proof of 1.7 will suffice for this.

1.10. <u>Lemma</u>. For all $\alpha > 1$, $J_\alpha(\mathbb{R})$ satisfies Σ_1 selection.

<u>Proof</u>. Let $R \subseteq J_\alpha(\mathbb{R}) \times [\omega\alpha]^{<\omega}$, and

$$R(a, F) \text{ iff } \exists b(J_\alpha(\mathbb{R}) \models \varphi[a, F, b, p]) ,$$

where φ is Σ_0 and $p \in J_\alpha(\mathbb{R})$. Define

$$\begin{aligned} S(a, F) \text{ iff } \exists \gamma < \omega\alpha \ [&\exists b \in S_\gamma(\mathbb{R})\varphi(a, F, b, p) \\ &\wedge \forall \delta < \gamma \ \forall b \in S_\delta(\mathbb{R}) \sim \varphi(a, F, b, p) \\ &\wedge \forall G \in S_\gamma(\mathbb{R}) \ \forall b \in S_\gamma(\mathbb{R}) \\ &(G <_{BK} F \Rightarrow \sim\varphi(a, G, b, p))] . \end{aligned}$$

By Lemma 1.1, S is $\Sigma_1(J_\alpha(\mathbb{R}), \{p\})$, and clearly S uniformizes R.

Lemma 1.10 fails for $\alpha = 1$ since there is a Σ^1_1 subset of $\mathbb{R} \times \{0, 1\}$ with no Σ^1_1 uniformization.

Before going on to Σ_n selection, we make some simple use of our ability to construct Σ_1 Skolem hulls.

1.11. <u>Lemma</u>. Suppose that $\alpha > 1$, and for no $\beta < \alpha$ do we have $J_\beta(\mathbb{R}) \prec_1^{\mathbb{R}} J_\alpha(\mathbb{R})$. Then

(a) There is a $\Sigma_1(J_\alpha(\mathbb{R}))$ partial map $h : \mathbb{R} \xrightarrow{\text{onto}} J_\alpha(\mathbb{R})$,

(b) If $\beta < \alpha$, then there is a total map $h : \mathbb{R} \xrightarrow{\text{onto}} J_\beta(\mathbb{R})$ such that $h \in J_\alpha(\mathbb{R})$.

<u>Proof</u>. (a) Let $H = \text{Hull}_1^\alpha(\mathbb{R})$. From the proofs of 1.7 and 1.10, we see that $J_\alpha(\mathbb{R})$ has a Σ_1 Skolem function which is (<u>lightface</u>) $\Sigma_1(J_\alpha(\mathbb{R}))$. By 1.9 then, $H \prec_1 J_\alpha(\mathbb{R})$ and H is the image of $\mathbb{R}$ under a $\Sigma_1(H)$ map $\tilde{h}$. Let $\pi : H \cong J_\beta(\mathbb{R})$ be the collapse map. Since $\pi \restriction (\mathbb{R} \cup \{\mathbb{R}\})$ is the identity, $J_\beta(\mathbb{R}) \prec_1^{\mathbb{R}} J_\alpha(\mathbb{R})$, and therefore $\beta = \alpha$. But then $\pi''\tilde{h}$ is the desired h.

(b) Suppose $\beta < \alpha$. By increasing β if necessary we may assume that for some Σ_1 formula φ and real x, $J_{\beta+1}(\mathbb{R}) \models \varphi[x]$ but $J_\beta(\mathbb{R}) \not\models \varphi[x]$. Let

$$S_{\omega\beta+n}(\mathbb{R}) \models \varphi[x] .$$

Now rudimentary functions are simple ([3], Lemma 1.2), and so we can find a Σ_0 formula ψ such that for any transitive M and $x \in M$,

$$\psi(M, x) \Longleftrightarrow S^n(M) \models \varphi[x] .$$

Clearly, there is a first order formula θ such that for any transitive M and $x \in M$

$$(M \models \theta[x]) \iff \psi(M, x) .$$

Let θ be Σ_n. Let $H = \mathrm{Hull}_{n+1}(\mathbb{R})$. Then $H \prec_n J_\beta(\mathbb{R})$. Let $\pi : H \cong J_\gamma(\mathbb{R})$ be the collapse map. Since $\pi(x) = x$ and $\pi(\mathbb{R}) = \mathbb{R}$, $J_\gamma(\mathbb{R}) \models \theta[x]$, and so $\gamma = \beta$. It is easy to get a $\underset{\sim}{\Sigma}_\omega(H)$ map of $\mathbb{R}$ onto H, which then collapses to a $\underset{\sim}{\Sigma}_\omega(J_\beta(\mathbb{R}))$ map of $\mathbb{R}$ onto $J_\beta(\mathbb{R})$. This map belongs to $J_\alpha(\mathbb{R})$.

Let Θ be the least ordinal not the surjective image of $\mathbb{R}$, and $\underset{\sim}{\delta}^2_1$ the least ordinal not the image of $\mathbb{R}$ under a surjection f such that $\{\langle x, y\rangle \mid f(x) \leq f(y)\}$ is $\underset{\sim}{\Delta}^2_1$. A Skolem hull argument like those just given shows that $\Theta^{L(\mathbb{R})} = \mu\alpha[P(\mathbb{R}) \cap L(\mathbb{R}) \subseteq J_\alpha(\mathbb{R})]$. For $(\underset{\sim}{\delta}^2_1)^{L(\mathbb{R})}$ we have

1.12. <u>Lemma</u>. Let σ be least such that $J_\sigma(\mathbb{R}) \prec^{\mathbb{R}}_1 L(\mathbb{R})$. Then

(a) $(\underset{\sim}{\Sigma}^2_1)^{L(\mathbb{R})} = \underset{\sim}{\Sigma}_1(J_\sigma(\mathbb{R})) \cap P(\mathbb{R})$

(b) $(\underset{\sim}{\Delta}^2_1)^{L(\mathbb{R})} = J_\sigma(\mathbb{R}) \cap P(\mathbb{R})$

(c) $(\underset{\sim}{\delta}^2_1)^{L(\mathbb{R})} = \sigma$.

<u>Proof</u>. (a) Clearly, any $(\underset{\sim}{\Sigma}^2_1)^{L(\mathbb{R})}$ set of reals is $\Sigma_1(L(\mathbb{R}), \mathbb{R})$, and therefore $\Sigma_1(J_\sigma(\mathbb{R}), \mathbb{R})$. For the converse, let A be a $\underset{\sim}{\Sigma}_1(J_\sigma(\mathbb{R}))$ set of reals. By 1.11, A is $\Sigma_1(J_\sigma(\mathbb{R}), \mathbb{R})$. Let

$$\begin{aligned} x \in A &\iff J_\sigma(\mathbb{R}) \models \varphi[x, y] \\ &\iff L(\mathbb{R}) \models \varphi[x, y] \end{aligned}$$

where φ is Σ_1. Since $J_\sigma(\mathbb{R})$ is the surjective image of $\mathbb{R}$ under a map in $L(\mathbb{R})$,

$$\begin{aligned} x \in A \iff \exists E \in L(\mathbb{R})\ (&\mathfrak{M} = (\mathbb{R}, E) \text{ is a wellfounded extensional} \\ &\text{model of } V = L(\mathbb{R}) \wedge \mathbb{R}^{\mathfrak{M}} \text{ collapses to } \mathbb{R} \\ &\wedge \mathfrak{M} \models \varphi[x^{\mathfrak{M}}, y^{\mathfrak{M}}]) , \end{aligned}$$

where we use $x^{\mathfrak{M}}$ and $y^{\mathfrak{M}}$ for the inverse images of x and y under the collapse of $\mathfrak{M}$. Thus A is $(\underset{\sim}{\Sigma}^2_1)^{L(\mathbb{R})}$.

(b) By 1.11, $J_\sigma(\mathbb{R}) \prec_1 L(\mathbb{R})$, and so $J_\sigma(\mathbb{R})$ is admissible. If A is $(\underset{\sim}{\Delta}^2_1)^{L(\mathbb{R})}$, then A is $\underset{\sim}{\Delta}_1(J_\sigma(\mathbb{R}))$ by (a), and so $A \in J_\sigma(\mathbb{R})$ by admissibility.

(c) $(\underset{\sim}{\delta}^2_1)^{L(\mathbb{R})} \leq \sigma$ by (b) and the fact that $J_\sigma(\mathbb{R})$ is admissible. But if $\beta < \sigma$, then 1.11 (b) guarantees a map $h : \mathbb{R} \xrightarrow{\text{onto}} \beta$ in $J_\sigma(\mathbb{R})$. Since $\{\langle x, y\rangle \mid h(x) \leq h(y)\} \in J_\sigma(\mathbb{R})$, it is $(\underset{\sim}{\Delta}^2_1)^{L(\mathbb{R})}$ and therefore $\beta < (\underset{\sim}{\delta}^2_1)^{L(\mathbb{R})}$. Thus $\sigma \leq (\underset{\sim}{\delta}^2_1)^{L(\mathbb{R})}$.

Lemma 1.2 indicates an analogy between the pointclass $(\underset{\sim}{\Sigma}_1^2)^{L(\mathbb{R})}$ and the class $\Sigma_2^1 \cap P(\omega) = (\Sigma_2^1)^L \cap P(\omega)$. This analogy is strengthened by the result of [10] that

$$(\underset{\sim}{\Sigma}_2^1)^{L(\mathbb{R})} = \mathfrak{D}^{\mathbb{R}}\underset{\sim}{\Pi}_1^1 ,$$

while of course $\Sigma_2^1 \cap P(\omega) = \mathfrak{D}\,\Pi_1^1$. One further evidence of the analogy is that if $A \in (\underset{\sim}{\Delta}_1^2)^{L(\mathbb{R})}$, then $A \leq_W B$ for some $B \subseteq \mathbb{R}$ such that $\{B\}$ is $\underset{\sim}{\Pi}_1^1$. ($A \leq_W B$ iff $\exists f$ (f continuous $\wedge$ $A = f^{-1}(B)$)). To see this, let $A \in J_\alpha(\mathbb{R})$ where for some formula $\varphi(v)$ and real x, $J_{\alpha+1}(\mathbb{R}) \models \varphi[x]$ but $J_\alpha(\mathbb{R}) \models \sim\varphi[x]$. Take $B = \{\langle\theta, y\rangle \mid y \in \mathbb{R} \wedge J_{\alpha+1}(\mathbb{R}) \models \theta[y]\}$. It is easy to show that $\{B\}$ is $\Pi_1^1(x)$, and $A \leq_W B$.

We proceed to the Σ_n selection theorem. Our proof is a transcription of Jensen's proof for L, as simplified by S. Friedman (cf. [15]).

1.13. Definition. For any $\alpha \in OR$ and $n \in \omega$

$$\rho_n^\alpha = \text{the least } \beta \text{ such that there is a } \underset{\sim}{\Sigma}_n(J_\alpha(\mathbb{R})) \text{ partial map } f : J_\beta(\mathbb{R}) \xrightarrow{\text{onto}} J_\alpha(\mathbb{R})$$

if $n \geq 1$, and $\rho_n^\alpha = \alpha$ if $n = 0$.

By Lemma 1.4, there is in fact a $\underset{\sim}{\Sigma}_n(J_\alpha(\mathbb{R}))$ partial map from $\omega\rho_n^\alpha \times \mathbb{R}$ onto $J_\alpha(\mathbb{R})$. We shall write "$\rho_n^\alpha = \mathbb{R}$" to mean $\rho_n^\alpha = 1$.

1.14. Lemma. Suppose $n \geq 1$ and $J_\alpha(\mathbb{R})$ satisfies Σ_n selection. Then

$$\rho_n^\alpha = \text{the least } \beta \text{ such that some } \underset{\sim}{\Sigma}_n(J_\alpha(\mathbb{R})) \text{ subset of } J_\beta(\mathbb{R}) \text{ is not a member of } J_\alpha(\mathbb{R})\,.$$

Proof. Let β be as on the r. h. s. If $h : J_{\rho_n^\alpha}(\mathbb{R}) \xrightarrow{\text{onto}} J_\alpha(\mathbb{R})$ is partial Σ_n, then $\{a \in \text{dom } h \mid a \notin h(a)\}$ witnesses that $\beta \leq \rho_n^\alpha$. Now let $p \in J_\alpha(\mathbb{R})$ be such that $J_\alpha(\mathbb{R})$ has a Σ_n Skolem function which is $\Sigma_n(J_\alpha(\mathbb{R}), \{p\})$ and there is a $\Sigma_n(J_\alpha(\mathbb{R}), \{p\})$ subset of $J_\beta(\mathbb{R})$ not in $J_\alpha(\mathbb{R})$. Let

$$H = \text{Hull}_n^\alpha(J_\beta(\mathbb{R}) \cup \{p\})\,.$$

Then $H \prec_n J_\alpha(\mathbb{R})$ and there is a $\underset{\sim}{\Sigma}_n(H)$ partial map h of $J_\beta(\mathbb{R})$ onto H, by 1.9. Let $\pi : H \cong J_\delta(\mathbb{R})$ be the collapse. Then $\delta = \alpha$, as our new subset of $J_\beta(\mathbb{R})$ at α is constructed at δ. Thus $\pi''h$ witnesses that $\rho_n^\alpha \leq \beta$.

The next lemma gives the basic quantifier complexity reduction behind the proof of the Σ_n selection theorem.

1.15. <u>Lemma</u>. Let $n \geq 1$ and assume $J_\alpha(\mathbb{R})$ satisfies Σ_m selection for all $m \leq n$. Let Q be a $\underset{\sim}{\Sigma}_n(J_\alpha(\mathbb{R}))$ relation and f a $\underset{\sim}{\Sigma}_n(J_\alpha(\mathbb{R}))$ partial function. Define S by

$$S(a, \beta) \text{ iff } \beta < \rho_n^\alpha \wedge \forall b \in J_\beta(\mathbb{R})$$
$$(b \in \text{dom } f \Rightarrow Q(a, \beta, f(b))) .$$

Then S is $\underset{\sim}{\Sigma}_{n+1}(J_\alpha(\mathbb{R}))$.

<u>Proof</u>. We proceed by induction on n. Suppose $Q(a, b, c) \Longleftrightarrow \exists d P(a, b, c, d)$, where P is $\underset{\sim}{\Pi}_{n-1}(J_\alpha(\mathbb{R}))$. Let $\rho_n = p_n^\alpha$ and $\rho_{n-1} = \rho_{n-1}^\alpha$. Let $h : J_{\rho_{n-1}}(\mathbb{R}) \xrightarrow{\text{onto}} J_\alpha(\mathbb{R})$ be a partial $\underset{\sim}{\Sigma}_{n-1}(J_\alpha(\mathbb{R}))$ map.

<u>Case 1</u>. For some $\delta < \rho_n$ and total $\underset{\sim}{\Sigma}_n(J_\alpha(\mathbb{R}))$ map $g : J_\delta(\mathbb{R}) \to \rho_{n-1}$, the range of g is cofinal in ρ_{n-1}.

<u>Proof</u>. Suppose $S(a, \beta)$. Then if $b \in J_\beta(\mathbb{R})$ and $b \in \text{dom}(f)$, we can find $c \in J_{\rho_{n-1}}(\mathbb{R})$ so that $P(a, \beta, f(b), h(c))$. Moreover $c \in J_{g(d)}(\mathbb{R})$ for some $d \in J_\delta(\mathbb{R})$. Let

$$H(b, d) \Longleftrightarrow b \in J_\beta(\mathbb{R}) \cap \text{dom } f \wedge d \in J_\delta(\mathbb{R})$$
$$\wedge \forall e \, \forall \eta \, ((e = f(b) \wedge \eta = g(d))$$
$$\Rightarrow \exists c \in J_\eta(\mathbb{R}) \, P(a, \beta, e, h(c))) .$$

We have just seen that $S(a, \beta) \Rightarrow \forall b \, \exists d \, H(b, d)$. Our induction hypothesis implies (or if $n = 1$, it is trivial that) the consequent of the final implication defining H is $\underset{\sim}{\Pi}_n$. Thus H is $\underset{\sim}{\Sigma}_n \wedge \underset{\sim}{\Pi}_n$, and being bounded in $J_{\rho_n}(\mathbb{R})$, $H \in J_\alpha(\mathbb{R})$. But then $S(a, \beta)$ is equivalent to the existence of such an H in $J_\alpha(\mathbb{R})$; that is,

$$S(a, \beta) \Longleftrightarrow \beta < \rho_n \wedge \exists H \in J_\alpha(\mathbb{R})$$
$$(\forall b \in J_\beta(\mathbb{R}) \, (b \in \text{dom } f \Rightarrow \exists d \; H(b, d))$$
$$\wedge \forall b \in J_\beta(\mathbb{R}) \, \forall d \, \forall \eta \, \forall e$$
$$(H(b, d) \wedge \eta = g(d) \wedge e = f(b) \Rightarrow$$
$$\exists c \in J_\eta(\mathbb{R}) P(a, \beta, e, h(c)))) .$$

Using the induction hypothesis again, we have that S is $\underset{\sim}{\Sigma}_{n+1}(J_\alpha(\mathbb{R}))$.

<u>Case 2</u>. Otherwise.

<u>Proof</u>. Suppose $S(a, \beta)$. Let $D = \text{dom } f \cap J_\beta(\mathbb{R})$; then $D \in J_\alpha(\mathbb{R})$ since $\beta < \rho_n$. Define

$$R(b, \eta) \Longleftrightarrow b \in D \wedge \eta < \rho_{n-1} \wedge \exists c \in J_\eta(\mathbb{R}) P(a, \beta, f(b), h(c)) ,$$

so that R is $\underset{\sim}{\Sigma}_n(J_\alpha(\mathbb{R}))$. By Σ_n selection for $J_\alpha(\mathbb{R})$, we have a total $\underset{\sim}{\Sigma}_n(J_\alpha(\mathbb{R}))$ map $g : D \to \rho_{n-1}$ which uniformizes R. By case hypothesis, the range of g is bounded in ρ_{n-1}. But then

$$S(a, \beta) \Longleftrightarrow \exists \eta < \rho_{n-1} \forall b \in J_\beta(\mathbb{R}) \forall d$$
$$(d = f(b) \Rightarrow \exists c \in J_\eta(\mathbb{R}) P(a, \beta, d, h(c))) .$$

By induction hypothesis (or obviously in the case $n = 1$) S is $\underset{\sim}{\Sigma}_{n+1}(J_\alpha(\mathbb{R}))$.

1.16. Theorem. Suppose $\alpha \in OR$, $n \geq 1$, and $\rho^\alpha_{n-1} \neq 1$. Then $J_\alpha(\mathbb{R})$ satisfies Σ_n selection.

Proof. Suppose $R(a, F) \Longleftrightarrow \exists b\, Q(a, F, b)$, where Q is $\underset{\sim}{\Pi}_{n-1}(J_\alpha(\mathbb{R}))$ and $R \subseteq J_\alpha(\mathbb{R}) \times [\omega\alpha]^{<\omega}$. Let $f : J_{\rho^\alpha_{n-1}}(\mathbb{R}) \xrightarrow{\text{onto}} J_\alpha(\mathbb{R})$ be a partial $\underset{\sim}{\Sigma}_{n-1}(J_\alpha(\mathbb{R}))$ map. Define

$$S(a, F) \Longleftrightarrow \exists \beta < \rho^\alpha_{n-1}$$
$$[\exists b \in J_{\beta+1}(\mathbb{R})\ (b \in \operatorname{dom} f \wedge f(b)_0 = F \wedge Q(a, f(b)_0, f(b)_1))$$
$$\wedge \forall b \in J_\beta(\mathbb{R})\ (b \in \operatorname{dom} f \Rightarrow \sim Q(a, f(b)_0, f(b)_1)))$$
$$\wedge \forall b \in J_{\beta+1}(\mathbb{R})\ (b \in \operatorname{dom} f \Rightarrow (F \leq_{BK} f(b)_0 \vee \sim Q(a, f(b)_0, f(b)_1)))] .$$

Here, of course, $f(b) = \langle f(b)_0, f(b)_1 \rangle$. Lemma 1.15 easily implies that S is $\underset{\sim}{\Sigma}_n(J_\alpha(\mathbb{R}))$, and clearly S uniformizes R. (By Lemma 1.10, we may assume $n \geq 2$, and then 1.4 easily implies that ρ^α_{n-1} is a limit ordinal, so that "$\forall b \in J_{\beta+1}(\mathbb{R})$" is a bounded quantification over $J_{\rho^\alpha_{n-1}}(\mathbb{R})$.)

We shall see in §4 that the hypothesis $\rho^\alpha_{n-1} \neq 1$ of the Σ_n selection theorem is essential. Of course, if $L(\mathbb{R}) = L$ we can weaken it to "$\alpha \neq 1$ or $n \neq 1$"; that is, we need only rule out $\Sigma_1(J_1(\mathbb{R}))$. On the other hand, if $\text{Det}(L(\mathbb{R}))$ then $J_1(\mathbb{R})$ satisfies Σ_n selection iff n is even, and a similar periodicity of order two takes over on arbitrary $J_\alpha(\mathbb{R})$ above the least n such that $\rho^\alpha_n = 1$.

§2. Scales on $\Sigma_1(J_\alpha(\mathbb{R}))$ sets. Our positive results on the existence of scales are refinements of Theorem 1 of [10], which builds directly on a scale construction due to Moschovakis ([12]). We shall describe briefly the slight generalization of this construction we need. Suppose that $\alpha \in OR$, and that for each $x \in \mathbb{R}$ we have a game G_x in which I's moves come from $\mathbb{R} \times \alpha$ while II's moves come from $\mathbb{R}$. Thus a typical run of G_x has the form

I	x_0, β_0		x_2, β_1	...
II		x_1		x_3

where $x_i \in \mathbb{R}$ and $\beta_i < \alpha$ for all i. Suppose that G_x is closed and continuously associated to x in the strong sense that for some $Q \subseteq (\omega^{<\omega})^{<\omega} \times \alpha^{<\omega}$,

$$\text{I wins } G_x \text{ iff } \forall n\, Q(\langle x \upharpoonright n, x_0 \upharpoonright n, \ldots, x_n \upharpoonright n\rangle, \langle \beta_0, \ldots, \beta_n\rangle) .$$

Thus by Gale-Stewart, one of the players in G_x has a winning quasi-strategy. (We are not assuming the Axiom of Choice, so we don't get full strategies.) Let

$P_k(x, u)$ iff u is a position of length k from which
I has a winning quasi-strategy

and

$$P(x) \text{ iff } P_0(x, \phi) .$$

If sufficiently many games are determined, then Moschovakis' construction yields a scale on P. The prewellordering $\leq_i$ induced by the $i^{\underline{th}}$ norm in this scale is definable from the P_k's for $k \leq i$ by means of recursive substitution, the Boolean operations, and quantification over $\mathbb{R} \times \alpha$. In particular, if $\alpha < \omega\gamma$ and $\langle P_k \mid k \leq i\rangle \in J_\gamma(\mathbb{R})$, then $\leq_i \in J_\gamma(\mathbb{R})$. The determinacy required to construct the Moschovakis scale is closely related to the definability of the scale constructed. In particular, if $\alpha < \omega\gamma$ and $P_k \in J_\gamma(\mathbb{R})$ for all $k < \omega$, then $\mathrm{Det}(J_\gamma(\mathbb{R}))$ suffices.

In the circumstances just described we call the map $x \longmapsto G_x$ a <u>closed game representation</u> of P. Which sets P admit closed game representations, and therefore scales? According to [10], every $(\underset{\sim}{\Sigma}_1^2)^{L(\mathbb{R})}$ set admits a closed game representation in $L(\mathbb{R})$. Our first theorem refines that result.

2.1. <u>Theorem</u>. If $\alpha > 1$ and $\mathrm{Det}(J_\alpha(\mathbb{R}))$, then the pointclass $\Sigma_1(J_\alpha(\mathbb{R}))$ has the scale property.

<u>Proof</u>. Let us first assume that α is a limit ordinal, and deal with the general case later. Let $\varphi_0(v)$ be a Σ_1 formula, and let

$$P(x) \text{ iff } J_\alpha(\mathbb{R}) \models \varphi_0[x] ,$$

for $x \in \mathbb{R}$. For $\beta < \alpha$, let

$$P^\beta(x) \text{ iff } J_\beta(\mathbb{R}) \models \varphi_0[x] .$$

Thus $P = \bigcup_{\beta<\alpha} P^\beta$. For each $\beta < \alpha$ we will construct a closed game representation $x \longmapsto G_x^\beta$ of P^β. Let

$P_k^\beta(x, u) \Longleftrightarrow u$ is a position of length k from which
I has a winning quasi-strategy in G_x^β .

We shall arrange that for each $P_k^\beta \in J_\alpha(\mathbb{R})$, and that the map $(\beta, k) \longmapsto P_k^\beta$ is $\Sigma_1(J_\alpha(\mathbb{R}))$. This will suffice for 2.1. For let $\{\varphi_k^\beta\}$ be the Moschovakis scale on P^β, and let $\leq_k^\beta$ be the prewellorder of $\mathbb{R}$ induced by φ_k^β. Then $\leq_k^\beta \in J_\alpha(\mathbb{R})$ by the remarks above. A simple inspection of [12] shows that those remarks are true uniformly in β, so that the map $(\beta, k) \longmapsto \leq_k^\beta$ is $\Sigma_1(J_\alpha(\mathbb{R}))$. We can then define a scale $\{\psi_k\}$ on P:

$$\psi_0(x) = \mu\beta P^\beta(x) ,$$

and

$$\psi_{k+1}(x) = \langle \psi_0(x), \varphi_k^{\psi_0(x)}(x) \rangle .$$

(In the definition of ψ_{k+1}, we use the lexicographic order to assign ordinals to pairs of ordinals.) It is easy to check that $\{\psi_k\}$ is a $\Sigma_1(J_\alpha(\mathbb{R}))$ scale on P.

So let β and x be given; we want to define G_x^β. Our plan is to force Player I in G_x^β to describe a countable model of $V = L(\mathbb{R}) + \varphi_0(x) + \forall\gamma\,(J_\gamma(\mathbb{R}) \not\models \varphi_0[x])$ which contains all the reals played by II, while using ordinals less than $\omega\beta$ to prove that his model is wellfounded. If $J_\beta(\mathbb{R}) \models \varphi_0[x]$, then I will be able to win G_x^β by describing an elementary submodel of $J_\gamma(\mathbb{R})$, where γ is least so that $J_\gamma(\mathbb{R}) \models \varphi_0[x]$. On the other hand, if I has a winning quasi-strategy in G_x^β, we will be able to piece together the models he describes in different runs of G_x^β according to his strategy, and thereby produce a model of the form $J_\gamma(\mathbb{R})$ so that $\gamma \leq \beta$ and $J_\gamma(\mathbb{R}) \models \varphi_0[x]$.

Player I describes his model in the language $\mathcal{L}$ which has, in addition to $\in$ and $=$, constant symbols $\underline{x}_i$ for $i < \omega$. He uses $\underline{x}_i$ to denote the ith real played in the course of G_x^β. Let us fix recursive maps

$$m, n : \{\theta \mid \theta \text{ is an } \mathcal{L}\text{-formula}\} \to \{2n \mid 1 \leq n < \omega\}$$

which are one-one, have disjoint recursive ranges, and are such that whenever $\underline{x}_i$ occurs in θ, $i < \min(m(\theta), n(\theta))$. These maps give stages sufficiently late in G_x^β for I to decide certain statements about his model.

Player I's description must extend the following $\mathcal{L}$-theory T. The axioms of T include

(1) Extensionality

(2) $V = L(\mathbb{R})$

$(3)_\varphi$ $\exists v\, \varphi(v) \Rightarrow \exists v\, (\varphi(v) \wedge \forall u \in v \sim \varphi(u))$

$(4)_i$ $\underline{x}_i \in \mathbb{R}$.

Of course, these axioms are true in any $J_\gamma(\mathbb{R})$ as long as the $\underline{x}_i$'s are interpreted by reals. The next axiom restricts the models of this form to at most one possibility

(5) $\varphi_0(\underline{x}_0) \wedge \forall\delta\, (J_\delta(\mathbb{R}) \not\models \varphi[\underline{x}_0])$.

Finally, T has axioms which guarantee that any model $\mathfrak{A}$ can be regarded as a definable closure of $\{\underline{x}_i^{\mathfrak{A}} \mid i < \omega\}$. Recall from 1.4 the uniformly definable maps $f_\gamma : [\omega\gamma]^{<\omega} \times \mathbb{R} \xrightarrow{\text{onto}} J_\gamma(\mathbb{R})$; let $\theta_0(v_0, v_1, v_2)$ be a Σ_1 formula which for all γ defines the graph of f_γ over $J_\gamma(\mathbb{R})$. Now, for any $\mathcal{L}$-formula $\varphi(v)$ of one free variable, T has axioms

$(6)_\varphi$ $\exists v\, \varphi(v) \Rightarrow \exists v\, \exists F(\varphi(v) \wedge \theta_0(F, \underline{x}_{m(\varphi)}, v))$

$(7)_\varphi$ $\exists v\, (\varphi(v) \wedge v \in \mathbb{R}) \Rightarrow \varphi(\underline{x}_{n(\varphi)})$.

This completes the list of axioms of T.

A typical run of G_x^β has the form

I	i_0, x_0, η_0		i_1, x_2, η_1	...
II		x_1		x_3

where for all k, $i_k \in \{0, 1\}$, $x_k \in \mathbb{R}$, and $\eta_k < \omega\beta$. If u is a position of length n, say $u = \langle\langle i_k, x_{2k}, \eta_k, x_{2k+1}\rangle \mid k < n\rangle$, then we let

$$T^*(u) = \{\theta \mid \theta \text{ is a sentence of } \mathcal{L} \wedge i_{n(\theta)} = 0\},$$

and if p is a full run of G_x^β,

$$T^*(p) = \bigcup_{n<\omega} T^*(p \restriction n).$$

Now let $p = \langle\langle i_k, x_{2k}, \eta_k, x_{2k+1}\rangle \mid k < \omega\rangle$ be a run of G_x; we say that p is a winning run for I iff

(a) $x_0 = x$

(b) $T^*(p)$ is a complete, consistent extension of T such that for all i, m, n, "$\underline{x}_i(\underline{n}) = \underline{m}$" $\in T^*(p)$ iff $x_i(n) = m$, and

(c) If φ and ψ are $\mathcal{L}$-formulae of one free variable, and "$\iota v\varphi(v) \in OR \wedge \iota v\psi(v) \in OR$" $\in T^*(p)$, then "$\iota v\varphi(v) \leq \iota v\psi(v)$" $\in T^*(p)$ iff $\eta_{n(\varphi)} \leq \eta_{n(\psi)}$.

In condition (c) we have used the "unique v" operator as an abbreviation: if θ and φ are $\mathcal{L}$-formulae, then $\theta(\iota v\varphi(v))$ abbreviates "$\exists v(\theta(v) \wedge \forall u\, (\varphi(u) \Leftrightarrow u = v))$".

It is clear that G_x is closed and continuously associated to x. In order to show that $x \longmapsto G_x$ is the desired closed game representation of P^β, we want to characterize the winning positions for I in G_x as those in which I has been "honest" about the minimal model of $\varphi_0(x)$. More precisely, let us call a position $u = \langle\langle i_k, x_{2k}, \eta_k, x_{2k+1}\rangle \mid k < n\rangle$ of length n <u>(β, x)-honest</u> iff $J_\beta(\mathbb{R}) \models \varphi_0[x]$, and if $\gamma \leq \beta$ is least such that $J_\gamma(\mathbb{R}) \models \varphi_0[x]$, then

(i) $n > 0 \Rightarrow x_0 = x$

(ii) if we let $I_u(\underline{x}_i) = x_i$ for $i < 2n$, then all axioms of $T^*(u) \cup T$ thereby interpreted in $(J_\gamma(\mathbb{R}), I_u)$ are true in this structure, and

(iii) If $\theta_0, \ldots, \theta_m$ enumerates those $\mathcal{L}$-formulae θ of one free variable such that $n(\theta) < n$ and

$$(J_\gamma(\mathbb{R}), I_u) \models \iota v\theta(v) \in OR$$

and if $\delta_i < \omega\gamma$ is such that

$$(J_\gamma(\mathbb{R}), I_u) \models \iota v\theta_i(v) = \delta_i ,$$

then the map

$$\delta_i \longmapsto \eta_{n(\theta_i)}$$

is well-defined and extendible to an order preserving map of $\omega\gamma$ into $\omega\beta$.

According to the following claim, honesty is the only rational policy for I in G^β_x.

<u>Claim</u>. Let u be a position in G^β_x. Then I has a winning quasi-strategy in G^β_x starting from u iff u is (β, x)-honest.

<u>Proof</u>. ($\Rightarrow$) Let u be a (β, x)-honest position of length n; then it is easy to check that $\exists i, x, \eta \forall y (u^\frown\langle i, x, \eta, y\rangle)$ is (β, x)-honest. [If $n = n(\theta)$ for some sentence θ of $\mathcal{L}$, put $i = 0$ iff $(J_\gamma(\mathbb{R}), I_u) \models \theta$, otherwise i is random. If $n = m(\varphi)$ for φ an $\mathcal{L}$ formula of one free variable, choose x so that $(J_\gamma(\mathbb{R}), I_{u^\frown\langle i,x,\eta,y\rangle}) \models$ axiom $(6)_\varphi$ of T. If $n = n(\varphi)$, choose x for the sake of $(7)_\varphi$, and otherwise let x be random. Finally, if $n = n(\varphi)$ for φ of one free variable, and $\delta < \omega\gamma$ is such that $(J_\gamma(\mathbb{R}), I_u) \models \iota v\varphi(v) = \delta$, then set $\eta = f(\delta)$ where f witnesses part (iii) of honesty for u. Otherwise, choose η randomly.] Moreover, no (β, x)-honest position is an immediate loss for I in G^β_x. Thus if u is (β, x)-honest, I can win G^β_x from u by keeping to (β, x)-honest positions.

($\Leftarrow$) The proof we give here is due to Hugh Woodin; our original proof followed more closely the corresponding proof in [10].

Let θ be the sentence we want to prove, that is, let θ = "For all β, x, and u, if I has a winning quasi-strategy in G^β_x starting from u, then u is (β, x)-honest." We show that $M \models \theta$ whenever M is a countable, transitive model of (a sufficiently large fragment of) ZF + DC. But then ZF + DC $\vdash \theta$, and we're done.

So let M be such a model, and let

$$M \models S \text{ is a winning quasi-strategy for}$$
$$\text{I in } G^\beta_x \text{ starting from u.}$$

In M, define

$$P = \{v \mid v \text{ is a position in } G^\beta_x \text{ extending } u \text{ and allowed by } S\},$$

and let $\mathbb{P} = (P, \supseteq)$. Let G be $\mathbb{P}$-generic over M, and let $p = \bigcup G$. We can think of p as a "generic run" of G^β_x according to S. In particular, p satisfies the requirements for being a winning run for I, since these requirements are closed.

Thus $T^*(p)$ is a complete, consistent extension of T. Let $\mathfrak{B} \models T^*(p)$, and let $\mathfrak{A}$ be the substructure of $\mathfrak{B}$ whose universe is

$$|\mathfrak{A}| = \{b \in |\mathfrak{B}| \mid \exists i \in \omega\ \exists \psi\ (\psi \text{ is a formula of } \mathcal{L} \text{ with no constant symbols but } \underline{x}_i \text{ and } b = \iota v(\mathfrak{B} \models \psi[v])\}.$$

Then by axioms (3) and (6) of T, $\mathfrak{A} \prec \mathfrak{B}$.

Let $p = \langle\langle i_k, x_{2k}, \eta_k, x_{2k+1}\rangle \mid k < \omega\rangle$. If $b \in OR^{\mathfrak{A}}$, and $b = \iota v(\mathfrak{B} \models \psi[v])$, then let

$$f(b) = \eta_{n(\psi)}.$$

By requirement (c) on I in G^β_x, f is a well-defined order-preserving map of $OR^{\mathfrak{A}}$ into $\omega\beta$. Thus we may assume that $\mathfrak{A} = (J_\gamma(\mathbb{R}^{\mathfrak{A}}), I)$ for some $\gamma \leq \beta$, where by requirement (b) on I in G^β_x, $I(\underline{x}_i) = x_i$. By axioms (7) of T, $\mathbb{R}^{\mathfrak{A}} = \{x_i \mid i < \omega\}$; on the other hand, since p is a generic run and II is free to play whatever reals he pleases, $\{x_i \mid i < \omega\} = \mathbb{R}^M$.

Thus $\mathfrak{A} = (J_\gamma(\mathbb{R}^M), I)$, where $I = \bigcup_k I_{p \restriction k}$. Moreover, by axiom (5) of T and requirement (a) on I in G^β_x, γ is least such that $J_\gamma(\mathbb{R}) \models \varphi_0[x]$. It is now easy to verify conditions (i)-(iii) of (β, x)-honesty for u in M[G]; the map f defined above witnesses (iii). Since honesty is absolute, u is (β, x)-honest in M.

This proves the claim.

Notice that the empty position is (β, x)-honest iff $J_\beta(\mathbb{R}) \models \varphi_0[x]$. The claim therefore implies that I has a winning quasi-strategy in G^β_x iff $J_\beta(\mathbb{R}) \models \varphi_0[x]$. If we let

$$P^\beta_n(x, u) \iff \text{I has a winning quasi-strategy in } G^\beta_x \text{ from the length } n \text{ position } u$$

$$\iff u \text{ has length } n \text{ and is } (\beta, x)\text{-honest},$$

then it is easy to see that $P^\beta_n \in J_{\beta+1}(\mathbb{R})$ for all β, n, and in fact that the map $(\beta, n) \longmapsto P^\beta_n$ is $\Sigma_1(J_\alpha(\mathbb{R}))$. (To see that $P^\beta_n \in J_{\beta+1}(\mathbb{R})$, notice that the sentences which come up in evaluating the honesty of positions of length n have quantifier complexity bounded by some recursive function of n. Also, notice that if there is any order preserving map as required by part (iii) of honesty, then there is a $\undertilde{\Sigma}_1(J_\beta(\mathbb{R}))$ such map.) Thus the claim gives us 2.1 in the case that is a limit.

Minor modifications make this proof work for arbitrary α. Again, let

$$P(x) \Longleftrightarrow J_\alpha(\mathbb{R}) \models \varphi_0[x] ,$$

where φ_0 is Σ_1. Using the simplicity of the rudimentary function S as in the proof of 1.11 (b), we can find first order formulae $\psi_n(v)$ for $n < \omega$ so that for all $\beta \in OR$ and $x \in \mathbb{R}$,

$$J_\beta(\mathbb{R}) \models \psi_n[x] \text{ iff } S_{\omega\beta+n}(\mathbb{R}) \models \varphi_0[x] .$$

For each $\beta < \alpha$ and $n < \omega$ we can define a game $G_x^{\beta,n}$ so that I wins $G_x^{\beta,n}$ iff $\exists \gamma \leq \beta\ (J_\gamma(\mathbb{R}) \models \psi_n[x])$ iff $S_{\omega\beta+n}(\mathbb{R}) \models \varphi_0[x]$. The rules of $G_x^{\beta,n}$ are exactly those of G_x^β except that axiom (5) of T is replaced by

$$(5)'\ \psi_n(\underline{x}_0) \wedge \forall\delta\ (J_\delta(\mathbb{R}) \models -\psi_n(\underline{x}_0)) .$$

We can define (β, n, x)-honesty and prove the analogue of our claim just as before. This analogous claim yields 2.1 in the general case at once.

Of course, the existence of a $\Sigma_1(J_\alpha(\mathbb{R}))$ set of reals universal for $\Sigma_1(J_\alpha(\mathbb{R}), \mathbb{R})$ sets of reals immediately gives the boldface version of 2.1.

2.2. <u>Corollary</u>. If $\alpha > 1$ and $\mathrm{Det}(J_\alpha(\mathbb{R}))$, then the boldface pointclass $\Sigma_1(J_\alpha(\mathbb{R}), \mathbb{R})$ has the scale property.

The hypothesis that $\alpha > 1$ is necessary in 2.1 and 2.2, since $\Sigma_1(J_1(\mathbb{R})) \cap P(\mathbb{R}) = \Sigma_1^1$, and Σ_1^1 doesn't have the scale property.

Theorem 2.1 immediately implies the special cases from which it was abstracted.

2.3. <u>Corollary</u>. If $\mathrm{Det}(L(\mathbb{R}))$, then

(a) The pointclass of inductive sets has the scale property ([11]),

(b) The pointclass Σ_ω^* has the scale property; in particular, every coinductive set admits a Σ_ω^* scale ([12]),

(c) The pointclass $(\Sigma_1^2)^{L(\mathbb{R})}$ has the scale property ([10]).

<u>Proof</u>. Let

$$\kappa^{\mathbb{R}} = \mu\alpha[J_\alpha(\mathbb{R}) \text{ is admissible}] .$$

Then inductive $= \Sigma_1(J_{\kappa^{\mathbb{R}}}(\mathbb{R}))$ and $\Sigma_\omega^* = \Sigma_1(J_{\kappa^{\mathbb{R}}+1}(\mathbb{R}))$, so we have (a) and (b). Let

$$\sigma = \mu\alpha[J_\alpha(\mathbb{R}) <_1 L(\mathbb{R})] .$$

Then by Lemma 1.12, $(\Sigma_1^2)^{L(\mathbb{R})} = \Sigma_1(J_\sigma(\mathbb{R}))$, so we have (c).

We shall use 2.1 together with some further work to give a complete description of those levels of the Levy hierarchy for $L(\mathbb{R})$ (that is, those pointclasses of the form $\underset{\sim}{\Sigma}_n(J_\gamma(\mathbb{R}))$) which have the scale property. For this, it is convenient to have the following definition.

2.2. Definition. Let $\alpha, \beta \in OR$ and $\alpha \leq \beta$. The interval $[\alpha, \beta]$ is a Σ_1-gap iff

(i) $J_\alpha(\mathbb{R}) \prec_1^{\mathbb{R}} J_\beta(\mathbb{R})$
(ii) $\forall\alpha' < \alpha \ (J_{\alpha'}(\mathbb{R}) \not\prec_1^{\mathbb{R}} J_\alpha(\mathbb{R}))$
(iii) $\forall\beta' > \beta \ (J_\beta(\mathbb{R}) \not\prec_1^{\mathbb{R}} J_{\beta'}(\mathbb{R}))$.

That is, a Σ_1-gap is a maximal interval of ordinals in which no new Σ_1 facts about elements of $\mathbb{R} \cup \{V_{\omega+1}\}$ are verified. If $[\alpha, \beta]$ is a Σ_1-gap, we say α begins the gap and β ends it. Notice that we allow $\alpha = \beta$. We shall also allow $[(\underset{\sim}{\delta}_1^2)^{L(\mathbb{R})}, \Theta^{L(\mathbb{R})}]$ as a Σ_1-gap.

2.3. Lemma. The Σ_1-gaps partition $\Theta^{L(\mathbb{R})}$.

Proof. Given $\gamma < \Theta^{L(\mathbb{R})}$, let $\alpha \leq \gamma$ be least so that $J_\alpha(\mathbb{R}) \prec_1^{\mathbb{R}} J_\gamma(\mathbb{R})$, and let $\beta = \sup\{\delta < \Theta^{L(\mathbb{R})} \mid J_\gamma(\mathbb{R}) \prec_1^{\mathbb{R}} J_\delta(\mathbb{R})\}$. Then $[\alpha, \beta]$ is a Σ_1-gap, and $\gamma \in [\alpha, \beta]$. Any two distinct Σ_1-gaps are disjoint because we have restricted the parameters involved to $\mathbb{R} \cup \{V_{\omega+1}\}$.

Let us work our way now through an arbitrary Σ_1-gap, looking for levels of the Levy hierarchy having the scale property. If α begins a Σ_1-gap, then by Lemma 1.11 there is a partial $\Sigma_1(J_\alpha(\mathbb{R}))$ map from $\mathbb{R}$ onto $J_\alpha(\mathbb{R})$, and therefore $\underset{\sim}{\Sigma}_1(J_\alpha(\mathbb{R})) = \Sigma_1(J_\alpha(\mathbb{R}), \mathbb{R})$.

2.4. Corollary. If α begins a Σ_1 gap, $\alpha > 1$, and $\mathrm{Det}(J_\alpha(\mathbb{R}))$, then $\underset{\sim}{\Sigma}_1(J_\alpha(\mathbb{R}))$ has the scale property.

The next classes to consider are the $\underset{\sim}{\Sigma}_n(J_\alpha(\mathbb{R}))$, $n > 1$, for α which begin a Σ_1-gap. The scale property here depends upon the admissibility of $J_\alpha(\mathbb{R})$.

2.5. Lemma. Suppose α begins a Σ_1-gap and $J_\alpha(\mathbb{R})$ is not admissible. Then for all $n \geq 1$

$$\underset{\sim}{\Sigma}_{n+1}(J_\alpha(\mathbb{R})) = \exists^{\mathbb{R}}(\underset{\sim}{\Pi}_n(J_\alpha(\mathbb{R})))$$

and

$$\underset{\sim}{\Pi}_{n+1}(J_\alpha(\mathbb{R})) = \forall^{\mathbb{R}}(\underset{\sim}{\Sigma}_n(J_\alpha(\mathbb{R}))) .$$

Proof. The two conclusions are of course equivalent. Let S be $\underset{\sim}{\Sigma}_{n+1}(J_\alpha(R))$; say

$$S(u) \Longleftrightarrow \exists v\, P(u, v)$$

where P is $\underset{\sim}{\Pi}_n(J_\alpha(R))$. Let

$$f : R \xrightarrow{\text{onto}} J_\alpha(R)$$

be a partial $\underset{\sim}{\Sigma}_1(J_\alpha(R))$ map; there is such a map since α begins a Σ_1 gap. Then clearly

$$S(u) \Longleftrightarrow \exists x \in R\, (x \in \text{dom } f \wedge \forall v\, (v = f(x) \Rightarrow P(u, v))) .$$

If $n \geq 2$, this implies $S \in \exists^R (\underset{\sim}{\Pi}_n(J_\alpha(R)))$, as desired. For $n = 1$, we need to know that "$x \in \text{dom } f$" $\in \exists^R (\underset{\sim}{\Pi}_1(J_\alpha(R)))$. This is a direct consequence of inadmissibility. For by inadmissibility we have a total $\underset{\sim}{\Sigma}_1(J_\alpha(R))$ map

$$h : D \to \omega\alpha$$

such that $D \in J_\alpha(R)$ and h has range cofinal in $\omega\alpha$. Since α begins a Σ_1 gap, an easy Skolem hull argument gives a total

$$g : R \xrightarrow{\text{onto}} D$$

such that $g \in J_\alpha(R)$. Let $k = h \circ g$. Let

$$Q(u) \Longleftrightarrow J_\alpha(R) \models \varphi[u, p] ,$$

where φ is Σ_1, be any $\underset{\sim}{\Sigma}_1(J_\alpha(R))$ set. Then

$$Q(u) \Longleftrightarrow \exists x \in R\, (S_{k(x)}(R) \models \varphi[u, p])$$

$$\exists x \in R\, \forall S\, \forall \gamma\, [(\gamma = k(x) \wedge S = S_\gamma(R))$$

$$\Rightarrow S \models \varphi[u, p]]$$

so that $Q \in \exists^R(\Pi_1(J_\alpha(R)))$.

The second periodicity theorem, 2.4, and 2.5 yield at once

2.6. Corollary. Suppose α begins a Σ_1-gap, $\alpha > 1$, $J_\alpha(R)$ is not admissible, and $\text{Det}(J_{\alpha+1}(R))$. Then for all $n < \omega$, the classes $\underset{\sim}{\Sigma}_{2n+1}(J_\alpha(R))$ and $\underset{\sim}{\Pi}_{2n+2}(J_\alpha(R))$ have the scale property.

Martin [8] shows that at admissible α beginning a gap the scale property fails above $\Sigma_1(J_\alpha(R))$ in a strong way.

2.7. Theorem (Martin). Suppose α begins a Σ_1-gap, $J_\alpha(\mathbb{R})$ is admissible, and $\mathrm{Det}(J_{\alpha+1}(\mathbb{R}))$. Then there is a $\Pi_1(J_\alpha(\mathbb{R}))$ subset of $\mathbb{R} \times \mathbb{R}$ with no uniformization in $J_{\alpha+1}(\mathbb{R})$.

2.8. Corollary (Martin). If α begins a Σ_1-gap, $J_\alpha(\mathbb{R})$ is admissible, and $\mathrm{Det}(J_{\alpha+1}(\mathbb{R}))$, then none of the classes $\underset{\sim}{\Sigma}_n(J_\alpha(\mathbb{R}))$ or $\underset{\sim}{\Pi}_n(J_\alpha(\mathbb{R}))$, for $n > 1$, have the scale property.

We are ready to venture inside our Σ_1 gaps. There we shall find no new scales.

2.9. Theorem. Let $[\alpha, \beta]$ be a Σ_1-gap, and assume $\mathrm{Det}(J_{\alpha+1}(\mathbb{R}))$. Then there is a $\Pi_1(J_\alpha(\mathbb{R}))$ subset of $\mathbb{R} \times \mathbb{R}$ with no $\underset{\sim}{\Sigma}_1(J_\beta(\mathbb{R}))$ uniformization.

Proof. For $x, y \in \mathbb{R}$, let

$$C_\alpha(x, y) \Longleftrightarrow \exists \gamma < \omega\alpha \, \exists F \subseteq \gamma \ \ (F \text{ is finite} \\ \wedge \{y\} \text{ is } \Sigma_1(S_\gamma(\mathbb{R}), \{x, F\}))\ .$$

By Lemma 1.1, C_α is $\Sigma_1(J_\alpha(\mathbb{R}))$. Suppose for a contradiction that R is a $\underset{\sim}{\Sigma}_1(J_\beta(\mathbb{R}))$ relation uniformizing $\sim C_\alpha$. By Lemma 1.4, we can fix an $x \in \mathbb{R}$ and Σ_0 formula φ so that for some finite $F \subseteq \omega\beta$

$$R(u, v) \Longleftrightarrow \exists w \ (J_\beta(\mathbb{R}) \models \varphi[u, v, w, x, F])\ .$$

Now $\{y \mid C_\alpha(x, y)\}$ has a wellorder definable over $J_\alpha(\mathbb{R})$, and so by $\mathrm{Det}(J_{\alpha+1}(\mathbb{R}))$ is countable. Thus we have a unique $y \in \mathbb{R}$ so that $R(x, y)$. Let $\theta(u, v)$ be the Σ_1 formula

$$\exists \gamma \ \exists F \subseteq \gamma \ \exists S \ (F \text{ is finite} \wedge S = S_\gamma(\mathbb{R}) \\ \wedge \forall z \in \mathbb{R} \ (z = v \Longleftrightarrow \exists w \in S \ \varphi(u, z, w, u, F)))\ .$$

Then $J_\beta(\mathbb{R}) \models \theta[x, y]$, so $J_\alpha(\mathbb{R}) \models \theta[x, y]$. But inspecting θ, we see that this means $C_\alpha(x, y)$, a contradiction.

2.10. Corollary. If $[\alpha, \beta]$ is a Σ_1-gap and $\alpha < \gamma < \beta$, then none of the classes $\underset{\sim}{\Sigma}_n(J_\gamma(\mathbb{R}))$ or $\underset{\sim}{\Pi}_n(J_\gamma(\mathbb{R}))$, for $n < \omega$, have the scale property.

Theorem 2.9 extends in a simple way an example due to Solovay. For $x, y \in \mathbb{R}$, let $OD(x, y)$ iff y is ordinal definable from x. Solovay showed that if $\forall x \exists y \sim OD(x, y)$, then $\sim OD$ has no uniformization ordinal definable from a real. [If R is such a uniformization, ordinal definable from x, and $R(x, y)$, then $OD(x, y)$, a contradiction.] Let $\delta = (\underset{\sim}{\delta}^2_1)^{L(\mathbb{R})}$ and $\Theta = \Theta^{L(\mathbb{R})}$. By the reflection theorem and the stability of δ, $OD^{L(\mathbb{R})}$ is just the set C_δ defined in 2.9. In

$L(\mathbb{R})$, every set is ordinal definable from a real, so Solovay's example interpreted in $L(\mathbb{R})$ gives 2.9 for the gap $[\delta, \Theta]$ (as noticed by Kechris and Solovay). Theorem 2.9 comes from simply localizing this example. It is interesting that the non-uniformizable $\Pi_1(J_\alpha(\mathbb{R}))$ relation referred to in Martin's 2.7 is just $\sim C_\alpha$; all counterexamples to uniformization in $L(\mathbb{R})$ come from localizing a single example. The next easy proposition sheds more light on the C_α's.

2.11. <u>Proposition</u>. Suppose $J_\alpha(\mathbb{R})$ is admissible and $\mathrm{Det}(J_\alpha(\mathbb{R}))$. Let C_α be the set defined in the proof of 2.9. Then for all x, $\{y \mid C_\alpha(x, y)\}$ is the largest countable $\Sigma_1(J_\alpha(\mathbb{R}), \{x\})$ set of reals.

<u>Proof</u> (Sketch). Let P be countable and $\Sigma_1(J_\alpha(\mathbb{R}), \{x\})$. Since $\Sigma_1(J_\alpha(\mathbb{R}), \{x\})$ has the scale property, P is Suslin via a tree on $\omega \times \alpha$ which is $\Delta_1(J_\alpha(\mathbb{R}), \{x\})$ (here we use admissibility). Since P is countable, P is Suslin via $T \restriction \delta$ for some $\delta < \alpha$. Taking the appropriate derivatives of $T \restriction \delta$, we eventually isolate any $y \in P$ at some stage $\beta < \alpha$. But then $\{y\}$ is $\Sigma_1(J_\beta(\mathbb{R}), \{x, F\})$ for the appropriate F.

Let us return to the question of scales in Σ_1-gaps. We have only one question left, but its answer is sufficiently long-winded to warrant a new section.

§3. <u>Scales at the end of a Σ_1-gap</u>. The results of §2 leave open the question whether any of the classes $\underset{\sim}{\Sigma}_n(J_\beta(\mathbb{R}))$ have the scale property when β ends a Σ_1 gap. Notice that there are new pointclasses here, that is, $\rho_n^\beta = \mathbb{R}$ for some $n < \omega$. This follows at once from part (b) of Lemma 1.11. It turns out that the scale property for these classes hinges on a reflection property of β somewhat subtler than the "Σ_1 reflection on reals" involved in 2.1 and 2.4.

For $a \in J_\alpha(\mathbb{R})$, let $\Sigma^n_{a,\alpha}$ be the Σ_n-type realized by a in $J_\alpha(\mathbb{R})$, that is

$$\Sigma^n_{a,\alpha} = \{\theta(v) \mid \theta \text{ is either } \Sigma_n \text{ or } \Pi_n \wedge J_\alpha(\mathbb{R}) \models \theta[a]\} .$$

3.1. <u>Definition</u>. An ordinal β is <u>strongly Π_n-reflecting</u> iff every Σ_n type realized in $J_\beta(\mathbb{R})$ is realized in $J_\alpha(\mathbb{R})$ for some $\alpha < \beta$; that is, iff

$$\forall b \in J_\beta(\mathbb{R})\ \exists \alpha < \beta\ \exists a \in J_\alpha(\mathbb{R})\ (\Sigma^n_{b,\beta} = \Sigma^n_{a,\alpha}) .$$

3.2. <u>Definition</u>. Let $[\alpha, \beta]$ be a Σ_1-gap. We call $[\alpha, \beta]$ <u>strong</u> iff β is strongly Π_n-reflecting, where n is least such that $\rho_n^\beta = \mathbb{R}$. Otherwise, $[\alpha, \beta]$ is <u>weak</u>.

Martin's proof of Theorem 2.7 easily gives the following generalization.

3.3. Theorem. Let $[\alpha, \beta]$ be a strong Σ_1-gap, and assume $\mathrm{Det}(J_{\alpha+1}(\mathbb{R}))$. Then there is a $\Pi_1(J_\alpha(\mathbb{R}))$ relation (namely, the $\sim C_\alpha$ of 2.9) which has no uniformization in $J_{\beta+1}(\mathbb{R})$.

3.4. Corollary. If $[\alpha, \beta]$ is a strong Σ_1-gap and $\mathrm{Det}(J_{\alpha+1}(\mathbb{R}))$, then none of the classes $\utilde{\Sigma}_n(J_\beta(\mathbb{R}))$ or $\utilde{\Pi}_n(J_\beta(\mathbb{R}))$, for $n < \omega$, have the scale property.

Thus at strong gaps $[\alpha, \beta]$, the scale property first re-appears on $\utilde{\Sigma}_1(J_{\beta+1}(\mathbb{R}))$. At weak gaps it re-appears on $\utilde{\Sigma}_n(J_\beta(\mathbb{R}))$ where n is least so that $\rho_n^\beta = \mathbb{R}$; that is, as soon as possible given 2.9. Before we show this, let us consider examples of the two different kinds of gap.

3.5. Example. Let $[\alpha, \beta]$ be the first Σ_1 gap such that $\beta \geq \alpha + \omega_1$. A Σ_1 Skolem hull argument shows easily that $\beta = \alpha + \omega_1$ and $\rho_1^\beta = \mathbb{R}$. Since $\mathrm{cof}(\beta) > \omega$, any Σ_1-type realized in $J_\beta(\mathbb{R})$ is realized in some $J_\delta(\mathbb{R})$ for $\delta < \beta$. Thus β is strongly Π_1 reflecting.

3.6. Example. Let β be least such that $\rho_1^\beta \neq \mathbb{R}$, and let $\alpha < \beta$ be least so that $J_\alpha(\mathbb{R}) \prec_1^{\mathbb{R}} J_\beta(\mathbb{R})$. The minimality of β implies that $[\alpha, \beta]$ is a Σ_1 gap, and that $\rho_2^\beta = \mathbb{R}$. Now for any n, let θ_n be a Σ_2 sentence so that for all γ,

$$J_\gamma(\mathbb{R}) \models \theta_n \text{ iff } \exists \alpha_0 \ldots \exists \alpha_n (\alpha_0 < \ldots < \alpha_n < \gamma \text{ and } J_{\alpha_0}(\mathbb{R}) \prec_1 \ldots \prec_1 J_{\alpha_n}(\mathbb{R}) \prec_1 J_\gamma(\mathbb{R})) .$$

Every Σ_2-type realized in $J_\beta(\mathbb{R})$ contains all of the θ_n's, but no Σ_2-type realized in any $J_\delta(\mathbb{R})$ for $\delta < \beta$ includes all of the θ_n's. Thus β is not strongly Π_2 reflecting.

There are simpler weak gaps than that given by 3.6. For example, the first gap of the form $[\alpha, \alpha + 1]$ is weak. However, 3.6 seems to better illustrate the general situation. Indeed, the reader might gain some feeling for the proof of our next theorem by trying to show that for the β described in example 3.6, $\utilde{\Sigma}_2(J_\beta(\mathbb{R}))$ has the scale property.

3.7. Theorem. Let $[\alpha, \beta]$ be a weak Σ_1 gap, and suppose $\mathrm{Det}(J_\alpha(\mathbb{R}))$. Then if n is least so that $\rho_n^\beta = \mathbb{R}$, the class $\utilde{\Sigma}_n(J_\beta(\mathbb{R}))$ has the scale property.

Proof. Let α, β, and n satisfy the hypotheses. The case $n = 1$ is somewhat special, so we assume first that $n > 1$.

We shall follow the proof of 2.1 in outline, but some new complications arise in this situation. In order to eventually obtain a $\utilde{\Sigma}_n(J_\beta(\mathbb{R}))$ scale, we must regard $J_\beta(\mathbb{R})$ as the union of a canonical sequence of Σ_{n-1} substructures of

$J_\beta(\mathbb{R})$, each of which collapses to some $J_\delta(\mathbb{R})$ with $\delta < \beta$. Our non-reflecting Σ_n type gives such a sequence.

Let $h : J_\beta(\mathbb{R}) \times \mathbb{R} \to J_\beta(\mathbb{R})$ be a Σ_{n-1} Skolem function. We want to standardize the parameter from which such a function is Σ_{n-1} definable. Fix θ_{sk} a Σ_{n-1} formula, w_1 a real, and F' a finite subset of $\omega\beta$ such that

$$h(u, x) = v \text{ iff } J_\beta(\mathbb{R}) \models \theta_{sk}[u, x, v, w_1, F'] .$$

Now fix $F \leq_{BK} F'$ to be the BK-least finite subset of $\omega\beta$ so that

$$\{\langle u, x, v\rangle \mid J_\beta(\mathbb{R}) \models \theta_{sk}[u, x, v, w_1, F]\}$$

is a Σ_{n-1} Skolem function for $J_\beta(\mathbb{R})$.

We next standardize the parameter satisfying a non-reflected Σ_n type. Let $b \in J_\beta(\mathbb{R})$ be such that $\Sigma^n_{b,\beta}$ is not realized in any $J_\gamma(\mathbb{R})$ for $\gamma < \beta$. Let $b = f(G', w_2)$, where f is $\Sigma_1(J_\beta(\mathbb{R}))$, G' is a finite subset of $\omega\beta$, and $w_2 \in \mathbb{R}$. Set

$$\Sigma = \Sigma^n_{\langle G', w_2\rangle, \beta} .$$

Clearly Σ is not realized in $J_\gamma(\mathbb{R})$ for any $\gamma < \beta$. Now fix $G \leq_{BK} G'$ to be BK-least so that

$$\Sigma = \Sigma^n_{\langle G, w_2\rangle, \beta} .$$

We can now define our canonical sequence $\langle H_i \mid i < \omega\rangle$ of Σ_{n-1} hulls. Simultaneously we shall define sequences $\langle K_i \mid i < \omega\rangle$ of finite subsets of $\omega\beta$ and $\langle \theta_i \mid i < \omega\rangle$ of Σ_n formulae in Σ. Let $K_0 = \emptyset$. Given $K_0, \ldots, K_i$, let

$$\begin{aligned} H_i &= \mathrm{Hull}^\beta_{n-1}(\{F, G, K_0 \ldots K_i\} \cup \mathbb{R}) \\ &= \{a \mid \{a\} \text{ is } \Sigma_{n-1}(J_\beta(\mathbb{R}), \{F, G, K_0 \ldots K_i\} \cup \mathbb{R})\} . \end{aligned}$$

Because we have thrown F into H_i, $H_i \prec_{n-1} J_\beta(\mathbb{R})$ and H_i is the image of $\mathbb{R}$ under a partial $\underset{\sim}{\Sigma}_{n-1}(H_i)$ function. Since $\rho^\beta_{n-1} \neq \mathbb{R}$, H_i must collapse to some $J_\gamma(\mathbb{R})$ for $\gamma < \beta$. But then $\pi(\langle G, w_2\rangle)$ does not realize Σ in $J_\gamma(\mathbb{R})$, where π is the collapse map, and therefore $\langle G, w_2\rangle$ does not realize Σ in H_i. Since $H_i \prec_{n-1} J_\beta(\mathbb{R})$, there must be a Σ_n formula $\theta(v)$ in Σ so that $H_i \not\models \theta[\langle G, w_2\rangle]$. Let

$$\begin{aligned} \theta_i = {} & \text{the least } \Sigma_n \text{ formula } \theta(v) \in \Sigma \\ & \text{such that } H_i \not\models \theta[\langle G, w_2\rangle] , \end{aligned}$$

where by "least" we mean least in some fixed ordering $<_\Sigma$ of Σ of order type ω. Now every element of $J_\beta(\mathbb{R})$ is Σ_1 definable from a real and a finite subset of

$\omega\beta$, so that for some finite $K \subseteq \omega\beta$, $\mathrm{Hull}^{\beta}_{n-1}(\{F, K\} \cup \mathbb{R}) \models \theta_i[\langle G, w_2\rangle]$. Let K_{i+1} be the BK-least finite subset K of $\omega\beta$ such that

$$\mathrm{Hull}^{\beta}_{n-1}(\{F, G, K_0, \ldots, K_i, K\} \cup \mathbb{R}) \models \theta_i[\langle G, w_2\rangle] .$$

This completes the definitions of the H_i's, K_i's, and θ_i's.

We record some simple properties of these sequences in a claim.

Claim 1.

(a) $\forall i\ (H_i \prec_{n-1} H_{i+1} \prec_{n-1} J_\beta(\mathbb{R}))$.

(b) $\forall i\ (H_i \not\models \theta_i[\langle G, w_2\rangle] \wedge H_{i+1} \models \theta_i[\langle G, w_2\rangle])$.

(c) $\forall i\ \exists \gamma < \beta\ (H_i \cong J_\gamma(\mathbb{R}))$.

(d) $\bigcup_{i<\omega} H_i = J_\beta(\mathbb{R})$.

Proof. (a) and (b) are obvious, and (c) was proved in the course of defining θ_i.

For (d), notice that

$$\bigcup_{i<\omega} H_i \models \theta[\langle G, w_2\rangle], \qquad \forall \theta \in \Sigma ,$$

because the θ_i's were chosen least in an ordering of type ω. Thus if π collapses $\bigcup_{i<\omega} H_i$ to a transitive set, then

$$\pi : \bigcup_{i<\omega} H_i \cong J_\beta(\mathbb{R}) ,$$

as otherwise $\pi(\langle G, w_2\rangle)$ would realize Σ in some $J_\gamma(\mathbb{R})$ for $\gamma < \beta$. We shall show $\pi(F) = F$, $\pi(G) = G$, and $\pi(K_i) = K_i$ for all i. Of course $\pi(x) = x$ for $x \in \mathbb{R}$. Since $\bigcup_{i<\omega} H_i = \mathrm{Hull}^{\bigcup_i H_i}_{n-1}(\{F, G, K_0, K_1, \ldots\} \cup \mathbb{R})$, it follows that $J_\beta(\mathbb{R}) = \mathrm{Hull}^{\beta}_{n-1}(\pi''\{F, G, \ldots\} \cup \mathbb{R})$, so $J_\beta(\mathbb{R}) = \mathrm{Hull}^{\beta}_{n-1}(\{F, G, \ldots\} \cup \mathbb{R})$, that is, $J_\beta(\mathbb{R}) = \bigcup_{i<\omega} H_i$.

To see that $\pi(F) = F$, notice that θ_{sk} defines a Σ_{n-1} Skolem function from $\langle F, w_1\rangle$ over $\bigcup_i H_i$, and thus from $\langle \pi(F), w_1\rangle$ over $J_\beta(\mathbb{R})$. But $\pi(F) \leq_{BK} F$, while F is BK-least so that θ_{sk} defines a Σ_{n-1} Skolem function from $\langle F, w_1\rangle$ over $J_\beta(\mathbb{R})$. Thus $\pi(F) = F$.

Similarly, $\langle \pi(G), w_2\rangle$ realizes Σ in $J_\beta(\mathbb{R})$ and $\pi(G) \leq_{BK} G$, so the BK-minimality of G implies $\pi(G) = G$.

Finally, $\pi(K_i) = K_i$ by induction on i. Certainly $\pi(K_0) = K_0$. If $\pi(K_j) = K_j$ for $j \leq i$, then

$$\pi''H_{i+1} = \mathrm{Hull}^{\beta}_{n-1}(\{F, G, K_0 \ldots K_i, \pi(K_{i+1})\} \cup \mathbb{R}) .$$

Since $\pi''H_{i+1} \models \theta_i[\langle G, w_2\rangle]$, $\pi(K_{i+1}) \leq_{BK} K_{i+1}$, and K_{i+1} is BK-least so that

$\mathrm{Hull}^{\beta}_{n-1}(\{F, G, K_0 \ldots K_{i+1}\} \cup \mathbb{R}) \models \theta_i[\langle G, w_2\rangle]$, we have $\pi(K_{i+1}) = K_{i+1}$.

This proves Claim 1.

Let P be a $\underset{\sim}{\Sigma}_n(J_\beta(\mathbb{R}))$ set of reals. We want a closed game representation of P simple enough to yield a $\underset{\sim}{\Sigma}_n(J_\beta(\mathbb{R}))$ scale on P. By part (d) of Claim 1, we can fix $e \in \omega$ so that P is $\Sigma_n(J_\beta(\mathbb{R}), \{F, G, K_0 \ldots K_e\} \cup \mathbb{R})$. Let

$$P(x) \text{ iff } J_\beta(\mathbb{R}) \models \varphi_0[x, a]$$

where φ_0 is Σ_n, and a is a parameter of the form $\langle F, G, K_0, \ldots, K_e, w_3\rangle$ for $w_3 \in \mathbb{R}$. For $i \geq e$, let

$$P^i(x) \text{ iff } H_i \models \varphi_0[x, a] .$$

We shall construct closed game representations $x \longmapsto G^i_x$ of the P^i's in such a way that if

$$P^i_k(x, u) \text{ iff } u \text{ is a winning position for I in } G^i_x \text{ of length } k ,$$

then P^i_k is $\underset{\sim}{\Sigma}_\omega(H_{\max(i,k)})$. Suppose we have done this. Inspecting [12] again, we see that this gives a scale $\{\varphi^i_k\}_{k\in\omega}$ on P^i so that the prewellordering $\leq^i_k$ of $\mathbb{R}$ induced by φ^i_k is $\underset{\sim}{\Sigma}_\omega(H_{\max(i,k)})$. But $H_{\max(i,k)}$ collapses to some $J_\gamma(\mathbb{R})$ with $\gamma < \beta$, and $\leq^i_k$ is fixed by the collapse map, so $\leq^i_k \in J_\beta(\mathbb{R})$. Similarly, each $P^i \in J_\beta(\mathbb{R})$. But now $\rho^\beta_n = \mathbb{R}$, so any countable subset of $J_\beta(\mathbb{R})$ is $\underset{\sim}{\Sigma}_n(J_\beta(\mathbb{R}))$, so the map $(i, k) \longmapsto (P^i, \leq^i_k)$ is $\underset{\sim}{\Sigma}_n(J_\beta(\mathbb{R}))$. We then have the obvious scale $\{\psi_i\}_{i<\omega}$ on P:

$$\psi_0(x) = \mu i[P^i(x)] ,$$

and

$$\psi_{k+1}(x) = \langle \psi_0(x), \varphi^{\psi_0(x)}_k \rangle ,$$

where again ψ_{k+1} is defined using the lexicographic order. Since $(i, k) \longmapsto (P^i, \leq^i_k)$ is $\underset{\sim}{\Sigma}_n(J_\beta(\mathbb{R}))$, the scale $\{\psi_i\}_{i<\omega}$ is $\underset{\sim}{\Sigma}_n(J_\beta(\mathbb{R}))$.

It suffices, then, to construct the representations $x \longmapsto G^i_x$ in such a way that P^i_k is first-order definable over $H_{\max(i,k)}$ for all i, k. Our plan is to force I in G^i_x to describe the truth in $J_\beta(\mathbb{R})$ about F, G, and the K_j's. Since

$$J_\beta(\mathbb{R}) = \mathrm{Hull}^{\beta}_{n-1}(\{F, G\} \cup \{K_j \mid j < \omega\} \cup \mathbb{R})$$

it is enough that I describe only Σ_{n-1} truths; moreover, this restriction is important if P^i_k is to be first-order definable over $H_{\max(i,k)}$. For this reason we also restrict I to playing at move k only ordinals in H_k and formulae involving no K_j for $j > k$. Our main problem is how, within these restrictions, to prevent I from describing the Σ_{n-1} truths about some parameters different from F, G, or the K_j's.

Player I's description is given in the language $\mathcal{L}$, which has $\in$, $=$, constant symbols $\underline{x}_i$ for $i \in \omega$, and constant symbols $\underline{F}$, $\underline{G}$, and $\underline{K}_i$ for $i \in \omega$. If φ is an $\mathcal{L}$-formula containing no constants $\underline{K}_i$ for $i > m$, we say φ has support m.

Let $\mathcal{B}_{n-1}$ be the class of boolean combinations of Σ_{n-1} formulae of $\mathcal{L}$. I will actually describe the truth of $\mathcal{B}_{n-1}$ formulae. We shall use the "unique v" operator applied to Σ_{n-1} formulae to abbreviate formulae in $\mathcal{B}_{n-1}$. Let σ, τ, and φ be Σ_{n-1} formulae, and let ψ be Π_{n-1}. The reader can easily check that $\varphi(\iota v\sigma(v))$, $\psi(\iota v\sigma(v))$, and $\psi(\iota v\sigma(v, \iota u\tau(u)))$ can be considered abbreviations of $\mathcal{B}_{n-1}$ formulae. These are the sorts of abbreviations we shall use.

For expository reasons, we allow I to play finitely many sentences and finitely many reals in a single move of G^i_x. A typical run of G^i_x then has the form

I	T_0, s_0, η_0, m_0		T_1, s_1, η_1, m_1	...
II		y_0		y_1

where T_k is a finite set of sentences in $\mathcal{B}_{n-1}$, all of which have support k, $s_k \in \mathbb{R}^{<\omega^k}$, $\eta_k \in OR \cap H_k$, $m_k \in \omega$ and $m_k > k$, and $y_k \in \mathbb{R}$. The roles of these objects have been explained, with the exception of the m_k's. We shall explain their role shortly.

Given the run of G^i_x displayed above, set

$$\langle x_i \mid i < \omega\rangle = s_0{}^\frown\langle y_0\rangle{}^\frown s_1{}^\frown\langle y_1\rangle{}^\frown\cdots,$$

and

$$T^* = \bigcup_{k<\omega} T_k .$$

We say this run is a win for I just in case it meets the following requirements. Let $n : \mathcal{B}_{n-1} \xrightarrow{1-1} \omega$ be such that any $\theta \in \mathcal{B}_{n-1}$ has support $n(\theta)$ and involves no $\underline{x}_j$ for $j \geq n(\theta)$.

1. $s_0(0) = x$, $s_0(1) = w_1$, $s_0(2) = w_2$, and $s_0(3) = w_3$.
2. (a) T^* is consistent,
   (b) if $\theta \in \mathcal{B}_{n-1}$ and θ is a sentence, then either $\theta \in T_{n(\theta)}$ or $\sim\theta \in T_{n(\theta)}$,
   (c) if $\theta \in T_k$, then θ has support k and involves no constant $\underline{x}_j$ for $j \geq \mathrm{dom}(s_0{}^\frown\langle y_0\rangle{}^\frown\cdots{}^\frown\langle y_{k-1}\rangle{}^\frown s_k)$,

(d) The Axiom of Extensionality is in T_0,

(e) if $\tau(v)$ is Σ_{n-1} and $\exists u(u = \iota v\tau(v)) \in T_k$, then "$\exists\alpha(\iota v\tau(v) \in J_\alpha(R))$" $\in T_{k+1}$,

(f) $\underline{x}_k(\underline{n}) = \underline{m} \in T^*$ iff $x_k(n) = m$.

Requirement 2(e) could be replaced by "$V = L(R) \in T^*$" in the case $n > 2$.

Next we require I to verify that $\varphi_0(x, a)$ holds in H_i. Let $\varphi_0 = \exists v\, \psi_0$, where ψ_0 is Π_{n-1}.

3. For some Σ_{n-1} formula $\tau(v)$ with support i,
$\psi_0(\underline{x}_0, \langle \underline{F}, \underline{G}, \underline{K}_0 \ldots \underline{K}_e, \underline{x}_3\rangle, \iota v\tau(v)) \in T_i$.

Player I must verify that any real he describes is one of the x_k's:

4. If $\tau(v)$ is Σ_{n-1} and "$\iota v\tau(v) \in R$" $\in T_k$, then for some j, "$\iota v\tau(v) = \underline{x}_j$" $\in T_{k+1}$.

Player I must verify that the model he is describing is well-founded:

5. If $\sigma(v)$ and $\tau(v)$ are Σ_{n-1}, and "$\iota v\sigma(v) \in OR \wedge \iota v\tau(v) \in OR$" $\in T^*$, then "$\iota v\sigma(v) \leq \iota v\tau(v)$" $\in T^*$ iff $\eta_{n(\sigma)} \leq \eta_{n(\tau)}$.

Finally, I must verify that $\underline{F}$ denotes F, $\underline{G}$ denotes G, and $\underline{K}_j$ denotes K_j for $j < \omega$. In order to do so, I must verify certain Σ_n and Π_n sentences which come up as he plays. Now in order to verify $\forall v\theta(v)$, where θ is Σ_{n-1}, I must simply refrain from putting $\sim\theta(\iota v\sigma(v))$ into T^* for any Σ_{n-1} $\sigma(v)$. This is a closed requirement on I's play. Dually, in order to verify $\exists v\theta(v)$, where θ is Π_{n-1}, I must put $\theta(\iota v\sigma(v))$ into T^* for some $\Sigma_{n-1}\sigma(v)$. Unfortunately, this is an open requirement on I's play. In simple cases, such as requirement 3, we can bound in advance the move at which I is expected to verify his Σ_n sentence, so that the requirement becomes closed. In view of our restriction on sentences in T_k to those with support k, this amounts to bounding in advance the hull which is to satisfy I's Σ_n sentence. In some cases we can't do this, and so we require I himself to provide the bound. This is the role of m_k; it is I's prediction at move k of the later move at which he will discharge an accrued obligation to verify certain Σ_n sentences. (In 7(c)(ii) below, m_k plays a similar but slightly different role.)

One may object that this violates the spirit of our restrictions on the first k moves: since the m_j for $j < k$ can refer to arbitrarily large hulls, P^i_k won't be first-order definable over $H_{\max(i,k)}$. The Coding Lemma (of all things!) will overcome this objection.

Requirements 6(a) and (b) force I to assert that θ_{sk} defines a Σ_{n-1} Skolem function from $\underline{F}$ and $\underline{x}_1$, while 6(c) forces him to assert that nothing $<_{BK} \underline{F}$ has this property of $\underline{F}$.

6. (a) The sentence

$$\forall v, w, u, x\, [(\theta_{sk}(u, x, v, \underline{F}, \underline{x}_1) \wedge \theta_{sk}(u, x, w, \underline{F}, \underline{x}_1)) \Rightarrow v = w]$$

is in T_0,

(b) if $\theta(v_0, v_1)$ and $\tau(u)$ are Σ_{n-1}, and $\exists v\, \theta(v, \iota u\tau(u)) \in T_k$, then for some j

$$\exists v(\theta(v, \iota u\tau(u)) \wedge \theta_{sk}(\iota u\tau(u), \underline{x}_j, v, \underline{F}, \underline{x}_1))$$

is in T_{k+1},

(c) if "$\iota v\sigma(v) <_{BK} \underline{F}$" $\in T_k$, then either the sentence

$$\exists v, w, u, x\, (\theta_{sk}(u, x, v, \iota v\sigma(v), \underline{x}_1) \wedge$$
$$\theta_{sk}(u, x, w, \iota v\sigma(v), \underline{x}_1) \wedge v \neq w)$$

is in T_{m_k}, or for some Σ_{n-1} formulae $\theta(v_0, v_1)$ and $\tau(u)$, the sentence

$$\exists v\, \theta(v, \iota u\tau(u)) \wedge$$
$$\sim \exists x\, \exists v\, (\theta_{sk}(\iota u\tau(u), x, v, \iota v\sigma(v), \underline{x}_1) \wedge \theta(v, \iota u\tau(u)))$$

is in T_{m_k}.

A similar requirement will fix the meaning of $\underline{G}$. Recall the Σ_n formula $\theta_j(v)$ in Σ used to define K_{j+1}, and the order $<_\Sigma$ of Σ.

7. (a) If $\forall u\, \theta(u, v) \in \Sigma$, where θ is Σ_{n-1}, and if $\tau(u)$ is any Σ_{n-1} formula, then

$$\sim\theta\,(\iota u\tau(u), \langle \underline{G}, \underline{x}_2\rangle) \notin T_k$$

(b) if $\exists u\, \theta(u, v) \in \Sigma$, where θ is Π_{n-1}, and $\exists u\, \theta(u, v) <_\Sigma \theta_k$, then for some Σ_{n-1} formula $\tau(u)$

$$\theta(\iota u\tau(u),\ \underline{G}, \underline{x}_2\ ) \in T_k\ ,$$

(c) if "$\iota v\sigma(v) <_{BK} \underline{G}$" $\in T_k$, where $\sigma(v)$ is Σ_{n-1}, then either

(i) there is a Σ_{n-1} formula $\theta(u, v)$ such that $\forall u\, \theta(u, v) \in \Sigma$, but for some Σ_{n-1} $\tau(u)$,

$$\sim\theta\,(\iota u\tau(u), \langle \iota v\sigma(v), \underline{x}_2\rangle) \in T_{m_k}$$

or

(ii) there is a Π_{n-1} formula $\theta(u, v)$ such that $\exists u\, \theta(u, v)$ is one of the first m_k elements of Σ under $<_\Sigma$, and for all Σ_{n-1} $\tau(u)$ and all $j \in \omega$

$$\theta(\iota u\tau(u), \langle \iota v\sigma(v), \underline{x}_2\rangle) \notin T_j\ .$$

Finally, we have a requirement fixing the meaning of $\underline{K}_j$. Recall the sequence $\langle \theta_j \mid j < \omega \rangle$ of Σ_n formulae in Σ. Let $\theta_j = \exists u\, \theta'_j$, where θ'_j is Π_{n-1}.

8. (a) "$\underline{K}_0 = \emptyset$" $\in T_0$

(b) For any $j > 0$, there is a Σ_{n-1} $\tau(u)$ with support j such that

$$\theta'_{j-1}(\iota u \tau(u), \langle \underline{G}, \underline{x}_2 \rangle) \in T_j ,$$

(c) For any $j > 0$, and any k, if

$$\text{"}\iota v \sigma(v) <_{BK} \underline{K}_j\text{"} \in T_k ,$$

then for no Σ_{n-1} $\tau(u, v)$ with support $j - 1$ and no ℓ do we have

$$\theta'_{j-1}(\, u\tau(u, i v \sigma(v)), \langle \underline{G}, \underline{x}_2 \rangle) \in T_\ell .$$

This completes the description of the payoff for I in G^i_x.

We next want to characterize the winning positions for I in G^i_x as the honest ones. Let

$$u = \langle \langle T_k, s_k, \eta_k, m_k, y_k \rangle \mid k < s \rangle$$

be a position of length s. Let

$$\langle x_j \mid j < \ell \rangle = s_0 ^\frown \langle y_0 \rangle ^\frown \cdots ^\frown s_k ^\frown \langle y_k \rangle ,$$

and let

$$I_u(\underline{x}_j) = x_j \quad \text{for } j < \ell ,$$

$$I_u(\underline{F}) = F, \qquad I_u(\underline{G}) = G ,$$

and

$$I_u(\underline{K}_j) = K_j \quad \text{for all } j < \omega .$$

Let us call u reasonable if it is not an immediate loss for I because of $\langle T_k \mid k < s \rangle$, in the sense that all assertions in 1-4 and 6-8 about the T_k's hold so far for the T_k with $k < s$. [Take for example 7(c). If u is unreasonable on its account, there must be a $k < s$ such that $m_k < s$, and a formula "$\iota v \sigma(v) <_{BK} \underline{G}$" in T_k such that 7(c)(i) fails for T_{m_k} while there are enough $j < s$ to witness the failure of 7(c)(ii).]

Finally, we say that u is (i, x)-honest if the following conditions are met:

1. $H_i \models \varphi_0[x, a]$.
2. u is reasonable.
3. $s > 0 \Rightarrow s_0(0) = x \wedge s_0(1) = w_1 \wedge s_0(2) = w_2 \wedge s_0(3) = w_3$.
4. $(J_\beta(\mathbb{R}), I_u) \models \bigcup_{k<s} T_k$.

5. if $\langle \sigma_0, \ldots, \sigma_m \rangle$ enumerates those Σ_{n-1} formulae σ such that $n(\sigma) < s$ and

$$(J_\beta(R), I_u) \models \iota v\sigma(v) \in OR ,$$

and if $\delta_i < \omega\beta$ is such that

$$(J_\beta(R), I_u) \models \iota v\sigma_i(v) = \delta_i$$

then the map

$$\delta_i \longmapsto \eta_{n(\sigma_i)}$$

is well-defined and extendible to an order-preserving map of $OR \cap H_s$ into $OR \cap H_s$.

The final two requirements for (i, x)-honesty guarantee that I has made commitments m_k, $k < s$, which can be kept.

6. If $k < s$ and "$\iota v\sigma(v) <_{BK} \underline{F}$" $\in T_k$, then either

$$\exists v, w, u, x\, (\Theta_{sk}(u, x, v, \iota v\sigma(v), \underline{x}_1) \wedge$$
$$\Theta_{sk}(u, x, w, \iota v\sigma(v), \underline{x}_1) \wedge v \neq w)$$

is true in (H_{m_k}, I_u), or for some Σ_{n-1} formula $\Theta(v_0, v_1)$

$$\exists u[\, \exists v\, \Theta(v, u) \wedge \sim \exists x\, \exists v\, (\Theta(v, u)$$
$$\wedge\, \Theta_{sk}(u, x, v, \iota v\sigma(v), \underline{x}_1))]$$

is true in (H_{m_k}, I_u).

7. If $k < s$ and "$\iota v\sigma(v) <_{BK} \underline{G}$" $\in T_k$, then either there is a Π_n formula $\Theta(v) \in \Sigma$ such that

$$(H_{m_k}, I_u) \models \sim \Theta(\langle \iota v\sigma(v), \underline{x}_2 \rangle) ,$$

or there is a Σ_n formula $\Theta(v) \in \Sigma$ which is one of the first m_k elements of Σ such that

$$(J_\beta(R), I_u) \models \sim \Theta(\langle \iota v\sigma(v), \underline{x}_2 \rangle) .$$

Because of conditions 6 and 7 in the definition of (i, x)-honesty, the following claim requires a proof.

<u>Claim 2</u>. $\{(x, u) \mid u$ is an (i, x)-honest position of length $s\}$ is $\underset{\sim}{\Sigma}_\omega(H_{\max(i,s)})$.

<u>Proof</u>. Conditions 1, 2, and 3 are trivially $\underset{\sim}{\Sigma}_\omega(H_{max(i,s)})$. Condition 4 of honesty is first order over $H_{max(i,s)}$ because of our restrictions on the sentences in T_k for $k < s$. Condition 5 is first order because if there is any map as required by 5, there is a $\underset{\sim}{\Sigma}_1(H_s)$ such map, and H_s is $\underset{\sim}{\Sigma}_{n+5}$ over $H_{max(i,s)}$.

We turn now to condition 7; since the proof that 6 is $\underset{\sim}{\Sigma}_\omega(H_{max(i,s)})$ is similar but simpler, we shall omit it. Let

$$A = \{\langle K, \theta, m\rangle \mid K <_{BK} G \wedge K \in H_s \wedge \theta \in \Sigma \wedge (H_m, \in) \models \theta[\langle K, w_2\rangle]\} .$$

It suffices to show that A is $\underset{\sim}{\Sigma}_\omega(H_s)$; from this it easily follows that condition 7 is $\underset{\sim}{\Sigma}_\omega(H_{max(i,s)})$.

Since there is a $\underset{\sim}{\Sigma}_{n-1}(H_s)$ partial map of $\mathbb{R}$ onto H_s, there is a $\underset{\sim}{\Sigma}_n(H_s)$ total map

$$f : \mathbb{R} \xrightarrow{\text{onto}} \{K \in H_s \mid K <_{BK} G\} \times \Sigma .$$

For $y, z \in \mathbb{R}$, let

$$y \leq^* z \text{ iff } f(y)_0 <_{BK} f(z)_0 \vee (f(y)_0 = f(z)_0 \wedge f(y)_1 \leq_\Sigma f(z)_1) .$$

Thus $\leq^*$ is a $\underset{\sim}{\Sigma}_n(H_s)$ prewellorder of $\mathbb{R}$. Since $H_s \cong J_\gamma(\mathbb{R})$ for some $\gamma < \omega$, $\leq^* \in J_\beta(\mathbb{R})$. Now for $m < \omega$, let

$$A_m = \{\langle K, \theta\rangle \mid \langle K, \theta, m\rangle \in A\} .$$

Now A_m is $\underset{\sim}{\Sigma}_\omega(H_{max(m,s)})$, so $f^{-1}(A_m)$ is $\underset{\sim}{\Sigma}_\omega(H_{max(m,s)})$, and so $f^{-1}(A_m) \in J_\beta(\mathbb{R})$ for all m. By the Coding Lemma ([14]), $f^{-1}(A_m)$ is $\underset{\sim}{\Sigma}^1_1(\leq^*)$ for each m. [Since $\leq^*$, $f^{-1}(A_m) \in J_\beta(\mathbb{R})$, $\text{Det}(J_\beta(\mathbb{R}))$ suffices for this application of the Coding Lemma. Now $J_\alpha(\mathbb{R}) \prec_1^{\mathbb{R}} J_\beta(\mathbb{R})$, so if there is a non-determined game in $J_\beta(\mathbb{R})$, there is a non-determined game in $J_\alpha(\mathbb{R})$. However, we have assumed $\text{Det}(J_\alpha(\mathbb{R}))$.] Since $\underset{\sim}{\Sigma}^1_1(\leq^*)$ is closed under countable unions, we have $B \in \underset{\sim}{\Sigma}^1_1(\leq^*)$, where

$$B = \bigcup_m (f^{-1}(A_m) \times \{m\}) .$$

But then B is $\underset{\sim}{\Sigma}_\omega(H_s)$. Since

$$(K, \theta, m) \in A \text{ iff } \exists y \in \mathbb{R}\, (\langle y, m\rangle \in B \wedge f(y)_0 = K \wedge f(y)_1 = \theta)$$

we have that A is $\underset{\sim}{\Sigma}_\omega(H_s)$, as desired. This proves Claim 2.

Our original proof of Claim 2 required more than $\mathrm{Det}(J_\alpha(\mathbb{R}))$, since we had applied the Coding Lemma directly to A. Kechris had the idea of applying the Coding Lemma to the A_m's individually. This reduces the determinacy needed in the hypotheses of 3.7 to $\mathrm{Det}(J_\alpha(\mathbb{R}))$, a reduction which is crucial for the use of 3.7 in Kechris-Woodin [7].

The next claim finishes our proof of 3.7 in the case that $n > 1$.

Claim 3. For all i, x, and u, u is (i, x)-honest iff u is a winning position for I in G^i_x.

Proof (Sketch). ($\Rightarrow$) It is enough to show that whenever v is (i, x)-honest, then $\exists\, T, s, \eta, m\ \forall y\ (v^\frown\langle T, s, \eta, m, y\rangle$ is (i, x)-honest). For if so, then I can win G^i_x by keeping to (i, x)-honest positions. So let v be an (i, x)-honest position of length k. Because v satisfies 2 and 4 of honesty, I has told as much of the truth about $J_\beta(\mathbb{R})$ in his first k moves as we required him to tell. Because v satisfies 6 and 7 of honesty, I has made predictions m_i for $i < k$ which he can fulfill. Thus I can choose T and s to insure continued satisfaction of 2 and 4 by the new position; that is, he can tell as much more of the truth as he must. (If $k = i$, we need that v satisfies 1 as well.) Continued satisfaction of 1 and 3 are trivial to insure, while continued satisfaction of 6 and 7 can be insured by choosing m_k large enough. Finally, I must choose η so as to insure continued satisfaction of 5. Now let v determine the map

$$g(\delta_i) = \eta_{n(\sigma_i)}, \qquad \text{for } 0 \leq i \leq m ,$$

as in 5 of honesty for v. We may assume that $i \leq j \Rightarrow \delta_i \leq \delta_j$. It is enough to see that g can be extended to an order-preserving map of $OR \cap H_{k+1}$ into $OR \cap H_{k+1}$. Now by 5 for v and the fact that $H_k \prec_1 J_\beta(\mathbb{R})$,

$$H_k \models \exists f : [\delta_i, \delta_{i+1}] \xrightarrow{op} [g(\delta_i), g(\delta_{i+1})]$$

for all $i < m$. Since $H_k \prec_1 H_{k+1}$,

$$H_{k+1} \models \exists f : [\delta_i, \delta_{i+1}] \xrightarrow{op} [g(\delta_i), g(\delta_{i+1})]$$

for all $i < m$. Similarly,

$$H_{k+1} \models \exists f : [0, \delta_0] \xrightarrow{op} [0, g(\delta_0)] .$$

Thus in order to obtain the desired extension of g, it is enough to show that there is an order-preserving $f : (OR \cap H_{k+1}) - \delta_m \longrightarrow (OR \cap H_{k+1}) - g(\delta_m)$. But such an f fails to exist just in case for some $n < \omega$

$$H_{k+1} \models (g(\delta_m) - \delta_m) \cdot n \text{ doesn't exist} ,$$

as the following diagram makes clear:

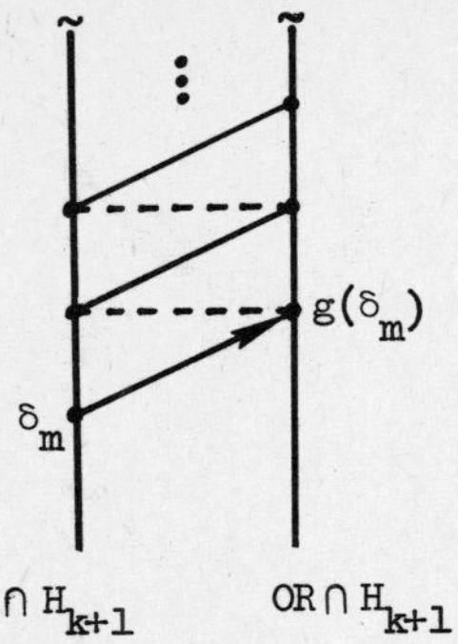

This is a Π_1 property of H_{k+1}, and so would pass downward to H_k, which is impossible since g can be extended over H_k. Thus there is such an f.

($\Leftarrow$) Let S be a winning strategy for I starting from u. As in 2.1, let $\langle\langle T_k, s_k, \eta_k, m_k, y_k\rangle \mid k < \omega\rangle$ be a "generic play" according to S, and let $\langle x_k \mid k < \omega\rangle = s_0{}^\frown\langle y_0\rangle{}^\frown s_1{}^\frown\langle y_1\rangle \cdots$. Since $\bigcup_k T_k$ is consistent, it has a model $\mathfrak{B}$. By payoff requirements 6(a) and (b), $\mathfrak{A} < \mathfrak{B}$ where

$$\mathfrak{A} = \mathrm{Hull}^{\mathfrak{B}}(\{\underline{x}_j^{\mathfrak{B}} \mid j < \omega\} \cup \{\underline{F}^{\mathfrak{B}}, \underline{G}^{\mathfrak{B}}\} \cup \{\underline{K}_j^{\mathfrak{B}} \mid j < \omega\}) .$$

If $\mathfrak{A} \models \iota v\sigma(v) \in OR$, then set

$$f(\iota v\sigma(v)^{\mathfrak{A}}) = \eta_{n(\sigma)} .$$

By payoff requirement 5, f is well-defined and order-preserving. Thus

$$\mathfrak{A} \cong (J_\gamma(R^{\mathfrak{A}}), I)$$

for some $\gamma \leq \beta$ and interpretation of constants I. By payoff requirements 7(a) and (b), $\mathfrak{A}$ realizes Σ, so $\gamma = \beta$. By 2(f), 4, and genericity, $R^{\mathfrak{A}} = R^M$, where M is our ground model. Clearly

$$I(\underline{x}_j) = x_j \quad \text{for } j < \omega ,$$

while payoff requirements 6, 7, and 8 insure that

$$I(\underline{F}) = F, \quad I(\underline{G}) = G ,$$

and

$$I(\underline{K}_j) = K_j \quad \text{for } j < \omega .$$

Therefore

$$\mathfrak{A} = (J_\beta(\mathbb{R}), I)$$

where I is the natural interpretation. It is easy to verify conditions 1 through 4 and conditions 6 and 7 of (i, x)-honesty for u at this point. For condition 5, notice that the order-preserving map f defined above actually maps any H_s into H_s (since I is the natural interpretation). Thus f witnesses the satisfaction of condition 5.

Of course, Claim 3 applied to the empty position implies that $x \longmapsto G^i_x$ is a closed game representation of P^i. Claims 2 and 3 imply that the sets P^i_s are simply definable enough to yield $\underset{\sim}{\Sigma}_n(J_\beta(\mathbb{R}))$ scale on P.

We are left with the proof of the theorem in the case n = 1. Since the basic plan in this case is the same as the plan in the case n > 1, we shall just sketch a few of the details.

Assume first that β is a limit ordinal and n = 1. We shall need no analogue of F. Let G be a finite subset of β and w_1 a real such that Σ, the Σ_1-type of $\langle G, w_1\rangle$ over $J_\beta(\mathbb{R})$, is not realized in $J_\gamma(\mathbb{R})$ for any $\gamma < \beta$. Further, let G be BK-least so that $\langle G, w_1\rangle$ realizes Σ. Define $\beta_i < \beta$ and $\theta_i(v) \in \Sigma$ by: $\beta_0 = 0$, and

$$\theta_i = \text{least } \Sigma_1 \text{ formula } \theta(v) \in \Sigma \text{ such that } J_{\beta_i}(\mathbb{R}) \not\models \theta[\langle G, w_1\rangle]$$

and

$$\beta_{i+1} = \text{least } \beta > \beta_i \text{ such that } J_\beta(\mathbb{R}) \models \theta_i[\langle G, w_1\rangle] .$$

The β_i's are analogues of the K_i's. Let also

$$H_i = \text{Hull}_\omega^{\beta_{i+1}}(\{G, \beta_0 \ldots \beta_i\} \cup \mathbb{R}) .$$

Of course $H_i \cong J_\gamma(\mathbb{R})$ for some $\gamma < \beta$, moreover, H_i is the image of $\mathbb{R}$ under a $\underset{\sim}{\Sigma}_1(H_{i+1})$ map. As before, $\bigcup_{i<\omega} H_i = J_\beta(\mathbb{R})$.

Let

$$P(x) \Longleftrightarrow J_\beta(\mathbb{R}) \models \varphi[x] ,$$

where φ is Σ_1 and we have dropped the parameter for simplicity. Let

$$P^i(x) \Longleftrightarrow H_i \models \varphi[x] .$$

Again, it is enough to construct representations $x \longmapsto G^i_x$ of the P^i in such a way that the corresponding P^i_k are first-order over $H_{\max(i,k)}$. For this, it is convenient to have in I's language $\mathcal{L}$, besides the $\underline{x}_i$, constants $\underline{\beta}_i$ and $\underline{J_{\beta_i}(R)}$ for $i < \omega$. In G^i_x, I must play a Σ_0-complete, consistent set of Σ_0 sentences of $\mathcal{L}$, mentioning at move k no sentences involving $\underline{\beta}_j$ or $\underline{J_{\beta_j}(R)}$ for $j > k$. For each j, I must play at move j the Σ_0 sentence "$\underline{J_{\beta_j}(R)} \models V = L(R)$". At move $i + 1$, I must assert that some object definable over $J_{\beta_{i+1}}(R)$ from $G, \beta_0,,.\beta_i$, and some real (which he has played) witnesses that $\varphi(\underline{x}_0)$. I must play ordinals $\eta_k \in H_k$ to prove well-foundedness. Finally, I must prove that he is interpreting his constant symbols correctly; for $\underline{G}$ this involves commitments m_k made at move k as in the case $n > 1$. The Coding Lemma argument of Claim 2 goes through because each H_k is the image of R under a map in $J_\beta(R)$. Claim 3 is proved as in the case $n > 1$.

Finally, we have the case $\beta = \gamma + 1$ and $n = 1$. Let $\langle G, w\rangle$ realize over $J_\beta(R)$ a non-reflected Σ_1-type. Now $G = h(p, J_\gamma(R))$ for some rud function h and $p \in J_\gamma(R)$. Using the simplicity of rud functions (as in 2.1), we can recursively associate to any Σ_1 formula θ and any $n \in \omega$ a formula $\psi_{\theta,n}$ so that for all α and all $a \in J_\alpha(R)$,

$$J_\alpha(R) \models \psi_{\theta,n}[a] \text{ iff } S_{\omega\alpha+n}(R) \models \theta[h(a, J_\alpha(R))] .$$

It follows that if Σ is the full elementary type realized by $\langle p, w\rangle$ over $J_\gamma(R)$, then Σ is not realized in $J_\delta(R)$ for any $\delta < \gamma$. Again, we may replace p by some finite $G \subseteq \omega\gamma$ (changing w) and take G to be BK-least so that $\langle G, w\rangle$ realizes Σ over $J_\gamma(R)$. Then $J_\gamma(R) = \text{Hull}(\{G\} \cup R)$.

Let

$$P(x) \Longleftrightarrow J_\beta(R) \models \varphi[x]$$

where φ is Σ_1 and we have ignored the parameter. Let

$$P^i(x) \Longleftrightarrow S_{\omega\gamma+i}(R) \models \varphi[x] .$$

As above, we can find first-order formulae ψ_i such that

$$P^i(x) \Longleftrightarrow J_\gamma(R) \models \psi_i[x] .$$

We construct the desired representation $x \longmapsto G^i_x$ of P^i as follows. Let $\mathcal{L}$ have $\in, =, \underline{x}_i$ for $i < \omega$, and $\underline{G}$. In G^i_x player I must produce a (fully) complete, consistent $\mathcal{L}$ theory containing $\psi_i(\underline{x}_0)$, while using ordinals less than $\omega\gamma$ to prove well-foundedness. At move k he can put only Σ_k sentences into his theory. His theory must be "Skolemized", and whenever he describes a real he must place it bodily on the board, as before. Finally, I must prove that he is interpreting $\underline{G}$ by G; this again involves commitments of the form m_k. The Coding Lemma argument of Claim 2 goes through because for each k we have a map $f : \mathbb{R} \to J_\gamma(\mathbb{R})$ such that $f \in J_\beta(\mathbb{R})$ and $G \in f''\mathbb{R}$ and $f''\mathbb{R} \prec_k J_\gamma(\mathbb{R})$. Claim 3 is proved as in the case $n > 1$.

This completes the proof of 3.7.

In each case in the proof of 3.7, we expressed our arbitrary $\underset{\sim}{\Sigma}_n(J_\beta(\mathbb{R}))$ set of reals as a countable union of sets belonging to $J_\beta(\mathbb{R})$. Thus we have

3.8. __Corollary__. If $[\alpha, \beta]$ is a weak Σ_1-gap, and n is least such that $\rho_n^\beta = \mathbb{R}$, then $P \in \underset{\sim}{\Sigma}_n(J_\beta(\mathbb{R})) \cap P(\mathbb{R})$ iff $P = \bigcup_i P^i$, where $P^i \in J_\beta(\mathbb{R})$ for all i.

3.9. __Corollary__. If $[\alpha, \beta]$ is a weak Σ_1-gap, and n is least such that $\rho_n^\beta = \mathbb{R}$, then each of the classes $\underset{\sim}{\Sigma}_{n+2k}(J_\beta(\mathbb{R}))$ and $\underset{\sim}{\Pi}_{n+2k+1}(J_\beta(\mathbb{R}))$, for $k < \omega$, has the scale property.

__Proof__. By the second periodicity theorem, it is enough to show that

$$\underset{\sim}{\Sigma}_{n+k+1}(J_\beta(\mathbb{R})) \cap P(\mathbb{R}) = \exists^{\mathbb{R}}(\underset{\sim}{\Pi}_{n+k}(J_\beta(\mathbb{R}))) \cap P(\mathbb{R}) .$$

The proof of this is like the proof of Lemma 2.5, the key fact being that

$$\Sigma_n(J_\beta(\mathbb{R})) \cap P(\mathbb{R}) \subseteq \exists^{\mathbb{R}}(\Pi_n(J_\beta(\mathbb{R})) .$$

This fact follows trivially from 3.8.

Our description of the levels of the Levy hierarchy for $L(\mathbb{R})$ having the scale property is now complete. It is interesting that the negative results on scales in $L(\mathbb{R})$ (2.8, 2.10, and 3.4) all come from negative results on uniformization (2.7, 2.9, and 3.3). In $L(\mathbb{R})$, the best way to uniformize arbitrary relations in a given Levy class is by means of scales on those relations, as we claimed in the introduction.

It follows easily from Wadge's Lemma that if $A \in \underset{\sim}{\Pi}_1(J_\alpha(\mathbb{R})) - \underset{\sim}{\Sigma}_1(J_\alpha(\mathbb{R}))$ and A admits a $\underset{\sim}{\Sigma}_n(J_\beta(\mathbb{R}))$ scale, then so does every $\underset{\sim}{\Pi}_1(J_\alpha(\mathbb{R}))$ set. This observation and an inspection of our results above yield the promised characterization of the smallest Levy class $\underset{\sim}{\Sigma}_n(J_\beta(\mathbb{R}))$ at which a scale on a given set A is definable. Let $\langle\beta, n\rangle$ be lexicographically least so that A admits a $\underset{\sim}{\Sigma}_n(J_\beta(\mathbb{R}))$ scale; then $A \in \underset{\sim}{\Sigma}_n(J_\beta(\mathbb{R}))$ and either $\underset{\sim}{\Sigma}_n(J_\beta(\mathbb{R}))$ or $\underset{\sim}{\Sigma}_{n+1}(J_\beta(\mathbb{R}))$ has the scale property. Our results on the scale property therefore characterize $\langle\beta, n\rangle$ by means of reflection properties.

§4. Suslin cardinals. We shall use the results of §2 and §3 to extend some work of [5] and [6] (see also [16]) in the global theory of boldface pointclasses. To simplify the exposition, we assume ZF + AD + DC throughout this section.

The results of [5] and [6] are formulated in terms of a hierarchy slightly finer and considerably more general than the Levy hierarchy for $L(\mathbb{R})$. We now define this hierarchy. Let $\mathrm{Sep}(\Gamma)$, $\mathrm{PWO}(\Gamma)$, and $\mathrm{Scale}(\Gamma)$ mean, respectively, that Γ has the separation, prewellordering, and scale properties. If Γ and Λ are nonselfdual boldface pointclasses, let

$$\{\Gamma, \Gamma\} <_w \{\Lambda, \Lambda\} \quad \text{iff} \quad \Gamma \subseteq \Lambda \cap \breve{\Lambda} .$$

Wadge and Martin have shown that $<_w$ is a wellorder. The set of pairs $\{\Gamma, \breve{\Gamma}\}$ such that $\exists^{\mathbb{R}}\Gamma \subseteq \Gamma$ or $\exists^{\mathbb{R}}\breve{\Gamma} \subseteq \breve{\Gamma}$ has order type Θ under $<_w$; for $0 \leq \alpha < \Theta$, let P_α be the αth such pair. Now let

$$\boldsymbol{\Sigma}^1_\alpha = \text{unique } \Gamma \in P_\alpha\ [\exists^{\mathbb{R}}\Gamma \subseteq \Gamma \wedge (\forall^{\mathbb{R}}\Gamma \not\subseteq \Gamma \vee \mathrm{Sep}(\Gamma))] .$$

Since exactly one of $\mathrm{Sep}(\Gamma)$ and $\mathrm{Sep}(\breve{\Gamma})$ holds for every nonselfdual boldface Γ, $\boldsymbol{\Sigma}^1_\alpha$ is well defined. Let

$$\boldsymbol{\Pi}^1_\alpha = (\boldsymbol{\Sigma}^1_\alpha)^{\smallsmile} ,$$

$$\boldsymbol{\Delta}^1_\alpha = \boldsymbol{\Sigma}^1_\alpha \cap \boldsymbol{\Pi}^1_\alpha ,$$

and

$$\boldsymbol{\delta}^1_\alpha = \sup\{\lambda \mid \exists f\ (f : \mathbb{R} \xrightarrow{\text{onto}} \lambda \wedge \{\langle x,y\rangle \mid f(x) \leq f(y)\} \in \boldsymbol{\Delta}^1_\alpha)\} .$$

The sequence $\langle \boldsymbol{\Sigma}^1_\alpha \mid \alpha < \Theta\rangle$ is just the natural extension of the projective hierarchy cofinally through all boldface pointclasses. In particular, $\boldsymbol{\Sigma}^1_0 = \boldsymbol{\Sigma}^0_1$, and for $1 \leq n < \omega$, $\boldsymbol{\Sigma}^1_n$ is just the class usually given that name. For $\alpha < \Theta^{L(\mathbb{R})}$, $\boldsymbol{\Sigma}^1_\alpha$ is essentially the αth level of the restriction to sets of reals of the Levy hierarchy for $L(\mathbb{R})$, as we now show.

First, more terminology. If Γ is a boldface pointclass, then

$$\bigcup_\alpha \Gamma = \{ \bigcup_{\beta<\alpha} A_\beta \mid \forall \beta < \alpha\ (A_\beta \in \Gamma)\}$$

and

$$\bigcap_\alpha \Gamma = (\bigcup_\alpha \Gamma)^{\smallsmile} .$$

We can classify limit ordinals λ according to the closure properties of $\boldsymbol{\Sigma}^1_\lambda$:

$$\lambda \text{ is type I } \iff \bigcap_2 \underset{\sim}{\Sigma}^1_\lambda \subseteq \underset{\sim}{\Sigma}^1_\lambda \wedge \bigcap_\omega \underset{\sim}{\Sigma}^1_\lambda \not\subseteq \underset{\sim}{\Sigma}^1_\lambda ,$$

$$\lambda \text{ is type II } \iff \bigcap_2 \underset{\sim}{\Sigma}^1_\lambda \subseteq \underset{\sim}{\Sigma}^1_\lambda ,$$

$$\lambda \text{ is type III } \iff \bigcap_\omega \underset{\sim}{\Sigma}^1_\lambda \subseteq \underset{\sim}{\Sigma}^1_\lambda \wedge \forall^{\mathbb{R}} \underset{\sim}{\Sigma}^1_\lambda \not\subseteq \underset{\sim}{\Sigma}^1_\lambda ,$$

$$\lambda \text{ is type IV } \iff \forall^{\mathbb{R}} \underset{\sim}{\Sigma}^1_\lambda \subseteq \underset{\sim}{\Sigma}^1_\lambda .$$

The type of λ is just the type, in the sense of [6], of the projective-like hierarchy immediately above $\underset{\sim}{\Sigma}^1_\lambda$. Some facts from [6] and [16]: λ is of type I iff $\mathrm{cof}(\lambda) = \omega$ iff $\underset{\sim}{\Sigma}^1_\lambda = \bigcup_\omega (\bigcup_{\alpha<\lambda} \underset{\sim}{\Sigma}^1_\alpha)$. If λ is of type II, III, or IV, then $\underset{\sim}{\Delta}^1_\lambda = \bigcup_{\alpha<\lambda} \underset{\sim}{\Sigma}^1_\alpha$. If λ is of type I, then $\mathrm{PWO}(\underset{\sim}{\Sigma}^1_\lambda)$; otherwise (provided $\lambda < \Theta^{L(\mathbb{R})}$) $\mathrm{PWO}(\underset{\sim}{\Pi}^1_\lambda)$. (At this point we should warn the reader that our definition of the extended projective hierarchy differs slightly from that of [5]. The difference is that the classes $\underset{\sim}{\Sigma}^1_\lambda$ for λ of type I or II are omitted from the hierarchy of [5].)

Let $\langle \delta_\alpha \mid \alpha < \Theta^{L(\mathbb{R})} \rangle$ enumerate in increasing order those ordinals δ such that $\rho^\delta_n = \mathbb{R}$ for some n. Let n_α be the least n such that $\rho^{\delta_\alpha}_n = \mathbb{R}$.

Lemma 4.1. For all $\alpha < \Theta^{L(\mathbb{R})}$,

(a) $\mathrm{PWO}(\underset{\sim}{\Sigma}_{n_\alpha}(J_{\delta_\alpha}(\mathbb{R})))$,

and

(b) $\underset{\sim}{\Sigma}_{n_\alpha}(J_{\delta_\alpha}(\mathbb{R})) \cap P(\mathbb{R}) \subseteq \bigcup_{\delta_\alpha} (J_{\delta_\alpha}(\mathbb{R}) \cap P(\mathbb{R}))$.

Proof. Let $\delta = \delta_\alpha$, $n = n_\alpha$, and $\rho = \rho^\delta_{n-1}$. Fix a partial $\Sigma_{n-1}(J_\delta(\mathbb{R}))$ map $f : J_\rho(\mathbb{R}) \xrightarrow{\text{onto}} J_\delta(\mathbb{R})$. Suppose

$$x \in A \iff J_\delta(\mathbb{R}) \models \exists v \varphi[x] ,$$

where φ is Π_{n-1}, and for $x \in A$ let

$$\psi(x) = \mu\gamma < \rho \; [\exists u \in J_\gamma(\mathbb{R})(u \in \mathrm{dom}\, f \wedge J_\delta(\mathbb{R}) \models \varphi[x, f(u)])] .$$

By 1.15, ψ is a $\underset{\sim}{\Sigma}_n(J_\delta(\mathbb{R}))$ norm, so we have (a). Fix $\gamma < \rho$. Then $\mathrm{dom}\, f \cap J_\gamma(\mathbb{R}) \in J_\delta(\mathbb{R})$, from which it easily follows that $\{x \mid \psi(x) \leq \gamma\} \in J_\delta(\mathbb{R})$. This proves (b). ⊣

Theorem 4.2. For all $\alpha < \Theta^{L(\mathbb{R})}$,

(a) if $\omega\alpha$ is of type I, then for all $k < \omega$

$$\underset{\sim}{\Sigma}^1_{\omega\alpha+k} = \underset{\sim}{\Sigma}_{n_\alpha+k}(J_{\delta_\alpha}(\mathbb{R})) ;$$

(b) if $\omega\alpha$ is of type II or III, then for all $k < \omega$

$$\utilde{\Sigma}^1_{\omega\alpha+k+1} = \utilde{\Sigma}_{n_\alpha+k}(J_{\delta_\alpha}(\mathbb{R})) \; ;$$

(c) if $\omega\alpha$ is of type IV, then

$$\utilde{\Pi}^1_{\omega\alpha} = \utilde{\Sigma}_{n_\alpha}(J_{\delta_\alpha}(\mathbb{R})) \; ,$$

and for $k < \omega$ such that $k \neq 0$,

$$\utilde{\Sigma}^1_{\omega\alpha+k+1} = \utilde{\Sigma}_{n_\alpha+k}(J_{\delta_\alpha}(\mathbb{R})) \; .$$

Proof. By induction on α. Set $\Lambda = \bigcup_{\beta<\omega\alpha} \utilde{\Sigma}^1_\beta = J_{\delta_\alpha}(\mathbb{R}) \cap P(\mathbb{R})$.

(a) Since $\utilde{\Sigma}^1_{\omega\alpha} = \bigcup_\omega \Lambda$, $\utilde{\Sigma}^1_{\omega\alpha} \subseteq \utilde{\Sigma}_{n_\alpha}(J_{\delta_\alpha}(\mathbb{R}))$. Let $A \in \utilde{\Sigma}_{n_\alpha}(J_{\delta_\alpha}(\mathbb{R})) \cap P(\mathbb{R})$. By 4.1 (b) and the fact that $\mathrm{cof}(\omega\alpha) = \omega$

$$A = \bigcup_{n<\omega} \Big(\bigcup_{\beta<\delta_\alpha} A^n_\beta \Big) \; ,$$

where for some fixed $\gamma_n < \omega\alpha$, $A^n_\beta \in \utilde{\Sigma}^1_{\gamma_n}$ for all $\beta < \delta_\alpha$. By Kechris [5], we may assume that $\utilde{\Sigma}^1_{\gamma_n}$ is closed under wellordered unions, so that $A \in \bigcup_\omega \Lambda = \utilde{\Sigma}^1_{\omega\alpha}$, as desired. Thus (a) is true for $k = 0$, and for $k > 0$ follows easily.

(b) $\utilde{\Sigma}^1_{\omega\alpha} \neq \utilde{\Sigma}_{n_\alpha}(J_{\delta_\alpha}(\mathbb{R})) \cap P(\mathbb{R})$, by 4.1 (a) and the fact that $\mathrm{PWO}(\utilde{\Pi}^1_{\omega\alpha})$. Thus $\utilde{\Sigma}^1_{\omega\alpha+1} \subseteq \utilde{\Sigma}_{n_\alpha}(J_{\delta_\alpha}(\mathbb{R})) \cap P(\mathbb{R})$. But $\utilde{\Sigma}^1_{\omega\alpha+1}$ is closed under wellordered unions by Kechris' theorem, so 4.1 (b) gives $\utilde{\Sigma}_{n_\alpha}(J_{\delta_\alpha}(\mathbb{R})) \cap P(\mathbb{R}) \subseteq \utilde{\Sigma}^1_{\omega\alpha+1}$. Thus (b) is true for $k = 0$, and for $k > 0$ again follows easily.

(c) The Coding lemma implies $\bigcup_{\omega\alpha} \utilde{\Pi}^1_{\omega\alpha} \subseteq \utilde{\Pi}^1_{\omega\alpha}$, which with the proof of 4.1 (b) implies the first statement. The second statement is immediate once we notice that $\utilde{\Sigma}^1_{\omega\alpha+1} = \bigcup_2 (\utilde{\Sigma}^1_{\omega\alpha} \cup \utilde{\Pi}^1_{\omega\alpha})$. ⊣

If T is a tree on $\omega \times \kappa$, then $[T]$ is the set of infinite branches of T, and $p([T])$ the projection of this set, that is,

$$p([T]) = \{x \in \mathbb{R} \mid \exists f\, \forall n\, ((x \restriction n, f \restriction n) \in T)\} \; .$$

We let

$$S(\kappa) = \{p([T]) \mid T \text{ is a tree on } \omega \times \kappa\} \; ,$$

and call the members of $S(\kappa)$ κ-Suslin sets. If $k > \omega$, then the κ-Suslin sets are precisely those admitting scales all of whose norms map into κ. We say κ is Suslin iff $S(\kappa) - \bigcup_{\alpha<\kappa} S(\alpha) \neq \emptyset$. Suslin ordinals are cardinals. We shall locate the pointclasses $S(\kappa)$ among the $\utilde{\Sigma}^1_\nu$'s, and the Suslin cardinals among the $\utilde{\delta}^1_\nu$'s.

Let ν_α be the αth ordinal ν such that either $\text{Scale}(\underset{\sim}{\Sigma}^1_\nu)$ or $\text{Scale}(\underset{\sim}{\Pi}^1_\nu)$. By the second periodicity theorem, $\nu_{\alpha+1} = \nu_\alpha + 1$ as long as ν_α is not a limit ordinal of type IV. If ν_α is of type IV, then $\nu_{\alpha+1}$ is a limit ordinal of type I by the results of 2 and 3.

Theorem 4.3. Let $\lambda < (\underset{\sim}{\Sigma}^2_1)^{L(\mathbb{R})}$ be a limit ordinal, and let $\nu = \sup\{\nu_\alpha \mid \alpha < \lambda\}$. Then

(a) If ν is type I, then for all $n < \omega$

$$\text{Scale}(\underset{\sim}{\Sigma}^1_{\nu+2n}), \quad \text{Scale}(\underset{\sim}{\Pi}^1_{\nu+2n+1}),$$

$$S(\kappa_{\lambda+n}) = \underset{\sim}{\Sigma}^1_{\nu+n+1},$$

$$\kappa_{\lambda+2n+1} = \underset{\sim}{\delta}^1_{\nu+2n+1} = (\kappa_{\lambda+2n})^+,$$

$$\text{cof}(\kappa_{\lambda+2n}) = \omega ;$$

(b) if ν is type II or III, then for all $n < \omega$

$$\text{Scale}(\underset{\sim}{\Pi}^1_{\nu+2n}), \quad \text{Scale}(\underset{\sim}{\Sigma}^1_{\nu+2n+1}),$$

$$S(\kappa_{\lambda+n}) = \underset{\sim}{\Sigma}^1_{\nu+n+1},$$

$$\kappa_{\lambda+2n+2} = \underset{\sim}{\delta}^1_{\nu+2n+2} = (\kappa_{\lambda+2n+1})^+,$$

$$\text{cof}(\kappa_{\lambda+2n+1}) = \omega ;$$

(c) if ν is type IV, then

$$\text{Scale}(\underset{\sim}{\Pi}^1_\nu), \quad S(\kappa_\lambda) = \underset{\sim}{\Pi}^1_\nu, \quad \kappa_\lambda = \underset{\sim}{\delta}^1_\nu,$$

and if $\mu = \nu_{\lambda+1}$, then for all $n < \omega$

$$\text{Scale}(\underset{\sim}{\Sigma}^1_{\mu+2n}), \quad \text{Scale}(\underset{\sim}{\Pi}^1_{\mu+2n+1}),$$

$$S(\kappa_{\lambda+n+1}) = \underset{\sim}{\Sigma}^1_{\mu+n+1},$$

$$\kappa_{\lambda+2n+2} = \underset{\sim}{\delta}^1_{\mu+2n+1} = (\kappa_{\lambda+2n+1})^+,$$

$$\text{cof}(\kappa_{\lambda+2n+1}) = \omega .$$

Proof (Sketch). By induction on λ.

(a) By 4.2 and the results of 2 and 3, $\underset{\sim}{\Sigma}^1_\nu = \underset{\sim}{\Sigma}_1(J_\alpha(\mathbb{R}))$ for some α beginning a Σ_1 gap. Thus $\text{Scale}(\underset{\sim}{\Sigma}^1_\nu)$, and by induction $\kappa_\lambda = \sup\{\underset{\sim}{\delta}^1_{\nu_\beta} \mid \beta < \lambda\}$. Thus $\kappa_\lambda < \underset{\sim}{\delta}^1_{\nu+1}$, and $S(\kappa_\lambda) \subseteq \underset{\sim}{\Sigma}^1_{\nu+1}$. But $\underset{\sim}{\Sigma}^1_{\nu+1} \subseteq S(\kappa_\lambda)$ since $\underset{\sim}{\Sigma}^1_\nu \subseteq S(\kappa_\lambda)$ and $S(\kappa_\lambda)$ is closed under $\exists^{\mathbb{R}}$ and countable intersection. Thus $S(\kappa_\lambda) = \underset{\sim}{\Sigma}^1_{\nu+1}$. The remaining assertions follow by arguments like those for the projective hierarchy ($\underset{\sim}{\Sigma}^1_\nu$ being analogous to $\underset{\sim}{\Sigma}^1_0$).

(b) By 4.2 and the results of §2 and §3, $\underset{\sim}{\Sigma}^1_{\nu+1} = \underset{\sim}{\Sigma}_1(J_\alpha(\mathbb{R}))$ for some α beginning a Σ_1 gap. Thus $\text{Scale}(\underset{\sim}{\Sigma}^1_{\nu+1})$, and $\kappa_\lambda = \sup\{\delta^1_{\nu_\beta} \mid \beta < \lambda\} = \underset{\sim}{\delta}^1_\nu$. We have $S(\kappa_\lambda) = \underset{\sim}{\Sigma}^1_{\nu+1}$ for the same reasons as in (a). The remaining assertions follow by the arguments for the projective hierarchy, except for $\text{Scale}(\underset{\sim}{\Pi}^1_\nu)$. This can be proved using the ideas of the proof of $\text{PWO}(\underset{\sim}{\Pi}^1_\nu)$; cf. [5], Theorem 3.1 (iii), and [16], Theorem 3.1.

(c) By 4.2 and 2 and 3, $\underset{\sim}{\Pi}^1_\nu = \underset{\sim}{\Sigma}_1(J_\alpha(\mathbb{R}))$, where α begins a Σ_1 gap. This proves the first set of assertions. The second set follows from our analysis of the ends of gaps in 2 and 3, and the arguments for the projective hierarchy ($\underset{\sim}{\Sigma}^1_\mu$ is analogous to $\underset{\sim}{\Sigma}^1_0$). ⊣

Corollary 4.4. (a) The sequences $\langle \kappa_\alpha \mid \alpha \leq (\underset{\sim}{\delta}^2_1)^{L(\mathbb{R})} \rangle$ and $\langle \nu_\alpha \mid \alpha \leq (\underset{\sim}{\delta}^2_1)^{L(\mathbb{R})} \rangle$ are continuous at limits.

(b) For any $\kappa \leq (\underset{\sim}{\delta}^2_1)^{L(\mathbb{R})}$, either $S(\kappa)$ or its dual has the scale property.

As a final application of §2 and §3, we prove

Corollary 4.5. In $L(\mathbb{R})$, the reliable cardinals are precisely the Suslin cardinals.

This result should be stamped "Made in Los Angeles". The reader who cares can untangle some of the credits for it from what follows.

Recall that an ordinal λ is reliable iff there is a scale $\{\varphi_i\}$ on a set $A \subseteq \mathbb{R}$ so that

$$\varphi_i : A \to \lambda$$

for all i, and

$$\lambda = \{\varphi_0(x) \mid x \in A\} .$$

The interest of the notion of reliability stems from results of [1] and [13], which to date have been proved only for reliable ordinals.

One direction of 4.5 is easy.

Lemma 4.6. Every Suslin cardinal is reliable.

Proof. Clearly ω is reliable. Let κ be an uncountable Suslin cardinal, let $A \in S(\kappa) - \bigcup_{\alpha<\kappa} S(\alpha)$, and let $\{\varphi_i\}$ be a scale on A all of whose norms map onto (perhaps improper) initial segments of κ. Let

$$B = \{x \in \mathbb{R} \mid \lambda i.x(i + 1) \in A\} ,$$

and for $x \in B$,

$$\psi_0(x) = \varphi_{x(0)}(\lambda i.x(i+1)) ,$$

$$\psi_{n+1}(x) = \varphi_n(\lambda i.x(i+1)) .$$

Then $\{\psi_i\}$ is a scale on B all of whose norms map into κ. Since $A \notin \bigcup_{\alpha<} S(\alpha)$, $\kappa = \{\psi_0(x) \mid x \in B\}$. $\dashv$

There are reliable ordinals which are not cardinals (cf. [1]). The first step toward showing that in $L(\mathbb{R})$ all reliable cardinals are Suslin, and the realization that it is a first step, are due to Kechris.

Lemma 4.7 (Kechris). Let λ be reliable, and let $\kappa = \sup\{\gamma \leq \lambda \mid \gamma$ is Suslin$\}$. Then there is a strictly increasing sequence of sets in $S(\kappa)$ of length λ.

Proof. Let $\{\varphi_i\}$ be a scale on A witnessing the reliability of λ. For $\alpha < \lambda$, let

$$A_\alpha = \{x \in A \mid \varphi_0(x) \leq \alpha\} .$$

Then $\alpha < \beta \Rightarrow A_\alpha \subsetneq A_\beta$; moreover, each A_α is in $S(\lambda)$ via a subtree of the tree of $\{\varphi_i\}$. But $S(\lambda) = S(\kappa)$ by the definition of κ. $\dashv$

The next and most substantial step toward 4.5 is due to Martin (exploiting an idea of Jackson).

Theorem 4.8 ([9]). For $1 \leq n < \omega$, there is no strictly increasing sequence of $\utilde{\Sigma}^1_{2n}$ sets of length $(\utilde{\delta}^1_{2n-1})^+$.

By 4.7 and 4.8 and standard facts about the Suslin cardinals below $\utilde{\delta}^1_\omega$, every reliable cardinal below $\utilde{\delta}^1_\omega$ is Suslin. But Martin's argument gives more than 4.8, and in fact, with some care we can prove

Theorem 4.9. Suppose $\alpha < \Theta^{L(\mathbb{R})}$ and Scale($\utilde{\Pi}^1_\alpha$). Then there is no strictly increasing sequence of $\exists^{\mathbb{R}}\utilde{\Pi}^1_\alpha$ sets of length $(\utilde{\delta}^1_\alpha)^+$.

Proof. We shall assume that the reader is familiar with the proof of 4.8 given in [9]. Our only problem in extending that proof is to show that the coding of ordinals below $(\utilde{\delta}^1_{2n-1})^+$ which it employs generalizes suitably.

Since Scale($\utilde{\Pi}^1_\alpha$), every $\exists^{\mathbb{R}}\utilde{\Pi}^1_\alpha$ set has a scale whose norms map into $\utilde{\delta}^1_\alpha$. Let $U \subseteq \mathbb{R}^3$ be universal $\exists^{\mathbb{R}}\utilde{\Pi}^1_\alpha$, and $\{\varphi_i\}$ a scale on U mapping into $\utilde{\delta}^1_\alpha$. Define a tree T on $\omega \times \omega \times \utilde{\delta}^1_\alpha$ by

$$T = \{(s, t, u) \mid \ell h(s) = \ell h(t) = \ell h(u) \wedge$$

$$\exists x \supseteq s\ \exists y \supseteq t\ [\bigwedge_{i<\ell h(s)} U(x, (y)_i, (y)_{i+1}) \wedge$$

$$\bigwedge_{i<\ell h(s)} \varphi_{(i)_0}(x, (y)_{(i)_1}, (y)_{(i)_1+1}) = U(i)]\} .$$

Here we let $\langle \cdot, \cdot \rangle$ be a bijection of $\omega \times \omega$ onto ω, and for any i, $i = \langle (i)_0, (i)_1 \rangle$, and for any y, i, n, $(y)_i(n) = y(\langle i, n \rangle)$. For $x \in \mathbb{R}$, let

$$T_x = \{(t, u) \mid (x \restriction \ell h(t), t, u) \in T\} .$$

T_x is the "Kunen tree" associated to x. If U_x is a wellfounded relation, then T_x is a wellfounded tree, and $|U_x| \leq |T_x|$ (where $|R|$ denotes the rank of the relation R). In order to carry out the argument of [9] in the present situation, we need only verify

(a) If $f : \mathbb{R} \to \mathbb{R}$ is continuous and $\forall x \in \mathbb{R}\ \forall \beta < \underset{\sim}{\delta}^1_\alpha\ (\forall \gamma < \beta\ \ T_{(x)_0} \restriction$ is wellfounded $\Rightarrow T_{f(x)} \restriction \beta$ is wellfounded), then $\exists \delta < (\underset{\sim}{\delta}^1_\alpha)^+\ \forall x \in \mathbb{R}\ \ T_{(x)_0}$ is wellfounded $\Rightarrow |T_{f(x)}| < \delta$;

(b) If $f : \mathbb{R} \to \mathbb{R}$ is continuous and $\forall x \in \mathbb{R}\ \forall \beta < \underset{\sim}{\delta}^1_\alpha\ (T_x \restriction \beta$ is wellfounded $T_{f(x)} \restriction \beta$ is wellfounded) then there are unboundedly many $\delta < (\underset{\sim}{\delta}^1_\alpha)^+$ such that $\forall x \in \mathbb{R}$

$$|T_x| < \delta \Rightarrow |T_{f(x)}| < \delta ;$$

(c) For all sufficiently large $\delta < (\underset{\sim}{\delta}^1_\alpha)^+$, $\{x \mid |T_x| < \delta\}$ is not $\underset{\sim}{\Pi}^1_{\alpha+1}$.

(If R is a tree on $\omega^k \times \alpha$, then $R \restriction \gamma = \{(\sigma_1 \ldots \sigma_k, \tau) \in R \mid \mathrm{range}(\tau) \subseteq \gamma\}$.) Now (c) can be proved exactly as in [9]. Since (a) and (b) have similar proofs, we shall just prove (b).

<u>Lemma 4.10</u>. Suppose $\alpha < \Theta^{L(\mathbb{R})}$, α is a successor, and $\mathrm{Scale}(\underset{\sim}{\Pi}^1_\alpha)$. Let R be a tree on $\omega \times \beta$, where $\beta < \underset{\sim}{\delta}^1_\alpha$, and let $f : \mathbb{R} \to \mathbb{R}$ be continuous and such that $\forall x\ (R_x$ is wellfounded $\Rightarrow R_{f(x)}$ is wellfounded). Then there is a club $C \subseteq \underset{\sim}{\delta}^1_\alpha$ so that for $\delta \in C$, $\forall x\ (|R_x| < \delta \Rightarrow |R_{f(x)}| < \delta)$.

<u>Proof</u>. By a result of Martin, $\bigcup_\delta \underset{\sim}{\Delta}^1_\alpha \subseteq \underset{\sim}{\Delta}^1_\alpha$ for all $\delta < \underset{\sim}{\delta}^1_\alpha$ (cf. [4], Theorem 3.7). It follows (as in [4], Theorem 3.8) that for all $u \in \beta^{<\omega}$, and $\delta < \underset{\sim}{\delta}^1_\alpha$,

$$\{x \mid |R^u_x| < \delta\} \in \underset{\sim}{\Delta}^1_\alpha ,$$

where R^u_x is the subtree of R_x below u; the proof is by induction on δ.

It is enough to show that $\forall\delta < \underset{\sim}{\delta}^1_\alpha \; \exists\eta < \underset{\sim}{\delta}^1_\alpha \; \forall x \in \mathbb{R} \; (|R_x| < \delta \Rightarrow |R_{f(x)}| < \eta)$. So fix $\delta < \underset{\sim}{\delta}^1_\alpha$, and let $A = \{x \mid |R_x| < \delta\}$. Now our hypotheses on α and 4.3 together imply that $\underset{\sim}{\Sigma}^1_\alpha = S(\gamma)$ for some $\gamma < \underset{\sim}{\delta}^1_\alpha$. Thus we can fix a tree Q on $\omega \times \gamma$ so that $A = p([Q])$. Define a tree P on $\omega \times \gamma \times \omega \times \beta$ by

$$P = \{(s, t, u, v) \mid \ell h(s) = \ell h(t) = \ell h(u) = \ell h(v) \wedge$$
$$(s, t) \in Q \wedge (u, v) \in R \wedge \exists x \supseteq s \; \exists y \supseteq u \; (f(x) = y)\} .$$

Then P is wellfounded, and if $x \in A$ then $R_{f(x)}$ can be embedded in P. Thus $\eta = |P|$ is as desired. ⊣

We can now complete the proof of (b) in the case α is a successor as in [9]. Since α is a successor and $\mathrm{Scale}(\underset{\sim}{\Pi}^1_\alpha)$, we have $\cup_\omega \underset{\sim}{\Pi}^1_\alpha \subseteq \underset{\sim}{\Pi}^1_\alpha$. But then the ω-club subsets of $\underset{\sim}{\delta}^1_\alpha$ generate a normal ultrafilter μ on $\underset{\sim}{\delta}^1_\alpha$. Set $\kappa = \underset{\sim}{\delta}^1_\alpha$. Since $\underset{\sim}{\Sigma}^1_\alpha = S(\gamma)$ for some $\gamma < \kappa$, we can apply Kunen's argument ([4], Theorem 14.3) to show that ${}^\kappa\kappa/\mu$ has order type $^{+}$. For $\beta < \kappa$ let

$$C_\beta = \{\delta < \kappa \mid \forall x \; (|T_x \upharpoonright \beta| < \delta \Rightarrow |T_{f(x)} \upharpoonright \beta| < \delta)\} ,$$

so that C_β is club in κ by 4.10. Let $D = \{\delta < \kappa \mid \forall\beta < \delta \; (\delta \in C_\beta)\}$, and let

$$C = \{[h]_\mu \mid . \; h : k \to D \wedge [h]_\mu \geq \; \} .$$

Clearly C is unbounded in κ^+. Suppose $|T_x| < \delta$, where $\delta \in C$. Let $\delta = [h]$, where $h : \kappa \to D$. Then

$$|T_x \upharpoonright \beta| < h(\beta) \qquad (\mu\text{-a.e.}) ,$$

so

$$|T_{f(x)} \upharpoonright \beta| < h(\beta) \qquad (\mu\text{-a.e.}) ,$$

so

$$|T_{f(x)}| < [h]_\mu = \delta ,$$

as desired. ⊣

Finally, let α be a limit. Since $\mathrm{Scale}(\underset{\sim}{\Pi}^1_\alpha)$, 4.3 and 4.2 tell us that $\exists^{\mathbb{R}} \underset{\sim}{\Pi}^1_\alpha = \underset{\sim}{\Sigma}_1(J_\rho(\mathbb{R}))$ for some ρ such that $\mathrm{cof}(\rho) > \omega$. Let $\lambda = \mathrm{cof}(\rho)$, and let $\mathfrak{u}$ be the supercompactness measure on $P_{\omega_1}(\lambda)$ given by Theorem 6.2.1 of [2]. Let μ be the measure on λ defined by

$$\mu(A) = 1 \;\text{ iff }\; \{X \mid \sup X \in A\} \in \mathfrak{u} .$$

Then μ is weakly normal (that is, if $h(\beta) < \beta$ a.e., then $\exists\gamma < \lambda$ ($h(\beta) < \gamma$ a.e.)) and $\mu(A) = 1$ for every ω-club A. Let

$$g : \lambda \xrightarrow{\text{cofinal}} \rho$$

be strictly increasing, continuous, have range cofinal in ρ, and be such that for all $\beta < \lambda$, $J_{g(\beta)}(\mathbb{R}) \not\prec_1^{\mathbb{R}} J_{g(\beta)+1}(\mathbb{R})$. For $\beta < \lambda$, let also

$$h(\beta) = \sup\{|\leq| \mid \leq \text{ is a prewellorder of } \mathbb{R} \text{ in } J_{g(\beta)}(\mathbb{R})\} .$$

Then h is strictly increasing, continuous, and has range cofinal in $\underset{\sim}{\delta}^1_\alpha$.

The following claim is the crucial new ingredient we need in the case α is a limit ordinal. The proof of part (2), its non-trivial part, is due to Kechris.

<u>Claim</u>. (1) $[h]_\mu = \underset{\sim}{\delta}^1_\alpha$,
(2) $[\lambda\beta.h(\beta)^+] = (\underset{\sim}{\delta}^1_\alpha)^+$.

<u>Proof</u>. (1) $\underset{\sim}{\Delta}^1_\alpha = \bigcup_{\beta<\alpha} \underset{\sim}{\Delta}^1_\beta = J_\rho(\mathbb{R}) \cap P(\mathbb{R})$, and therefore every $\underset{\sim}{\Delta}^1_\alpha$ set is γ-Suslin for some $\gamma < \underset{\sim}{\delta}^1_\alpha$. If $\underset{\sim}{\delta}^1_\alpha$ is regular (i.e. $\lambda = \underset{\sim}{\delta}^1_\alpha$), then the proof of Theorem 3.2 of [16] implies that $\bigcup_\omega \underset{\sim}{\Pi}^1_\alpha \subseteq \underset{\sim}{\Pi}^1_\alpha$, so that the ω-club subsets of $\underset{\sim}{\delta}^1_\alpha$ generate a normal ultrafilter, which must then be μ. But then $[h]_\mu = \underset{\sim}{\delta}^1_\alpha$. So suppose $\lambda < \underset{\sim}{\delta}^1_\alpha$. The proof of Theorem 6.2.1 of [2] shows in this case that $\mu \in J_\rho(\mathbb{R})$. Thus if $\gamma < \underset{\sim}{\delta}^1_\alpha$, then ${}^\lambda\gamma/\mu$ has order type less than $\underset{\sim}{\delta}^1_\alpha$. By the weak normality of μ, $[h]_\mu$ is just the supremum of these order types for $\gamma < \underset{\sim}{\delta}^1_\alpha$. Thus $[h]_\mu = \underset{\sim}{\delta}^1_\alpha$.

(2) That $[\lambda\beta.h(\beta)^+]_\mu \geq (\underset{\sim}{\delta}^1_\alpha)^+$ is easy to see. For the other inequality, let $\ell : \lambda \to \underset{\sim}{\delta}^1_\alpha$ and $\ell(\beta) < h(\beta)^+$ for all β. We shall construct a wellfounded tree W on $\underset{\sim}{\delta}^1_\alpha$ so that for μ-a.e. $\beta < \lambda$, $\ell(\beta) \leq |W \restriction h(\beta)|$. This implies that

$$[\ell]_\mu \leq [\lambda\beta.|W \restriction h(\beta)|]_\mu = |[\lambda\beta.W \restriction h(\beta)]_\mu| .$$

But by (a) of our claim, $[\lambda\beta.W \restriction h(\beta)]_\mu$ is a wellfounded tree on $\underset{\sim}{\delta}^1_\alpha$. Thus $[\ell]_\mu < (\underset{\sim}{\delta}^1_\alpha)^+$, as desired.

In order to construct W, we construct first a tree R on $\omega \times \omega \times \omega \times \underset{\sim}{\delta}^1_\alpha$ so that for all limit ordinals $\beta < \lambda$,

$$(*) \qquad h(\beta)^+ = \sup\{|R_{xy} \restriction h(\beta)| \mid R_{xy} \text{ is wellfounded}\} .$$

For this, let

$$V = \{\langle x, y\rangle \mid J_\beta(\mathbb{R}) \models \sigma_{x(0)}[\lambda i.x(i+1), y]\} ,$$

where σ_i is the ith Σ_1 formula of two free variables, and let θ be a Σ_1 formula so that (identifying $\mathbb{R}^2$ with $\mathbb{R}$)

$$V(x) \text{ iff } J_\rho(\mathbb{R}) \models \theta[x] .$$

Let also

$$L(x, y) \text{ iff } \exists\delta < \rho\ (J_\delta(\mathbb{R}) \models \theta[x] \wedge \neg\theta[y]) .$$

Since L is $\Sigma_1(J_\rho(\mathbb{R}))$, Theorem 2.1 gives us a scale $\{\varphi_i\}$ on L. Notice that the scale $\{\varphi_i\}$ we get from 2.1 has the property that if $L(x, y)$, then $\varphi_0(x, y) = \mu\delta[J_\delta(\mathbb{R}) \models \theta[x]]$, and for all $i > 0$, $\varphi_i(x, y)$ is the order type of a prewellorder of $\mathbb{R}$ which is $\underset{\sim}{\Sigma}_\omega(J_{\varphi_0(x,y)}(\mathbb{R}))$. Let

$$M(x, y) \text{ iff } \forall n < \omega\ L((x)_n, y) ,$$

and let $\{\psi_i\}$ be the scale on M given by

$$\psi_{\langle m,n\rangle}(x, y) = \varphi_m((x)_n, y) .$$

Finally, let

$$R = \{(s, t, u, v) \mid \ell h(s) = \ell h(t) = \ell h(u) = \ell h(v) \wedge$$
$$\exists x \supseteq s\ \exists y \supseteq t\ \exists z \supseteq u\ (M(x, \langle y, z\rangle) \wedge \bigwedge_{i<\ell h(s)} v(i) = \psi_i(x, \langle y, z\rangle))\} .$$

In order to show (*), fix a limit ordinal $\beta < \lambda$, and fix an $x \in \mathbb{R}$ so that $(x)_n \in V$ for all n, and in fact

$$g(\beta) = \mu\gamma\ [\forall n < \omega\ (J_\gamma(\mathbb{R}) \models \theta[(x)_n])] .$$

We can find such an x by the universality of V and the fact that g grows as fast as it does. Then for all y, z,

$$M(x, \langle y, z\rangle) \text{ iff } \langle y, z\rangle \in p([R_x]) ,$$
$$\text{iff } \langle y, z\rangle \in p([R_x \restriction h(\beta)]) ,$$

where the second equivalence follows from our observations on the "local" nature of $\{\varphi_i\}$. Thus for any y, R_{xy} is wellfounded iff $R_{xy} \restriction h(\beta)$ is wellfounded. Now $\{\langle y, z\rangle \mid M(x, \langle y, z\rangle)\}$ is clearly universal for the class $\Pi_1(J_{g(\beta)}(\mathbb{R}), \mathbb{R})$, and so if we let

$$y \in N \text{ iff } \exists z\ M(x, \langle y, z\rangle) ,$$

then N is in the class $\exists^{\mathbb{R}}\Pi_1(J_{g(\beta)}(\mathbb{R}), \mathbb{R})$, but not its dual. From 4.2 we see easily that $\exists^{\mathbb{R}}\Pi_1(J_{g(\beta)}(\mathbb{R}), \mathbb{R})$ is a class of the form $\underset{\sim}{\Sigma}^1_{\gamma+1}$, where Scale($\underset{\sim}{\Pi}^1_{\gamma+1}$) and, by 4.3, $\underset{\sim}{\delta}^1_{\gamma+1} = h(\beta)^+$. Now for $\eta < \underset{\sim}{\delta}^1_{\gamma+1}$, $\{y \mid |R_{xy} \restriction h(\beta)| < \eta\}$ is $\underset{\sim}{\Delta}^1_{\gamma+1}$, by the argument given in the proof of 4.10. Since $N = \{y \mid R_{xy}$ is wellfounded$\} = \{y \mid R_{xy} \restriction h(\beta)$ is wellfounded$\}$, and N is not $\underset{\sim}{\Delta}^1_{\gamma+1}$, we have (*).

We can now complete the proof of (2) of the claim in the case $\underset{\sim}{\delta}^1_\alpha$ is singular, that is, $\lambda < \underset{\sim}{\delta}^1_\alpha$. Fix a norm $\varphi : \mathbb{R} \xrightarrow{\text{onto}} \lambda$ so that the induced prewellordering $\leq_\varphi$ is in $J_\rho(\mathbb{R})$. By the Coding lemma there is a relation P such that $\forall w \exists x, y\ P(w, x, y)$, and for $\beta < \lambda$ a limit

$$(\varphi(w) = \beta \wedge P(w, x, y)) \Rightarrow (R_{xy} \text{ is wellfounded} \wedge \ell(\beta) < |R_{xy} \restriction h(\beta)|) ,$$

and $P \in J_\rho(\mathbb{R})$. Thus P is γ-Suslin for some $\gamma < \underset{\sim}{\delta}^1_\alpha$; let $P = p([S])$, where S is a tree on $\omega^3 \times \gamma$ and $\gamma < \underset{\sim}{\delta}^1_\alpha$. Finally, let

$$W = \{(q, r, s, t, u) \mid \ell h(q) = \ell h(r) = \cdots = \ell h(u)$$
$$\wedge (q, r, s, t) \in S \wedge (r, s, u) \in R\} .$$

Then W is wellfounded. If $\beta < \lambda$ is a limit ordinal, $\gamma < h(\beta)$, $\varphi(w) = \beta$, and $P(w, x, y)$, then the tree $R_{xy} \restriction h(\beta)$ can be embedded into $W_{wxy} \restriction h(\beta)$. Since $|R_{xy} \restriction h(\beta)| > \ell(\beta)$, $|W \restriction h(\beta)| > \ell(\beta)$, as desired.

We shall just outline the proof of Claim (b) in the case that $\underset{\sim}{\delta}^1_\alpha$ is regular, Fix a complete $\underset{\sim}{\Pi}^1_\alpha$ set A, and a $\underset{\sim}{\Pi}^1_\alpha$-scale $\{\tau_i\}$ on A. Since $\bigcup_\omega \underset{\sim}{\Pi}^1_\alpha \subseteq \underset{\sim}{\Pi}^1_\alpha$, (by [16]), any $\underset{\sim}{\Sigma}^1_\alpha$ subset of A is bounded in τ_0 -- here we use the regularity of $\underset{\sim}{\delta}^1_\alpha$. Thus μ is just the ω-club measure on $\lambda = \underset{\sim}{\delta}^1_\alpha$, and $g(\beta) = h(\beta) = \beta$ for μ-a.e. β. Consider the following Solovay game.

I	w
II	z, x, y

II wins if $\exists n ((w)_n \notin A \wedge \forall m < n ((z)_m \in A))$ or $\forall n ((w)_n, (z)_n \in A)$ and, if $\beta = \sup\{\tau_0((w)_n), \tau_0((z)_n) \mid n < w\}$, then R_{xy} is wellfounded and $\ell(\beta) < |R_{xy} \restriction \beta|$. By boundedness and the property (*) of R, II has a winning strategy σ. We can now use σ and the scale $\{\tau_i\}$ to construct W just as we used the tree S to construct W in the case $\underset{\sim}{\delta}^1_\alpha$ is singular. Where we used before that S was a tree on an ordinal less than $\underset{\sim}{\delta}^1_\alpha$, we now use that for μ-a.e. β, whenever $\tau_0((w)_n) < \beta$ for all n, then $\tau_i((w)_n) < \beta$ for all i, n.

This proves the claim.

Finally, we prove boundedness property (b) of the Kunen tree coding in the case α is a limit. Suppose $f : \mathbb{R} \to \mathbb{R}$ is continuous and $\forall \eta < \underset{\sim}{\delta}^1_\alpha\ \forall x \in \mathbb{R}\ (T_x \restriction \eta$ is wellfounded $\Rightarrow T_{f(x)}$ is wellfounded). Now for μ-a.e. $\beta < \lambda$, $h(\beta)^+ = \underset{\sim}{\delta}^1_{\gamma+1}$ for some γ of type I such that $\text{Scale}(\underset{\sim}{\Pi}^1_{\gamma+1})$. By Lemma 4.10, the set

$$C_\beta = \{\delta < h(\beta)^+ \mid \forall x \in \mathbb{R}\ (|T_x \restriction h(\beta)| < \delta \Rightarrow T_{f(x)} \restriction h(\beta) < \delta)\}$$

is club in $h(\beta)^+$ for such β. Let

$$C' = [\lambda\beta . C_\beta]_\mu - \underset{\sim}{\delta}^1_\alpha ,$$

so that by our claim C' is club in $(\underset{\sim}{\delta}^1_\alpha)^+$. Now it is easy to show that there is a function $F : (\underset{\sim}{\delta}^1_\alpha)^\alpha \to (\underset{\sim}{\delta}^1_\alpha)^+$ so that whenever $\beta < (\underset{\sim}{\delta}^1_\alpha)^+$ and W is a tree on $\underset{\sim}{\delta}^1_\alpha$ so that $|W| = \beta$, then $|[\lambda\gamma \cdot W \upharpoonright h(\gamma)]_\mu| = F(\beta)$. (If μ is normal, then F is the identity.) Let

$$C = C' \cap \{\beta \mid F''\beta \subseteq \beta\} .$$

Suppose $[p]_\mu \in C$ and $|T_x| < [p]$. Then $F(|T_x|) < [p]_\mu$, so $|T_x \upharpoonright h(\beta)| < p(\beta)$ for μ-a.e. β, so $|T_{f(x)} \upharpoonright h(\beta)| < p(\beta)$ for μ-a.e. β, so $F(|T_{f(x)}|) < [p]$. But it is easy to check that $\beta \leq F(\beta)$ for all β, so in fact $|T_{f(x)}| < [p]$. Thus C is as demanded by boundedness requirement (b), and we have (modulo [9], of course) proved 4.9.

From 4.9 and 4.7 we have at once the following theorem.

Theorem 4.11. Let $\lambda < \Theta^{L(\mathbb{R})}$ be a reliable ordinal. Then there is a Suslin cardinal κ so that $\kappa \leq \lambda < \kappa^+$.

Proof. Let $\kappa = \sup\{\gamma \leq \lambda \mid \gamma$ is Suslin$\}$. By 4.4, κ is Suslin. If $S(\kappa) = \underset{\sim}{\Pi}^1_\alpha$ where α is type IV, then Scale($\underset{\sim}{\Pi}^1_\alpha$) and $\kappa = \underset{\sim}{\delta}^1_\alpha$ by 4.3. But $\exists^{\mathbb{R}} \underset{\sim}{\Pi}^1_\alpha = \underset{\sim}{\Pi}^1_\alpha$, so 4.7 and 4.9 combined tell us that $\lambda < \kappa^+$. Otherwise, inspection of 4.3 shows that $S(\kappa) = \underset{\sim}{\Sigma}^1_{\alpha+1} = \exists^{\mathbb{R}} \underset{\sim}{\Pi}^1_\alpha$ for some α. If Scale($\underset{\sim}{\Pi}^1_\alpha$), then $\kappa = \underset{\sim}{\delta}^1_\alpha$ by 4.3, and again 4.7 and 4.9 imply that $\lambda < \kappa^+$. If Scale($\underset{\sim}{\Pi}^1_\alpha$) fails, then Scale($\underset{\sim}{\Pi}^1_{\alpha+1}$) must hold, and $\underset{\sim}{\delta}^1_{\alpha+1} = \kappa^+$ is Suslin. By the definition of κ, $\lambda < \kappa^+$. ⊣

Theorem 4.11 clearly implies that all reliable cardinals below $\Theta^{L(\mathbb{R})}$ are Suslin. This completes the proof of Corollary 4.5.

REFERENCES

[1] H.S. Becker, Thin collections of sets of projective ordinals and analogs of L, Annals of Mathematical Logic, vol. 19, 1980, pp. 205-241.

[2] L.A. Harrington and A.S. Kechris, On the determinacy of games on ordinals, Annals of Mathematical Logic, vol. 20, 1981, pp. 109-154.

[3] R.B. Jensen, The fine structure of the constructible hierarchy, Annals of Mathematical Logic, vol. 4, 1972, pp. 229-308.

[4] A.S. Kechris, AD and projective ordinals, Cabal Seminar 76-77, Lecture Notes in Mathematics, vol. 689, Springer Verlag, 1978.

[5] A.S. Kechris, Souslin cardinals, κ-Souslin sets, and the scale property in the hyperprojective hierarchy, Cabal Seminar 77-79, Lecture Notes in Mathematics, vol. 839, Springer-Verlag, 1981.

[6] A.S. Kechris, R.M. Solovay, and J.R. Steel, The axiom of determinacy and the prewellordering property, Cabal Seminar 77-79, Lecture Notes in Mathematics, vol. 839, Springer-Verlag, 1981.

[7] A.S. Kechris and W.H. Woodin, The equivalence of determinacy and partition properties, to appear.

[8] D.A. Martin, The largest countable this, that, or the other, this volume.

[9] S. Jackson and D.A. Martin, Pointclasses and wellordered unions, this volume.

[10] D.A. Martin and J.R. Steel, The extent of scales in $L(\mathbb{R})$, this volume.

[11] Y.N. Moschovakis, Inductive scales on inductive sets, Cabal Seminar 76-77, Lecture Notes in Mathematics, vol. 689, Springer-Verlag, 1978.

[12] Y.N. Moschovakis, Scales on coinductive sets, this volume.

[13] Y.N. Moschovakis, Ordinal games, Cabal Seminar 77-79, Lecture Notes in Mathematics, vol. 839, Springer-Verlag, 1981.

[14] Y.N. Moschovakis, Descriptive Set Theory, Studies in Logic, vol. 100, North Holland, Amsterdam, 1980.

[15] S.G. Simpson, A short course in admissible recursion theory, Generalized recursion theory II, Studies in Logic, vol. 94, North Holland, Amsterdam, 1978, pp. 355-390.

[16] J.R. Steel, Closure properties of pointclasses, Cabal Seminar 77-79, Lecture Notes in Mathematics, vol. 839, Springer-Verlag, 1981.

THE REAL GAME QUANTIFIER PROPAGATES SCALES

Donald A. Martin
Department of Mathematics
University of California
Los Angeles, California 90024

§0. Introduction. Moschovakis [1980, 6E] shows, assuming sufficient determinacy, that the game quantifier (for games on the integers) propagates scales. In Moschovakis [1983] he shows how to put scales on sets defined by applying the real game quantifier to a closed matrix. In this paper we show, from appropriate determinacy hypotheses, that the real game quantifier, and the real or integer game quantifier of any fixed countable length, propagate scales.

Throughout the paper, we work in ZF + DC.

If α is an ordinal, a game type of length α is a function $g:\alpha \to \{0,1\}$. If g is a game type of length α and $Y \subseteq {}^{\omega}\alpha$, the game G_g^Y is played as follows: White and Black produce in α moves, an element y of ${}^{\omega}\alpha$. White chooses $y(\beta)$ if $g(\beta) = 0$ and Black chooses $y(\beta)$ if $g(\beta) = 1$. White wins just in case $y \in Y$.

If $Y \subseteq {}^{\omega}(\gamma + \alpha)$ and g is a game type of length α,

$$g(Y) = \left\{x \in {}^{\omega}\gamma\text{: White has a winning strategy for } G_g^{Y_x}\right\}$$

where $Y_x = \{z\colon x^\frown z \in Y\}$. (Note that $g(Y)$ depends in general on γ, though our notation does not indicate this dependence.)

The notion of a scale on a subset of ${}^{\omega}\alpha$ is defined just like that of a scale on a subset of ${}^{\omega}\omega$, using the product topology on ${}^{\omega}\alpha$.

The purpose of this paper is to show that, if $Y \subseteq {}^{\omega}(\gamma + \alpha)$, α is countable, Y admits a scale, and g is a game type of length α, then $g(Y)$ admits a scale. We shall need to assume determinacy for certain games of length $\leq 4\alpha$ ($=\alpha$ if α is a limit ordinal), the games $G_g^{Y_x}$ and certain other games whose winning conditions are related to the norms on Y. Our definability results for the scale on $g(Y)$ will be natural generalizations of those of Moschovakis, the principal difference being that we need to use a well-ordering of a subset of ω of order type α.

Our plan is to break down Moschovakis' scale-propagation technique into three basic pieces and then to assemble the pieces in a routine fashion. The three pieces are:

(1) Infimum norms. Here we show how to go from a sequence of norms on Y to a sequence of norms on Y to a sequence of norms on $g(Y)$ when $\alpha = 1$ and $g(0) = 0$ (i.e., for one-move games where White moves). If the original sequence is a scale, the resulting sequence will be a scale, but our main result is similar to Moschovakis Infimum Lemma for existential real quantification.

(2) Supremum norms. This is like (1) except that $g(0) = 1$ (Black moves). Our method is similar to that of Moschovakis' Supremum Lemma.

(3) Game norms. If g is any game type and φ is any norm on Y, a natural generalization of a result of Moschovakis [1980] gives a norm $g(\varphi)$ on $G(Y)$, granted appropriate determinacy.

In §1 we discuss infimum norms, in §2 we discuss supremum norms, in §3 we discuss game norms, and in §4 we assemble a scale on $g(Y)$ from a scale on Y, infimum norms, supremum norms, game norms, and a well-ordering of a subset of ω of order type α. In §5 we define canonical winning strategies. In §6 we prove results about propagation of the scale property.

§1. Infimum norms. Let $Y \subseteq {}^{\omega}(\gamma + 1)$. Suppose $\bar{\varphi} = (\varphi_i : i \in \omega)$ is a putative scale on Y, i.e. $\varphi_i : Y \to$ Ordinals for each i and $\varphi_{i+1}(y_1) \leq \varphi_{i+1}(y_2)$ implies $\varphi_i(y_1) \leq \varphi_i(y_2)$ for all $y_1, y_2 \in Y$ and every $i \in \omega$ (The last clause is merely for convenience, and differs from Moschovakis' definition [1980].)

Let $k \in \omega$. We define a putative scale $\psi = (\psi_i : i \in \omega) = \mathrm{Inf}_k\, \bar{\varphi}$ on $g(Y)$, where $g : 1 \to \{0,1\}$ and $g(0) = 0$. Our definition will have the property that each ψ_i is defined solely from $\varphi_0, \ldots, \varphi_i$.

For $i \leq k$, let $\psi_i(x) = \inf\{\varphi_i(x^\frown n) : n \in \omega \,\&\, x^\frown n \in Y\}$. For $i > k$ let

$$\psi_i(x) = \inf\{\langle \varphi_k(x^\frown n),\, n,\, \varphi_i(x^\frown n)\rangle : n \in \omega \,\&\, x^\frown n \in Y\}$$

where $\langle\ ,\ ,\ \rangle$ embeds the lexicographic order on an appropriate $\kappa \times \omega \times \kappa$ into the ordinals.

For each $i \in \omega$ and $x_1, x_2 \in {}^{\omega}\gamma$ we now define the game for verifying $x_1 \in g(Y)$ and either $x_2 \notin g(Y)$ or $\psi_i(x_1) \leq \psi_i(x_2)$.

$i \leq k$. II plays n_1 and I plays n_2. II wins just in case $x_1^\frown n_1 \in Y$ and either $x_2^\frown n_2 \notin Y$ or $\varphi_i(x_1^\frown n_1) \leq \varphi_i(x_2^\frown n_2)$.

$i > k$. II Plays n_1, I Plays n_2, I plays $j_1 < 3$, and II plays $j_2 \leq j_1$. If $j_1 = j_2 = 2$, II wins if $x_1^\frown n_1 \in Y$ and either $x_2^\frown n_2 \notin Y$ or $\varphi_i(x_1^\frown n_1) \leq \varphi_i(x_2^\frown n_2)$. If $j_1 = j_2 = 1$, II wins if $n_1 \leq n_2$. If $j_1 = j_2 = 0$, II wins if $x_1^\frown n_1 \in Y$ and either $x_2^\frown n_2 \notin Y$ or $\varphi_k(x_1^\frown n_1) \leq \varphi_k(x_2^\frown n_2)$. If $1 = j_2 < j_1$, II wins if $n_1 < n_2$. If $0 = j_2 < j_1$, II wins if $x_1^\frown n_1 \in Y$ and either $x_2^\frown n_2 \notin Y$ or $\varphi_k(x_1^\frown n_1) < \varphi_k(x_2^\frown n_2)$.

It is easily checked that II has a winning strategy for this game just in case $x_1 \in g(Y)$ and either $x_2 \notin g(Y)$ or $\psi_i(x_1) \leq \psi_i(x_2)$.

There is a similar game for verifying $x_1 \in g(Y)$ and either $x_2 \notin g(Y)$ or $\psi_i(x_1) < \psi_i(x_2)$. We omit the definition.

A <u>standard play</u> of the game for verifying $x_1 \in g(Y)$ and either $x_2 \notin g(Y)$ or $\psi_i(x_1) \leq \psi_i(x_2)$ is a play of the game in which either $i \leq k$ or $i > k$ and $j_1 = j_2 \neq 1$. The <u>terminal condition</u> of a standard play is $(x_1{}^\frown n_1, x_2{}^\frown n_2, k)$ if $i > k$ and $j_1 = 0$ and $(x_1{}^\frown n_1, x_2{}^\frown n_2, i)$ if $i \leq k$ or $j_1 = 2$. Notice that II wins a standard play just in case $x_1{}^\frown n_1 \in Y$ and either $x_2{}^\frown n_2 \notin Y$ or $\varphi_j(x_1{}^\frown n_1) \leq \varphi_j(x_2{}^\frown n_2)$ where j is the third component of the terminal condition.

<u>Lemma 1.1.</u> Let $\bar\varphi$ be a putative scale on Y. Let $\bar\psi = \mathrm{Inf}_k\, \bar\varphi$. Let $f : \omega \to \omega$ be such that

(a) $f(i) \leq f(i + 1)$ for all i;
(b) $f(i) \leq i$ for all i;
(c) range (f) is unbounded.

Let $x_i \in g(Y)$ for each $i \in \omega$, where $g : 1 \to \{0,1\}$ and $g(0) = 0$. Assume that $\psi_{f(i)}(x_i) \geq \psi_{f(i)}(x_{i+1})$ for all i. Let σ_i be a winning strategy for II for the game for verifying $x_{i+1} \in g(Y)$ and either $x_i \notin g(Y)$ or $\psi_{f(i)}(x_{i+1}) \leq \psi_{f(i)}(x_i)$, for each $i \in \omega$.

There are numbers $n, n_1, n_2, \ldots$ such that $\lim_i n_i = n$ and, for every n_0 with $x_0{}^\frown n_0 \in Y$, there is an $f^* : \omega \to \omega$ satisfying (a), (b), and (c), and

(d) $f^*(i) \leq f(i)$ for all i,
(e) $f^*(i) \geq \min\{f(i), k\}$,

and such that $(x_{i+1}{}^\frown n_{i+1}, x_i{}^\frown n_i, f^*(i))$ is the terminal condition of a standard play consistent with σ_i for each i.

<u>Proof.</u> Let n_{i+1} be II's first move is given by σ_i, for $i = 0,1,2,\ldots$ Let n_0 be such that $x_0{}^\frown n_0 \in Y$.

Assume inductively that $x_i{}^\frown n_i \in Y$. Since (n_{i+1}, n_i) is consistent with σ_i for $f(i) \leq k$ and $(n_{i+1}, n_i, 0, 0)$ is consistent with σ_i for $i > k$ we have $x_{i+1}{}^\frown n_{i+1} \in Y$. For $f(i) \leq k$ we have also $\varphi_{f(i)}(x_{i+1}{}^\frown n_{i+1}) \leq \varphi_{f(i)}(x_i{}^\frown n_i)$. For $f(i) > k$, we have $\varphi_k(x_{i+1}{}^\frown n_{i+1}) \leq \varphi_k(x_i{}^\frown n_i)$.

Since the $\varphi_k(x_i{}^\frown n_i)$ are non-increasing for $f(i) > k$, there must be an i_0 with $f(i_0) > k$ such that $i \geq i_0$ implies $\varphi_k(x_i{}^\frown n_i) = \varphi_k(x_{i_0}{}^\frown n_{i_0})$.

Let $i \geq i_0$. The play $(n_{i+1}, n_i, 1, 0)$ is inconsistent with σ_i, since it is a win for I. Thus $(n_{i+1}, n_i, 1, 1)$ is consistent with σ_i. This means $n_{i+1} \leq n_i$. Let then i_1 be large enough that $i_1 \geq i_0$ and $n_i = n_{i_1}$ for all $i \geq i_1$. Let $n = n_{i_1}$. Let $f^*(i) = f(i)$ if $f(i) \leq k$ or $i \geq i_1$. Let $f^*(i) = k$ otherwise. We have already seen that $(n_{i+1}, n_i, f^*(i))$ is the terminal condition of a standard play consistent with σ_i when $i < i_1$. Suppose $i \geq i_1$. Consider the play $(n_{i+1}, n_i, 2, j_2)$ consistent with σ_i. $j_2 = 0$ or $j_2 = 1$ would lose for II, so $j_2 = 2$. Hence $(n_{i+1}, n_i, f(i))$ is the terminal condition of a standard play consistent with σ.

Lemma 1.1 is similar to Moschovakis' [1983] Infimum Lemma. Our situation is simpler than his in that our moves are integers instead of reals, but our situation is complicated by the fact that we are going to use our lemma in studying long games, so we need the σ_i and the functions f and f^* for bookkeeping and avoiding the axiom of choice.

Let us say that a putative scale $\bar{\varphi} = (\varphi_i : i \in \omega)$ on Y is <u>i-lsc</u>. if whenever $(y_j : j \in \omega)$ converges to y, each $y_j \in Y$, and the norms $\varphi_i(y_j)$ are all eventually constant as j increases, then $y \in Y$ and $\varphi_i(y) \leq \lim_j \varphi_i(y_j)$. Note that $\bar{\varphi}$ is a scale just in case $\bar{\varphi}$ is i-lsc for every $i \in \omega$.

<u>Lemma 1.2</u>. Let $\bar{\varphi}$ be a putative scale on Y, let $\bar{\psi} = \mathrm{Inf}_k\, \bar{\varphi}$, and let $i_0 \in \omega$. If $\bar{\varphi}$ is i_0-lsc so is $\bar{\psi}$.

<u>Proof</u>. Let $x_j \in g(Y)$ for $j \in \omega$ and suppose $(x_j : j \in \omega)$ converges to x and the norms $\psi_i(x_j)$ are eventually constant. By thinning the sequence if necessary, we may assume that $\psi_i(x_i) = \lim_j \psi_i(x_j)$ for each i. Let n_i be such that $x_i{}^\frown n_i \in Y$ and $\varphi_i(x_i{}^\frown n_i)$, for $i \leq k$, or $\langle \varphi_k(x_i{}^\frown n_i), n_i, \varphi_i(x_i{}^\frown n_i)\rangle$, for $i > k$, is as small as possible. If $i \leq k$, we have that $\varphi_i(x_i{}^\frown n_i) = \varphi_i(x_j{}^\frown n_j)$ for all $j \geq i$, since the minimality of $\varphi_j(x_j{}^\frown n_j)$ implies that of $\varphi_i(x_j{}^\frown n_j)$. If $i > k$, we have then that $\varphi_k(x_i{}^\frown n_i) = \varphi_k(x_k{}^\frown n_k)$. Hence, for all $j > k, n_j = n_{k+1} = n$. Hence $(x_i{}^\frown n_i : i \in \omega)$ converges to $(x^\frown n)$. If $j > i > k$, we must have $\varphi_i(x_i{}^\frown n_i) = \varphi_i(x_j{}^\frown n_j)$. Thus all norms $\varphi_i(x_j{}^\frown n_j)$ are eventually constant. It follows that $x^\frown n \in Y$ and $\varphi_{i_0}(x^\frown n) \leq \lim_j \varphi_{i_0}(x_j{}^\frown n_j)$. Thus $x \in g(Y)$. If $i_0 \leq k$, then $\psi_{i_0}(x) \leq \varphi_{i_0}(x^\frown n) \leq \lim_j \varphi_{i_0}(x_j{}^\frown n_j) = \lim_j \psi_{i_0}(x_j)$. If $i > k$, it suffices to note that $\inf\{\varphi_k(x^\frown m) : m \in \omega\} \leq \varphi_k(x^\frown n) \leq \lim_j \varphi_k(x_j{}^\frown n_j)$ and so that

$$\psi_{i_0}(x) = \inf\{\langle \varphi_k(x^\frown m), m, \varphi_i(x^\frown m)\rangle : m \in \omega\}$$

$$\leq \langle \varphi_k(x^\frown n), n, \varphi_{i_0}(x^\frown n)\rangle \leq \lim_j \langle \varphi_k(x_j{}^\frown n_j), n_j, \varphi_{i_0}(x_j{}^\frown n_j)\rangle \quad .$$

§2. <u>Supremum norms</u>. Let $Y \subseteq {}^{\omega}(\gamma + 1)$. Suppose $\bar{\varphi} = (\varphi_i : i \in \omega)$ is a putative scale on Y. Let $k \in \omega$. We define a putative scale $\bar{\psi} = (\psi_k : k \in \omega) = \mathrm{Sup}_k\, \bar{\varphi}$ on $g(Y)$, where $g : 1 \to \{0,1\}$ and $g(0) = 1$.

If $i \leq k$, let

$$\psi_i(x) = \sup\{\varphi_i(x^\frown n) : n \in \omega\} \ .$$

If $i > k$, let

$$\psi_i(x) = \langle \psi_k(x), \varphi_{p_0}(x^\frown m_0), \varphi_{p_1}(x^\frown m_1), \ldots, \varphi_{p_{i-k-1}}(x^\frown m_{i-k-1})\rangle \ .$$

where $n \mapsto (p_n, m_n)$ is a bijection between ω and $\omega \times \omega$ with $p_n \leq n$ for all n, with $\langle\ ,\ldots,\ \rangle$ an appropriate embedding of the lexicographic ordering into the ordinals.

$\bar{\psi}$ is clearly a putative scale.

For each i and x_1, x_2, we define the game for verifying that $x_1 \in g(Y)$ and either $x_2 \notin g(Y)$ or $\psi_i(x_1) \leq \psi_i(x_2)$.

For $i \leq k$, I plays n_1 and II plays n_2. II wins if $x_1 ^\frown n_1 \in Y$ and either $x_2 ^\frown n_2 \notin Y$ or $\varphi_i(x_1 ^\frown n_1) \leq \varphi_i(x_2 ^\frown n_2)$.

If $i > k$, I plays $j_1 \leq i-k$ and II plays $j_2 \leq j_1$. If $j_1 = j_2 = 0$, I plays n_1 II Plays n_2, and II wins $\Longleftrightarrow x_1 ^\frown n_1 \in Y$ and either $x_2 ^\frown n_2 \notin Y$ or $\varphi_k(x_1 ^\frown n_1) \leq \varphi_k(x_2 ^\frown n_2)$. If $j_1 = j_2 = n+1$, II wins $\Longleftrightarrow x_1 ^\frown m_n \in Y$ and either $x_2 ^\frown m_n \notin Y$ or $\varphi_{p_n}(x_1 ^\frown m_n) \leq \varphi_{p_n}(x_2 ^\frown m_n)$. The cases $j_1 > j_2 = 0$ and $j_1 > j_2 = n+1$ are similar, except that "$\leq$"^{n} replaces "$\leq$", and II plays n_2 before I plays n_1 when $j_2 = 0$.

There is a similar game for verifying that $x_1 \in g(Y)$ and either $x_2 \notin g(Y)$ or $\psi_i(x_1) < \psi_i(x_2)$.

A standard play of our game for verifying $x_1 \in g(Y)$ and either $x_2 \notin g(Y)$ or $\psi_i(x_1) \leq \psi_i(x_2)$ is a play in which $i \leq k$ or $j_1 = j_2$. The terminal condition of a standard play is $(x_1 ^\frown n_1, x_2 ^\frown n_2, i)$ if $i \leq k$, $(x_1 ^\frown n_1, x_2 ^\frown n_2, k)$ if $i > k$ and $j_1 = 0$, and $(x_1 ^\frown m_n, x_2 ^\frown m_n, p_n)$ if $j_1 = n+1$.

Lemma 2.1. Let $\bar{\varphi}$ be a putative scale on Y. Let $\bar{\psi} = \mathrm{Sup}_k\, \varphi$. Let f satisfy (a), (b), and (c) as in Lemma 1.1. Let $x_i \in g(Y)$ for each $i \in \omega$ (where $g : 1 \to \{0,1\}$ and $g(0) = 1$). Assume that $\psi_{f(i)}(x_i) \geq \psi_{f(i)}(x_{i+1})$ for all i. Let σ_i, $i = 0,1,\ldots$, be winning strategies for the game for verifying $x_{i+1} \in g(Y)$ and either $x_i \notin g(Y)$ or $\psi_{f(i)}(x_{i+1}) \leq \psi_{f(i)}(x_i)$. For each $n \in \omega$ there are numbers $n_0, n_1, \ldots$ such that $\lim_i n_i = n$ and there is a function $f^* : \omega \to \omega$ such that f^* satisfies (a), (b), (c), (d), and (e) of Lemma 1.1 and such that $(x_{i+1} ^\frown n_{i+1}, x_i ^\frown n_i, f^*(i))$ is the terminal condition of a standard play according to σ_i for each i.

Proof. Since the norms $\psi_i(x_j)$ are eventually non-decreasing as j increases, they are eventually constant.

Let $j(i)$ be given by $p_j(i) = i$ and $m_{j(i)} = n$. For each $t > k$, let $u(t)$ be such that $u(t) > u(t-1)$ if $t > k+1$, $f(u(t)) > k + j(t)$, and $\psi_{k+j(t)}(x_j) = \psi_{k+j(t)}(x_{u(t)})$ for all $j \geq u(t)$.

Set $n_i = n$ for all $i \geq u(k+1)$. For $u(t+1) > i \geq u(t)$, let I Play $j_1 = j(t) + 1$ against σ_i. This is legal, since $f(i) \geq f(u(t)) > k + j(t) = k + j_1 - 1$ so $f(i) - k \geq j_1$. Since $\psi_{k+j(t)}(x_{i+1}) = \psi_{k+j(t)}(x_i)$, $\psi_k(x_i) = \psi_k(x_{i+1})$ and $\varphi_{p_j}(x_i ^\frown m_j) = \varphi_{p_j}(x_{i+1} ^\frown m_j)$ for all $j < j(t)$. Thus σ_i, a winning strategy, cannot call for II to play $j_2 < j(t) + 1$. Hence $j_2 = j(t) + 1$ and the terminal condition of this standard play is

$$(x_{i+1}{}^\frown m_{j(t)},\ x_i{}^\frown m_{j(t)},\ p_{j(t)}) = (x_{i+1}{}^\frown n,\ x_i{}^\frown n,\ t)$$

Assume that n_{i+1} is defined, where $k < f(i)$ and $i < u(k+1)$. Let I play $j_1 = 0$ against σ_i. II must respond with $j_2 = 0$. Let I play n_{i+1} as his n_1. Let n_i be II's response via σ_i. The terminal condition of this standard play is then

$$(x_{i+1}{}^\frown n_{i+1},\ x_i{}^\frown n_i, k)\ .$$

Suppose n_{i+1} is defined and $f(i) \leq k$. Let I play n_{i+1} and II play n_i as given by σ_i. The terminal condition of this standard play is

$$(x_{i+1}{}^\frown n_{i+1},\ x_i{}^\frown n_i,\ f(i))$$

Now define $f^*(i) = f(i)$ if $f(i) \leq k$, $f^*(i) = k$ if $k < f(i)$ and $i < u(k+1)$, and $f^*(i) = t$ if $u(t) \leq i < u(t+1)$. Since $f(u(t)) > k + j(t) > m_{j(t)} = t$, $f^*(i)$ is always $\leq f(i)$.

Lemma 2.1 has the same relation to Moschovakis[1] [1983]. Supremum Lemma as Lemma 1.1 has to the Infimum Lemma.

Lemma 2.2. Let $\bar{\varphi}$ be a putative scale on Y and let $\bar{\psi} = \sup_k \varphi$. For $i_0 \in \omega$, if $\bar{\varphi}$ is i_0-lsc, so is $\bar{\psi}$.

Proof. Let $(x_j : j \in \omega)$ converge to x with each $x_i \in g(Y)$ and let the norms $\psi_i(x_j)$ be eventually constant. Let $n \in \omega$. For each i, the norms $\varphi_i(x_j{}^\frown n)$ are eventually constant as j increases. Thus $x^\frown n \in Y$ and $\varphi_{i_0}(x^\frown n) \leq \lim_j \varphi_{i_0}(x_j{}^\frown n)$. Since $\bar{\varphi}$ is a putative scale, $\varphi_i(x^\frown n) \leq \lim_j \varphi_i\ (x_j{}^\frown n)$ for each $i \leq i_0$. Since this holds for all n, it follows that $x \in g(Y)$ and, from the definition of ψ_{i_0}, that $\psi_{i_0}(x) \leq \lim_j \psi_{i_0}(x_j)$. provided that $\psi_i(x) \leq \lim_j \psi_i(x_j)$ for all i such that $i \leq k$ and $i \leq i_0$. Fix such an i and let $\psi_i(x_{j_0}) = \psi_i(x_j)$ for all $j \geq j_0$. We must show that $\sup\{\varphi_i(x^\frown n) : n \in \omega\} \leq \sup\{\varphi_i(x_{j_0}{}^\frown n) : n \in \omega\}$. We already know, since $i \leq i_0$, that $\varphi_i(x^\frown n) \leq \lim_j \varphi_i(x_j{}^\frown n)$. Let $j^* \geq j_0$ be such that $\varphi_i(x_{j*}{}^\frown n) \geq \varphi_i(x^\frown n)$. For some m, $\varphi_i(x_{j_0}{}^\frown m) \geq \varphi_i(x_{j*}{}^\frown n)$, and the proof is complete.

§3. Game norms. Let $Y \subseteq {}^\omega(\gamma + \alpha)$ and let $g : \alpha \to \{0,1\}$. Let φ be a norm on Y. We define a norm $\psi = g(\varphi)$ on $g(Y)$. Let $x_1, x_2 \in {}^\omega\gamma$. We consider a game of length 2α, played as follows. If White moves at β, i.e. if $g(\beta) = 0$, then II moves at 2β and chooses $z_1(\beta)$, and I moves at $2\beta + 1$ and chooses $z_2(\beta)$. If Black moves at β, then I chooses $z_1(\beta)$ at 2β and II chooses $z_2(\beta)$

$2\beta + 1$. II wins the game if $x_1{}^\frown z_1 \in Y$ and either $x_2{}^\frown z_2 \notin Y$ or $\varphi(x_1{}^\frown z_1) \leq \varphi(x_2{}^\frown z_2)$.

(Our game, and the lemmas that follow are obvious generalizations of Moschovakis [1980, 6E].)

We define $x_1 \preccurlyeq x_2$ to hold if and only if $x_1, x_2 \in g(Y)$ and II has a winning strategy for the associated game. If we can show that $\preccurlyeq$ is a prewellordering, then our norm ψ is essentially defined. Note: We shall let ψ have range an initial segment of the ordinals.

<u>Lemma 3.1</u>. Assume all the games used in defining $\leq$ are determined. There is no infinite sequence $x_0, x_1, x_2, \ldots$ of members of $g(Y)$ such that $x_i \not\preccurlyeq x_{i+1}$ for all i.

<u>Proof</u>. Suppose such a sequence exists. Let σ be a strategy for White witnessing that $x_0 \in g(Y)$. Let σ_i, for each i, be a winning strategy for I for witnessing that $x_i \not\preccurlyeq x_{i+1}$. We shall define $z_i \in {}^\omega\alpha$ such that z_0 is a play according to σ and (z_i, z_{i+1}) is a play according to σ_i, for each i. Suppose each $z_i \upharpoonright \beta$ is defined for some $\beta < \alpha$, with $z_0 \upharpoonright \beta$ consistent with σ and each $(z_i \upharpoonright \beta, z_{i+1} \upharpoonright \beta)$ consistent with σ_i. If White moves at β, let $z_0(\beta)$ be given by σ and, inductively, let $z_{i+1}(\beta)$ be given by σ_i. If Black moves at β, let $z_i(\beta)$ be given by σ_i.

Since z_0 is according to σ, $x_0{}^\frown z_0 \in Y$. Since (z_i, z_{i+1}) is according to σ_i, we have by induction that $z_{i+1} \in Y$ and $\varphi(x_{i+1}{}^\frown z_{i+1}) < \varphi(x_i{}^\frown z_i)$. This is a contradiction.

<u>Lemma 3.2</u>. Assume that all games used in defining $\preccurlyeq$ are determined. $\preccurlyeq$ is a prewellordering of $g(Y)$.

<u>Proof</u>. $x \preccurlyeq x$ since otherwise $x_i = x$ contradicts Lemma 3.1. Similarly, a failure of $(x \preccurlyeq x'$ or $x' \preccurlyeq x)$ contradicts Lemma 3.1. If $x_1 \preccurlyeq x_2$ and $x_2 \preccurlyeq x_3$, then composing the two strategies in the obvious way gives $x_1 \preccurlyeq x_3$. Lemma 3.1 implies directly that the relation $(x_1 \preccurlyeq x_2 \;\&\; x_2 \not\preccurlyeq x_1)$ is well-founded.

For any x_1 and x_2 we call the associated game <u>the game for verifying</u> $x_1 \in g(Y)$ <u>and either</u> $x_2 \notin g(Y)$ <u>or</u> $\psi(x_1) \leq \psi(x_2)$. Note that II has a winning strategy for the game if and only if $x_1 \in g(Y)$ and either $x_2 \notin g(Y)$ or $\psi(x_1) \leq \psi(x_2)$.

We now define <u>the game for verifying</u> $x_1 \in g(Y)$ <u>and either</u> $x_2 \notin g(Y)$ <u>or</u> $\psi(x_1) < \psi_2(x_2)$. This game is played just as the one for $x_2 \in g(Y)$ and either $x_1 \notin g(Y)$ or $\psi(x_2) \leq \psi(x_1)$ except that I wins if and only if $x_1{}^\frown z_1 \in Y$ and either $x_2{}^\frown z_2 \notin Y$ or $\varphi(x_1{}^\frown z_1) < \varphi(x_2{}^\frown z_2)$. <u>Note</u>. Unlike the games of §1 and §2, it is a winning strategy for I, not for II, which "verifies" the relation in question.

Lemma 3.3. Assume all relevant games are determined. I has a winning strategy for the game for verifying $x_1 \in g(Y)$ and either $x_2 \notin g(Y)$ or $\psi(x_1)<\psi(x_2)$ if and only if $x_1 \in g(Y)$ and either $x_2 \notin g(Y)$ or $\psi(x_1) < \psi(x_2)$.

Proof. If I has a strategy for this game clearly $x_1 \in g(Y)$. If $x_2 \in g(Y)$ and $\psi(x_2) \leq \psi(x_1)$, we can play II's strategy witnessing this against I's strategy and get a contradiction.

Suppose $x_1 \in g(Y)$ and either $x_2 \notin g(Y)$ or $\psi(x_1) < \psi(x_2)$ and suppose also that II has a winning strategy for the game. We can compose this strategy with II's strategy witnessing $x_1 \preccurlyeq x_1$, getting a strategy witnessing that $x_2 \preccurlyeq x_1$.

§4. The main construction. Let $Y \subseteq {}^{\omega}(\gamma + \alpha)$ and let $g : \alpha \to \{0,1\}$ with α countable. Let $\bar{\varphi} = (\varphi_i : i \in \omega)$ be a putative scale on Y. Let $\beta \mapsto k_\beta$ be a one-one function from α into ω.

For $\beta \leq \alpha$, let $g_\beta(\delta) = g(\beta + \delta)$ for $\beta + \delta < \alpha$. Let $X_\beta = g_\beta(Y)$. $X_\beta \subseteq {}^{\omega}(\gamma + \beta)$. For $\beta_1 < \beta_2 \leq \alpha$, let $g_{\beta_1\beta_2} = g_{\beta_1} \upharpoonright \delta$ where δ is the least ordinal such that $\beta_1 + \delta = \beta_2$. Note that $X_{\beta_1} = g_{\beta_1,\beta_2}(X_{\beta_2})$.

We define a relation $\lhd$ on $\omega \times (\alpha + 1)$ as follows:

If $\beta < \alpha$ and $k_\beta < n$, then $(m,\beta + 1) \lhd (n,\beta)$ for all $m \leq n$. If $\beta < \alpha$ and $k_\beta \geq n$, let δ be minimal such that $\delta = \alpha$ or $k_\delta < n$. $(n,\delta) \lhd (n,\beta)$. Otherwise $\lhd$ never obtains.

Lemma 4.1. $\lhd$ is a well-founded relation.

Proof. Suppose $(n_0,\beta_0) \rhd (n_1,\beta_1) \rhd \dots$. Since $n_0 \geq n_1 \geq n_2 \geq \dots$, $n_i = n$ for all $i \geq$ some i_0. Since $\beta_0 < \beta_1 < \dots$, let $\beta < \sup\{\beta_i : i \in \omega\}$ be such that $k_\delta \geq n$ for all δ such that $\beta \leq \delta < \sup\{\beta_i : i \in \omega\}$. Let i be such that $i \geq i_0$ and $\beta_i \geq \beta$. Since $n_i = n$ and $k_\delta \geq n$ for all δ with $\beta_i \leq \delta \leq \beta_{i+1}$, $(n_{i+1},\beta_{i+1}) \ntriangleleft (n_i,\beta_i)$.

The construction which follows will depend on certain games' being determined. Assume for the rest of this section that all relevant games are determined.

By induction on $\lhd$ we define norms ψ_i^β on X_β for $i \in \omega$ and $\beta \leq \alpha$.

Let $\psi_i^\alpha = \varphi_i$ for all $i \in \omega$.

If $k_\beta < i$ and $g(\beta) = 0$, let $\psi_i^\beta = (\mathrm{Inf}_{k_\beta}\, \overline{\psi^{\beta+1}})_i$. This is well defined since it depends only on $\psi_0^{\beta+1},\dots,\psi_i^{\beta+1}$.

If $k_\beta < i$ and $g(\beta) = 1$, let $\psi_i^\beta = (\sup_{k_\beta}\, \overline{\psi^{\beta+1}})_i$.

If $k_\beta > i$, let $\psi_i^\beta = g_{\beta,\delta}(\psi_i^\delta)$ where $\delta > \beta$ is minimal such that $\delta = \alpha$ or $k_\delta < i$.

Lemma 4.2. Let $\beta < \alpha$ and $x_1, x_2 \in X_\beta$. For all $i \in \omega$, if $g(\beta) = 0$ then $\psi_i^\beta(x_1) \le \psi_i^\beta(x_2) \iff (\mathrm{Inf}_{k_\beta}\ \overline{\psi^{\beta+1}})_i(x_1) \le (\mathrm{Inf}_{k_\beta}\ \overline{\psi^{\beta+1}})_i(x_2)$, and if $g(\beta) = 1$ then $\psi_i^\beta(x_1) \le \psi_i^\beta(x_2) \iff (\mathrm{Sup}_{k_\beta}\ \overline{\psi^{\beta+1}})_i(x_1) < (\mathrm{Sup}_{k_\beta}\ \overline{\psi^{\beta+1}})_i(x_2)$.

Proof. For $i > k_\beta$ it is immediate from the definition that ψ_i^β is identical with the corresponding infimum or supremum norm. Assume then that $i \le k_\beta$. For some $\delta > \beta$, $\psi_i^\beta = g_{\beta,\delta}(\psi_i^\delta)$. It is easily seen that $g_{\beta,\delta}(\psi_i^\delta) = g_{\beta,\beta+1}(g_{\beta+1,\delta}(\psi_i^\delta)) = g_{\beta,\beta+1}(\psi_i^{\beta+1})$. Since $i \le k_\beta$, the verification game for $g_{\beta,\beta+1}(\overline{\psi_i^{\beta+1}})$ is identical with that for $(\mathrm{Inf}_{k_\beta}(\overline{\psi^{\beta+1}}))_i$ or $(\mathrm{Sup}_{k_\beta}(\overline{\psi^{\beta+1}}))_i$, depending on whether $g(\beta) = 0$ or $g(\beta) = 1$.

Note. ψ_i^β may not be identical with the corresponding infimum or supremum norm when $i \le k_\beta$, simply because the infimum and supremum norms need not have range an initial segment of the ordinals. Had we wished, we could have modified the definitions to make this so.

Lemma 4.3. If $k_\delta \ge i$ for all δ such that $\beta_1 \le \delta < \beta_2$ then $\psi_i^{\beta_1} = g_{\beta_1,\beta_2}\psi_i^{\beta_2}$.

By induction on $\lhd$ we define games for verifying $x_1 \in X_\beta$ and either $x_2 \notin X_\beta$ or $\psi_i^\beta(x_1) \le \psi_i^\beta(x_2)$, and also games for verifying that $x_1 \in X_\beta$ and $x_2 \notin X_\beta$ or $\psi_i^\beta(x_1) < \psi_i^\beta(x_2)$.

If $k_\beta \ge i$, let $\delta > \beta$ be minimal such that $\delta = \alpha$ or $k_\delta < i$. First play the game given by the fact that $\psi_i^\beta = g_{\beta,\delta}\ \psi_i^\delta$ (reversing the roles of I and II in the $<$ case). If (z_1, z_2) is a play of this game, then II wins just in case $x_1{}^\frown z_1 \in X_\delta$ and either $(x_2{}^\frown z_2) \notin X_\delta$ or $\psi^\delta(x_1{}^\frown z_1)\ (\le)\ \psi^\delta(x_2{}^\frown z_2)$. Now play the game for verifying this fact.

If $k_\beta < i$, we first play the finite game given by the fact that ψ_i^β is an infimum or supremum norm defined from $\overline{\psi_i^{\beta+1}}$. The winning conditions for this game involve an inequality on $\psi_{i'}^{\beta+1}$ for some $i' \le i$ or else a numerical inequality. In the latter case, the game terminates. In the former case, continue by playing the appropriate game.

A standard play of the game for verifying $x_1 \in X_\beta$ and either $x_2 \notin X_\beta$ or $\psi_i^\beta(x_1) \le \psi_i^\beta(x_2)$ is a play in which all the plays of constituent games for infimum or supremum norms are standard. The terminal condition of a standard play is defined in the obvious way. We also speak of a standard partial play of such a game and of the terminal condition of such a partial play.

Lemma 4.4. If $\overline{\varphi}$ is 0-lsc so is $\overline{\psi^0}$.

Proof. Let $(x_j : j \in \omega)$ converge to x and suppose that each $x_j \in g(Y)$ and that the $\psi_i^0(x_j)$ are eventually constant as j increases. Thinning the sequence if necessary, we may suppose that $\psi_i(x_i) = \psi_i(x_j)$ for each $j \ge i$. For a contradiction, let σ be a winning strategy for I for the game for verifying

$x_0 \in g(Y)$ and either $x \notin g(Y)$ or $(g(\varphi_0))(x_0) < g((\varphi_0))(x)$. Note that $g(\varphi_0) = \psi_0^0$. Let σ_i be a winning strategy for II for the game for verifying $x_{i+1} \in g(Y)$ and either $x_i \notin g(Y)$ or $\psi_i^0(x_{i+1}) \leq \psi_i^0(x_i)$.

We shall define $z, z_0, z_1, \ldots,$ each $\in {}^{\alpha}\omega$, functions $f^\beta : \omega \to \omega$ for $\beta \leq \alpha$, and standard plays $w_0, w_1, \ldots,$ with each w_i consistent with σ_i, such that the following conditions hold:

(i) $(z_i : i \in \omega)$ converges to z.

(ii) (z, z_0) is a play consistent with σ.

(iii) For each $\alpha \leq \beta$ and each $i \in \omega$, the terminal condition of the corresponding part of the plan w_i is

$$(x_{i+1}{}^\frown z_{i+1} \restriction \beta,\ x_i{}^\frown z_i \restriction \beta,\ f^\beta(i)) .$$

(iv) $f^0(i) = i$ for all i.

(v) $f^\beta(i) \leq f^\beta(i+1)$ for all $\beta \leq \alpha$ and all $i \in \omega$.

(vi) Range f^β is unbounded for all $\beta \leq \alpha$.

(vii) $f^{\beta+1}(i) \geq \min\{f^\beta(i), k_\beta\}$.

(viii) For each i, $f^\beta(i)$ is a non-increasing function of β.

(ix) If $\lambda \leq \alpha$ is a limit ordinal, then $f^\lambda(i) = \inf\{f^\beta(i) : \beta < \lambda\}$ for all $i \in \omega$.

Suppose $z^\beta = z \restriction \beta$, $z_0^\beta = z_0 \restriction \beta, \ldots,$ f^β, and the appropriate parts, w_i^β, of the w_i are defined. Suppose $(z_i^\beta : i \in \omega)$ converges to z^β, (z^β, z_0^β) is consistent with σ, the w_i^β are consistent with σ_i, and (iii), (v) and (vi) hold at β, and $f^\beta(i) \leq i$ for all $i \in \omega$.

Suppose first that $g(\beta) = 0$. (iii), (v), (vi) and $f^\beta(i) \leq i$ guarantee that the hypotheses of Lemma 1.1 are satisfied, with Y replaced by $X_{\beta+1}$, k replaced by k_β, with (the appropriate fragments of) the σ_i, and with $f(i) = f^\beta(i)$. Let $n, n_1, n_2, \ldots$ be as given by Lemma 1.1. Let n_0 be the move given by σ at $(z^\beta{}^\frown n, z_0^\beta)$. Let f^* be as given by Lemma 1.1. Let $f^{\beta+1}(i) = f^*(i)$. Let $w_i^{\beta+1}$ be w_i^β followed by the play given by Lemma 1.1. Let $z(\beta) = n$ and $z_i(\beta) = n_i$. Lemma 1.1 guarantees that $(z_i^{\beta+1} : i \in \omega)$ converges to $z^{\beta+1}$, the $w_i^{\beta+1}$ are consistent with the σ_i, (iii), (v), and (vi) hold at $\beta + 1$, $f^{\beta+1}(i) \leq f^\beta(i) \leq i$ for all i, and $f^{\beta+1}(i) \geq \min\{f^\beta(i), k_\beta\}$.

Next suppose that $g(\beta) = 1$. We proceed just as in the first case, using Lemma 2.1 in place of Lemma 1.1. We let n be the move given by σ at (z^β, z_0^β). We omit the details.

Finally suppose that $\lambda < \alpha$ is a limit ordinal, that z^λ, the z_i^λ, the w_i^λ, and the f^β for $\beta < \lambda$, are defined. Suppose $(z_i^\beta : i \in \omega)$ converges to z^β for each $\beta < \lambda$, (z^β, z_0^β) is consistent with σ for $\beta < \lambda$, the w_i^β are consistent with σ_i for $\beta < \lambda$, (iv) holds, (iii), (v), (vi), and (vii) hold for $\beta < \lambda$, (ix) holds for limit ordinals $< \lambda$, and each $f^\beta(i)$ is non-increasing as a function of

$\beta < \lambda$.

It follows that $(z_i^\lambda : i \in \omega)$ converges to z^λ, (z^λ, z_0^λ) is consistent with σ, and each w_i^λ is consistent with σ_i. Define f_i^λ by (ix). It is readily seen that (iii) and (v) hold at λ. To see that (vi) holds at λ, let $n \in \omega$. We must show $f^\lambda(i) \geq n$ for some i. Let $\beta < \lambda$ be such that $k_\delta \geq n$ for every δ such that $\beta \leq \delta < \lambda$. Let i be such that $f^\beta(i) \geq n$. For $\beta \leq \delta < \lambda$, $f^{\delta+1}(i) \geq \min\{f^\delta(i), k_\delta\}$ by (v) and (vii). Thus induction and (ix) give that $f^\delta(i) \geq n$ for all $\delta < \lambda$. Hence $f^\lambda(i) \geq n$.

We have given the construction and verified that it has the required properties. Now let us prove the lemma. Since (z, z_0) is a play consistent with σ, $z_0 \in Y$ and either $z \notin Y$ or $\varphi_0(x_0{}^\frown z_0) < \varphi_0(x^\frown z)$. Since each w_i is consistent with σ_i and the terminal condition of w_i is $(x_{i+1}{}^\frown z_{i+1}, x_i{}^\frown z_i, f^\alpha(i))$, it follows by induction that $x_i{}^\frown z_i \in Y$ and $\varphi_{f^\alpha(i)}(x_{i+1}{}^\frown z_{i+1}) \leq \varphi_{f^\alpha(i)}(x_i{}^\frown z_i)$ for each $i \in \omega$. Since $(z_i : i \in \omega)$ converges to z and $(x_i : i \in \omega)$ converges to x, it follows, since $\overline{\varphi}$ is 0-loc, that $x^\frown z \in Y$ and $\varphi_0(x^\frown z) \leq \lim_i \varphi_0(x_i{}^\frown z_i) \leq \varphi_0(x_0{}^\frown z_0)$. This is a contradiction.

<u>Theorem 4.5</u>. If $\overline{\varphi}$ is a scale, so is $\overline{\psi^0}$.

<u>Proof</u>. Let (n,β) be $\lhd$-minimal such that $\overline{\varphi^\beta}$ is not n-lsc. If $k_\beta < n$, it follows from Lemmas 1.2 and 2.2 that $\overline{\varphi^{\beta+1}}$ is not n-lsc. Since $(n, \beta+1) \lhd (n,\beta)$ this is a contradition. If $k_\beta \geq n$, let $\gamma > \beta$ be the least ordinal such that $\gamma = \alpha$ or $k_\gamma < n$. Apply Lemma 4.4 with Y replaced by X_γ, g replaced by $g_{\beta,\gamma}$, and $(\varphi_i : i \in \omega)$ replaced by $(\psi_{n+i}^\gamma : i \in \omega)$. Since $\overline{\psi^\beta}$ is not n-lsc, gives that $\overline{\psi^\gamma}$ is not n-lsc. Since $(n,\gamma) \lhd (n,\beta)$, we have a contradiction.

§5. <u>Canonical winning strategies</u>. Let g, Y, $\beta \mapsto k_\beta$, $\overline{\varphi}$, $(\overline{\psi^\beta} : \beta \leq \alpha)$, and $(X_\beta : \beta \leq \alpha)$ be as in §4, with $\overline{\varphi}$ a scale on Y. Let $x \in g(Y)$. We define the <u>canonical strategy for White</u> as follows:

At position z^β with $g(\beta) = 0$, if $x^\frown z^\beta \notin X^\beta$, White plays $z(\beta) = 0$. If $x^\frown z^\beta \in X_\beta$, I plays the smallest n such that $x^\frown z^{\beta\frown} n \in X_{\beta+1}$ and, for all m such that $x^\frown z^{\beta\frown} m \in X_{\beta+1}$, $\psi_{k_\beta}^{\beta+1}(x^\frown z^{\beta\frown} n) \leq \psi_{k_\beta}^{\beta+1}(x^\frown z^{\beta\frown} m)$.

<u>Theorem 5.1</u>. Assume all games involved in the definition of the $\overline{\psi^\beta}$ are determined. The canonical strategy is a winning strategy.

<u>Proof</u>. We prove the following stronger fact. Let z^β be any position consistent with the canonical strategy. Let $n \in \omega$ and $\gamma \geq \beta$ be such that, for all δ with $\beta \leq \delta < \gamma$, $k_\delta \geq n$. We define a game $G(z^\beta, n, \gamma)$ whose length is 2ρ, where ρ is the least ordinal such that $\beta + \rho = \gamma$. Let $z_1^\beta = z_2^\beta = z^\beta$. For $\beta \leq \delta < \gamma$,

if $g(\delta) = 1$, I plays $z_1(\delta)$ and then II plays $z_2(\delta)$. If $\beta \leq \delta < \gamma$, and $g(\delta) = 0$, II must play $z_1(\delta)$ = the move given by the canonical strategy at z_1^δ; then I plays $z_2(\delta)$. II wins just in case $x^\frown z_1^\gamma \in X_\gamma$ and either $x^\frown z_2^\gamma \notin X_\gamma$ or $\psi_n^\gamma(x^\frown z_1^\gamma) \leq \psi_n^\gamma(x^\frown z_2^\gamma)$.

We shall prove that II has a winning strategy for $G(z^\beta, n, \gamma)$. Suppose this is false for some (γ, β, n) and choose the lexicographically least (γ, β, n). Choose some z^β witnessing this fact.

We first show that γ is a limit ordinal $> \beta$. If $\gamma = \beta$, then $z^\beta \notin X_\beta$. Let $\gamma' = \beta$, let $n' = 0$, and let $\beta' = 0$. I can win $G(z^0, n', \gamma')$ by playing $z_1(\delta) = z^\beta(\delta)$ and playing $z_2(\delta)$ according to some strategy witnessing $x \in X_0$. This contradicts the minimality of (γ, β, n) unless $\beta = 0$, which is clearly impossible. If $\beta < \gamma = \delta + 1$, let II play a winning strategy for $G(z^\beta, n, \delta)$. By the definition of the canonical strategy, the play must reach a winning position (z_1^δ, z_2^δ) in $G(z^\beta, n, \gamma)$.

Let us then consider the case γ is a limit ordinal $> \beta$. We choose a winning strategy τ for I for $G(z^\beta, n, \gamma)$ as follows. If some z^γ extending z^β and consistent with the canonical strategy is such that $x^\frown z^\gamma \notin X_\gamma$, let I play $z_1(\delta) = z^\gamma(\delta)$. By the argument of the last paragraph, z^β is won for White, so let I play a $z_2(\delta)$ according to a winning strategy for White. If every position z^γ extending z^β which is consistent with the canonical strategy satisfies $x^\frown z^\gamma \in X_\gamma$, let I play an arbitrary winning strategy for $G(z^\beta, n, \gamma)$. Note that, in either case, every play (z_1^γ, z_2^γ) consistent with τ satisfies $x^\frown z_2^\gamma \in X_\gamma$ and either $x^\frown z_1^\gamma \notin X_\gamma$ or $\psi_n^\gamma(x^\frown z_1^\gamma) > \psi_n^\gamma(x^\frown z_2^\gamma)$.

Let $z_0^\beta = z^\beta$ and $\beta_0 = \beta$. Suppose inductively we have defined $\beta_0 < \cdots < \beta_i < \gamma$ and $z_0^{\beta_j}, z_1^{\beta_j}, \ldots, z_j^{\beta_j}, z^{\beta_j}$ for each $j \leq i$, and strategies σ_i for $j < i$ such that

(a) $z_j^{\beta_j} \subseteq z_j^{\beta_{j+1}} \subseteq \cdots \subseteq z_j^{\beta_i}$;

(b) $z^{\beta_0} \subseteq z^{\beta_1} \subseteq \cdots \subseteq z^{\beta_i}$;

(c) $(z^{\beta_i}, z_0^{\beta_i})$ is consistent with τ;

(d) $z_j^{\beta_j} \in X_{\beta_j}$ for all $j \leq i$;

(e) For all $j < i$, $\psi_{n+j}^{\beta_{j+1}}(x^\frown z_{j+1}^{\beta_{j+1}}) \leq \psi_{n+j}^{\beta_{j+1}}(x^\frown z_j^{\beta_{j+1}})$, σ_j is a winning strategy for II witnessing this fact, and $(z_{j+1}^{\beta_i}, z_j^{\beta_i})$ is consistent with σ_j. Furthermore, $k_\delta \geq n + j$ for all δ such that $\beta_j \leq \delta < \gamma$.

(f) $x^\frown z^{\beta_i} \in X_{\beta_i}$ and $\psi_{n+1}^{\beta_i}(x^\frown z^{\beta_i}) \leq \psi_{n+i}^{\beta_i}(x^\frown z_i^{\beta_i})$.

Let $\beta_{i+1} > \beta_i$ be such that $\beta_{i+1} < \gamma$ and $k_\delta \geq n + i + 1$ for all δ such that $\beta_{i+1} \leq \delta < \gamma$. By the minimality of γ, let τ_i be a winning strategy for II for $G(z^{\beta_i}, n + i, \beta_{i+1})$. We get $z_0^{\beta_{i+1}}, \ldots, z_{i+1}^{\beta_{i+1}}$ as follows: $(z_j^{\beta_{i+1}}, z_{j+1}^{\beta_{i+1}})$ will be consistent with σ_j, for all $j < i$. $(z_{i+1}^{\beta_{i+1}}, z_i^{\beta_{i+1}})$ will be consistent with some strategy for II witnessing (f). (In particular, $z_i^{\beta_{i+1}}$ will extend z^{β_i}.)

$(z^{\beta_{i+1}}, z_{i+1}^{\beta_{i+1}})$ will be consistent with τ_i. $(z^{\beta_{i+1}}, z_0^{\beta_{i+1}})$ will be consistent with τ. Since τ is a strategy for I and all other strategies are for II, the reader will easily check that the $z_0^{\beta_{i+1}}, \ldots, z_{i+1}^{\beta_{i+1}}, z^{\beta_{i+1}}$ are determined uniquely, once the strategy witnessing (f) is chosen. Since $\psi_{n+i}^{\beta_{i+1}}(x^\frown z_{i+1}^{\beta_{i+1}}) \leq \psi_{n+i}^{\beta_{i+1}}(x^\frown z_i^{\beta_{i+1}})$, we may complete the construction by choosing σ_i witnessing this fact.

Now let $z^\gamma \supseteq z^{\beta_i}$ for each i and let $z_j^\gamma \supseteq z_j^{\beta_i}$ for each $i \geq j$. Since the σ_j are winning strategies, we have $x^\frown z_{j+1}^\gamma \in X_\gamma$ and $\psi_{n+j}^\gamma(x^\frown z_{j+1}^\gamma) \leq \psi_{n+j}^\gamma(x^\frown z_j^\gamma)$ for each j. By the properties of τ, we have $x^\frown z_0^\gamma \in X_\gamma$ and $\psi_n^\gamma(x^\frown z_0^\gamma) < \psi_n^\gamma(x^\frown z^\gamma)$. Since $(x^\frown z_j^\gamma : j \in \omega)$ converges to $x^\frown z^\gamma$ (note that $z_{i+1} \restriction \beta_i) = z \restriction \beta_i)$, we must have $x^\frown z^\gamma \in X^\gamma$ and $\psi_n^\gamma(x^\frown z^\gamma) \leq \lim_j \psi_n^\gamma(x^\frown z_j^\gamma) \leq \psi_n^\gamma(x^\frown z_0^\gamma)$. This is a contradiction.

§6. Definability. For pointclasses Γ and game types g, we wish to define a pointclass $g(\Gamma)$ and prove theorems such as $\text{Scale}(\Gamma) \Rightarrow \text{Scale}\ g(\Gamma)$. Since Γ is to be a pointclass in the sense of Moschovakis [1980], we shall define $g(\Gamma)$ only under the following assumption:

(*) There is a well-ordering R of a subset A of ω of order type α, where $g : \alpha \to \{0,1\}$, such that, if $\beta \mapsto k_\beta$ is the isomorphism between $(\alpha,<)$ and (A,R), and $\hat{g}(k_\beta) = g(\beta)$, then Γ is closed under preimages by functions recursive in $(R,\hat{g})$.

Let $\beta \to k_\beta$ be a one-one function from an ordinal α into ω. Let R be the induced well-ordering of a subset of ω. For $z \in {}^\omega\omega$, let $z_R^* \in {}^\alpha\omega$ be given by $z_R^*(\beta) = z(k_\beta)$.

Let $g : \alpha \to \{0,1\}$. Let γ be a pointclass satisfying (*). For simplicity we shall assume that pointclasses are collections of sets each of which is a subset of ${}^\gamma\omega$, where $\gamma < \omega^2$ but γ is not fixed.

$X \subseteq {}^\gamma\omega$ belongs to $g(\Gamma)$ just in case there is an R witnessing that Γ satisfies (*) and a $Y \in \Gamma$ such that, for the associated $\beta \mapsto k_\beta$,

$$X = (\{x^\frown z_R^* : z \in Y\}) .$$

Theorem 6.1. Let α be a countable limit ordinal. Let $g : \alpha \to \{0,1\}$ be such that, for every limit ordinal $\lambda < \alpha$ and every $n \in \omega$, $\{g(\lambda + m) : m \geq n\} = \{0,1\}$. (In other words, White and Black both make infinitely many moves in every ω-block.) Suppose Γ satisfies (*) and that all games involved in defining $g(\Gamma)$ are determined.

If $\text{Scale}(\Gamma)$ then $\text{Scale}(g(\Gamma))$.

Proof. Let Y, R witness that $X \in g(\Gamma)$. Let $\beta \mapsto k_\beta$ be the associated function. Let $\bar{\varphi}$ be a scale on Y as given by $\text{Scale}(\Gamma)$. Let $\varphi_i^*(x^\frown z_R^*) = \varphi_i(x^\frown z)$.

$\overline{\varphi}^*$ is a scale on Y^* with $X = g(Y^*)$. Let $\overline{\psi^\beta}$, $\beta \leq \alpha$ be defined as in §4. We must show that the relations

(a) $x_1 \in X$ & $(x_2 \notin X$ or $\psi_i^0(x_1) \leq \psi_i^0(x_2))$
(b) $x_1 \in X$ & $(x_2 \notin X$ or $\psi_i^0(x_1) < \psi_i^0(x_2))$

are in Γ. (We take these relations to be subsets of ${}^{(\omega+\omega+1)}\omega$.) Since the two cases are similar, we consider only $\leq$.

By inserting dummy moves where necessary, we can make (a) hold just in case II wins $G(x_1,x_2,i)$, where $G(x_1,x_2,i)$ is a game of type g. The winning conditions of a play of this game are of one of the forms

$$n_1 \leq n_2$$

$$n_1 < n_2$$

$$x_1{}^\frown z_1 \in Y \;\&\; (x_2{}^\frown z_2 \notin Y \text{ or } \varphi_j(x_1{}^\frown z_1) \leq \varphi_j(x_2{}^\frown z_2))$$

$$x_1{}^\frown z_1 \in Y \;\&\; (x_2{}^\frown z_2 \notin Y \text{ or } \varphi_j(x_1{}^\frown z_1) < \varphi_j(x_2{}^\frown z_2))$$

where z_1, z_2 and n_1, n_2 or j are determined from the play in the obvious way.

We must find an R' and a Y' which witnesses that the property that II has a winning strategy for $G(x_1,x_2,i)$ belongs to $g(\Gamma)$.

We do this as follows: If the β-th move in the new game corresponds to picking $z^*_{1_R}(\delta)$, we let $k'_\beta = 4k_\delta$. If it corresponds to $z^*_{2_R}(\delta)$, we let $k'_\beta = 4k_\delta + 2$. We let the other non-dummy moves corresponding to δ be κ^1_δ, κ^2_δ, and $p^3_{k_\delta}$ and $p^4_{k_\delta}$ if needed, where p_i is the $i+1$-st prime. We let the dummy moves corresponding to β be $p^5_{k_\delta}$, $p^6_{k_\delta}$, etc. The corresponding R' and $\hat{g}'$ are recursive in $(R,\hat{g})$. If $z^*_{R'}$ is a play of the game, z_1 and z_2 are clearly recursive uniformly in $(R,\hat{g})$ and z. We need to be able to find which of the four winning conditions hold and the value of j or (n_1,n_2) as a function of (x_1,x_2,i,z) recursive in $(R,\hat{g})$. Note that, given $m = k_\beta$ and $k \in \omega$, we can find effectively from R the q such that $q = k_\delta$, where δ is the least ordinal $\geq \beta$ such that $k_\delta < k$ (if it exists, and we can determine whether it exists). Repeating this procedure at most i times, beginning with $k = i$ and $m = k_0$, we can find all the places in the game $G(x_1,x_2,i)$ where the numbers j_1, j_2 as in §1 and §2 are played. This allows us to compute the desired information. We omit the details.

<u>Corollary 6.2</u>. Let g be the type of the real game, i.e., let $g : \omega^2 \to \{0,1\}$ with $g(\beta) = 0 \iff \beta = \omega \cdot 2k + n$ for some $k,n \in \omega$. Assume that all integer games of length ω^2 in which White moves at exactly the even ordinals and whose payoffs are in Γ (in the obvious sense) are determined. If Scale(Γ) then Scale($g(\Gamma)$).

<u>Proof</u>. Let Y witness $X \in g(\Gamma)$. As in the proof of Theorem 6.1, we get a scale on X which belongs to $g'(\Gamma)$, where $g' : \omega^2 \to \{0,1\}$ and $g'(\beta) = 0 \iff \beta$ is even. But $g'(\Gamma) = g(\Gamma)$. To see this, replace the g' game by a g game as follows: Replace each ω-block by two ω-blocks. White plays a strategy for the next ω moves of the original game and then Black chooses a play consistent with that strategy.

We could prove more complicated definability theorems by letting the φ_i belong to different classes Γ_i. We could also prove a generalization of Corollary 6.2 for real games of arbitrary countable length. We could also prove definability results for our canonical strategies. Since there are no ideas involved beyond those already presented and those of Moschovakis [1980] and [1983], we shall do none of this.

REFERENCES

Y.N. Moschovakis [1980], Descriptive Set Theory, Studies in Logic, Vol. 100, North-Holland Publishing Co., Amsterdam, 1980.

Y.N. Moschovakis [1983], Scales on coinductive sets, this volume.

SOME CONSISTENCY RESULTS IN ZFC USING AD

W. Hugh Woodin[1]
Department of Mathematics
California Institute of Technology
Pasadena, California 91125

For the most part the uses of the axiom of determinacy (AD) have been to settle natural questions that arise about sets under its influence, i.e. (certain) sets of reals. This combined with the fact that to assume AD requires restricting ones attention to a fragment of the universe in which the axiom of choice fails, would seem to indicate that AD has little to offer in the way of solutions to problems in more conventional set theory. Set theorists as a rule ignore constraints of definability in choosing objects for their amusement nor do they wish to abandon the axiom of choice.

Recently, however, there have been several applications of AD to obtain consistency results in ZFC (see [SVW] or [W]). These methods revolve around starting with a model of ZF + AD and constructing a forcing extension in which ZFC holds, the hope being that enough of the influence of AD will extend to produce some desired property in the generic extension.

We shall be concerned with the results obtained by Steel and Van Wesep [SVW], they show the consistency of ZFC together with ω_2 is the second uniform indiscernible and the nonstationary ideal on ω_1 is ω_2-saturated. The latter solves a well known problem within the theory of saturated ideals. For their result, Steel and Van Wesep needed to assume the consistency of $ZF + AD + \mathbb{R}\text{-}AC$, where $\mathbb{R}\text{-}AC$ is the axiom of choice for families indexed by the reals. This theory is substantially stronger than ZF + AD. We reduce the assumption needed to just the consistency of ZF + AD. As we have suggested their method is to start with a model of $ZF + AD + \mathbb{R}\text{-}AC$ and then construct a forcing extension in which ZFC holds. The forcing is mild enough so that in the generic extension ω_2 is the second uniform indiscernible and the nonstationary ideal on ω_1 is ω_2-saturated. Basically we simply show that $ZF + AD + V = L(\mathbb{R})$ suffices to carry out their forcing arguments. We also isolate a single combinatorial principle which in ZFC implies both that ω_2 is the second uniform indiscernible and that the nonstationary ideal on ω_1 is ω_2-saturated. This principle we show holds in the generic extension. The point of this is twofold. First it offers a means to those uninterested in AD for mining the combinatorial riches of this model and second it suggests that a theory weaker than AD may suffice for these consistency results.

---

[1] Research partially supported by NSF Grant MCS 80-21468.

Finally we extend the results of [SVW] to show that if ZF + AD is consistent then so is ZFC + MA + ¬CH + the nonstationary ideal on ω_1 is ω_2-saturated. Actually we show something stronger, namely as in [SVW], $P(\omega_1)/NS$ can have a very simple form (even in the presence of MA + ¬CH).

We shall for the most part be working in ZF + DC + AD throughout this paper. Notation for the most part will be as in [SVW] and we assume familiarity with the elementary aspects of set theory in the context of AD, as presented in [SVW].

§1. As usual we define the reals as elements of ω^ω. Giving ω the discrete topology naturally induces a product topology on ω^ω, it is with respect to this topology that we define the notion of category. Of course ω^ω with this topology is homeomorphic to the space of irrationals so in addition we have naturally a notion of Lebesgue measure on ω^ω.

Suppose $\alpha < \beta$ are countable ordinals. Let T_α denote the space $(\omega + \alpha)^\omega$ and let $T_{\alpha,\beta} = \prod_{\alpha \le \delta < \beta} T_\delta$. The spaces $T_\alpha, T_{\alpha,\beta}$ with their natural topologies are each homeomorphic to ω^ω (for T_α choose the topology induced by the discrete topology on $\omega + \alpha$, for $T_{\alpha,\beta}$ choose the product topology). Suppose $f \in T_{\alpha,\beta}$. Thus f is a function with domain $[\alpha,\beta) = \{\delta \mid \alpha \le \delta < \beta\}$ such that for each $\delta \in [\alpha,\beta)$, $f(\delta) \in T_\delta$. We denote $f(\delta)$ by f_δ. Note that f_δ is also a function, $f_\delta : \omega \to \omega + \delta$. We are using notation from [SVW].

We define the basic notion of forcing introduced in [SVW].

<u>Definition 1.1</u>. Let P denote the set of conditions defined as follows. A condition is a pair $(f,\vec{X})$ such that $f \in T_{0,\alpha}$ for some $\alpha < \omega_1$, and $\vec{X} = \langle X_\beta \mid \beta < \omega_1 \rangle$, where:

(1) For each $\delta < \alpha$, $f_\delta : \omega \to \omega + \delta$ is onto.

(2) For each $\beta < \omega_1$, $X_\beta \subseteq T_{0,\beta}$ such that if $\beta > \alpha$ then $\{h \in T_{\alpha,\beta} \mid f^\frown h \in X_\beta\}$ is comeager in $T_{\alpha,\beta}$.

(3) (coherence) For each δ, β, $\alpha \le \delta < \beta$;

(a) for $g \in X_\beta$, if $f \subseteq g$ then $g_\delta : \omega \to \omega + \delta$ is onto and $g \restriction \delta \in X_\delta$.

(b) for $g \in X_\delta$, if $f \subseteq g$ then $\{h \in T_{\delta,\beta} \mid g^\frown h \in X_\beta\}$ is comeager in $T_{\delta,\beta}$.

Define the order on P by:

$$(g,\vec{Y}) \le (f,\vec{X}) \text{ iff } f \subseteq g, \ g \in \bigcup X_\beta \text{ and for each } \beta < \omega_1, \ Y_\beta \subseteq X_\beta .$$

Let $\mathbb{P}$ denote the partial order $(P,\le)$.

Let $P' = \{(f,\vec{X}) \mid (f,\vec{X})$ satisfies (1) and (2) in the definition of $P\}$. Let $\mathbb{P}'$ denote P' together with the obvious order extending the order on P.

If we assume that every set of reals has the property of Baire then $\mathbb{P}$ is dense in $\mathbb{P}'$. Specifically if $(f,\vec{X}) \in P'$ then there is a refinement $\vec{Y}$ of $\vec{X}$

such that $(f,\vec{Y}) \in P$, $(f,\vec{Y}) \leq (f,\vec{X})$ in $\mathbb{P}'$. The construction of $\vec{Y}$ is straightforward using the following observations:

(1) (Kuratowski, Ulam) Suppose $A \subseteq \mathbb{R} \times \mathbb{R}$ is comeager, then $\{x \mid A_x$ is comeager$\}$ is comeager. $A_x = \{y \mid (x,y) \in A\}$. This is simply the 'Fubini' theorem for category.

(2) Suppose $\langle A_\zeta : \zeta < \lambda \rangle$ is a wellordered sequence of comeager sets of reals. Then $\bigcap A_\zeta$ is comeager.

Note that (1) is true in ZF + DC, (2) is true in ZF + DC if in addition one assumes that every set of reals has the property of Baire. Under this additional assumption the converse of (1) is true as well.

Thus assuming that every set of reals has the property of Baire, forcing with $\mathbb{P}$ is equivalent to forcing with $\mathbb{P}'$. All of this is noted in [SVW] to which we refer the reader for more details.

We now define a set of conditions Q_{ω_1}; $t \in Q_{\omega_1}$ iff t is a function with domain a finite subset of ω_1, such that for $\delta \in \operatorname{dom} t$, $t(\delta) \in (\omega + \delta)^{<\omega}$. Order the elements of Q_{ω_1} by $t_1 \leq t_2$ iff $t_2 \subseteq t_1$. Let $\mathbb{Q}_{\omega_1}$ denote the corresponding partial order $(Q_{\omega_1}, \leq)$.

Suppose α, β are countable ordinals with $\alpha < \beta$. Each element, t, of Q_{ω_1} naturally defines a basic open set in $T_{\alpha,\beta}$. We denote this by $[t]_{\alpha,\beta}$. Thus $[t]_{\alpha,\beta} = \{f \in T_{\alpha,\beta} \mid t(\delta) \subseteq f_\delta$ for each $\delta \in \operatorname{dom} t \cap [\alpha,\beta)\}$. Frequently when the subscripts are implicit from context we denote $[t]_{\alpha,\beta}$ by $[t]$. Of course $[t]_{\alpha,\beta}$ depends only on $t \restriction [\alpha,\beta)$. Let $Q_{\alpha,\beta} = \{t \in Q_{\omega_1} \mid \operatorname{dom} t \subseteq [\alpha,\beta)\}$ and let $\mathbb{Q}_{\alpha,\beta}$ denote the associated partial order. Note that the completion of $\mathbb{Q}_{\alpha,\beta}$ is isomorphic to the algebra of regular open subsets of $T_{\alpha,\beta}$.

Though most of our considerations are within the theory of ZF + AD + DC, some of the theorems we prove require less. Much in the spirit of [SVW] we indicate weaker hypotheses when it is easy to do so. We abbreviate 'all sets of reals have the property of Baire' by 'C'.

The following lemmas are proved in [SVW] and require only ZF + DC + C. We include them with proofs for the sake of completeness.

Lemma 1.2 (ZF + DC + C). Assume $D \subseteq P$ is open and dense in $\mathbb{P}$. Suppose $(f,\vec{X}) \in P$ with $f \in T_{0,\alpha}$ and for $\beta > \alpha$ let $S_\beta = \{h \in T_{\alpha,\beta} \mid f^\frown h \in X_\beta$ and $(f^\frown h,\vec{Y}) \in D$ for some $\vec{Y}\}$. Then for sufficiently large β, S_β is comeager in $T_{\alpha,\beta}$.

Proof. We are given that every set of reals has the property of Baire, hence if S_β is not comeager in $T_{\alpha,\beta}$ there must exist $t \in Q_{\alpha,\beta}$ such that $S_\beta \cap [t]$ is meager. We claim that for each $t \in Q_{\omega_1}$ there exist β_t such that for $\beta > \beta_t$, $S_\beta \cap [t]$ is not meager in $T_{\alpha,\beta}$. If not then for each β such that $S_\beta \cap [t]$ is meager let $Z_\beta = \{f^\frown h \mid h \notin S_\beta \cap [t]\}$. Let $Z_\beta = X_\beta$ otherwise. $(f,\vec{Z}) \in P'$ so choose $(f,\vec{Y}) \in P$ with $Y_\beta \subseteq Z_\beta$ for each β. Choose $g \in Y_\beta$ such that $t \in Q_{\alpha,\beta}$ and

and $g \in [t]_{\alpha,\beta}$. Then $(f^\frown g, \vec{Y}) \in P$ and has no extension in D.

Thus for each $t \in Q_{\omega_1}$ there exists β_t such that for $\beta > \beta_t$, $S_\beta \cap [t]$ is meager in $T_{\alpha,\beta}$. Choose a limit ordinal γ, $\alpha < \gamma < \omega_1$, such that for $\beta < \gamma$ and $t \in Q_{\alpha,\beta}$, $\beta_t < \gamma$. Hence $S_\gamma \cap [t]$ is not meager for each $t \in Q_{\alpha,\gamma}$ and therefore S_γ is comeager in $T_{\alpha,\gamma}$. Finally this implies that S_β is comeager in $T_{\alpha,\beta}$ for $\beta > \gamma$. ⊣

Lemma 1.3 (ZF + DC + C). $\mathbb{P}$ is (ω,∞) distributive.

Proof. Let $\langle D_n \mid n < \omega\rangle$ be a sequence of open dense sets in $\mathbb{P}$. Suppose $(f,\vec{X}) \in P$ with $f \in T_{0,\alpha}$. It suffices to find $(g,\vec{Y}) \leq (f,\vec{X})$ such that $(g,\vec{Y}) \in D_n$ for each n. For each n and β, $\alpha < \beta < \omega_1$, define S^n_β relative to D_n and $(f,\vec{X})$ as S_β is defined in Lemma 1.2. By Lemma 1.2 there exists $\bar{\beta} < \omega_1$ such that for each n, $S^n_{\bar{\beta}}$ is comeager in $T_{\alpha,\bar{\beta}}$. Choose $h \in \bigcap_n S^n_{\bar{\beta}}$. Using DC choose for each n, $\vec{Y}_n$ such that $(f^\frown h, \vec{Y}_n) \leq (f,\vec{X})$ and $(f^\frown h, \vec{Y}_n) \in D_n$. For each $\beta < \omega_1$ let $Y_\beta = \bigcap_n (\vec{Y}_n)_\beta$. $(f^\frown h, \vec{Y})$ is as required. ⊣

We will now use AD to prove two additional theorems regarding $\mathbb{P}$. First we establish some mild coding, we need to code both elements of $T_{\alpha,\beta}$ and simple subsets of $T_{\alpha,\beta}$ by reals in a reasonably effective fashion.

Let H_{ω_1} denote all sets with countable transitive closure, i.e. H_{ω_1} is the set of all hereditarily countable sets. Suppose $a \in H_{\omega_1}$ and let $TC(a)$ denote the transitive closure of a. The set a can be easily coded by coding the countable structure $\langle TC(a), a, \epsilon\rangle$. Thus we choose as codes for a, structures $\langle \omega, A, E\rangle$ (where $A \subseteq \omega$, $E \subseteq \omega \times \omega$) that are isomorphic to $\langle TC(a), a, \epsilon\rangle$. Codes of a we regard in an obvious fashion as reals, this simply amounts to identifying elements of $P(\omega) \times P(\omega \times \omega)$ with reals.

This method of encoding defines a partial map π of $\mathbb{R}$ onto H_{ω_1} with domain a Π^1_1 set of reals. Further π chosen in this manner is easily seen to be Δ_1 over H_{ω_1}. Recall that a subset of H_{ω_1} is Σ_1 over H_{ω_1} if it is definable in the structure $\langle H_{\omega_1}, \epsilon\rangle$ by a Σ_1-formula without parameters, similarly a subset is Π_1 over H_{ω_1} if it can be defined by a Π_1-formula without parameters. A subset of H_{ω_1} is Δ_1 if it is both Σ_1 and Π_1 over H_{ω_1}. A subset is $\utilde{\Sigma}_1$ if it is definable by a Σ_1-formula with parameters from H_{ω_1}, similarly for $\utilde{\Pi}_1$, $\utilde{\Delta}_1$. In light of the fact that the coding map π is Δ_1, it follows that a set is $\utilde{\Sigma}_1$ iff it can be defined by a Σ_1-formula with real parameters.

Fix $B \subseteq \mathbb{R} \times H_{\omega_1}$ that is Σ_1 over H_{ω_1} and universal for $\utilde{\Sigma}_1$ subsets of H_{ω_1}, i.e. such that for each $A \subseteq H_{\omega_1}$, $\utilde{\Sigma}_1$ over H_{ω_1}, there is a real, x, with $A = B_x$ where $B_x = \{a \in H_{\omega_1} \mid \langle x,a\rangle \in B\}$.

Note that $T_{\alpha,\beta} \subseteq H_{\omega_1}$ and therefore π defines a coding of the spaces $T_{\alpha,\beta}$. Also it is easily verified that borel subsets of $T_{\alpha,\beta}$ are $\utilde{\Delta}_1$ over H_{ω_1}. Hence borel subsets of $T_{\alpha,\beta}$ are coded by reals via the universal set B, i.e. $x \in \mathbb{R}$

codes $A \subseteq T_{\alpha,\beta}$ provided that $A = B_x$. We can assume by judicious choice of B that for $x \in \text{dom}\,\pi$, $B_x = \{\pi(x)\}$, therefore we view the coding induced by B as extending π.

We are perhaps being overly technical, we are simply coding Σ_1 subsets of H_{ω_1} by reals via a universal set with the additional requirement that the coding of ordinals be a standard one, the latter for boundedness considerations.

Finally conditions $(f,\vec{X}) \in P$ we regard as subsets of H_{ω_1} by identifying $(f,\vec{X})$ with the corresponding set, $\{\langle f,g\rangle \mid g \in \bigcup X_\alpha\}$. Let $\Sigma_1(H_{\omega_1}) = \{A \subseteq H_{\omega_1} \mid A$ is Σ_1 over $H_{\omega_1}\}$.

Theorem 1.4 (ZF + AD + DC). $P \cap L(\mathbb{R})$ is dense in $\mathbb{P}$. In fact $P \cap \Sigma_1(H_{\omega_1})$ is dense in $\mathbb{P}$.

Proof. Suppose $(f,\vec{X})$ is a condition in P with $f \in T_\alpha$. For each $\beta < \omega_1$, $\beta > \alpha$, let $X_{\alpha,\beta} = \{h \in T_{\alpha,\beta} \mid f^\frown h \in X_\beta\}$. Thus $X_{\alpha,\beta}$ is comeager in $T_{\alpha,\beta}$. We construct a sequence $\langle Z_{\alpha,\beta} \mid \beta < \omega_1,\ \beta > \alpha\rangle$ in $L(\mathbb{R})$ such that for each $\beta > \alpha$, $Z_{\alpha,\beta}$ is comeager in $T_{\alpha,\beta}$ and $Z_{\alpha,\beta} \subseteq X_{\alpha,\beta}$. To find $\langle Z_{\alpha,\beta} \mid \beta < \omega_1,\ \beta > \alpha\rangle$ consider the following Solovay-type game:

I	II
n_0	m_0
n_1	m_1
$\vdots$	$\vdots$
x	y

Player II wins iff whenever x codes a countable ordinal $\gamma > \alpha$ then y codes a sequence $\langle Z_{\alpha,\beta} \mid \beta < \beta_0,\ \beta > \alpha\rangle$ where $\beta_0 \geq \gamma$ and each $Z_{\alpha,\beta}$ is comeager in $T_{\alpha,\beta}$, $Z_{\alpha,\beta} \subseteq X_{\alpha,\beta}$. View y as coding $\bigcup Z_{\alpha,\beta}$. Thus II wins iff whenever x codes a countable ordinal $\gamma > \alpha$ i.e. $B_x = \{\gamma\}$, $\gamma > \alpha$, then for all $\beta \leq \gamma$, $\beta > \alpha$, $B_y \cap T_{\alpha,\beta}$ is comeager and contained in $X_{\alpha,\beta}$.

By standard arguments player I cannot have a winning strategy since any strategy for I can be forced to produce a real that fails to code an ordinal or must always produce codes for ordinals bounded by some fixed countable ordinal γ_0. In the latter case II can easily defeat the strategy by playing a code for a sequence $\langle Z_{\alpha,\beta}\rangle$ of length longer than γ_0. Implicit in this is the fact that each $X_{\alpha,\beta}$ contains a comeager borel set.

Thus since by AD the game is determined, II must have a winning strategy. Let s be such a strategy (which we regard as a real) and for $x \in \mathbb{R}$ let $s(x)$ denote the response by s to I playing x.

Let $Z_{\alpha,\beta} = \{h \in T_{\alpha,\beta} \mid h \in B_{s(x)}$ for some real x, a code of $\gamma > \beta\}$. Thus for each $\beta < \omega_1$, $\beta > \alpha$, $Z_{\alpha,\beta}$ is comeager in $T_{\alpha,\beta}$ and is a subset of $X_{\alpha,\beta}$. Let $Z_\beta = \{f^\frown h \mid h \in Z_{\alpha,\beta}\}$. $(f,\vec{Z}) \in P'$ and clearly $(f,\vec{Z}) \in L(\mathbb{R})$. Thus $(f,\vec{Z})$

can be refined in $L(\mathbb{R})$ to $(f,\vec{Z}')$ with $(f,\vec{Z}') \in P$. Finally it follows that $(f,\vec{Z}') \leq (f,\vec{X})$ in $\mathbb{P}$.

It is apparent from the definition of $(f,\vec{Z})$ that $(f,\vec{Z})$ is Σ_1 over H_{ω_1}. We claim that the refinement $(f,\vec{Z}')$ can be chosen so that $(f,\vec{Z}')$ is Σ_1 over H_{ω_1}.

To see this simply let $Z'_{\alpha,\beta} = \{h \in T_{\alpha,\beta} \mid h$ is generic over $L[s,f]$ for the partial order $\mathbb{Q}_{\alpha,\beta}\}$ for $\beta < \omega_1$, $\beta > \alpha$. Let $Z'_\beta = \{f^\frown h \mid h \in Z'_{\alpha,\beta}\}$. It is routine to verify that $(f,\vec{Z}') \in P$. Further it follows that $(f,\vec{Z}')$ is Σ_1 over H_{ω_1}, the key parameter is $\langle s,f\rangle^\#$. ⊣

Theorem 1.4 is in effect a coding lemma for certain elements of P. We now want to consider terms for subsets of ω_1 in the forcing language for $\mathbb{P}$.

Suppose $A \subseteq \mathbb{Q}_{\omega_1} \times \omega_1$. Associated to A is a term for a subset of ω_1. Toward defining this term suppose G is generic over V for $\mathbb{P}$. The generic object G defines in a natural fashion an ω_1 sequence of reals which may be regarded as an element of T_{0,ω_1}, generalizing our notation slightly. This sequence is defined from G by $\bigcup\{f \mid (f,\vec{X}) \in G$ for some $\vec{X}\}$. The entire generic object is easily recovered from this sequence hence we identify G with this sequence. Using the set A we can define a subset of ω_1 in $V[G]$ as follows. $\alpha \in S$ iff $\langle t,\alpha\rangle \in A$ for some $t \in \mathbb{Q}_{\omega_1}$ with $G \in [t]$. Denote the corresponding term in $V^{\mathbb{P}}$ by τ_A.

<u>Lemma 1.5</u> (ZF + DC + C). Suppose $(f,\vec{X}) \in P$ and τ is a term for a subset of ω_1, $\tau \in V^{\mathbb{P}}$. Then for some $A \subseteq \mathbb{Q}_{\omega_1} \times \omega_1$ and condition $(f,\vec{Y}) \in P$, $(f,\vec{Y}) \leq (f,\vec{X})$ and $(f,\vec{Y}) \Vdash \tau = \tau_A$.

<u>Proof</u>. Fix α such that $f \in T_{0,\alpha}$.

Define A as follows. $\langle t,\gamma\rangle \in A$ iff for some $\beta > \alpha$, $t \in \mathbb{Q}_{\alpha,\beta}$ and the set, $\{h \in T_{\alpha,\beta} \mid h \in [t]$ and for some $\vec{Y}$, $(f^\frown h,\vec{Y}) \Vdash \gamma \in \tau\}$, is comeager in $[t]$.

For $\beta > \alpha$ let $Z_{\alpha,\beta} = \{h \in T_{\alpha,\beta} \mid f^\frown h \in X_\beta$ and for all $\vec{Y}$, $\gamma < \omega_1$, if $(f^\frown h,\vec{Y}) \Vdash$ '$\gamma \in \tau$' then $h \in [t]$ for some $t \in \mathbb{Q}_{\alpha,\beta}$ with $\langle t,\gamma\rangle \in A\}$. $Z_{\alpha,\beta}$ is comeager in $T_{\alpha,\beta}$ for each $\beta > \alpha$, therefore let $Z_\beta = \{f^\frown h \mid h \in Z_{\alpha,\beta}\}$. Choose a refinement $\vec{Y}$ of $\vec{Z}$ such that $(f,\vec{Y}) \in P$. Hence $(f,\vec{Y}) \leq (f,\vec{X})$ and $(f,\vec{Y}) \Vdash \tau = \tau_A$. ⊣

The generic object for $\mathbb{P}$ can also be viewed as a filter over $\mathbb{Q}_{\omega_1}$, i.e. if G is generic for $\mathbb{P}$ we can identify G with $\{t \in \mathbb{Q}_{\omega_1} \mid G \in [t]\}$ which is a filter over $\mathbb{Q}_{\omega_1}$. Lemma 1.5 simply says that terms for subsets of ω_1 in $V^{\mathbb{P}}$ correspond to terms for subsets of ω_1 in $V^{\mathbb{Q}_{\omega_1}}$.

Subsets of $\mathbb{Q}_{\omega_1} \times \omega_1$ are in essence subsets of ω_1. By an early theorem of Solovay, assuming AD every subset of ω_1 is constructible from a real. This in the presence of sharps is equivalent to saying that every subset of ω_1 is Σ_1 over H_{ω_1} and so assuming AD, every $A \subseteq \mathbb{Q}_{\omega_1} \times \omega_1$ is Σ_1 over H_{ω_1} and constructible from a real.

Theorem 1.6 (ZF + AD + DC). Suppose $(f,X) \in P$ and $\tau \in V^{\mathbb{P}}$ is a term for a subset of ω_1. Then there is a condition $(f,Y) \leq (f,X)$, $(f,Y) \in \Sigma_1(H_{\omega_1})$ and a set $A \subseteq Q_{\omega_1} \times \omega_1$, $A \in \Sigma_1(H_{\omega_1})$, such that $(f,Y) \Vdash \tau = \tau_A$.

Proof. Immediate by the preceding remarks, Theorem 1.4 and Lemma 1.5. ⊣

Steel and Van Wesep prove in [SVW] that if $V \models$ 'ZF + AD + $\mathbb{R}$-AC' and G is generic over V for $\mathbb{P}$ then in the generic extension V[G], ω_1-DC holds. As usual we view G as an ω_1 sequence of reals. By Theorem 1.4 G is generic over $L(\mathbb{R})$ for $\mathbb{P}$ defined in $L(\mathbb{R})$ (i.e. $\mathbb{P} \cap L(\mathbb{R})$) and by Theorem 1.6, V[G] and $L(\mathbb{R}, G)$ have the same subsets of ω_1 so that in fact $L(\mathbb{R}, G)$ is just $L(P(\omega_1))$ defined in V[G]. Hence $L(\mathbb{R}, G) \models \omega_1$-DC. But this is a statement in $L(\mathbb{R})$ about forcing with $\mathbb{P}$ and one would hope that this is provable about $L(\mathbb{R})$ just assuming AD, i.e. this should be a theorem of ZF + AD + V = $L(\mathbb{R})$.

We now define two choice principles which we will show hold in $L(\mathbb{R})$ assuming AD.

For each infinite countable ordinal α let $S_\alpha \subseteq \mathbb{R}$ be the set of those reals coding α. The set S_α naturally defines a Baire space homeomorphic to a comeager subset of α^ω, i.e. S_α with its natural topology is homeomorphic to $\{f \in \alpha^\omega \mid f : \omega \to \alpha \text{ is onto}\}$. Note that for $\alpha \neq \beta$, $S_\alpha \cap S_\beta = \emptyset$.

Definition 1.7. (1) Let **$\mathbb{R}$-AC denote the axiom: For each $f : \mathbb{R} \to V$ there is a choice function $g : \mathbb{R} \to V$ such that $\{x \in \mathbb{R} \mid g(x) \in f(x) \text{ or } f(x) = \emptyset\}$ is comeager.

(2) Let **ω_1-AC denote the axiom: For each $f : \mathbb{R} \to V$ there is a choice function $g : \mathbb{R} \to V$ such that $\{x \in \mathbb{R} \mid g(x) \in f(x) \text{ or } f(x) = \emptyset\} \cap S_\alpha$ is comeager in S_α for every infinite countable ordinal α.

Lemma 1.8 (ZF + AD + DC). Assume V = $L(\mathbb{R})$. Then **$\mathbb{R}$-AC and **ω_1-AC hold.

Proof. Let $f : \mathbb{R} \to V$ be given, we seek the appropriate choice function. Since V = $L(\mathbb{R})$ every set is ordinal definable from a real and therefore we can assume $f : \mathbb{R} \to P(\mathbb{R})$. But then f is in essence a subset of $\mathbb{R} \times \mathbb{R}$ and in this case **$\mathbb{R}$-AC is simply uniformization on a comeager set, a well known consequence of AD.

We now find a choice function as required by **ω_1-AC. Fix $f : \mathbb{R} \to P(\mathbb{R})$ as given, assume $f(x) \neq \emptyset$ for every $x \in \mathbb{R}$ and let $A \subseteq \mathbb{R} \times \mathbb{R}$ be the corresponding subset of $\mathbb{R} \times \mathbb{R}$. We seek a function $g : \mathbb{R} \to \mathbb{R}$ such that $\{x \in \mathbb{R} \mid g(x) \in f(x)\} \cap S_\alpha$ is comeager in S_α for all infinite ordinals $\alpha < \omega_1$. To find g it suffices to find $H \subseteq A$, H a Σ^1_2 set, i.e. simple, such that $\{x \in \mathbb{R} \mid H_x \neq \emptyset\} \cap S_\alpha$ is comeager in S_α for every $\alpha > \omega$, $\alpha < \omega_1$, where for $x \in \mathbb{R}$, $H_x = \{y \in \mathbb{R} \mid \langle x,y \rangle \in H\}$. Since H is a Σ^1_2 set we can find (and this is a theorem of ZF) a function $g : \mathbb{R} \to \mathbb{R}$ such that for all $x \in \mathbb{R}$, $H_x = \emptyset$ or $g(x) \in H_x$. Clearly g is as required.

To construct H consider the following integer game which is a Solovay game:

$$\begin{array}{cc} \text{I} & \text{II} \\ n_0 & m_0 \\ n_1 & m_1 \\ \vdots & \vdots \\ z & w \end{array}$$

Player II wins if whenever z codes an ordinal α, w codes a Σ^1_2 set $H \subseteq A$ such that $\{x \in \mathbb{R} \mid H_x \neq \emptyset\} \cap S_\gamma$ is comeager in S_γ for all infinite $\gamma \leq \alpha$. More precisely, using the Σ_1-universal set $B \subseteq \mathbb{R} \times H_{\omega_1}$, II wins if whenever z codes an ordinal α, i.e. $B_z = \{\alpha\}$, $B_w \subseteq A$ and $\{x \in \mathbb{R} \mid (B_w)_x \neq \emptyset\} \cap S_\gamma$ is comeager in S_γ for all infinite $\gamma \leq \alpha$. Note that Σ^1_2 subsets of $\mathbb{R} \times \mathbb{R}$ are Σ_1 over H_{ω_1} and conversely. Equivalently we could have used a universal Σ^1_2 subset of $\mathbb{R} \times (\mathbb{R} \times \mathbb{R})$ for the decoding of w.

By $**\mathbb{R}$-AC and standard arguments player I cannot have a winning strategy in this game, the situation is similar to that in the proof of Theorem 1.4. Hence by AD player II has a winning strategy. Let s be a winning strategy for player II and for $z \in \mathbb{R}$ let $s(z)$ denote the response by s to player I playing z. Define $H \subseteq \mathbb{R} \times \mathbb{R}$ by $\{\langle x,y\rangle \in \mathbb{R} \times \mathbb{R} \mid \langle x,y\rangle \in B_{s(z)}$ for some $z \in \mathbb{R}\}$. Clearly $H \subseteq A$ and by the definition of H, H is Σ_1 over H_{ω_1} in particular $H \subseteq \mathbb{R} \times \mathbb{R}$ is Σ^1_2. Finally $\{x \in \mathbb{R} \mid H_x \neq \emptyset\} \cap S_\gamma$ is comeager in S_γ for all infinite $\gamma < \omega_1$. ⊣

We remark that assuming ZF + AD + DC, Lemma 1.8 is true in a more general context than $V = L(\mathbb{R})$, more precisely Lemma 1.8 holds whenever $V = L(P(\mathbb{R}))$ and Θ is regular. Recall $\Theta = \sup\{\zeta \in \mathrm{OR} \mid$ there is an onto map $h : \mathbb{R} \to \zeta\}$.

For all countable ordinals $\alpha < \beta$ let $T^*_{\alpha,\beta}$ denote the comeager subset of $T_{\alpha,\beta}$ defined by $\{h \in T_{\alpha,\beta} \mid$ for all $\gamma \in \mathrm{dom}\, h$, $h(\gamma) : \omega \to \omega + \gamma$ is onto$\}$. Note that for all $\beta > \omega_1$, $T^*_{\alpha,\beta}$ is definably homeomorphic to S_β.

<u>Lemma 1.9</u> (ZF + DC + C + $**\mathbb{R}$-AC). Suppose $(f,\vec{X}) \in P$ and $(f,\vec{X}) \Vdash \exists Z\, \varphi(Z)$ where $\varphi(Z)$ is a formula in the forcing language for $\mathbb{P}$. Then there is a refinement $\vec{Y}$ of $\vec{X}$ and a term $\tau \in V^{\mathbb{P}}$ such that $(f,\vec{Y}) \Vdash \varphi(\tau)$.

<u>Proof</u>. Fix α such that $f \in T_{0,\alpha}$.

Clearly the set $\{(g,\vec{Y}) \in P \mid (g,\vec{Y}) \Vdash \varphi(\tau)$ for some term $\tau \in V^{\mathbb{P}}\}$ is open and dense below $(f,\vec{X})$. Choose by Lemma 1.2 an infinite countable ordinal $\beta > \alpha$ such that the set $\{h \in T_{\alpha,\beta+1} \mid$ there is a term $\tau \in V^{\mathbb{P}}$ and a refinement $\vec{Y}$ of $\vec{X}$ such that $(f^\frown h,\vec{Y}) \Vdash \varphi(\tau)\}$ is comeager in $T_{\alpha,\beta+1}$. The Baire space $T^*_{\alpha,\beta+1}$ is homeomorphic to S_β which in turn is homeomorphic to a comeager subset of $\mathbb{R}$.

Hence by $**\mathbb{R}$-AC there is a choice function g such that $\operatorname{dom} g \subseteq T_{\alpha,\beta+1}$ is comeager in $T_{\alpha,\beta+1}$ and for $h \in \operatorname{dom} g$, $g(h)$ is a pair $\langle \tau_h, \vec{Y}_h \rangle$ such that $(f^\frown h, \vec{Y}_h) \in P$ and $(f^\frown h, \vec{Y}_h) \Vdash \varphi(\tau_h)$. Using the function g it is routine to define τ, $\vec{Y}$ as desired. ⊣

Lemma 1.9 asserts that the forcing language for $\mathbb{P}$ is in a weak sense full. This has as an immediate corollary that forcing with $\mathbb{P}$ preserves DC. To prove that the forcing language is full requires $\mathbb{R}$-AC, this is the method in [SVW].

<u>Lemma 1.10</u> (ZF + DC + C + $**\mathbb{R}$-AC). Suppose $V[G]$ is a generic extension of V obtained by forcing with $\mathbb{P}$. Then $V[G] \models$ DC.

<u>Proof</u>. Immediate by Lemma 1.8. ⊣

Using $**\omega_1$-AC we can improve Lemma 1.10 and show that in addition $V[G] \models \omega_1$-AC.

<u>Lemma 1.11</u> (ZF + DC + C + $**\omega_1$-AC). Suppose $V[G]$ is a generic extension of V obtained by forcing with $\mathbb{P}$. Then $V[G] \models \omega_1$-AC.

<u>Proof</u>. Let τ_F be a term in $V^{\mathbb{P}}$ for a function F with domain ω_1. For each $\gamma < \omega_1$ let $\tau_F(\gamma)$ denote the corresponding term for $F(\gamma)$. We assume that for each $\gamma < \omega_1$, $\tau_F(\gamma)$ is a term for a nonempty set.

Fix $(f, \vec{X}) \in P$ with $f \in T_{0,\alpha}$. For infinite $\beta < \omega_1$, $\beta > \alpha$, the Baire space $T^*_{\alpha,\beta+1}$ is definably homeomorphic to S_β. Hence by Lemma 1.9 and $**\omega_1$-AC there is a choice function g such that for infinite $\beta > \omega_1$ with $\beta > \alpha$, $\operatorname{dom} g \cap T_{\alpha,\beta+1}$ is comeager in $T_{\alpha,\beta+1}$ and for $h \in \operatorname{dom} g \cap T_{\alpha,\beta+1}$, $g(h)$ is a pair $\langle \tau_h, \vec{Y}_h \rangle$ such that $(f^\frown h, \vec{Y}_h) \leq (f, \vec{X})$ and $(f^\frown h, \vec{Y}_h) \Vdash \tau_h \in \tau_F(\beta)$.

By a routine argument using the 'Fubini' theorem for category etc., there is a refinement $\vec{Y}$ of $\vec{X}$ such that $\bigcup_\beta Y_{\beta+1} \subseteq \operatorname{dom} g$ and further for $h \in \bigcup_\beta Y_{\beta+1}$, $(f^\frown h, \vec{Y}_h) \geq (f, \vec{Y})$. Finally g defines a term τ_H for a function H that is forced by $(f, \vec{Y})$ to be a choice function for F, i.e. $(f, \vec{Y}) \Vdash \tau_H$ is a choice function for τ_F.

Hence $V[G] \models \omega_1$-AC. Of course we have used implicitly that ω_1 is not collapsed by forcing with $\mathbb{P}$. ⊣

Since AD implies that $**\omega_1$-AC holds in $L(\mathbb{R})$ we have shown, assuming AD, that forcing with $\mathbb{P}$ over $L(\mathbb{R})$ recovers ω_1-AC. We prove that forcing with $\mathbb{P}$ over $L(\mathbb{R})$ yields ω_1-DC in the generic extension. To do this we need deeper consequences of AD about $L(\mathbb{R})$.

A subset $A \subseteq \mathbb{R} \times \mathbb{R}$ can be uniformized if there is a choice function $g : \mathbb{R} \to \mathbb{R}$ such that for all $x \in \mathbb{R}$, $g(x) \in A_x$ or $A_x = \emptyset$.

We work now in $L(\mathbb{R})$. Recall that a set of reals is $\utilde{\Sigma}^2_1$ if it can be defined by a Σ^2_1-formula with real parameters (equivalently if it is Σ_1 over $L_\Theta(\mathbb{R})$

allowing real parameters).

The theorem of AD about $L(\mathbb{R})$ that we need is due to D. Martin and J. Steel (see [MMS]) and implies that assuming AD, sets of reals $\underset{\sim}{\Sigma}^2_1$ in $L(\mathbb{R})$ can be uniformized in $L(\mathbb{R})$. Our use of this will be by exploiting a trick due to R. Solovay (see Appendix B in [KSS]), the key idea of which is the following. Suppose λ is an ordinal. Then there is an ordinal $\alpha < \Theta$ such that $L_\alpha(\mathbb{R})$ is elementarily equivalent to $L_\lambda(\mathbb{R})$ and such that every set of reals in $L_\alpha(\mathbb{R})$ is $\underset{\sim}{\Sigma}^2_1$ in $L(\mathbb{R})$, in fact we can assume something stronger, that the structure $\langle L_\alpha(\mathbb{R}), \varepsilon\rangle$ is isomorphic to a structure $\langle \mathbb{R}, E\rangle$ with $E \subseteq \mathbb{R} \times \mathbb{R}$, E a $\underset{\sim}{\Sigma}^2_1$ set. This is a theorem of 'ZF + DC' and is proved by a reflection argument using the facts that Θ is a regular cardinal and that every set is ordinal definable from a real. It is a variant of a 'basis' theorem for $\underset{\sim}{\Sigma}^2_1$ subsets of $P(\mathbb{R})$, i.e. every $\underset{\sim}{\Sigma}^2_1$ subset of $P(\mathbb{R})$ contains a $\underset{\sim}{\Delta}^2_1$ set of reals, this is also a theorem of 'ZF + DC' (+ $V = L(\mathbb{R})$) due to R. Solovay.

<u>Theorem 1.12</u> (ZF + AD + DC + $V = L(\mathbb{R})$). Suppose $V[G]$ is a generic extension of V obtained by forcing with $\mathbb{P}$. Then $V[G] \models \omega_1$-DC.

<u>Proof</u>. As usual we identify G with the corresponding ω_1-sequence of reals. Suppose $(f,\vec{X}) \in P$ and $(f,\vec{X})$ belongs to the generic filter defined by G. We abuse notation slightly and indicate this by $(f,\vec{X}) \in G$.

By Lemma 1.10, $L(\mathbb{R}, G) \models$ DC. Hence to show that $L(\mathbb{R}, G) \models \omega_1$-DC it suffices to show that in $L(\mathbb{R}, G)$ any ω-closed subtree of $\mathbb{R}^{<\omega_1}$ has an unbounded branch, which for nontrivial subtrees is simply to require that there exists an ω_1-branch. Fix $\tau \in L(\mathbb{R})^{\mathbb{P}}$ a term for such a tree $T \in L(\mathbb{R}, G)$. We find in $L(\mathbb{R})$ a set $D \subseteq \mathbb{R} \times \mathbb{R}$ such that any uniformization function of D in $L(\mathbb{R})$ yields an ω_1-branch through T in $L(\mathbb{R}, G)$.

Let $A = \{\langle f,g\rangle \mid f \in T^*_{0,\beta+1}$ for some β, $g \in \mathbb{R}^{<\omega_1}$, $\operatorname{dom} g \subseteq \beta$, and for some $x \in \mathbb{R}$ and $\vec{Y}$, $(f,\vec{Y}) \Vdash g^\frown x \in \tau\}$. For $\langle f,g\rangle \in A$ let $F(\langle f,g\rangle) = \{\langle x,\vec{Y}\rangle \mid x \in \mathbb{R}$, $(f,\vec{Y}) \in P$ and $(f,\vec{Y}) \Vdash g^\frown x \in \tau\}$. This defines a function $F : A \to L(\mathbb{R})$.

Suppose H is a choice function for F, i.e. $H : A \to L(\mathbb{R})$ with $H(\langle f,g\rangle) \in F(\langle f,g\rangle)$ for all $\langle f,g\rangle \in A$. Regard H as a pair of functions H_1, H_2 thus $H(\langle f,g\rangle) = \langle H_1(\langle f,g\rangle), H_2(\langle f,g\rangle)\rangle$. From H one can define in $L(\mathbb{R}, G)$ a function $H^* : T \to \mathbb{R}$ such that for all $g \in T$, $g^\frown H^*(g) \in T$. To find H^* work in $L(\mathbb{R}, G)$. For each $g \in T$ choose the least ordinal γ such that if $f = G \restriction \gamma$ then $\langle f,g\rangle \in A$ and $(f, H_2(\langle f,g\rangle)) \in G$. Define $H^*(g) = H_1(\langle f,g\rangle)$. Since $H \in L(\mathbb{R})$, H^* is defined for all $g \in T$.

Using the function H^* it is easy to construct an ω_1-branch through T in $L(\mathbb{R}, G)$.

Note, this is a key point, that A is isomorphic to a subset of $\mathbb{R}$ (i.e. the set A is in 1 to 1 correspondence with a set of reals) hence finding H reduces in a canonical fashion to uniformizing a certain subset of $\mathbb{R} \times \mathbb{R}$ which we take as D.

Suppose $L(\mathbb{R}, G) \models$ 'ω_1-DC fails'. By the obvious homogeneity of $\mathbb{P}$, $[[\omega_1\text{-DC fails}]]_{\mathbb{P}} = 1$. This is a statement of $L(\mathbb{R})$ hence by the remarks preceding the statement of this theorem there is an ordinal $\xi < \Theta$ such that $L_\xi(\mathbb{R}) \models$ '$ZF^- + AD + DC + [[\omega_1\text{-DC fails}]]_{\mathbb{P}} = 1$' and with the additional property that every set of reals in $L_\xi(\mathbb{R})$ is Σ^2_1 in $L(\mathbb{R})$. ZF^- refers to ZF-replacement, i.e. a large enough fragment of ZF. Fix ξ with these properties.

It is an immediate consequence of Theorem 1.4 that $\mathbb{P} \cap L_\xi(\mathbb{R})$ is dense in $\mathbb{P}$. Therefore $L_\xi(\mathbb{R}, G)$ is a generic extension of $L_\xi(\mathbb{R})$ via forcing with $\mathbb{P}$ and so $L_\xi(\mathbb{R}, G) \models$ 'ω_1-DC fails'. Choose $T \in L_\xi(\mathbb{R}, G)$ such that in $L_\xi(\mathbb{R}, G)$, T is an ω-closed subtree of $\mathbb{R}^{<\omega_1}$ with no unbounded branch. Fix $\tau \in L_\xi(\mathbb{R})^{\mathbb{P}}$ a term for T and proceeding as above define in $L_\xi(\mathbb{R})$, A, F and D. But then D is Σ^2_1 in $L(\mathbb{R})$ hence in $L(\mathbb{R})$ the set D can be uniformized. This yields in $L(\mathbb{R})$ a choice function for F and this in turn produces a branch through T in $L(\mathbb{R}, G)$. Finally by Theorem 1.6 this branch, being in essence a subset of ω_1, must lie in $L_\xi(\mathbb{R}, G)$, a contradiction. $\dashv$

Having established that in the presence of AD forcing with $\mathbb{P}$ over $L(\mathbb{R})$ yields ω_1-DC, the consistency results of [SVW] easily follow. Before indicating how we offer a different perspective of forcing with $\mathbb{P}$.

Lemma 1.5 clearly suggests an intimate relationship between forcing with $\mathbb{P}$ and forcing with $\mathbb{Q}_{\omega_1}$, we will prove a theorem clarifying this but first we state a theorem due independently to H. Friedman.

<u>Theorem 1.13</u> (ZF + DC + C + $**\mathbb{R}$-AC). Suppose c is a Cohen real over V. Then there is a generic elementary embedding:

$$j : V \to M \subseteq V[c] \quad \text{where} \quad c \in M .$$

<u>Proof</u>. Since every set of reals has the property of Baire the complete boolean algebra generated by the Cohen conditions is isomorphic to $P(\mathbb{R})/\mathcal{I}$ where $\mathcal{I}$ denotes the ideal of meager sets. A standard generic ultrapower argument using $**\mathbb{R}$-AC instead of the full axiom of choice yields j. $\dashv$

In the special case of $V = L(\mathbb{R})$ the generic elementary embedding of Theorem 1.13 is easily (and uniquely) defined. Suppose c is a Cohen real over $L(\mathbb{R})$ and let $\mathbb{R}_c$ denote the set of reals as defined in $V[c]$. Each set in $L(\mathbb{R})$ is ordinal definable from a real, this plus the observation that $j \restriction OR$ must be the identity uniquely defines j. If φ is a formula that defines in $L(\mathbb{R})$ a set A, φ with parameters $x \in \mathbb{R}$, $\gamma \in OR$, then φ defines $j(A)$ in $L(\mathbb{R}_c)$.

Forcing with $\mathbb{Q}_{\omega_1}$ is equivalent as far as adding reals to forcing with $\mathbb{C}_{\omega_1}$, the partial order of Cohen conditions for adding ω_1 reals (if one assumes the axiom of choice $\mathbb{C}_{\omega_1}$ and $\mathbb{Q}_{\omega_1}$ are in fact isomorphic). Suppose $G \subseteq \mathbb{Q}_{\omega_1}$ is generic over $L(\mathbb{R})$. Then Theorem 1.13 easily extends to yield a generic elementary embedding, $j : L(\mathbb{R}) \to L(\mathbb{R}_G)$ where $\mathbb{R}_G$ denotes the set of reals in $V[G]$.

Theorem 1.14 (ZF + DC + C + $**\mathbb{R}$-AC). Assume $V = L(\mathbb{R})$ and suppose $G \subseteq \mathbb{Q}_{\omega_1}$ is generic over $L(\mathbb{R})$ for the partial order $\mathbb{Q}_{\omega_1}$. Let $\mathbb{R}_G$ denote the set of reals in $V[G]$. Then G defines a generic filter over $L(\mathbb{R}_G)$ for the partial order $\mathbb{P}$ defined in $L(\mathbb{R}_G)$.

Proof. For sets of ordinals $s \in L(\mathbb{R}_G)$ let $\mathbb{R}_s$ denote the set of reals in $L(\mathbb{R}, s)$. The embedding $j : L(\mathbb{R}) \to L(\mathbb{R}_G)$ induces unique elementary embeddings $j_1 : L(\mathbb{R}) \to L(\mathbb{R}_s)$, $j_2 : L(\mathbb{R}_s) \to L(\mathbb{R}_G)$ with $j = j_2 \circ j_1$. We denote j_2 by j_s.

Let P_G denote P defined in $L(\mathbb{R}_G)$ and let $\mathbb{P}_G$ denote the corresponding partial order. Similarly define P_s, $\mathbb{P}_s$ for sets of ordinals $s \in L(\mathbb{R}_G)$, i.e. P_s denotes P defined in $L(\mathbb{R}_s)$, etc.

In $L(\mathbb{R}, G)$, G clearly defines a filter over $\mathbb{P}_G$. Suppose $D \subseteq P_G$ is dense and open in $\mathbb{P}_G$, $D \in L(\mathbb{R}_G)$. Thus D is ordinal definable from a real x_0, so fix a formula φ with parameters only x_0 and possibly some ordinal, such that φ defines D. Choose a countable ordinal α large enough so that $x_0 \in L(\mathbb{R}, G \upharpoonright \alpha)$ where $G \upharpoonright \alpha = G \cap \mathbb{Q}_{0,\alpha}$. The restriction $G \upharpoonright \alpha$ canonically defines $f \in T_{0,\alpha}$ which we interpret as a set of ordinals. Hence φ defines in $L(\mathbb{R}_f)$ a dense, open set $D_f \subseteq P_f$. Further $j_f(D_f) = D$.

Work in $L(\mathbb{R}_f)$ and choose by Lemma 1.2 $\beta > \alpha$ such that the set $A = \{h \in T_{\alpha,\beta} \mid (f^\frown h, \vec{Y}) \in D_f$ for some sequence $\vec{Y}\}$ is comeager in $T_{\alpha,\beta}$.

Let h be the element of $T_{\alpha,\beta}$ defined by $G \cap \mathbb{Q}_{\alpha,\beta}$. Let $g = f^\frown h$ which as usual we can regard as a set of ordinals. The embedding $j_f : L(\mathbb{R}_f) \to L(\mathbb{R}_G)$ induces unique elementary embeddings $j_1 : L(\mathbb{R}_f) \to L(\mathbb{R}_g)$, $j_g : L(\mathbb{R}_g) \to L(\mathbb{R}_G)$ with $j_f = j_g \circ j_1$. Further $h \in j_1(A)$ hence in $L(\mathbb{R}_g)$ there exists a sequence $\vec{Y}_g$ such that $(g, \vec{Y}_g) \in j_1(D_f) = D_g$.

For each countable ordinal $\gamma > \beta$ let h_γ denote that element of $T_{\beta,\gamma}$ determined by $G \cap \mathbb{Q}_{\beta,\gamma}$. It follows that for each γ, $g^\frown h_\gamma \in j_g((\vec{Y}_g)_\gamma)$ hence $j_g((g, \vec{Y}_g))$ belongs to the filter over $\mathbb{P}_G$ defined by G. But $j_g((g, \vec{Y}_g)) \in j_g(D_g) = D$. ⊣

The assumption $V = L(\mathbb{R})$ is not necessary in Theorem 1.14 though the proof is perhaps more straightforward under this additional assumption.

An immediate corollary to Theorem 1.14 is that forcing with $\mathbb{P}$ does not collapse any cardinals. This is also proved in [SVW], though by a different argument. One can also use Theorem 1.14 to reprove the (ω, ∞) distributivity of $\mathbb{P}$.

As noted in [SVW] if one replaces the spaces $T_{\alpha,\beta}$ by $\prod_{\alpha \leq \delta < \beta} \omega^\omega$ then there is a partial order $\mathbb{P}^*$ analogous to $\mathbb{P}$. Forcing with $\mathbb{P}^*$ over a ground model of 'ZF + DC + C + $**\mathbb{R}$-AC' adds an ω_1-sequence of distinct reals without adding any new reals or collapsing any cardinals. Further there is an obvious variant of Theorem 1.14 for $\mathbb{P}^*$ using C_{ω_1} instead of Q_{ω_1}. The main point of this is that for arbitrary cardinals λ it indicates a procedure for generalizing the definition of $\mathbb{P}^*$ in order to define $\mathbb{P}^*_\lambda$, an (ω,∞) distributive partial order with the property that forcing with $\mathbb{P}^*_\lambda$ adds a λ-sequence of distinct reals without collapsing any cardinals. We are of course assuming ZF + DC + C + $**\mathbb{R}$-AC, for the sake of simplicity assume as well that $V = L(\mathbb{R})$. Define $\mathbb{P}^*_\lambda$ so that if G is a generic λ-sequence of Cohen reals over $L(\mathbb{R})$ then G defines a generic filter over $L(\mathbb{R}_G)$ for $j(\mathbb{P}^*_\lambda)$ i.e. $\mathbb{P}^*_\lambda$ defined in $L(\mathbb{R}_G)$. The desired properties of $\mathbb{P}^*_\lambda$ will then follow. We leave the details to the reader.

Thus assuming ZF + DC + C + $**\mathbb{R}$-AC it is possible to create in a generic fashion arbitrarily long wellordered sequences of distinct reals without introducing new reals or collapsing any cardinals.

§2. Throughout this section we work in ZF + AD + DC.

Assume G is generic over $L(\mathbb{R})$ for $\mathbb{P}$. Then by Theorem 1.12, $L(\mathbb{R}, G) \models \omega_1$-DC. We use this model to prove two consistency results in ZF + ω_1-DC. The point is that some of the combinatorial consequences of AD carry over to $L(\mathbb{R}, G)$, in fact we will show that a form of determinacy holds in $L(\mathbb{R}, G)$.

The first result concerns infinitary partition properties. Suppose κ is an infinite cardinal and that $S \subseteq \kappa$. For $\alpha \leq \kappa$ let $[S]^\alpha$ denote the set of all subsets of S with ordertype α. We use the standard notation $\kappa \to (\kappa)^\alpha_\beta$ to indicate that for any map $e : [\kappa]^\alpha \to \beta$ there is a subset S of κ, S of ordertype κ, such that $e \restriction [S]^\alpha$ is constant.

Assuming AD + DC there exist many cardinals κ such that $\kappa \to (\kappa)^\kappa_2$ (see [KKMW]. Thus granting the consistency of ZF + AD + DC it is impossible to refute in ZF + DC the existence of uncountable cardinals κ for which $\kappa \to (\kappa)^\kappa_2$. If however one assumes ω_1-DC, many partition properties must fail for all uncountable cardinals as is implicit in the following theorem.

Theorem 2.1 (ZF + DC). Assume there is an uncountable sequence of distinct reals. Then for all uncountable cardinals κ, $\kappa \not\to (\kappa)^{\omega_1}_2$.

Proof. A slight refinement of the results in [KKMW], using similar methods, shows that if for some uncountable cardinal κ, $\kappa \to (\kappa)^{\omega_1}_2$ then every ω_1-Souslin set of reals is determined. Recall that a set $A \subseteq \omega^\omega$ is ω_1-Souslin if there is a tree T on $\omega \times \omega_1$ such that $A = \{x \in \omega^\omega \mid$ for some $f \in \omega_1^\omega$, $\langle x,f\rangle$ defines an infinite branch through $T\}$.

Using an ω_1-sequence of distinct reals it is straightforward to construct an ω_1-Souslin set of reals that is not determined. ⊣

In contrast to Theorem 2.1 we show that it is possible in the presence of ω_1-DC for there to exist uncountable cardinals κ for which $\kappa \to (\kappa)^\alpha_{<\kappa}$ for all

$\alpha < \omega_1$.

Theorem 2.2 ($ZF + AD + DC + V = L(\mathbb{R})$). Suppose $V[G]$ is a generic extension of V obtained by forcing with $\mathbb{P}$. Let κ be any regular cardinal of V, $\kappa \geq \omega_2$, such that in V, $\kappa \to (\kappa)^{\alpha}_{<\kappa}$ for all $\alpha < \omega_1$. Then in $V[G]$, $\kappa \to (\kappa)^{\alpha}_{<\kappa}$ for all $\alpha < \omega_1$.

Proof. We work for the moment in $L(\mathbb{R})$ and isolate the relevant combinatorial occurrence responsible for the theorem. We will proceed by successive approximation.

Since $\kappa \to (\kappa)^{\xi}_{<\kappa}$ for all $\xi < \omega_1$, for each $\alpha < \omega_1$ there is a canonical κ-complete measure defined over $[\kappa]^{\alpha}$, i.e. $S \subseteq [\kappa]^{\alpha}$ has measure 1 if for some $C \subseteq \kappa$, C closed and unbounded in κ, $[C]^{\alpha} \subseteq S$.

(1) Suppose $F : [\kappa]^{\alpha} \to P(\mathbb{R})$ is a function such that the set $\{s \in [\kappa]^{\alpha} \mid F(s)$ is comeager$\}$ is of measure 1. Then there is a closed set $C \subseteq \kappa$, unbounded in κ, and a comeager set $A \subseteq \mathbb{R}$, such that for $x \in [C]^{\alpha}$, $A \subseteq F(s)$.

Proof of (1). Using $**\mathbb{R}$-AC define a partial function $H : \mathbb{R} \to P(\kappa)$ with comeager domain, such that for $x \in \text{dom } H$, $H(x)$ is closed and unbounded in κ, $[H(x)]^{\alpha} \subseteq \{s \in [\kappa]^{\alpha} \mid x \in F(s)\}$ or $[H(x)]^{\alpha} \subseteq \{s \in [\kappa]^{\alpha} \mid x \notin F(s)\}$. Let $C = \{\gamma < \kappa \mid$ the set $\{x \in \mathbb{R} \mid \gamma \in H(x)\}$ is comeager$\}$ and let $A = \{x \in \mathbb{R} \mid C \subseteq H(x)\}$. Since the wellordered intersection of comeager sets is comeager it follows that C is closed and unbounded in κ and that A is comeager.

(2) Suppose $F : [\kappa]^{\alpha} \to L(\mathbb{R})$ is a function such that on a set of measure 1, $F(s) \cap T_{0,\zeta+1}$ is comeager in $T_{0,\zeta+1}$ for each $\zeta < \omega_1$. Then there is a closed, unbounded set $C \subseteq \kappa$ and a set A, $A \cap T_{0,\zeta+1}$ comeager in $T_{0,\zeta+1}$ for each $\zeta < \omega_1$, such that for $s \in [C]^{\alpha}$, $A \subseteq F(s)$.

Proof of (2). Using $**\omega_1$-AC define a function $H : \text{dom } H \to P(\kappa)$ such that $\text{dom } H \cap T_{0,\zeta+1}$ is comeager in $T_{0,\zeta+1}$ for each $\zeta < \omega_1$ and such that for $f \in \text{dom } H$, $H(f)$ is closed, unbounded in κ, $[H(f)]^{\alpha} \subseteq \{s \in [\kappa]^{\alpha} \mid f \in F(s)\}$ or $[H(f)]^{\alpha} \subseteq \{s \in [\kappa]^{\alpha} \mid f \notin F(s)\}$. For each $\zeta < \omega_1$ let $C_{\zeta} = \{\gamma < \kappa \mid$ the set $\{f \in T_{0,\zeta+1} \mid \gamma \in H(f)\}$ is comeager in $T_{0,\zeta+1}\}$ and let $A_{\zeta} = \{f \in T_{0,\zeta+1} \mid C_{\zeta} \subseteq H(f)\}$. Thus for each $\zeta < \omega_1$, C_{ζ} is closed, unbounded in κ, and $A_{\zeta} \cap T_{0,\zeta+1}$ is comeager in $T_{0,\zeta+1}$. Define $C = \bigcap_{\zeta} C_{\zeta}$, $A = \bigcup_{\zeta} A_{\zeta}$. Since $\kappa > \omega_2$, C is closed and unbounded in κ.

We shall actually need the following variant of (2) which is an immediate corollary of (2).

(3) Suppose $F : [\kappa]^{\alpha} \to P$ is a function such that on a set of measure 1, $F(s)$ is a condition of the form $(\emptyset, \vec{X})$. Then there is a closed, unbounded set $C \subseteq \kappa$ and a condition $(\emptyset, \vec{Y}) \in P$ such that for $s \in [C]^{\alpha}$, $(\emptyset, \vec{Y}) \leq F(s)$.

Proof of (3). Immediate using (2).

We use (3) to prove Theorem 2.2.

Since $\mathbb{P}$ is (ω,∞) distributive, for each $\alpha < \omega_1$, $[\kappa]^\alpha$ is the same computed in $L(\mathbb{R})$ or $L(\mathbb{R}, G)$. Suppose that for some $\alpha < \omega_1$, $\kappa \not\to (\kappa)^\alpha_{<\kappa}$ in $L(\mathbb{R}, G)$. Choose in $L(\mathbb{R}, G)$ $\lambda < \kappa$ and a map $e : [\kappa]^\alpha \to \lambda$ with no homogeneous set. Choose a term $\tau \in L(\mathbb{R})^{\mathbb{P}}$ for e, by Lemma 1.9 we can assume that for some sequence $\vec{X}$, $(\emptyset,\vec{X}) \Vdash$ 'τ defines a map $e : [\kappa]^\alpha \to \lambda$ with no homogeneous set'. For $s \in [\kappa]^\alpha$ let $\tau(s)$ denote the term corresponding to $e(s)$.

We again work in $L(\mathbb{R})$. Define a map $e^* : [\kappa]^\alpha \to \lambda \times Q_{\omega_1} \times \omega_1$ by $e^*(s) = \langle \gamma, t, \beta \rangle$ for the 'first' triple $\langle \gamma, t, \beta \rangle$ such that the set $\{f \in T_{0,\beta} \mid$ for some $\vec{Y}$, $(f,\vec{Y}) \Vdash \tau(s) = \gamma\}$ is comeager in $[t]$.

Since $\kappa \to (\kappa)^\zeta_{<\kappa}$ for all $\zeta < \omega_1$ there is a closed, unbounded subset C of κ and a triple $\langle \gamma_0, t_0, \beta_0 \rangle \in \lambda \times Q_{\omega_1} \times \omega_1$ such that for $s \in [C]^\alpha$, $e^*(s) = \langle \gamma_0, t_0, \beta_0 \rangle$.

By (3) we can find $C^* \subseteq C$, closed and unbounded, $f^* \in T_{0,\beta_0}$, and a sequence $\vec{Y}^*$, such that $(f^*,\vec{Y}^*) \in P$ and for $s \in [C^*]^\alpha$, $(f^*, \vec{Y}^*) \Vdash \tau(s) = \gamma$. Hence $(f^*,\vec{Y}^*) \Vdash$ "C^* is homogeneous for e", a contradiction. —|

Suppose S is a set of ordinals. Define an equivalence relation on reals by $x \sim_S y$ if $L[x,S] = L[y,S]$. For $x \in \mathbb{R}$ let $[x]_S$ denote the equivalence class of x. We call the equivalence classes S-degrees. This is a standard generalization of the familiar notion of Turing degrees. A cone of S-degrees is a set of S-degrees defined by $\{[x]_S \mid x_0 \in L[S,x]\}$ for some $x_0 \in \mathbb{R}$. We say that S-degree determinacy holds if for any set, A, of S-degrees either A contains a cone or the compliment of A contains a cone. In addition, to avoid (some) trivialities, we require that there must be no maximal S-degree. It is a fundamental theorem of AD, due to D. Martin, that Turing (degree) determinacy holds. Thus assuming AD, S-degree determinacy holds for any S.

Assuming ω_1-DC, S-degree determinacy must fail for any S such that the S-degrees are countable i.e. such that $[x]_S$ is countable for each $x \in \mathbb{R}$. This however does not rule out the possibility of S-degree determinacy if the corresponding degrees are uncountable.

Theorem 2.3 (ZF + AD + DC + V = $L(\mathbb{R})$). Assume $G \subseteq Q_{\omega_1}$ defines a generic filter over V for $\mathbb{P}$. Then in V[G], G-degree determinacy holds.

Proof. For reals x, y define $x \leq_L y$ if $x \in L[y]$ and let $[x]_L = \{z \in \mathbb{R} \mid L[z] = L[x]\}$. Thus $[x]_L$ is the constructible degree of x which we call the L-degree of x. Note that the corresponding form of (degree) determinacy, L-degree determinacy, is an immediate consequence of Turing determinacy hence of AD.

As usual we identify elements of $T^*_{0,\alpha+1}$ with reals. This is done in a natural fashion. Further suppose x, y are reals, we let $x * y$ denote the real coding the pair $\langle x,y\rangle$ in some canonical fashion.

We isolate in a series of claims the main technical lemma that we shall need.

(1) Suppose $F : \mathbb{R} \to P(\mathbb{R})$ is a function such that for $x \in \mathbb{R}$, $F(x)$ is comeager. Then for every $x_0 \in \mathbb{R}$ there is a real $x_1 \geq_L x_0$ and a comeager set $A \subseteq \mathbb{R}$ such that for any real $x \geq_L x_1$ and any $c \in A$, $L[x,c] = L[y,c]$ for some $y \in \mathbb{R}$, $y \geq_L x_0$, and $c \in F(y)$.

Proof of (1). Suppose not and assume the claim fails for some real, x_0. Then (in the worst case) by $**\mathbb{R}$-AC and L-degree determinacy there is a partial function $H : \mathbb{R} \to \mathbb{R}$ with comeager comain such that for $c \in \text{dom } H$ and any real $x \geq_L H(c)$ there is no real $y \in L[x,c]$ for which $L[y,c] = L[x,c]$, $y \geq_L x_0$, and $c \in F(y)$.

However since every set of reals has the property of Baire, the function H must agree with a borel function, h, on some comeager borel set D. Let y be a real coding x_0, h, and D. Choose $c \in F(y)$ such that c is a Cohen real over $L[y]$. Thus $c \in D$ and so $h(c) = H(c)$. Further $L[y,c] = L[y,h(c),c]$. Let $x = y * h(c)$. Therefore $H(c) \leq_L x$ but $L[x,c] = L[y,c]$ and $c \in F(y)$, a contradiction since $x_0 \leq_L y$.

We shall need an 'ω_1' form of (1).

(2) Suppose $F : \mathbb{R} \to P(\mathbb{R})$ is a function such that for $x \in \mathbb{R}$, $F(x) \cap T_{0,\alpha+1}$ is comeager in $T_{0,\alpha+1}$ for each $\alpha < \omega_1$. Then for every $x_0 \in \mathbb{R}$ there is a real $x_1 \geq_L x_0$ and a set A, $A \cap T_{0,\alpha+1}$ comeager in $T_{0,\alpha+1}$ for each $\alpha < \omega_1$, such that for any real $x \geq_L x_1$ and any $f \in A$, $L[x,f] = L[y,f]$ for some real y, $y \geq_L x_0$ and $f \in F(y)$.

Proof of (2). Fix x_0. Let $A_1 = \{\langle f,z\rangle \mid f \in T^*_{0,\alpha+1}$ for some $\alpha < \omega_1$, $z \in \mathbb{R}$ and for any $x \geq_L z$, $L[x,f] = L[y,f]$ for some $y \in \mathbb{R}$, $y \geq_L x_0$ and $f \in F(y)\}$.

Since the spaces $T^*_{0,\alpha+1}$ are each homeomorphic to a comeager set of reals, by (1), $\{f \mid \langle f,z\rangle \in A_1$ for some $z\} \cap T_{0,\alpha+1}$ is comeager in $T_{0,\alpha+1}$ for each $\alpha < \omega_1$.

The set A_1 defines a partial function $J : H_{\omega_1} \to P(\mathbb{R})$ (in the codes, a partial function $J^* : \mathbb{R} \to P(\mathbb{R})$). By the proof of Lemma 1.8 there is a choice (partial) function H, $\text{dom } H \cap T_{0,\alpha+1}$ comeager in $T_{0,\alpha+1}$ for each $\alpha < \omega_1$, such that H is Σ_1 over H_{ω_1} (in the codes, a Σ^1_2 partial function $H^* : \mathbb{R} \to \mathbb{R}$). Let x_1 be a real coding x_0 and a parameter sufficient to define H in a Σ_1 fashion. Let $A = \text{dom } H$. Suppose $f \in A$, $f \in T^*_{0,\alpha+1}$, and $x \geq_L x_1$. Then $L[x,f] = L[x,H(f),f] = L[z,f]$ where $z = x * H(f)$. The point being that by absoluteness considerations $H(f) \in L[x,f]$. Thus $z \geq_L H(f)$ and so $L[z,f] = L[y,f]$ for some $y \in \mathbb{R}$, $y \geq_L x_0$, and $f \in F(y)$. But then $L[x,f] = L[y,f]$ and therefore A and x_1 have the required properties.

It is a variant of (2) that is the main lemma that we will need.

(3) Suppose $F : \mathbb{R} \to P$ is a function that for $x \in \mathbb{R}$, $F(x)$ is a condition of the form $(\emptyset,\vec{X})$. Then for every real x_0, there is a real $x_1 \geq_L x_0$ and a condition $(\emptyset,\vec{Y}) \in P$ such that for any real $x \geq_L x_1$ and any condition $(f,\vec{Z}) \leq (\emptyset,\vec{Y})$, if $\operatorname{dom} f = \alpha + 1$ (i.e. $f \in T^*_{0,\alpha+1}$) for some $\alpha < \omega_1$ then there is a real $y \in L[x,f]$ for which $L[x,f] = L[y,f]$, $x_0 \leq_L y$ and $(f,\vec{X}) \leq F(y)$ for some sequence $\vec{X}$.

Proof of (3). Immediate using (2).

In addition to (3) we will also need:

(4) Fix $\alpha < \omega_1$. Suppose $F : \mathbb{R} \to P(T_{0,\alpha+1})$ is a function such that for $x \in \mathbb{R}$, $F(x)$ is comeager in $T_{0,\alpha+1}$. Then for every $x \in \mathbb{R}$ the set $\{c \in \mathbb{R} \mid$ for some $f,g \in T^*_{0,\alpha+1}$, $L[x,c,f] = L[x,g,f]$ and $f \in F(x*c) \cap F(x*g)\}$ is comeager in $\mathbb{R}$.

Proof of (4). Routine.

We continue with the proof of Theorem 2.3. Fix $\tau \in L(\mathbb{R})^{\mathbb{P}}$ a term for a set of G-degrees. Assume toward a contradiction, using Lemma 1.9, that for some sequence $\vec{X}$, $(\emptyset,\vec{X}) \Vdash$ 'τ defines a partition of the G-degrees neither piece of which contains a cone'.

We work in $L(\mathbb{R})$ and fix some additional notation. Let $\mathrm{RO}\,Q_{\omega_1}$ denote the set of elements of the complete boolean algebra generated by Q_{ω_1}. Similarly for $\alpha < \beta < \omega_1$ let $\mathrm{ROQ}_{\alpha,\beta}$ denote the set of elements in the completion of $Q_{\alpha,\beta}$. Thus in a natural fashion $\mathrm{ROQ}_{\omega_1} = \bigcup_{\alpha<\omega_1} \mathrm{ROQ}_{0,\alpha}$. We depart slightly from previous conventions and for $b \in \mathrm{ROQ}_{\omega_1}$ let $[b] = \{f \mid$ for some $\alpha < \beta < \omega_1$ and $t \in Q_{\alpha,\beta}$, $t \leq b$ and $f \in [t]_{\alpha,\beta}\}$. The difference is that for $t \in Q_{\omega_1}$, if $\operatorname{dom} t \not\subseteq \operatorname{dom} f$ then $f \notin [t]$.

For each $x \in \mathbb{R}$ define $b_x \in \mathrm{ROQ}_{\omega_1}$ such that for some sequence $\vec{Y}$ and all conditions $(f,\vec{Z}) \leq (\emptyset,\vec{Y})$, $(f,\vec{Z}) \Vdash$ '$[x]_G \in \tau$' iff $f \in [b_x]$. Clearly b_x depends only on $[x]_L$, hence assume with no loss of generality that for some real, y_0, and all $x \geq_L y_0$, $b_x \neq 0$. We claim the map $[x]_L \mapsto b_x$ is constant on a cone (of L-degrees). To show this we first note that on a cone, $b_x \in \mathrm{ROQ}_{0,\alpha}$ for some $\alpha < \omega_1^x$ i.e. for some α countable in $L[x]$. To see this suppose not. Then this must fail on a cone, hence for some real, z_0, and any $x \geq_L z_0$, $b_x \notin \mathrm{ROQ}_{0,\alpha}$ for any $\alpha < \omega_1^x$. For each $c \in \mathbb{R}$ let $e(c)$ be the least α such that $b_{c*z_0} \in \mathrm{ROQ}_{0,\alpha}$. Thus on a comeager set A, the range of e is bounded by some ordinal $\alpha_0 < \omega_1$. Define for each $c \in \mathbb{R}$, $F(c) \subseteq T_{0,\alpha_0+1}$ by $F(c) = \{h \in T^*_{0,\alpha_0+1} \mid h \in [b_{z_0*c}]$ iff for some sequence $\vec{Z}$, $(h,\vec{Z}) \Vdash$ '$[z_0*c]_G \in \tau$'$\}$. For $t \in Q_{0,\alpha_0+1}$ and $f \in T_{0,\alpha_0+1}$ let f_t denote the element of T_{0,α_0+1} obtained by perturbing f

by t, hence $f_t \in [t]$. For every $c \in \mathbb{R}$, $F(c)$ is comeager in T_{0,α_0+1}. By shrinking if necessary we can assume in addition that for $f \in F(c)$ and $t \in Q_{0,\alpha_0+1}$, $f_t \in F(c)$. Finally by (4), for some $f,g \in T^*_{0,\alpha_0+1}$ and some $c \in A$, $L[z_0, c, f] = L[z_0,g,f]$ and $f \in F(z_0 * c) \cap F(z_0 * g)$. Thus it follows that $b_{z_0*g} = b_{z_0*c}$ and so in particular $b_{z_0*g} \in ROQ_{0,\alpha_0+1}$ but $\omega_1^{z_0*g} > \alpha_0 + 1$ a contradiction since $z_0 * g \geq_L z_0$.

Hence on a cone, $b_x \in ROQ_{0,\alpha}$ for some $\alpha < \omega_1^x$. But then on a cone, $e(x) < \omega_1^x$. By standard arguments it follows that $e(x)$ is constant on a cone and so if $b_0 = \vee\{t \in Q_{\omega_1} \mid \text{on a cone } t \leq b_x\}$ then $b_x = b_0$ on a cone. Fix $x_0 \in \mathbb{R}$ such that for $x \geq_L x_0$, $b_x = b_0$.

For each $x \in \mathbb{R}$ let $F_x = \{f \mid f \in T_{0,\alpha}$ for some $\alpha < \omega_1$ and $f \in [b_x]$ iff for some sequence $\vec{Z}$, $(f,\vec{Z}) \Vdash [x]_G \in \tau\}$.

This defines in a natural fashion a map $F : \mathbb{R} \to P$ such that for $x \in \mathbb{R}$, $F(x)$ is a condition of the form $(\emptyset,\vec{X})$. By (3) there is a real, x_1, and a condition $(\emptyset,\vec{Y}) \in P$ such that for any real $x \geq_L x_1$ and any condition $(f,\vec{Z}) \leq (\emptyset,\vec{Y})$, if $\text{dom } f = \alpha + 1$ for some $\alpha < \omega_1$ then there is a real $y \in L[x,f]$ for which $L[x,f] = L[y,f]$, $x_0 \leq_L y$ and $(f,\vec{X}) \leq F(y)$ for some sequence $\vec{X}$. Assume by refining if necessary that $(\emptyset,\vec{Y}) \Vdash$ 'τ defines a partition of the G-degrees neither piece of which contains a cone'. Choose $g \in \bigcup_\beta Y_\beta$ such that $g \in [b_0]$. We claim that $(g,\vec{Y}) \Vdash$ 'τ contains the cone of G-degrees generated by x_1'. Suppose not and choose $x \geq_L x_1$ and a condition $(f,\vec{Z}) \leq (g,\vec{Y})$ with $\text{dom } f = \alpha + 1$ for some $\alpha < \omega_1$, such that $(f,\vec{Z}) \Vdash$ '$[x]_G \notin \tau$'. Therefore for some $y \in \mathbb{R}$, $L[x,f] = L[y,f]$, $y \geq_L x_0$ and $(f,\vec{X}) \leq F(y)$ for some sequence $\vec{X}$. Hence, since $f \in [b_0]$ and $y \geq_L x_0$, for some condition $(f,\vec{X}) \leq (f,\vec{Z})$, $(f,\vec{X}) \Vdash$ '$[y]_G \in \tau$'. But $L[f,x] = L[f,y]$ so $(f,\vec{X}) \Vdash$ '$[x]_G = [y]_G$' and hence $(f,\vec{X}) \Vdash$ '$[x]_G \in \tau$', a contradiction.

Thus $(g,\vec{Y}) \Vdash$ 'τ contains the cone generated by x_1' but $(g,\vec{Y}) \leq (\emptyset,\vec{Y})$ contradicting that $(\emptyset,\vec{Y}) \Vdash$ 'τ does not contain a cone'. $\dashv$

§3. Following the basic approach of [SVW] we show the consistency of 'ZFC + ω_2 is the second uniform indiscernible + the nonstationary ideal on ω_1 is ω_2-saturated' assuming the consistency of 'ZF + AD'. In fact were this our only goal we could easily finish by using Theorem 1.12 and the relevant proofs of [SVW] (the use of $\mathbb{R}$-AC in [SVW] is really only in establishing the appropriate version of Theorem 1.12). Our approach is slightly different than that of [SVW], we work through a combinatorial intermediary:

* For all $x \in \mathbb{R}$, $x^\#$ exists, and for some (filter) $G \subseteq Q_{\omega_1}$ and all $A \subseteq \omega_1$, $A \in L[x][G]$ for some $x \in \mathbb{R}$ with G generic over $L[x]$ for Q_{ω_1}.

Assuming AD, ω_2 is the second uniform indiscernible and the nonstationary ideal on ω_1 is ω_2-saturated (trivially since AD implies that the filter of closed, unbounded, subsets of ω_1 is an ultrafilter). In fact assuming AD something even stronger is true: For every $x \in \mathbb{R}$, $x^{\#}$ exists, and for all $A \subseteq \omega_1$, $A \in L[x]$ for some $x \in \mathbb{R}$. This of course must fail in ZFC, (*) is an attempt to find a version more palatable with the axiom of choice.

Theorem 3.1 (ZF + AD + DC + V = $L(\mathbb{R})$). Suppose $V[G]$ is a generic extension of V obtained by forcing with $\mathbb{P}$. Then $V[G] \models *$.

Proof. This theorem can be proved in a variety of ways. We use Theorems 1.13 and 1.14 and use the relevant notation.

Suppose $G_1 \subseteq Q_{\omega_1}$ defines a generic filter over V for $\mathbb{Q}_{\omega_1}$. Then by Theorem 1.14, $L(\mathbb{R}_{G_1})[G_1]$ is a generic extension of $L(\mathbb{R}_{G_1})$ for forcing with $\mathbb{P}_{G_1}$ ($\mathbb{P}$ defined in $L(\mathbb{R}_{G_1})$).

We show that (*) holds in $L(\mathbb{R}_{G_1})[G_1]$ and in fact that G_1 is the appropriate witness. It is easily verified that in $L(\mathbb{R}_{G_1})[G_1] \subseteq V[G_1]$, for every real, x, $x^{\#}$ exists. Hence to verify (*) it suffices to show that for all $A \subseteq \omega_1$, $A \in V[G_1]$, $A \in L[x][G_1]$ for some real, x, with G_1 $\mathbb{Q}_{\omega_1}$ generic over $L[x]$ for $\mathbb{Q}_{\omega_1}$. Fix $A \subseteq \omega_1$, $A \in V[G_1]$. Choose a term $\tau_A \in V^{\mathbb{Q}_{\omega_1}}$ for A. Working in V, let $S_A = \{\langle p,\alpha\rangle \mid p \in Q_{\omega_1},\ \alpha < \omega_1, \text{ and } p \Vdash \alpha \in \tau_A\}$. Hence $S_A \in L[x_0]$ for some real, $x_0 \in V$ and therefore $A \in L[x_0][G_1]$. But G_1 is generic over V for $\mathbb{Q}_{\omega_1}$ so G_1 is generic over $L[x_0]$ for $\mathbb{Q}_{\omega_1}$.

Thus (*) holds in $L(\mathbb{R}_{G_1})[G_1]$. Therefore by Theorem 1.13, forcing over $L(\mathbb{R})$ with $\mathbb{P}$ must yield $*$ in the generic extension. ⊣

By a recent theorem of A. Kechris (see [K]) ZF + AD implies DC in $L(\mathbb{R})$. Thus we obtain as a corollary to the previous theorem:

Theorem 3.2. Assume ZF + AD is consistent. Then so is ZFC + $*$.

Proof. By Theorem 3.1, if ZF + AD is consistent then so is ZF + ω_1-DC + $*$.

Suppose $V \models$ ZF + ω_1-DC + $*$. Then $L(P(\omega_1)) \models$ ZF + ω_1-DC + $*$. Forcing over $L(P(\omega_1))$ it is possible to recover the axiom of choice without adding new subsets of ω_1 (thereby preserving $*$) i.e. in $L(P(\omega_1))$ let $T = \{f : \delta \to P(\omega_1) \mid \delta < \omega_2\}$. Define for $f,g \in T$, $f \leq g$ iff $g \subseteq f$. Suppose $G \subseteq T$ is generic over $L(P(\omega_1))$ for the partial order $\langle T, \leq\rangle$. Then since ω_1-DC holds in $L(P(\omega_1))$, $L(P(\omega_1))$ and $L(P(\omega_1))[G]$ have the same subsets of ω_1 and so $L(P(\omega_1))[G] \models$ ZFC + $*$. ⊣

Before proceeding we generalize some notation. Suppose α, β are ordinals with $\alpha < \beta$. Define $Q_{\alpha,\beta}$ in the obvious fashion extending the definition in the case of α, β countable, i.e. $Q_{\alpha,\beta} = \{f \mid f$ is a function with $\operatorname{dom} f \subseteq [\alpha,\beta)$, dom f finite, and for $\delta \in \operatorname{dom} f$, $f(\delta) \in (\omega+\delta)^{<\omega}\}$. Let $\mathbb{Q}_{\alpha,\beta}$ denote the corresponding partial order. Let $RO\mathbb{Q}_{\alpha,\beta}$ denote the elements of the completion of $\mathbb{Q}_{\alpha,\beta}$, etc. We will on occasion denote $\mathbb{Q}_{0,\alpha}$ by $\mathbb{Q}_\alpha$.

Theorem 3.3 (ZFC). Assume $*$. Then ω_2 is the second uniform indiscernible and the nonstationary ideal on ω_1 is ω_2-saturated. In fact $P(\omega_1)/NS \cong RO\mathbb{Q}_{\omega_2}$.

Proof. To show that ω_2 is the second uniform indiscernible it suffices to show that for each ordinal, α, with $\omega_1 < \alpha < \omega_2$, α is collapsed to ω_1 inside some $L[x]$, i.e. that there is an onto function $h : \omega_1 \to \alpha$ with $h \in L[x]$ for some $x \in \mathbb{R}$. This is an immediate consequence of $*$ using the fact that $\mathbb{Q}_{\omega_1}$ is c.c.c.

The proof that $P(\omega_1)/NS \cong RO\mathbb{Q}_{\omega_2}$ is based upon the corresponding proof in [SVW].

For the remainder of this proof ω_1, ω_2 refer to the ω_1, ω_2 of V.

Define a map $I : P(\omega_1)/NS \cong RO\mathbb{Q}_{\omega_1,\omega_2}$ as follows. Fix $G_0 \subset \mathbb{Q}_{\omega_1}$ as given by $*$. Suppose $A \subseteq \omega_1$. Choose $x \in \mathbb{R}$, $A \in L[x][G_0]$ with G_0 generic over $L[x]$ for $\mathbb{Q}_{\omega_1}$. Choose a term $\tau \in L[x]^{\mathbb{Q}_{\omega_1}}$ for A. Regard τ as a subset of ω_1 (i.e. as a subset of $\mathbb{Q}_{\omega_1} \times \omega_1$). Let $j : L[x] \to L[x]$ be an elementary embedding with $j(\alpha) = \alpha$ for $\alpha < \omega_1$ and $j(\omega_1) = \omega_2$. Thus $j(\mathbb{Q}_{\omega_1}) = \mathbb{Q}_{\omega_2}$ and so $j(\tau)$ is a term in $L[x]^{\mathbb{Q}_{\omega_2}}$ for a subset of ω_2. Note that $\mathbb{Q}_{\omega_2} \cong \mathbb{Q}_{\omega_1} \times \mathbb{Q}_{\omega_1,\omega_2}$ and therefore in $L[x][G_0]$, $j(\tau)$ defines naturally a term $\tau^* \in L[x][G_0]^{\mathbb{Q}_{\omega_1,\omega_2}}$ for a subset of ω_2. Define $I(\tau) = [[\omega_1 \in \tau^*]]_{\mathbb{Q}_{\omega_1,\omega_2}}$ (i.e. $[[\omega_1 \in \tau^*]]$ computed in $L[x][G_0]^{\mathbb{Q}_{\omega_1,\omega_2}}$).

It is routine to verify that the map I is well defined and further that I defines a boolean isomorphism of $P(\omega_1)/NS$ into $RO\mathbb{Q}_{\omega_1,\omega_2}$. This suffices for showing that the nonstationary ideal on ω_1 is ω_2-saturated, with a little more work one can show the map I is onto, the basic observation is that terms in $V^{\mathbb{Q}_\alpha}$ for elements of $RO\mathbb{Q}_{\alpha,\beta}$ correspond canonically to elements of $RO\mathbb{Q}_\beta$. We briefly indicate the argument.

Fix $b \in \mathbb{Q}_{\omega_1,\omega_2}$. Hence $b \in RO\mathbb{Q}_{\omega_1,\lambda}$ for $\lambda < \omega_2$. Code b by a subset $A_b \subseteq \omega_1$, view A_b as coding a collapse of λ to ω_1 etc. (for instance pick $A_b \subseteq \omega_1$ coding a structure $\langle \omega_1, E, S\rangle \cong \langle L_\lambda, \in, S\rangle$ where $S = \{t \in \mathbb{Q}_{\omega_1,\lambda} \mid t \leq b\}$). Choose $x \in \mathbb{R}$ with $A_b \in L[x][G_0]$ and for which G_0 is generic over $L[x]$ for $\mathbb{Q}_{\omega_1}$. Hence $b \in L[x][G_0]$ and so pick a term $\tau \in L[x]^{\mathbb{Q}_{\omega_1}}$ for b. Working in L[x] define $w = \{p \cup q \mid p \in \mathbb{Q}_{\omega_1},\ q \in \mathbb{Q}_{\omega_1,\lambda},$ and $p \Vdash q \leq \tau\}$. Let $c \in RO\mathbb{Q}_\lambda$ be that element of $RO\mathbb{Q}_\lambda$ defined by w. Choose in L[x] a code $A_c \subseteq \omega_1$ of c. Thus it follows that if $\delta < \omega_1$ is an indiscernible for L[x] then $A_c \cap \delta$ codes an element c_δ of $RO\mathbb{Q}_{\omega_1}$. Define $A = \{\delta \mid c_\delta \in G_0^*$, the filter over $RO\mathbb{Q}_{\omega_1}$ generated by $G_0\}$. Finally it is straightforward to verify that $I(A) = b$.

The partial orders $\mathbb{Q}_{\omega_2}$, $\mathbb{Q}_{\omega_1,\omega_2}$ are isomorphic hence $P(\omega_1)/NS \cong RO\mathbb{Q}_{\omega_2}$. ⊣

Using Theorem 3.3 it is possible to deduce from $*$ some useful variants of $*$.

Assume $\models$ ZFC + $*$. Let $G_0 \subseteq \mathbb{Q}_{\omega_1}$ be as given by $*$. Suppose $G \subseteq \mathbb{Q}_{\omega_1,\omega_2}$ defines a filter over $\mathbb{Q}_{\omega_1,\omega_2}$, generic over V. Via the isomorphism $I : P(\omega_1)/NS \cong RO\mathbb{Q}_{\omega_1,\omega_2}$ as constructed in the proof of Theorem 3.3 there is a generic elementary embedding:

$$j : V \to M \subseteq V[G] \text{ with } M^{\omega} \subseteq M \text{ in } V[G] .$$

This is simply the embedding corresponding to the appropriate generic ultrapower of V.

$G_0 \times G$ defines in a natural fashion a filter, $G_0 \otimes G$, over $\mathbb{Q}_{\omega_2}$. It is routine to verify that $j(G_0) = G_0 \otimes G$.

For $\alpha < \omega_1$ let $(G_0)_\alpha = G_0 \cap \mathbb{Q}_{\alpha,\omega_1}$. We claim that for some $\alpha < \omega_1$ and all $A \subseteq \omega_1$, $A \in L[x][(G_0)_\alpha]$ for some $x \in \mathbb{R}$ with $(G_0)_\alpha$ generic over $L[x^\#]$ for $\mathbb{Q}_{\alpha,\omega_1}$ (or even with $(G_0)_\alpha$ generic over $HOD_x^{L(\mathbb{R})}$ for $\mathbb{Q}_{\alpha,\omega_1}$, where $HOD_x^{L(\mathbb{R})}$ denotes HOD_x computed in $L(\mathbb{R})$).

To see this note that $j(G_0)_{\alpha+1} = G \cap \mathbb{Q}_{\omega_1+1,\omega_2}$ if $\alpha = \omega_1^V$ and therefore has this property in M.

The partial orders $\mathbb{Q}_{\omega_1}$, $\mathbb{Q}_{\alpha,\omega_1}$ are isomorphic (in L) hence $(G_0)_\alpha$ may be viewed as a filter on $\mathbb{Q}_{\omega_1}$. Thus $*$ is equivalent to the variant: For all $x \in \mathbb{R}$, $x^\#$ exists, and for some (filter) $G \subseteq \mathbb{Q}_{\omega_1}$ and all $A \subseteq \omega_1$, $A \in L[x][G]$ for some $x \in \mathbb{R}$ with G generic over $L[x^\#]$ for $\mathbb{Q}_{\omega_1}$.

It is straightforward to show that ZFC + Martin's Axiom (ZFC + MA) refutes $*$. However a slight weakening of $*$ seems sufficient to avoid this difficulty.

$**$ For all $x \in \mathbb{R}$, $x^\#$ exists, and for all $A \subseteq \omega_1$, there exists a filter $G \subseteq \mathbb{Q}_{\omega_1}$ and a real, x, such that $A \in L[x][G]$ with G generic over $L[x]$ for $\mathbb{Q}_{\omega_1}$.

<u>Theorem 3.4</u>. Assume ZFC + $*$ is consistent. Then so is ZFC + MA + $\neg$CH + $**$.

<u>Proof</u>. Assume $V \models$ ZFC + $*$. Suppose P is a c.c.c. partial order of size $\aleph_1$. View P as given by an order $<_P$ on ω_1 i.e. $P = \langle \omega_1, <_P \rangle$. It suffices to show that for any P if $G_P \subseteq P$ is a filter, generic over V then $V[G_P] \models **$. (Actually one can show $V[G_P] \models *$.)

Suppose $S \subseteq \omega_1$, $S \in V[G_P]$. Fix a term $\tau_S \in V^P$ for S. We regard τ_S as a subset of $P \times \omega_1$ i.e. of $\omega_1 \times \omega_1$.

Invoking $*$ (more precisely its 'useful' variant) fix a filter $G_0 \subseteq \mathbb{Q}_{\omega_1}$ such that for all $C \subseteq \omega_1$, $C \in L[x][G_0]$ for some $x \in \mathbb{R}$ with G_0 generic over $L[x^\#]$ for $\mathbb{Q}_{\omega_1}$.

Choose $x \in \mathbb{R}$ such that $P, \tau_S \in L[x][G_0]$ and for which G_0 is generic over $L[x^{\#}]$ for $\mathbb{Q}_{\omega_1}$. Select in $L[x]$ a term τ for P and let $A = \mathbb{Q}_{\omega_1} * P$ the iteration defined in $L[x]$ using τ. The pair $\langle G_0, G_P\rangle$ defines a filter, $G_0 \otimes G_P$, on A that is generic over $L[x^{\#}]$. Further $S \in L[x^{\#}][G_0 \otimes G_P]$. Note that A is ω_1^V-c.c. in $L[x^{\#}]$ hence in $L[x^{\#}]$, $RO(A) \cong RO\mathbb{Q}_{\omega_1^V}$ (since $A \in L[x]$). Therefore $L[x^{\#}][G_0 \otimes G_p] = L[x^{\#}][G_1]$ for some $G_1 \subseteq \mathbb{Q}_{\omega_1}$, G_1 generic over $L[x^{\#}]$ for $\mathbb{Q}_{\omega_1}$. $S \in L[x^{\#}][G_1]$ and so we are done. $\dashv$

We now consider the problem of Martin's Axiom and the saturation of the non-stationary ideal on ω_1.

Theorem 3.5. Assume ZFC + * is consistent. Then so is ZFC + MA + $\neg$CH + $P(\omega_1)/NS \cong RO\mathbb{Q}_{\omega_2}$.

Proof. We fix some notation. Suppose P, Q are (separative) partial orders with $P \subseteq Q$ and P relatively complete in Q i.e. $RO(P)$ is a complete subalgebra of $RO(Q)$. Suppose $G \subseteq P$ is a filter, generic over V. Then we denote by G/P the quotient partial order Q/G computed in $V[G]$.

By a chain $\langle P_\alpha : \alpha < \lambda\rangle$ of partial orders we shall always mean a chain for which P_α is relatively complete in P_β for all $\alpha < \beta$, i.e. chains will correspond to iterations of forcing.

Assume $V_0 \models$ ZFC + *. We work in V_0 and assume $2^{\aleph_0} = 2^{\aleph_1} = \aleph_2$. Construct a chain of partial orders $\langle P_\alpha : \alpha < \omega_2\rangle$ such that each partial order P_α is c.c.c. and of size $\aleph_1$, $P_\beta = \bigcup_{\alpha<\beta} P_\alpha$ for limit $\beta < \omega_2$, and such that if $P = \bigcup_\alpha P_\alpha$ then $V_0^P \models$ MA. Assume each P_α is given as an order $<_\alpha$ on $\omega_1 \cdot \alpha$, i.e. $P_\alpha = \langle \omega_1 \cdot \alpha, <_\alpha\rangle$. Thus implicit in P_α is the chain $\langle P_\beta : \beta < \alpha\rangle$.

We claim $V_0^P \models P(\omega_1)/NS \cong RO\mathbb{Q}_{\omega_2}$.

Fix, by *, $G_0 \subseteq \mathbb{Q}_{\omega_1}$ such that for all $A \subseteq \omega_1$, $A \in L[x][G_0]$ for some $x \in \mathbb{R}$ with G_0 generic over $L[x]$ for $\mathbb{Q}_{\omega_1}$.

Suppose $G \subseteq \mathbb{Q}_{\omega_1,\omega_2}$ defines a filter on $\mathbb{Q}_{\omega_1,\omega_2}$ generic over V_0. Corresponding to G and the isomorphism $I : P(\omega_1)/NS \cong RO\mathbb{Q}_{\omega_1,\omega_2}$ as constructed in the proof of Theorem 3.3 there is a generic elementary embedding:

$$j : V_0 \to M \subseteq V_0[G] \quad \text{with} \quad M^\omega \subseteq M \quad \text{in} \quad V_0[G] .$$

Thus $j(G_0) = G_0 \otimes G$ the filter over $\mathbb{Q}_{\omega_2}$ determined by $G_0 \times G$.

Let G_P be a generic filter over V_0 for P.

The generic elementary embedding:

$$j : V_0 \to M \subseteq V_0[G]$$

lifts to define a generic elementary embedding:

$$j : V_0[G_P] \to N \subseteq V_0[G][G_{j(P)}]$$

where $G_{j(P)}$ is a filter generic over $V_0[G]$ for $j(P)$ and $N = M[G_{j(P)}]$.

To show that the nonstationary ideal is still ω_2-saturated in $V_0[G_P]$ one must show in effect that $j(P)$ is c.c.c. in $V_0[G]$. Of course $j(P)$ is c.c.c. in M but this a priori does not help to show that $j(P)$ is c.c.c. in $V_0[G]$ since we are only given that M is closed under ω sequences in $V_0[G]$.

In any case by standard arguments in the theory of saturated ideals, in $V_0[G_P]$, $P(\omega_1)/NS \cong B \subseteq RO(\mathbb{Q}_{\omega_1,\omega_2} * j(P)/j''P)$. Thus we must show that in $V_0[G_P]$, $RO(\mathbb{Q}_{\omega_1,\omega_2} * j(P)/j''P) \cong RO\mathbb{Q}_{\omega_1,\omega_2}$.

For each $\alpha < \omega_2$ let $G = G_P \cap P_\alpha$. Hence G_α is generic over V_0 for P_α and $G_P = \bigcup_\alpha G_\alpha$. We first consider (in V_0) $\mathbb{Q}_{\omega_1,\omega_2} * j(P_\alpha)$. The key point is that this iteration (which we view in canonical fashion as a partial order on $Q_{\omega_1,\omega_2} \times j(\omega_1 \cdot \alpha)$) is an element of $L[x,G_0]$ for some real, x. To see this work in V_0 and choose $x \in \mathbb{R}$ such that $P \in L[x][G_0]$ with G_0 generic over $L[x]$ for $\mathbb{Q}_{\omega_1}$. Choose a term for P_α in $L[x]$ and using this define the iteration $A_\alpha = \mathbb{Q}_{\omega_1} * \bar{P}_\alpha$ in $L[x]$. View A_α as a partial order $\langle Q_{\omega_1} \times \omega_1 \cdot \alpha, \overset{*}{<}_\alpha \rangle$. Using $x^\#$ one can compute in V_0 $j(A_\alpha) \in L[x]$. But $j(A_\alpha) = \mathbb{Q}_{\omega_2} * j(P_\alpha)$, i.e. $j(P_\alpha) = j(A_\alpha)/\mathbb{Q}_{\omega_2}$ the quotient computed in $L[x][G_0 \otimes G]$. Hence $\mathbb{Q}_{\omega_1,\omega_2} * j(P_\alpha) = j(A_\alpha)/G_0$ this quotient computed in $L[x][G_0]$.

Fix $\alpha < \omega_2$. Choose in V_0, $y \in \mathbb{R}$ such that $P_\alpha \in L[y][G_0]$ with G_0 generic over $L[y]$ for $\mathbb{Q}_{\omega_1}$ and with the additional property that α is collapsed to ω_1 in $L[y]$. Thus $\mathbb{Q}_{\omega_1,\omega_2} * j(P_\alpha) \in L[y][G_0]$. G_α is clearly generic over $L[y^\#][G_0]$ for P_α. We claim that $C_\alpha = \mathbb{Q}_{\omega_1,\omega_2} * j(P_\alpha)/j''P_\alpha$ can be computed in $L[y^\#,G_0][G_\alpha]$. The relevant observation is that $P_\alpha = \langle \omega_1 \cdot \alpha, <_\alpha \rangle$ and so the computation of $j''P_\alpha$ simply requires $j''(\omega_1 \cdot \alpha)$ which can be computed in $L[y^\#]$. Let $x = y^\#$. Thus C_α is of size $\omega_2^{V_0}$ in $L[x,G_0][G_\alpha]$. We claim that C_α is $\omega_2^{V_0}$ c.c. in $L[\langle x,G_0,G_\alpha\rangle^\#]$. This is because $j(P_\alpha) \in L[\langle x,G_0\rangle^\#][G]$ and is $\omega_2^{V_0}$ c.c. in $L[\langle x,G_0\rangle^\#][G]$ $(\langle x,G_0\rangle^\# \in M$, $j(P_\alpha)$ is $\omega_2^{V_0}$ c.c. in M). Chain conditions are preserved under factoring hence $j(P_\alpha)/j''P_\alpha$ is $\omega_2^{V_0}$ c.c. in $L[\langle x,G_0\rangle^\#][G][G_\alpha]$ and so it follows that C_α is $\omega_2^{V_0}$ c.c. in $L[\langle x,G_0\rangle^\#][G_\alpha]$. Finally $L[\langle x,G_0\rangle^\#][G_\alpha] = L[\langle x,G_0,G_\alpha\rangle^\#]$.

$\mathbb{Q}_{\omega_1,\omega_2} * j(P)$ defines in V_0 a chain $\mathbb{Q}_{\omega_1,\omega_2} * \langle D_\gamma : \gamma < \lambda \rangle$ where $\lambda = j(\omega_2)$ and for $\alpha < \omega_2$, $\mathbb{Q}_{\omega_1,\omega_2} * j(P_\alpha) = \mathbb{Q}_{\omega_1,\omega_2} * D_{j(\alpha)}$. For $\gamma < \lambda$ let $\alpha_\gamma = \{\alpha < \omega_2 \mid j(\alpha) < \gamma\}$. Let E_γ be the quotient $\mathbb{Q}_{\omega_1,\omega_2} * D_\gamma / j''P_{\alpha_\gamma}$ computed in $V_0[G_{\alpha_\gamma}]$. Thus $C_\alpha = E_{j(\alpha)}$.

We identify (in $V_0[G_P]$) the sequence $\langle E_\gamma : \gamma < \lambda \rangle$ with the corresponding chain of partial orders. Hence $E_\beta = \bigcup_{\gamma < \beta} E_\gamma$ for limit $\beta < \lambda$ and for each $\gamma < \lambda$, $E_\gamma \in L[S_\gamma]$ for some $S_\gamma \subseteq \omega_1$ with E_γ of size ω_2 in $L[S_\gamma]$ and E_γ $\omega_2^{V_0}$ c.c. in $L[S^\#]$ (S_γ is just a reasonable code of G_0, G_{α_γ} and an appropriate real from V_0). Thus using the indiscernibles of $L[S_\gamma]$, $E_\gamma = \bigcup_{\zeta < \beta} E_{\gamma,\zeta}$ where $\langle E_{\gamma,\xi} : \xi < \omega_2 \rangle$ is a chain of partial orders of size $\aleph_1$ and $E_{\gamma,\beta} = \bigcup_{\xi < \beta} E_{\gamma,\xi}$ for limits $\beta < \omega_2$.

Let $E = \bigcup_\gamma E_\gamma$. Hence $E = Q_{\omega_1,\omega_2} * j(P)/j''P$. By a suitable diagonalization argument, using $\mathrm{cof}(\lambda) = \omega_2$, $E = \bigcup_\xi F_\xi$ where $\langle F_\xi : \xi < \omega_2\rangle$ is a chain of partial orders each of size $\aleph_1$, $F_\beta = \bigcup_{\xi<\beta} F_\xi$ for limits $\beta < \omega_2$. Thus E is ω_2-saturated. Let $B = RO(E)$. Hence $P(\omega_1)/NS \cong B$ in $V_0[G_P]$. However because of the decomposition $E = \bigcup_{\gamma\in C} E_{\gamma,f(\gamma)}$ it follows that B is isomorphic to a complete subalgebra of ROQ_{ω_2} and further since forcing with E collapses ω_1, $B \cong ROQ_{\omega_2}$ and so in $V_0[G_P]$, $P(\omega_1)/NS \cong ROQ_{\omega_2}$. ⊣

Much in the spirit of [SVW] we summarize some of these independence results in the following theorem.

<u>Theorem 3.6</u>. The following are equiconsistent:
1) ZF + AD
2) ZFC + $AD^{L(\mathbb{R})}$ + ω_2 is the second uniform indiscernible + the nonstationary ideal on ω_1 is ω_2-saturated.
3) ZFC + $AD^{L(\mathbb{R})}$ + MA + $\neg$CH + ω_2 is the second uniform indiscernible + the nonstationary ideal on ω_1 is ω_2-saturated.

As we have indicated $*$ is a ZFC version of:

‡ For every $x \in \mathbb{R}$, $x^\#$ exists, and for all $A \subseteq \omega_1$, $A \in L[x]$ for some $x \in \mathbb{R}$.

We will show that assuming ZFC + $*$, $L(\mathbb{R}) \models$ ‡. In fact we will prove something stronger but first we isolate what is necessary in $L(\mathbb{R})$ in order for it to be possible to force over $L(\mathbb{R})$ to obtain a model of ZFC + $*$. Clearly the following in addition to ‡ will suffice:

1) C + $**\mathbb{R}$-AC.
2) Forcing with $\mathbb{P}$ yields ω_1-DC.

Let $\mathfrak{F}$ denote the filter over the reals generated by wellordered intersections of comeager subsets of $\mathbb{R}$. Assume $\mathfrak{F}$ is nontrivial ($\emptyset \notin \mathfrak{F}$) and let $\mathbb{P}^*$ denote the partial order defined as $\mathbb{P}$ is defined, using $\mathfrak{F}$ in place of the comeager filter. Upon examination of the relevant proofs it becomes apparent that (1) and (2) can be replaced by:

(1)' $\mathfrak{F}$ is nontrivial and for every set $A \subseteq \mathbb{R}$ there is a borel set B and a set $D \in \mathfrak{F}$ such that $A \cap D = B \cap D$. For any function $f : \mathbb{R} \to V$ there is a (partial) choice function $g : \mathbb{R} \to V$, dom $g \in \mathfrak{F}$, such that for $x \in \mathrm{dom}\, g$, $g(x) \in f(x)$ or $f(x) = \emptyset$.

(2)' Forcing with $\mathbb{P}^*$ yields ω_1-DC.

Note that (1)' simply asserts that every set of reals has the property of Baire relative to a set in $\mathfrak{F}$ (this is the '$\mathfrak{F}$' version of C) and that the appropriate version of $**\mathbb{R}$-AC holds. Also it is possible to isolate choice principles in the spirit of $**\omega_1$-AC that are equivalent to (2)' (in ZF + DC + (1)'), we leave the details of this to the curious reader.

Theorem 3.7. Assume ZFC + $*$. Then $L(\mathbb{R}) \models \ddagger$. In fact $L(\mathbb{R}) \models (1)' + (2)' + \ddagger$ and $L(P(\omega_1))$ is a generic extension of $L(\mathbb{R})$ via forcing with $\mathbb{P}^*$ as defined in $L(\mathbb{R})$.

Proof. Assume $V \models$ ZFC + $*$.

Fix $G_0 \subseteq \mathbb{Q}_{\omega_1}$ as given by $*$. Suppose $G \subseteq \mathbb{Q}_{\omega_1,\omega_2}$ defines a filter, generic over V for $\mathbb{Q}_{\omega_1,\omega_2}$. Let $I : P(\omega_1)/NS \cong RO\mathbb{Q}_{\omega_1,\omega_2}$ be the isomorphism as constructed in the proof of Theorem 3.3. Hence there is a generic elementary embedding:

$$j : V \to M \subseteq V[G] \quad \text{with} \quad M^\omega \subseteq M \quad \text{in} \quad V[G]\ .$$

Further $j(G_0) = G_0 \otimes G$ the filter over $\mathbb{Q}_{\omega_2}$ determined by $G_0 \times G$.

Let $L(\mathbb{R}_G)$ denote the $L(\mathbb{R})$ of $V[G]$. To show that $L(\mathbb{R}) \models (1)'$ it suffices to show that $L(\mathbb{R}_G) \models (1)'$ since $L(\mathbb{R}_G)$ is also the $L(\mathbb{R})$ of M. That $L(\mathbb{R}_G) \models (1)'$ is immediate via the following claim:

Claim. Assume $V_0 \models$ ZFC and that $V_0[g]$ is a generic extension of V_0 obtained by adding a generic ω_1-sequence of Cohen reals. Let $L(\mathbb{R}_g)$ denote the $L(\mathbb{R})$ of $V_0[g]$. Then $L(\mathbb{R}_g) \models (1)'$.

Proof. Routine.

Note that $V[G] = V[G_1][G_2]$ where $G_1 = G \cap \mathbb{Q}_{\omega_1+1}$, $G_2 = G \cap \mathbb{Q}_{\omega_1+1,\omega_2}$. Hence it follows by the claim that $L(\mathbb{R}_G) \models (1)'$ and so $L(\mathbb{R}) \models (1)'$.

Let $\mathbb{P}_0^*$ denote the partial order $\mathbb{P}^*$ as defined in $L(\mathbb{R})$. By an argument analogous to the proof of Theorem 1.14 it follows that $G_0 \otimes G$ defines a filter over $j(\mathbb{P}_0^*)$, generic over $L(\mathbb{R}_G)$.

Hence G_0 defines a filter over $\mathbb{P}_0^*$ that is generic over $L(\mathbb{R})$. By $*$ the generic extension $L(\mathbb{R})[G_0] = L(P(\omega_1))$. Thus $L(\mathbb{R})[G_0] \models \omega_1$-DC, this proves $L(\mathbb{R}) \models (2)'$.

Finally to show that $L(\mathbb{R}) \models \ddagger$, observe that since $L(\mathbb{R})[G_0] \models *$ an analysis in $L(\mathbb{R})$ of the forcing language for $\mathbb{P}_0^*$ will yield $\ddagger$. ⊣

Thus by Theorem 3.7, ZFC + $*$ is equiconsistent with ZF + DC + (1)' + (2)' + $\ddagger$. Observe that ZF + DC + (1)' + (2)' is equiconsistent with ZF (add ω_1 Cohen reals to L, the new $L(\mathbb{R})$ satisfies (1)' + (2)'). Thus ignoring possibly significant interference effects between (1)' + (2)' and $\ddagger$, the consistency strength of ZFC + $*$ could well approximate that of ZF + DC + $\ddagger$. We conjecture that ZF + AD $\vdash$ Con(ZF + DC + $\ddagger$), in fact we will be foolish enough to suggest a scenario for a proof. Assume ZF + AD + $V = L(\mathbb{R})$. The model that we are interested in is HOD. Fix $\lambda = \Theta^{L(\mathbb{R})}$. Suppose $G \subseteq Q_\lambda$ is generic over HOD for Q_λ. Let $L(\mathbb{R}_G)$ denote the $L(\mathbb{R})$ of $HOD[G]$. We conjecture that $L(\mathbb{R}_G) \models \ddagger$.

There is actually some evidence that $ZF + \underset{\sim}{PD} \vdash Con(ZF + DC + \ddagger)$, where $\underset{\sim}{PD}$ denotes the axiom of projective determinacy.

We use ZFC + $*$ to produce one final independence result. For each $n < \omega$ let $\Sigma_n(\omega_1)$ denote the class of those subsets of ω_1 that are definable over $\langle P(\omega_1), \epsilon\rangle$ by a Σ_n-formula without parameters. Similarly define the classes $\Pi_n(\omega_1)$. These classes are 'ω_1' versions of the more conventional classes $\Sigma^1_n(\omega)$, $\Pi^1_n(\omega)$. We work within ZFC and consider the question of which of the classes $\Sigma_n(\omega_1)$, $\Pi_n(\omega_1)$ $n < \omega$ have the prewellordering property.

Recall that assuming $\underset{\sim}{PD}$, for each $n < \omega$, $PWO(\Sigma^1_{2n}(\omega))$ and $PWO(\Pi^1_{2n+1}(\omega))$, see [M] for details.

<u>Theorem 3.8</u>. Assume ZFC + $*$ + $\underset{\sim}{PD}$. Then for each n, $PWO(\Sigma_{2n}(\omega_1))$ and $PWO(\Pi_{2n+1}(\omega_1))$.

<u>Proof</u> (sketch). Note that as a consequence of $*$, a set $S \subseteq \omega_1$ that is $\Sigma_k(\omega_1)$ defines in the codes a set $S^* \subseteq \mathbb{R}$ that is Σ^1_{k+2} and conversely. By $\underset{\sim}{PD}$ for each integer k, $PWO(\Sigma^1_{2k}(\omega^\omega))$ and $PWO(\Pi^1_{2k+1}(\omega^\omega))$, see [M] for the relevant details. Using 'generic' codes of countable ordinals it follows from this that for each $n < \omega$, $PWO(\Sigma_{2n}(\omega_1))$ and $PWO(\Pi_{2n+1}(\omega_1))$. $\dashv$

Let $\Pi_1(P(\omega_1))$ denote the class of those subsets of $P(\omega_1)$, Π_1 over $\langle P(\omega_1), \epsilon\rangle$. Assume ZFC + $*$ + $\underset{\sim}{PD}$. Then it can be shown that $PWO(\Pi_1(P(\omega_1)))$ fails as does $PWO(\Sigma_1(P(\omega_1)))$.

Many of the questions posed in [SVW] remain unanswered. We add one to the list. Does $ZF + DC + AD + V = L(A, \mathbb{R})$ $(A \subseteq \mathbb{R})$ suffice for Theorem 1.12? This may seem (and be) a rather technical question however it tests the power of AD. The point being that assuming $AD_{\mathbb{R}}$, the conclusion of Theorem 1.12 holds in $L(A, \mathbb{R})$ for each set of reals $A \subseteq \mathbb{R}$. It is therefore natural to ask if AD suffices to show this in $L(A, \mathbb{R})$.

We note that prior to the results here the consistency of Martin's Axiom with the existence of an ω_2-saturated ideal on ω_1 was open. This problem will surely fall to more conventional assumptions. The situation for the nonstationary ideal is less clear. Of course the model produced here (Theorem 3.5) has the additional feature that in it $P(\omega_1)/NS$ is essentially as simple as possible given that $MA + \neg CH$ holds (it is known that $MA + \neg CH$ refutes the existence of an $\aleph_1$-dense ideal).

The problem of Martin's Axiom and the existence of ω_2-saturated ideals is an instance of a more general question, can there exist a c.c.c. indestructible ω_2-saturated ideal? Note that assuming ZFC + $*$, '$P(\omega_1)/NS \cong ROQ_{\omega_2}$' is true in any c.c.c. extension via a partial order of size $\aleph_1$ (this because $*$ holds in any such forcing extension, see Theorem 3.4).

The consistency of ZFC + * can also be obtained from the consistency of ZF + DC + Every set of reals is Souslin + $\aleph_1$ is measurable. This fooolws via a recent result of A. Kechris which states that if $\aleph_1$ is measurable then a subset $A \subseteq \omega_1$ is constructible from a real if and only if the set of reals defined by A (in the codes) is Souslin and CoSouslin.

REFERENCES

A.S. Kechris [K], The axiom of determinacy implies dependent choices in L(R), Journal of Symbolic Logic, to appear.

A.S. Kechris, E.M. Kleinberg, Y.N. Moschovakis and H. Woodin [KKMW], The axiom of determinacy, strong partition relations and non-singular measures, Cabal Seminar 77-79, Lecture Notes in Mathematics, Vol. 839, Springer-Verlag, (1981), 75-100.

A.S. Kechris, R.M. Solovay and J.R. Steel [KSS], The axiom of determinacy and the prewellordering property, Cabal Seminar 77-79, ibid, 101-126.

D.A. Martin, Y.N. Moschovakis and J.R. Steel [MMS], The extent of definable scales, Bull. Amer. Math. Soc., 6(1982), 435-440.

Y.N. Moschovakis [M], Descriptive Set Theory, North Holland, (1980).

J.R. Steel and R. VanWesep [SVW], Two consequences of determinacy consistent with choice, Trans. Amer. Math. Soc.

H. Woodin [W], An $\aleph_1$ dense ideal on $\aleph_1$, in preparation.

INTRODUCTION TO Q-THEORY

Alexander S. Kechris(1)
Department of Mathematics
California Institute of Technology
Pasadena, California 91125

Donald A. Martin(2)
Department of Mathematics
University of California
Los Angeles, California 90024

Robert M. Solovay(3)
Department of Mathematics
University of California
Berkeley, California 94720

TABLE OF CONTENTS

---

(1) Research partially supported by NSF Grant MCS81-17804. The author is an A. P. Sloan Foundation Fellow.

(2) Research partially supported by NSF Grant MCS78-02989.

(3) Research partially supported by NSF Grant MCS79-06077.

Introduction

Working in the context of Projective Determinacy (PD), we introduce and study in this paper a countable Π^1_{2n+1} set of reals Q_{2n+1} and an associated real y^0_{2n+1} for each $n \geq 0$ (real means element of ω^ω in this paper). Our theory has analytical (descriptive set theoretic) as well as set theoretic aspects, strongly interrelated with each other.

In the analytical direction Q_{2n+1} can be thought of as a generalization of the concept of the hyperarithmetic $(= \Delta^1_1)$ reals, to all odd levels of the projective hierarchy, and y^0_{2n+1} as an analog of the Kleene $\mathcal{O}$; indeed $Q_1 = \{\alpha : \alpha \text{ is } \Delta^1_1\}$ and y^0_1 is (Turing equivalent to) the Kleene $\mathcal{O}$. So Q-theory at level $2n + 1$ can be understood as a version of hyperarithmetic theory for the $(2n + 1)$th level of this hierarchy. But also Q_{2n+1} for $2n + 1 \geq 3$ is the set of reals in an inner model of set theory, and in this and many other respects Q_{2n+1} for $2n + 1 \geq 3$ appears as an appropriate generalization to odd levels higher than two of the concept of the constructible reals. In that sense y^0_{2n+1} appears as an analog of $0^\#$.

At level 3, the set theoretic aspects become much more concrete. For instance Q_3 consists of exactly those reals which are ordinal definable in $L[x]$, for all sufficiently large (in the sense of constructibility degrees) reals x. Moreover there is some evidence leading to speculation that Q_3 should be intimately connected with the presently unravelling theory of inner models of large cardinals. In some sense, which we hope future work will make precise, it seems that Q_3 should be the set of reals in an "ultimate inner model for large cardinals", where "ultimate" refers here to large cardinals stopping just short of those implying $\underset{\sim}{\Delta}^1_2$-DETERMINACY. If this is so, Q-theory at this level would be another manifestation of the intricate interweaving of the theories of determinacy and large cardinal hypotheses.

We will now describe in somewhat more concrete terms these general ideas. We start with analytical aspects (see I below) and we conclude with the set theoretic ones (II below). (The reader can also consult the expository paper [K5] for a more detailed survey of the role of Q-theory in the current study of projective sets from PD.)

I. Analytical Aspects. The original motivation for the development of Q-theory came from a search for an appropriate generalization of the Kleene Basis Theorem for Σ^1_1 sets (see [Mo3, 4E.8]). We recall first the precise statement of the Kleene Basis Theorem. For any $k \geq 1$, let W^0_k denote the complete Π^1_k set of integers. Thus W^0_1 has the same Turing degree as Kleene's $\mathcal{O}$ (the set of Gödel numbers of notations for recursive ordinals). Kleene proved: The set of reals recursive in W^0_1 forms a basis for Σ^1_1 (i.e. if $A \subseteq \omega^\omega$ is a non-empty

Σ^1_1 set, then A contains some real recursive in W^0_1.)

Descriptive set theory (as developed under the assumption of PD; see [Mo3]) is permeated by a strong periodicity of order two. The odd (resp. even) levels of the projective hierarchy closely resemble one another. Thus it is natural to conjecture the following naive generalization of the Kleene Basis Theorem for all $n \geq 0$:

$$(NG_n) : \text{Reals } \Delta^1_{2n+1} \text{ in } W^0_{2n+1} \text{ form a basis for } \Sigma^1_{2n+1} \text{ sets of reals .}$$

Unfortunately (as the reader has probably guessed from our terminology) the "naive" NG_n is false for $n \geq 1$. With hindsight we can see that W^0_{2n+1} is not, in this context, the proper analogue of W^0_1 on level $2n + 1$, if $n \geq 1$.

To describe the "right" analogue of W^0_1 at level $2n + 1$, we need to review (see [K2]) the theory of Π^1_{2n+1}-singletons. (A real α is a Π^1_{2n+1}-singleton iff $\{\alpha\}$ is a Π^1_{2n+1} subset of ω^ω.) If α is a Π^1_{2n+1}-singleton and β has the same Δ^1_{2n+1}-degree as α, then β is a Π^1_{2n+1}-singleton. Restricted to the set of Δ^1_{2n+1}-degrees of Π^1_{2n+1}-singletons, the natural partial ordering of Δ^1_{2n+1}-degrees becomes a well-ordering. The least degree of a Π^1_{2n+1}-singleton is 0, the degree consisting of all Δ^1_{2n+1} reals. We call these the trivial Π^1_{2n+1}-singletons. There are non-trivial Π^1_{2n+1}-singletons. We let y^0_{2n+1} denote one of least possible Δ^1_{2n+1}-degree.

We prove (Theorem 5.6) that reals Δ^1_{2n+1} in y^0_{2n+1} form a basis for Σ^1_{2n+1}. The characterization of y^0_{2n+1} is only up to Δ^1_{2n+1}-degree. We give in §6 an intrinsic characterization of a Turing degree within this Δ^1_{2n+1} degree (due to Harrington). With this refinement, reals recursive in y^0_{2n+1} form a basis for Σ^1_{2n+1}. As one would hope, y^0_1 has the same Turing degree as W^0_1. Thus we can view Theorem 5.6 as the right generalization to odd levels greater than one, of the Kleene basis theorem.

We have already remarked that the naive generalization of the Kleene Basis Theorem is false at odd levels greater than one. (With hindsight, the key difference that makes level 1 behave differently from the odd levels greater than one is that the set WO of well-orderings of ω is Δ^1_3 but not Δ^1_1.) The refutations of NG_n (and of various extensions of NG_n) for $n \geq 1$ point to the notion of the hull of a Σ^1_{2n+1} set A. Let $\mathrm{Hull}_{2n+1}(A) = \{\alpha : (\forall \beta \in A)\ (\alpha \text{ is } \Delta^1_{2n+1} \text{ in } \beta)\}$. Clearly if A is non-empty, $\mathrm{Hull}_{2n+1}(A)$ is countable. One can show: (a) no set $\mathrm{Hull}_{2n+1}(A)$ (with A Σ^1_{2n+1} and non-empty) is a basis for Σ^1_{2n+1}; (b) if $n \geq 1$, then by judicious choice of A, one can make $\mathrm{Hull}_{2n+1}(A)$ have strong closure properties, e.g. closure under the Δ^1_{2n+1} jump, so that $W^0_{2n+1} \in \mathrm{Hull}_{2n+1}(A)$. It turns out that there is a maximal set of the form $\mathrm{Hull}_{2n+1}(A)$ (A as above). This maximal hull is precisely the set Q_{2n+1}.

For $n = 0$ it can be seen that $\mathrm{Hull}_1(A) = \{\alpha : \alpha \text{ is } \Delta^1_1\}$ for <u>all</u> non-empty Σ^1_1 sets A, so that $Q_1 = \{\alpha : \alpha \text{ is } \Delta^1_1\}$, and Q-theory can be understood as an appropriate generalization of hyperarithmetic theory to all odd levels of the projective hierarchy (it turns out of course that $Q_{2n+1} \supsetneq \{\alpha : \alpha \text{ is } \Delta^1_{2n+1}\}$, if $n \geq 1$, so that this theory diverges from the "naive" generalization of Δ^1_1 to Δ^1_{2n+1}). This aspect is developed in §§1-9 and §14 of our paper. In §§15, 16 we discuss versions of Q-theory appropriate for various point-classes beyond the projective hierarchy. Here we make contact with generalized recursion theory, especially Kleene recursion in ${}^3E$, and inductive definability.

II. <u>Set Theoretic Aspects</u>. Recall that, provided $\aleph^L_1 < \aleph_1$, $L \cap \omega^\omega$ is the maximal countable Σ^1_2 set of reals; see [So]. Kechris and Moschovakis have shown (assuming PD) that for each $n \geq 1$, there is a maximal countable Σ^1_{2n} set C_{2n}, and that C_{2n} is the set of reals in an inner model of ZFC (for instance $L(C_{2n})$). So C_2 is just $L \cap \omega^\omega$. In [K2] it is shown that for $n \geq 0$, there is a maximal countable Π^1_{2n+1} set C_{2n+1}. However C_{2n+1} is definitely not the set of reals of a transitive model of ZFC; in fact, it is not even downward closed under Turing reducibility. On the other hand Q_{2n+1} can be characterized as the maximal countable Π^1_{2n+1} set downward closed under Turing as well as Δ^1_{2n+1}-reducibility. Moreover for $n \geq 1$, Q_{2n+1} is the set of reals in an inner model of set theory. This is discussed in §§12, 13. Further results showing that the Q_{2n+1}'s and the C_{2n}'s for $n \geq 1$ fit into a natural sequence are given in §10.

We elaborate now on the speculative remarks on the theory of inner models for large cardinals (as developed in the work of Kunen [Ku], Dodd and Jensen [DJ] and mainly Mitchell [Mi1], [Mi2], [Mi3]).

First we remark that (working in the theory ZF + DC + $\underset{\sim}{\Delta}^1_2$-DETERMINACY) one can show that there is a Δ^1_3 real α such that any real appearing in a model considered in Mitchell's cited papers is recursive in α. (From Mitchell's work it follows that there is a Σ^1_3 set A and a Σ^1_3 well-ordering of A of order type at most $\aleph_1$, such that the reals in Mitchell's models form an initial segment of the well-ordering. From [Mo3, 6G.10, 6G.12] it follows that A is countable, and from [Mo3, 6E.5] that there is a Δ^1_3 real in which every element of A is recursive.)

One might hope that every Δ^1_3 real appears in a canonical inner model for some large cardinal axiom. More ambitiously, one might hope for an "ultimate Mitchell model" $\mathfrak{M}^\infty$ which contains all the Δ^1_3 reals. It follows from results in this paper, the assumptions of the last paragraph, and reasonable assumptions on $\mathfrak{M}^\infty$ that $Q_3 \subseteq \mathfrak{M}^\infty$. (This can be seen as follows: From §2 we have that if $\alpha \in Q_3$, then for some countable ordinal $\xi < \omega_1$, and <u>all</u> (real) codes w of ξ, $\alpha \in \Delta^1_3(w)$. Thus by relativization $\alpha \in \mathfrak{M}^\infty[w]$ for all such codes. Taking w to be generic we conclude that $\alpha \in \mathfrak{M}^\infty$.) It is tempting to conjecture that the reals of $\mathfrak{M}^\infty$ should be precisely Q_3 (it would then follow that in $\mathfrak{M}^\infty$, as in other Mitchell models, the reals have a Δ^1_3-good well-ordering.) One would also like $\mathfrak{M}^\infty$ to have all large cardinals compatible with $\mathfrak{M}^\infty \models \neg\,\Delta^1_2$-DETERMINACY (in the way that L has all the large cardinals compatible with $\neg\, 0^\#$).

One plausible candidate for $\mathfrak{M}^\infty$ is the ultraproduct

$$\prod_d \mathrm{HOD}^{L[d]}/\mu \ ,$$

where d varies over constructibility degrees and μ is the cone measure on constructibility degrees; see §13. This ultraproduct is understood as taken within $L(\omega^\omega)$ under the assumption that $L(\omega^\omega) \models$ AD. (Woodin has given the following alternative way of describing this model, working in ZF + DC only: Consider the notion of forcing consisting of all constructibly pointed perfect trees, i.e. those perfect trees which are constructible from every path through them. Let x be the real generic over $L(\omega^\omega)$ for this notion of forcing (in the Boolean extension of $L(\omega^\omega)$). Let

$$\mathfrak{M}^\infty = \mathrm{HOD}^{L[x]} .$$

Then $\mathfrak{M}^\infty \subseteq L(\omega^\omega)$ and if $L(\omega^\omega) \models$ AD Woodin shows that $\mathfrak{M}^\infty$ is equal to the above ultraproduct.) The reals in this model are (on the assumptions above) exactly Q_3. On the same assumptions, this model is known to have some large cardinals, for example it has hypermeasures of length ω.

The real y^0_3 should be related to $\mathfrak{M}^\infty$. In §11 we define a natural sequence of reals Y^0_k, $k \geq 2$. It is likely that $Y^0_{2n+1} \equiv_T y^0_{2n+1}$, for $n \geq 1$. Moreover $Y^0_2 \equiv_T 0^\#$ (see [Ma3]). This suggests that the Y^0_k, $k \geq 2$ and so the y^0_{2n+1}, $n \geq 1$, should be "sharps" of models of set theory. In particular, y^0_3 should be the "sharp" of $\mathfrak{M}^\infty$. It is worth noting that, at least in a weak sense, Y^0_3 is the "sharp" of the $\mathfrak{M}^\infty$ mentioned above; see §13. That model is L[P] for P a set of ordinals. If we take the ordinary $P^\#$ and consider sentences where the only parameters are indiscernibles, we get essentially Y^0_3.

It is interesting to note also that there is a plausible definition of an "ultimate core model" K^∞, starting from a strong enough large cardinal axiom.

Suppose there are cardinals κ and λ with $\kappa < \lambda$ and an elementary embedding $j : R(\lambda) \to R(\lambda)$ with critical point κ. Suppose also that if we iterate j in the obvious way (getting a directed system $\langle M_\eta : \eta \in OR\rangle$ and $\langle j_{\alpha,\beta} : \alpha \leq \beta \in OR\rangle$ with $j_{01} = j$, $M_0 = R(\lambda)$), then all the M_η's are well-founded.

By [Ma5], $\underset{\sim}{\Delta}^1_2$-DETERMINACY follows. To define K^∞, we define $K^\infty \cap R(\kappa)$ and use our system of $j_{\alpha\beta}$'s to stretch this to a definition of K^∞. (I.e., $x \in K^\infty$ iff $x \in j_{0\eta}(K^\infty \cap R(\kappa))$, for η large.) Let then $x \in R(\kappa)$. Find ν inaccessible with $x \in R(\nu)$, $\nu < \kappa$. Form the Boolean extension $V^{\mathcal{B}}$ of V that adjoins a generic map f of ω onto ν. In $V^{\mathcal{B}}$ there is a real w that canonically encodes f. Put $x \in K^\infty$ iff for some $z \in \Delta^1_3(w)$, z encodes x.

With this definition, it is not hard to show that $K^\infty \cap \omega^\omega = Q_3$. Moreover, one can relate K^∞ to the models considered in §13 and show that, for all sufficiently large constructibility degrees x,

$$K^\infty \cap R(\aleph_1^{L[x]}) = HOD^{L[x]} \cap R(\aleph_1^{L[x]}) .$$

(We should mention that Woodin has, using $L(\omega^\omega) \models AD$, constructed the same K^∞ simply by starting with the equation above and stretching by sharps.) What is missing (and seems to be quite difficult) is an analysis of K^∞ from below, i.e. some version of "K^∞ is the union of its mice" for some notion of mouse.

Acknowledgement. In concluding, the authors would like to acknowledge the contributions of L. Harrington, at the early stages, and H. Woodin at the later stages of the development of this theory. Their results are included here with their permission.

§1. Preliminaries.

1.1 We use in this paper standard terminology and notation in descriptive set theory, following in most instances that of Moschovakis' textbook [Mo3]. Our basic spaces will be ω, $\mathbb{R} = \omega^\omega$ (the reals, denoted by $\mathcal{N}$ in [Mo3]) and 2^ω. (Product) spaces are of the form $\mathcal{X} = X_1 \times \cdots \times X_k$, where X_i is a basic space, members of these spaces are called points and subsets of them pointsets. A pointclass is a collection of pointsets.

We will adhere to the following notational conventions throughout this paper. Letters e, i, j, k, ℓ, m, n denote always members of ω, $\alpha, \beta, \gamma, \delta, \varepsilon$ members of ω^ω and $\xi, \eta, \vartheta, \kappa, \lambda$ ordinals. As usual $f : X \twoheadrightarrow Y$ means that f is an onto map.

For any pointclass Γ,

$$\Gamma\text{ - DET}$$

is an abbreviation for the assertion that all games in Γ are determined. In particular,

$$PD \Longleftrightarrow \bigcup_n \Sigma^1_n\text{ - DET}$$

is Projective Determinacy and

$$AD \Longleftrightarrow \text{Power}(\omega^\omega)\text{ - DET}$$

the Axiom of (full) Determinacy.

Our basic theory is ZF + DC and all further hypotheses are always stated explicitly.

1.2 Assuming $\underset{\sim}{\Delta}^1_{2n}$-DET an extensive theory of $\Pi^1_{2n+1}(\Sigma^1_{2n+1},\Delta^1_{2n+1})$ and $\Sigma^1_{2n+2}(\Pi^1_{2n+2},\Delta^1_{2n+2})$ pointsets has been developed over the years. We will need throughout this paper a number of results from this theory, all of which can be found with appropriate references in the union of [Mo3] and [K2], with a few exceptions to be noted below. For the reader's convenience we will review quickly this material, recalling along the way some useful (standard) notation as well.

We assume from now on for the rest of this section $\underset{\sim}{\Delta}^1_{2n}$-DET. (For $n = 0$, $\Sigma^1_0 = \Sigma^0_1$, $\Pi^1_0 = \Pi^0_1$, $\Delta^1_0 = \Delta^0_1$, so $\underset{\sim}{\Delta}^1_0$-DET is just clopen determinacy which is provable in ZF + DC, thus no strong hypothesis is being made in this case.)

Let Γ be a pointclass, A a pointset. A norm on A is a map $\varphi : A \to \kappa$ from A into an ordinal κ. If $\varphi : A \twoheadrightarrow \kappa$ maps A onto κ we call A regular. Most of the time we will be using regular norms in the sequel. We call φ a Γ-norm if the two relations below

$$x \leq^*_\varphi y \Longleftrightarrow x \in A \wedge (\varphi(x) \leq \varphi(y)) ,$$

$$x <^*_\varphi y \Longleftrightarrow x \in A \wedge (\varphi(x) < \varphi(y))$$

are in Γ_1 where we put $\varphi(y) = \infty =$ an ordinal bigger than $\sup\{\varphi(x) : x \in A\}$, for all $y \notin A$. Finally we say that Γ is normed iff every pointset in Γ admits a Γ-norm.

The pointclasses Π^1_{2n+1}, Σ^1_{2n+2} are normed (First Periodicity Theorem; [Mo3], 6B.1). If A is a Π^1_{2n+1}-complete pointset, i.e. $A \in \Pi^1_{2n+1}$ and for every Π^1_{2n+1} pointset B there is total recursive function F such that $x \in B \Longleftrightarrow F(x) \in A$, then for every (regular) Π^1_{2n+1}-norm $\varphi : A \twoheadrightarrow \kappa$ the ordinal κ is always the same, i.e.

$$\kappa = \underset{\sim}{\delta}^1_{2n+1} = \sup\{\xi : \xi \text{ is the length of a } \underset{\sim}{\Delta}^1_{2n+1} \text{ prewellordering on } \omega^\omega\}$$

([Mo3], 4C.14).

Some corollaries of the fact that Π^1_{2n+1}, Σ^1_{2n+2} are normed are the following:

(i) Π^1_{2n+1} (Σ^1_{2n+2}) satisfy reduction and Σ^1_{2n+1} (Π^1_{2n+2}) satisfy separation ([Mo3], 4B.10-11).

(ii) Let $\mathfrak{D}^1_{2n+1}$ denote the set of Δ^1_{2n+1} reals and $\mathfrak{D}^1_{2n+1}(\beta)$ the relativized notion, i.e.

$$\alpha \in \mathfrak{D}^1_{2n+1}(\beta) \Longleftrightarrow \alpha \in \Delta^1_{2n+1}(\beta) .$$

Then there is a partial function $d : \omega \times \omega^\omega \times \omega \to \omega$ with Π^1_{2n+1} graph such that

$$\alpha \in \mathfrak{D}^1_{2n+1}(\beta) \Longleftrightarrow \exists i \, \forall m \, (\alpha(m) = d(i,\beta,m))$$

([Mo3], 4D.2, 4D.5 and 6B.2). From this we have the Bounded Quantification Theorem, i.e. for each $P(\alpha,\beta,x)$ in Π^1_{2n+1} the pointset

$$R(\beta,x) \iff \exists\alpha \in \Delta^1_{2n+1}(\beta)P(\alpha,\beta,x)$$

is also in Π^1_{2n+1} ([Mo3], 4D.3 and 6B.2). In particular the relation

$$\alpha \in \Delta^1_{2n+1}(\beta)$$

is Π^1_{2n+1}.

One technical comment that will be needed in §12: To prove that the collection of Π^1_{2n+1} relations on ω only are normed we only need

$$ZF + DC + \underset{\sim}{\Delta}^1_{2n-2} - DET + \Delta^1_{2n} - DET .$$

Again let Γ be a pointclass and A a pointset. A scale on A is a sequence $\overline{\varphi} = \{\varphi_n\}$ of norms on A such that

(i) If $x_i \in A$, $i = 0,1,\ldots$ and $x_i \to x$

and

(ii) For each n, and for all large enough i

$$\varphi_n(x_i) = \text{constant} = \lambda_n ,$$

then $x \in A$ and $\varphi_n(x) \leq \lambda_n$. We call $\{\varphi_n\}$ a Γ-scale if the pointsets

$$R(n,x,y) \iff x \leq^*_{\varphi_n} y ,$$

$$S(n,x,y) \iff x <^*_{\varphi_n} y ,$$

are in Γ. We say that Γ is scaled if every $A \in \Gamma$ admits a Γ-scale.

The pointclasses Π^1_{2n+1}, Σ^1_{2n+2} are scaled (Second Periodicity Theorem; [Mo3], 6C.3). Some corollaries of this fact are the following:

(i) The Uniformization Theorem for Π^1_{2n+1}, i.e. if $R(x,y)$ is Π^1_{2n+1} there is $R^*(x,y)$ in Π^1_{2n+1} such that $R^* \subseteq R$ and $\exists y\, R(x,y) \iff \exists!\, y\, R^*(x,y)$ ([Mo3], 6C.5).

(ii) The Basis Theorem for Π^1_{2n+1}, i.e. every nonempty Π^1_{2n+1} set contains a Π^1_{2n+1}-singleton, and a Basis Theorem for Σ^1_{2n+1}, i.e. every nonempty Σ^1_{2n+1} set contains a real recursive in some Π^1_{2n+1}-singleton.

We turn now to definability estimates for winning strategies. The basic theorem here is the Third Periodicity Theorem ([Mo3], 6E.1), which asserts that in every Σ^1_{2n} game in which Player I has a winning strategy, he actually has a Δ^1_{2n+1} winning strategy. We will also use the following consequences of this result:

(i) The Spector-Gandy Theorem for Π^1_{2n+1}, which asserts that every Π^1_{2n+1} pointset $P(x)$ can be written as

$$P(x) \Longleftrightarrow \exists \alpha \in \mathfrak{D}^1_{2n+1}(x) R(\alpha, x) ,$$

for some $R \in \Pi^1_{2n}$ ([Mo3], 6E.7).

(ii) Closure of Π^1_{2n+1} under quantification on ordinals $< \omega_1$: Let WO be the set of reals coding wellorderings of ω and for $w \in$ WO let $|w|$ the ordinal of the wellordering coded by w. Thus $|\ | : \mathrm{WO} \twoheadrightarrow \omega_1$. A pointset $P(w,x)$ is <u>invariant</u> on w if

$$w \in \mathrm{WO} \wedge v \in \mathrm{WO} \wedge |w| = |v| \Longleftrightarrow [P(w,x) \Longleftrightarrow P(v,x)] .$$

Then we have that if $P(w,x)$ is Π^1_{2n+1} and invariant on w,

$$R(x) \Longleftrightarrow \exists w\ P(w,x)$$

is also Π^1_{2n+1}. (This follows immediately from [K4], 1.2.)

(iii) Every thin (i.e. containing no perfect subset) Σ^1_{2n+1} set contains only Δ^1_{2n+1} reals (so in particular is countable). Also every non-empty Δ^1_{2n+1} thin set A can be written as $\{(\alpha)_n : n \in \omega\}$ for some $\alpha \in \mathfrak{D}^1_{2n+1}$ ([Mo3], 6E.5).

We will finally need some facts about Δ^1_{2n+1}-degrees and thin Π^1_{2n+1} sets.

The Δ^1_{2n+1}-<u>reducibility</u> $\leq_{2n+1}$ is defined in the usual way:

$$\alpha \leq_{2n+1} \beta \Longleftrightarrow \alpha \in \Delta^1_{2n+1}(\beta) .$$

Let

$$\alpha \equiv_{2n+1} \beta \Longleftrightarrow \alpha \leq_{2n+1} \beta \wedge \beta \leq_{2n+1} \alpha ,$$

be the relation of Δ^1_{2n+1}-<u>equivalence</u> and define the Δ^1_{2n+1}-<u>degree</u> of α by

$$[\alpha]_{2n+1} = \{\beta : \beta \equiv_{2n+1} \alpha\} .$$

If $d = [\alpha]_{2n+1}$ is a Δ^1_{2n+1}-degree its <u>jump</u> d' is defined by

$$d' = [W^{\alpha}_{2n+1}]_{2n+1} ,$$

where W^{α}_{2n+1} is a complete $\Pi^1_{2n+1}(\alpha)$ subset of ω. Letting for $d = [\alpha]_{2n+1}$, $e = [\beta]_{2n+1}$

$$d \leq e \Longleftrightarrow \alpha \leq_{2n+1} \beta ,$$

be the partial ordering of Δ^1_{2n+1}-degrees and

$$d < e \Longleftrightarrow d \leq e \wedge e \not\leq d ,$$

we clearly have

$$d < d' .$$

We pass finally to the structure of thin Π^1_{2n+1} sets. First there exists a largest thin Π^1_{2n+1} set which is denoted by

$$C_{2n+1}$$

([Mo3]), 6E.9). The set C_{2n+1} is closed under $\equiv_{2n+1}$ (i.e. if $\alpha \in C_{2n+1}$ and $\beta \equiv_{2n+1} \alpha$, then $\beta \in C_{2n+1}$), and the relation $\leq_{2n+1}$ is a prewellordering, when restricted to C_{2n+1} ([K2], 1B-1 and 1B-3). This means that the Δ^1_{2n+1}-degrees of elements of C_{2n+1} are wellordered under their natural partial ordering, and we let

$$d_0^{2n+1}, d_1^{2n+1}, \ldots, d_\xi^{2n+1}, \ldots ; \quad \xi < \rho_{2n+1}$$

be their increasing enumeration. Then also $d_{\xi+1}^{2n+1} = (d_\xi^{2n+1})'$ ([K2], 1B-6). Finally, there is a Δ^1_{2n+1}-good wellordering $<$ on C_{2n+1} which refines $\leq_{2n+1}$ i.e.

$$\alpha < \beta \Rightarrow \alpha \leq_{2n+1} \beta .$$

Here Δ^1_{2n+1}-good means that for each $\alpha \in C_{2n+1}$, $\{\beta : \beta \leq \alpha\}$ is countable and there are pointsets P, R in Π^1_{2n+1}, Σ^1_{2n+1} respectively such that

$$\alpha \in C_{2n+1} \Rightarrow [\{(\gamma)_n : n \in \omega\} = \{\beta : \beta < \alpha\} \Leftrightarrow P(\gamma,\alpha) \Leftrightarrow R(\gamma,\alpha)] ,$$

([K2], 1C-3).

The length of $<$ is again ρ_{2n+1} and this ordinal can be characterized as follows:

$$\rho_{2n+1} = \omega_1^{L(C_{2n+1})} .$$

Here $L(C_{2n+1})$ is the smallest inner model of ZF containing C_{2n+1} as an element. It can be shown that

$$L(C_{2n+1}) = L(C_{2n+2}) ,$$

where

$$C_{2n+2} = \{\alpha : \exists\beta \in C_{2n+1} \ (\alpha \text{ is recursive in } \beta)\} ,$$

so that if Σ^1_{2n+1}-DE holds, C_{2n+2} is the largest countable Σ^1_{2n+2} set. For $n = 0$, $C_2 = L \cap \omega^\omega$ and $L(C_2) = L$. We have that the reals in $L(C_{2n+2})$ are exactly C_{2n+2}, that $L(C_{2n+2})$ in Σ^1_{2n+2}-correct, i.e. Σ^1_{2n+2} formulas are absolute between this model and the real world, and that $L(C_{2n+2})$ satisfies AC + GCH, while its reals carry a Δ^1_{2n+2}-good wellordering (in $L(C_{2n+2})$). ([K2], 3C-1 and [K1], 2C-1).

Remark. In the references we gave about the results on C_{2n+1}, C_{2n+2}, sometimes theorems are stated with PD as a hypothesis instead of $\underset{\sim}{\Delta}^1_{2n}$-DET. However the proofs use only this latter hypothesis. For example in [K1], 2C-1 one sees beyond $\underset{\sim}{\Delta}^1_{2n}$-DET the extra assumption that all $\underset{\sim}{\Delta}^1_{2n+1}$ sets have the property of Baire. This however is now known to follow from $\underset{\sim}{\Delta}^1_{2n}$-DET; see for instance [K5], 5.3.1.

§2. Defining Q_{2n+1}. We assume throughout this section that $\underset{\sim}{\Delta}^1_{2n}$-DET holds.

Recall from §1 the important fact that Π^1_{2n+1} is closed under the "bounded" existential quantifier "$\exists\alpha \in \mathcal{D}^1_{2n+1}$", where $\mathcal{D}^1_{2n+1}$ is the set of Δ^1_{2n+1} reals. We isolate this property of the set $\mathcal{D}^1_{2n+1}$ in the following definition.

2.1 Definition. A set $B \subseteq \omega^\omega$ is called Π^1_{2n+1}-bounded if for every $P(\alpha,x)$ in Π^1_{2n+1} (here x varies over an arbitrary product space $\mathcal{X}$) the pointset

$$R(x) \iff \exists\alpha \in B\ P(\alpha,x)$$

is also Π^1_{2n+1}.

Thus $\mathcal{D}^1_{2n+1}$ is Π^1_{2n+1}-bounded, but ω^ω is not. We notice first some simple facts about this notion.

2.2 Proposition ($\underset{\sim}{\Delta}^1_{2n}$-DET). (i) Every Π^1_{2n+1}-bounded set in Π^1_{2n+1}.

(ii) Every Π^1_{2n+1}-bounded set B is contained in a Π^1_{2n+1}-bounded set $\overline{B}$ which is downward closed under $\leq_{2n+1}$, i.e. is an initial segment of Δ^1_{2n+1}-degrees.

(iii) Every Π^1_{2n+1}-bounded set is countable, in particular is contained in C_{2n+1}.

Proof. (i) Let B be Π^1_{2n+1}-bounded. Then $\alpha \in B \iff \exists\beta \in B(\alpha = \beta)$, so $B \in \Pi^1_{2n+1}$.

(ii) Given a Π^1_{2n+1}-bounded set B, let $\overline{B} = \{\alpha : \exists\beta \in B(\alpha \leq_{2n+1} \beta)\}$ be the downward closure of B under $\leq_{2n+1}$. To see that $\overline{B}$ is Π^1_{2n+1}-bounded note that

$$\exists\alpha \in \overline{B}\ P(\alpha,x) \iff \exists\beta \in B\ \exists\alpha \leq_{2n+1} \beta\ P(\alpha,x)\ .$$

(iii) First we show that every Π^1_{2n+1}-bounded set B is thin, so contained in C_{2n+1}. Assume not, towards a contradiction, and let $f : 2^\omega \to B$ be a continuous injection. Then if $P(\alpha,x)$ is Π^1_{2n+1} we have

$$R(x) \iff \exists\alpha \in 2^\omega P(\alpha,x) \iff \exists\beta \in B\ (\beta \in f[2^\omega] \wedge P(f^{-1}(\beta),x))$$

so R is $\underset{\sim}{\Pi}^1_{2n+1}$. But clearly R is a typical Σ^1_{2n+2} set, so we have a contradiction.

By (ii) we can assume without loss of generality that B is closed downwards under $\leq_{2n+1}$. So B is an initial segment of C_{2n+1} in the prewellordering $\leq_{2n+1} \restriction C_{2n+1}$. Thus if B is uncountable, towards a contradiction, it must be equal to C_{2n+2} since every proper initial segment of $\leq_{2n+1} \restriction C_{2n+1}$ is countable. By the Basis Theorem for Σ^1_{2n+1} there is a Π^1_{2n+1}-singleton $y \in C_{2n+1}$ such that the complement of C_{2n+1} (which is a nonempty Σ^1_{2n+1} set) contains a real x with $x \leq_{2n+1} y$. Thus we conclude that C_{2n+1} is not closed downward under $\leq_{2n+1}$, so $B \neq C_{2n+2}$, a contradiction. ⊣

We establish now the basic result about Π^1_{2n+1}-bounded sets.

2.3 Theorem ($\underset{\sim}{\Delta}^1_{2n}$-DET) There is a largest Π^1_{2n+1}-bounded set.

Proof. It is convenient to split here the proof in two cases, according as $n = 0$ or $n > 0$. In each case an alternative description of the maximal Π^1_{2n+1}-bounded set will be also given.

Case $n = 0$. We shall show that the largest Π^1_1-bounded set is actually the set $\mathcal{D}^1_1$ of all Δ^1_1 reals. Indeed, let B be Π^1_1-bounded and downward closed under $\leq_1$. Then $B \subseteq C_1$ and if $\{d_0, d_1, \ldots, d_\xi, \ldots\}_{\xi<\rho_1}$ is the increasing enumeration of Δ^1_1-degrees in C_1 ($\rho_1 = \omega^L_1$) then for some $\eta < \rho_1$, $B = \bigcup_{\xi<\eta} d_\xi$. Since $d_0 = \mathcal{D}^1_1$ it is enough to show that $\eta = 1$. If not, then $d_1 \subseteq B$. But the Kleene $\mathcal{O}$ (i.e. the complete Π^1_1 set of integers) is in d_1 and $\mathcal{O}$ is a basis for Σ^1_1, so some real $x \leq_1 \mathcal{O}$ belongs to $\omega^\omega - C_1$ therefore to $\omega^\omega - B$. But as $\mathcal{O} \in B$ and B is downward closed under $\leq_1$, $x \in B$, a contradiction.

Case $n > 0$. Let WO be the set of reals coding wellorderings of subsets of ω and for $w \in WO$ let $|w|$ be the ordinal of the associated wellordering. Clearly $|\ | : WO \twoheadrightarrow \omega_1$. We call w a code of $\xi = |w|$.

2.4 Definition. A real α is Δ^1_{2n+1} in a countable ordinal ξ if α is Δ^1_{2n+1} in every code w of ξ.

Put now

$$Q_{2n+1} = \{\alpha : \alpha \text{ is } \Delta^1_{2n+1} \text{ in a countable ordinal}\}$$

$$= \{\alpha : \exists \xi < \omega_1 \ \forall |w| = \xi \ (\alpha \in \Delta^1_{2n+1}(w))\} .$$

We shall prove that Q_{2n+1} is the largest Π^1_{2n+1}-bounded set.

First we show that $Q_{2n+1} \in \Pi^1_{2n+1}$. Indeed we have

$$\alpha \in Q_{2n+1} \Longleftrightarrow \exists \xi < \omega_1 \ \forall w \ [w \in WO \wedge |w| = \xi \Rightarrow \alpha \in \Delta^1_{2n+1}(w)] ,$$

which is Π^1_{2n+1} by the closure of Π^1_{2n+1} under existential quantification over ω_1. To verify that Q_{2n+1} is also Π^1_{2n+1}-bounded take a "test" Π^1_{2n+1} pointset $P(\alpha,x)$ and note that

$$R(x) \iff \exists\alpha \in Q_{2n+1}\ P(\alpha,x)$$

$$\iff \exists\xi < \omega_1\ \forall w\ [w \in WO \wedge |w| = \xi \Rightarrow$$

$$\exists\alpha \in \Delta^1_{2n+1}(w)\ [\alpha \in Q_{2n+1} \wedge P(\alpha,x)]] ,$$

so that $R \in \Pi^1_{2n+1}$.

Finally let B be an arbitrary Π^1_{2n+1}-bounded set. We shall prove that $B \subseteq Q_{2n+1}$. First we can assume without loss of generality that B is downward closed under $\leq_{2n+1}$. Since $B \subseteq C_{2n+1}$ this implies that if $<$ is a Δ^1_{2n+1}-good wellordering of C_{2n+1} which refines $\leq_{2n+1} \upharpoonright C_{2n+1}$, then B is a proper initial segment of $<$ as well. Let $\varphi : C_{2n+1} \twoheadrightarrow \rho_{2n+1}$ be the norm associated to $<$ (i.e. $\varphi(x) < \varphi(y) \iff x < y$). Since $<$ is Δ^1_{2n+1}-good, φ is a Π^1_{2n+1}-norm on C_{2n+1}. We shall prove that

$$\alpha \in B \Rightarrow \alpha \text{ is } \Delta^1_{2n+1} \text{ in } \varphi(\alpha) ,$$

which will complete our proof. Fix $w \in WO$ with $|w| = \varphi(\alpha)$. We will show that $\alpha \in \Delta^1_{2n+1}(w)$. Put

$$V(\beta) \iff \beta \leq \alpha \iff \varphi(\beta) \leq \varphi(\alpha) .$$

If we can show that $V \in \Sigma^1_{2n+1}(w)$, then we are done since V is countable so it contains only $\Delta^1_{2n+1}(w)$ reals, in particular $\alpha \in \Delta^1_{2n+1}(w)$ (since $\alpha \in V$). Since $|w| = \varphi(\alpha)$ we have

$$V(\beta) \iff \forall\gamma \in B\ [|w| \leq \varphi(\gamma) \Rightarrow \beta \leq \gamma] .$$

But Σ^1_{2n+1} is closed under universal quantification over B, so it is enough to compute that for $\gamma \in C_{2n+1}$, the relation

$$w \in WO \wedge |w| \leq \varphi(\gamma)$$

is Δ^1_{2n+1} (this means as usual that there are S_1, S_2 in Σ^1_{2n+1}, Π^1_{2n+1} resp. such that

$$\gamma \in C_{2n+1} \Rightarrow [w \in WO \wedge |w| \leq \varphi(\gamma) \iff S_1(w,\gamma) \iff S_2(w,\gamma)] .$$

Note that by its definition $\beta \leq \gamma$ is Δ^1_{2n+1}, for $\gamma \in C_{2n+1}$.) Let for $\gamma \in C_{2n+1}$,

$$I(\alpha,\gamma) \Longleftrightarrow \alpha \text{ enumerates the } <\text{-predecessors of } \gamma$$

$$\Longleftrightarrow \{(\alpha)_n : n \in \omega\} = \{\delta : \delta < \gamma\} .$$

Then since $<$ is a Δ^1_{2n+1}-good wellordering of C_{2n+1}, I is Δ^1_{2n+1}, for $\gamma \in C_{2n+1}$. If $I(\alpha,\gamma)$ let

$$n <_\alpha m \Longleftrightarrow (\alpha)_n < (\alpha)_m .$$

Then $<_\alpha$ is a wellordering of order type $\varphi(\gamma)$. We have now that for $\gamma \in C_{2n+1}$

$$w \in WO \wedge |w| \leq \varphi(\gamma) \Longleftrightarrow$$

$$w \in WO \wedge \exists\alpha [I(\alpha,\gamma) \wedge \text{the order type of } <_\alpha \text{ is } \geq |w|]$$

$$w \in WO \wedge \forall\alpha [I(\alpha,\gamma) \Rightarrow \text{the order type of } <_\alpha \text{ is } \geq |w|] ,$$

which proves the desired estimate. ⊣

We can now put down our official definition of Q_{2n+1}.

2.5 <u>Definition</u>. The set Q_{2n+1} is the largest Π^1_{2n+1}-bounded set of reals.

From the proof of 2.3 we have $Q_1 = \mathfrak{D}^1_1$, but we shall see later on that $\mathfrak{D}^1_{2n+1} \subsetneq Q_{2n+1}$ for $n > 0$.

<u>Remark</u>. It is now clear that the Π^1_{2n+1}-bounded sets are exactly the Π^1_{2n+1} subsets of Q_{2n+1}. Because if $B \subseteq Q_{2n+1}$ is such and $P(\alpha,x)$ is Π^1_{2n+1}, then

$$\exists\alpha \in B\, P(\alpha,x) \Longleftrightarrow \exists\alpha \in Q_{2n+1} [\alpha \in B \wedge P(\alpha,x)] .$$

§3. Σ^1_{2n+1}-<u>hulls</u>. We assume throughout this section that $\underset{\sim}{\Delta}^1_{2n}$-DET holds. We introduce here the useful technical concept of a Σ^1_{2n+1}-hull and we tie it up with the preceding notions.

3.1 <u>Definition</u>. Given $S \subseteq \omega^\omega$ we define its $2n+1$-<u>hull</u> by

$$\mathrm{Hull}_{2n+1}(S) = \{\alpha : \forall\beta \in S\, (\alpha \leq_{2n+1} \beta)\} .$$

A Σ^1_{2n+1}-<u>hull</u> is a set of the form $\mathrm{Hull}_{2n+1}(S)$, for some nonempty Σ^1_{2n+1} set S.

Note again the following simple facts.

3.2 Proposition ($\underset{\sim}{\Delta}^1_{2n}$-DET) (i) Every Σ^1_{2n+1}-hull is a downward closed under $\leq_{2n+1}$, Π^1_{2n+1} set.

(ii) Every Π^1_{2n+1}-bounded set is contained in a Σ^1_{2n+1}-hull.

Proof. (i) To see that $\text{Hull}_{2n+1}(S) = B$ is Π^1_{2n+1}-bounded, note that

$$\exists \alpha \in B\ P(\alpha,x) \Longleftrightarrow \forall \beta \in S\ \exists \alpha \leq_{2n+1} \beta\ [\alpha \in B \wedge P(\alpha,x)]$$

and observe that $B \in \Pi^1_{2n+1}$ by its definition.

(ii) Let B be Π^1_{2n+1}-bounded. Then B is countable, so

$$S = \{\alpha : \forall \beta \in B\ (\beta \text{ is recursive in } \alpha)\}$$

is a nonempty Σ^1_{2n+1} set. But clearly

$$B \subseteq \{\beta : \forall \alpha \in S\ (\beta \text{ is recursive in } \alpha)\} \subseteq \text{Hull}_{2n+1}(S)\ .$$

⊣

The following result is now immediate.

3.3 Theorem ($\underset{\sim}{\Delta}^1_{2n}$-DET). The set Q_{2n+1} is also the largest Σ^1_{2n+1}-hull. Moreover there is a Π^1_{2n} nonempty set P such that

$$Q_{2n+1} = \text{Hull}_{2n+1}(P)\ ,$$

and also

$$Q_{2n+1} = \{\alpha : \forall \beta \in P\ (\alpha \text{ is recursive in } \beta)\}\ ,$$

i.e. Q_{2n+1} is the recursive hull of P.

Proof. Let S be as in the proof of 3.2 (ii) for $B = Q_{2n+1}$. Let

$$\beta \in S \Longleftrightarrow \exists \gamma\ P(\langle \beta,\gamma \rangle)\ ,$$

where $P \in \Pi^1_{2n}$. Then obviously

$$Q_{2n+1} \subseteq \{\alpha : \forall x \in P\ (\alpha \text{ is recursive in } x)\} \subseteq \text{Hull}_{2n+1}(P)\ ,$$

but by the maximality of Q_{2n+1} we must also have

$$\text{Hull}_{2n+1}(P) \subseteq Q_{2n+1}\ ,$$

and the proof is complete.

⊣

§4. Relativization; Q_{2n+1}-degrees. We assume throughout this section that $\underset{\sim}{\Delta}^1_{2n}$-DET holds.

Given an arbitrary $\beta \in \omega^\omega$ everything said so far relativizes of course to β. A set $B \subseteq \omega^\omega$ is $\Pi^1_{2n+1}(\beta)$-bounded if for all $P(\alpha,x)$ in $\Pi^1_{2n+1}(\beta)$ the pointset $\exists\alpha \in B\ P(\alpha,x)$ is also $\Pi^1_{2n+1}(\beta)$. Equivalently, this means that for every $P(\alpha,x)$ in Π^1_{2n+1} the pointset $\exists\alpha \in B\ P(\alpha,x)$ is $\Pi^1_{2n+1}(\beta)$. This is because if $P(\alpha,x)$ is $\Pi^1_{2n+1}(\beta)$ then $P(\alpha,x) \Longleftrightarrow P'(\alpha,x,\beta)$, with $P' \in \Pi^1_{2n+1}$ and $\exists\alpha \in B\ P(\alpha,x) \Longleftrightarrow \exists\alpha \in B\ P'(\alpha,x,\beta)$ is $\Pi^1_{2n+1}(\beta)$. As before there exists a largest $\Pi^1_{2n+1}(\beta)$-bounded set which we denote by $Q_{2n+1}(\beta)$. Clearly this is $\Pi^1_{2n+1}(\beta)$ but more than that we have the following uniformity.

4.1 Proposition ($\underset{\sim}{\Delta}^1_{2n}$-DET). The pointset

$$R_{2n+1}(\alpha,\beta) \Longleftrightarrow \alpha \in Q_{2n+1}(\beta)$$

is Π^1_{2n+1}.

Proof. From the relativized proof of 2.3 we have that

$$R_1(\alpha,\beta) \Longleftrightarrow \alpha \in \Delta^1_1(\beta) ,$$

which is Π^1_1, and

$$R_{2n+1}(\alpha,\beta) \Longleftrightarrow \alpha \text{ is } \Delta^1_{2n+1} \text{ in some countable ordinal and } \beta ,$$

if $n > 0$. As in the proof of 2.3 this is also Π^1_{2n+1}. ⊣

By a similar argument we can see also the following uniformity.

4.2 Proposition ($\underset{\sim}{\Delta}^1_{2n}$-DET). If P is Π^1_{2n+1}, so is

$$R(x,\beta) \Longleftrightarrow \exists\alpha \in Q_{2n+1}(\beta)\ P(\alpha,\beta,x) .$$

The relativized $\Sigma^1_{2n+1}(\beta)$-hull operation is defined as follows:

$$\mathrm{Hull}^\beta_{2n+1}(S) = \{\alpha : \forall\gamma \in S\ (\alpha \leq_{2n+1} \langle\gamma,\beta\rangle)\} .$$

If $S \in \Sigma^1_{2n+1}(\beta)$ then we can find $S^* \in \Sigma^1_{2n+1}(\beta)$ such that $\mathrm{Hull}^\beta_{2n+1}(S) = \mathrm{Hull}_{2n+1}(S^*)$, so that in talking about $\Sigma^1_{2n+1}(\beta)$-hulls we can always use the absolute hull operation (applied to $\Sigma^1_{2n+1}(\beta)$ sets of course). To see that such S^* exists put

$$S^*(\delta) \Longleftrightarrow \exists\gamma\ (\delta = \langle\gamma,\beta\rangle \wedge S(\gamma)) .$$

Then $S^* \in \Sigma^1_{2n+1}(\beta)$ and

$$\text{Hull}^{\beta}_{2n+1}(S) = \text{Hull}_{2n+1}(S^*) .$$

Again $Q_{2n+1}(\beta)$ is the largest $\Sigma^1_{2n+1}(\beta)$-hull. We also have the following uniformity.

4.3 <u>Proposition</u> ($\underset{\sim}{\Delta}^1_{2n}$-DET). There is a Π^1_{2n} pointset $P_{2n}(\alpha,\beta)$ such that if $P_{2n}(\beta) = \{\alpha : P_{2n}(\alpha,\beta)\}$ we have

$$Q_{2n+1}(\beta) = \text{Hull}_{2n+1}(P_{2n}(\beta)) ,$$

and also

$$Q_{2n+1}(\beta) = \{x : \forall\alpha \in P_{2n}(\beta)\ (x \text{ is recursive in } \alpha)\} .$$

<u>Proof</u>. It is enough by the argument in 3.3 to find $S_{2n+1}(\alpha,\beta)$ in Σ^1_{2n+1} such that $Q_{2n+1}(\beta) = \text{Hull}_{2n+1}(S_{2n+1}(\beta))$ and also $Q_{2n+1}(\beta) = \{\gamma : \forall\alpha \in S_{2n+1}(\beta)(\gamma$ is recursive in $\alpha)\}$. The pointset S_{2n+1} is defined as follows:

$$S_{2n+1}(\alpha,\beta) \iff \forall\gamma \in Q_{2n+1}(\beta)\ (\gamma \text{ is recursive in } \alpha) . \qquad \dashv$$

We define now the notion of Q_{2n+1}-<u>reducibility</u> among reals as follows.

4.4 <u>Definition</u>. For α, $\beta \in \omega^\omega$ let

$$\alpha \leq^Q_{2n+1} \beta \iff \alpha \in Q_{2n+1}(\beta) .$$

Since $\Delta^1_{2n+1}(\beta) \subseteq Q_{2n+1}(\beta)$ clearly

$$\alpha \leq_{2n+1} \beta \Rightarrow \alpha \leq^Q_{2n+1} \beta .$$

For $n = 0$ we have

$$\leq_1 = \leq^Q_1$$

but later on we will see that this fails when $n > 0$.

Obviously $\alpha \leq^Q_{2n+1} \alpha$. To see that $\leq^Q_{2n+1}$ is a genuine notion of reducibility we need the following fact.

4.5 <u>Proposition</u> ($\underset{\sim}{\Delta}^1_{2n}$-DET). The relation $\leq^Q_{2n+1}$ is transitive.

<u>Proof</u>. Let $\alpha \leq^Q_{2n+1} \beta$, $\beta \leq^Q_{2n+1} \gamma$. We will show that $\alpha \leq^Q_{2n+1} \gamma$. For that define

$$R(\alpha') \iff \exists\beta' \in Q_{2n+1}(\gamma)\ (\alpha' \in Q_{2n+1}(\beta')) ,$$

so that

$$R = \bigcup \{Q_{2n+1}(\beta') : \beta' \in Q_{2n+1}(\gamma)\} .$$

Then $R \in \Pi^1_{2n+1}(\gamma)$ and $\alpha \in R$. If we can show that R is $\Pi^1_{2n+1}(\gamma)$-bounded then $R \subseteq Q_{2n+1}(\gamma)$ and we are done. To see this let $P(\alpha,x)$ be a test pointset in $\Pi^1_{2n+1}(\gamma)$. Then

$$\exists\alpha' \in R\, P(\alpha',x) \Leftrightarrow \exists\beta' \in Q_{2n+1}(\gamma)\, \exists\alpha' \in Q_{2n+1}(\beta')\, P(\alpha',x) ,$$

which is clearly $\Pi^1_{2n+1}(\gamma)$ and we are done. ⊣

4.6 <u>Definition</u>. For $\alpha, \beta \in \omega^\omega$ define

$$\alpha \equiv^Q_{2n+1} \beta \Leftrightarrow \alpha \leq^Q_{2n+1} \beta \wedge \beta \leq^Q_{2n+1} \alpha .$$

This is clearly an equivalence relation. The equivalence class of any $\alpha \in \omega^\omega$ is called the Q_{2n+1}-<u>degree</u> of α and is denoted by $[\alpha]^Q_{2n+1}$. So

$$[\alpha]^Q_{2n+1} = \{\beta : \beta \equiv^Q_{2n+1} \alpha\} .$$

There is of course an obvious partial ordering of Q_{2n+1}-degrees defined by

$$[\alpha]^Q_{2n+1} \leq [\beta]^Q_{2n+1} \Leftrightarrow \alpha \leq^Q_{2n+1} \beta .$$

The Q_1-degrees are exactly the Δ^1_1-degrees. For $n > 0$ every Q_{2n+1}-degree consists of a set of Δ^1_{2n+1}-degrees, which is infinite as we will see later on.

We know that C_{2n+1} is closed under $\equiv_{2n+1}$, i.e. it consists of Δ^1_{2n+1}-degrees. It will be useful to establish the analogous fact for Q_{2n+1}-degrees.

4.7 <u>Proposition</u> ($\underset{\sim}{\Delta}^1_{2n}$-DET). The largest thin Π^1_{2n+1} set C_{2n+1} is closed under $\equiv^Q_{2n+1}$, i.e. it consists of Q_{2n+1}-degrees.

<u>Proof</u>. Let

$$C = \{\alpha : \exists\beta \equiv^Q_{2n+1} \alpha\ (\beta \in C_{2n+1})\} .$$

Clearly $C \in \Pi^1_{2n+1}$, so it will be enough to show that C is thin. If not, let $P \subseteq C$ be perfect. Let

$$R(\alpha,\beta) \Leftrightarrow \alpha \in P \wedge \alpha \equiv^Q_{2n+1} \beta \wedge \beta \in C_{2n+1} .$$

Then $R \in \underset{\sim}{\Pi}^1_{2n+1}$, so by the Uniformization Theorem for Π^1_{2n+1}, there is $R^* \in \underset{\sim}{\Pi}^1_{2n+1}$ such that $R^* \subseteq R$ and

$$\forall\alpha \in P\, \exists!\, \beta\, R^*(\alpha,\beta) .$$

Let $f : P \to C_{2n+1}$ be the function with graph R^*. We claim that f is actually $\underset{\sim}{\Delta}^1_{2n+1}$ (it is obviously $\underset{\sim}{\Pi}^1_{2n+1}$). Indeed, for $\alpha \in P$,

$$f(\alpha) \neq \beta \Longleftrightarrow \exists \beta' \in Q_{2n+1}(\alpha)\ [R^*(\alpha,\beta') \wedge \beta' \neq \beta]\ ,$$

so that the complement of the graph of f is also $\underset{\sim}{\Pi}^1_{2n+1}$. Thus $f[P] \in \underset{\sim}{\Sigma}^1_{2n+1}$ and since it is a subset of C_{2n+1}, it is countable. But to each $\beta \in f[P]$ correspond at most countably many $\alpha \in P$ such that $f(\alpha) = \beta$, therefore P itself is countable, which is of course absurd. $\dashv$

Since $\leq_{2n+1} \subseteq \leq^Q_{2n+1}$ and $\leq_{2n+1} \restriction C_{2n+1}$ is a prewellordering, the same is true for $\leq^Q_{2n+1} \restriction C_{2n+1}$. This means that the Q_{2n+1}-degrees in C_{2n+1} are well-ordered in a transfinite increasing sequence

$$q^{2n+1}_0,\ q^{2n+1}_1,\ \ldots,\ q^{2n+1}_\xi,\ \ldots\ ; \qquad \xi < \rho_{2n+1}$$

under their usual partial ordering. The length of this hierarchy is exactly the same as the length of the hierarchy of Δ^1_{2n+1}-degrees

$$d^{2n+1}_0,\ d^{2n+1}_1,\ \ldots,\ d^{2n+1}_\xi,\ \ldots\ ; \qquad \xi < \rho_{2n+1}$$

in C_{2n+1}, i.e. $\rho_{2n+1} = \omega_1^{L(C_{2n+2})}$. This is immediate from the Σ^1_{2n+2}-correctness of $L(C_{2n+2})$.

Our next immediate goal will be to define an appropriate jump operation for Q_{2n+1}-degrees. The hierarchy of Δ^1_{2n+1}- and Q_{2n+1}-degrees in C_{2n+1} will be instrumental in what follows.

§5. The first non-trivial Π^1_{2n+1}-singleton.

Throughout this section we assume $\underset{\sim}{\Delta}^1_{2n}$-DET.

Recall that a real α is called a Π^1_{2n+1}-singleton if $\{\alpha\} \in \Pi^1_{2n+1}$. The concept of Π^1_{2n+1}-singleton is the key idea behind the definition of the jump operator for Q_{2n+1}-degrees. We shall need first some simple facts about it.

5.1 Proposition ($\underset{\sim}{\Delta}^1_{2n}$-DET). (i) The set of Π^1_{2n+1}-singletons is closed under $\equiv_{2n+1}$.

(ii) The relation $\leq_{2n+1}$ is a prewellordering on the set of Π^1_{2n+1}-singletons.

Proof. (i) Let $\{\alpha\} \in \Pi^1_{2n+1}$ and $\beta \equiv_{2n+1} \alpha$. Let $d : \omega \times \omega^\omega \times \omega \to \omega$ be a partial function with Π^1_{2n+1} graph such that

$$\gamma \leq_{2n+1} \delta \Longleftrightarrow \exists i\ [\forall n(d(i,\delta,n) = \gamma(n))]\ .$$

Choose i_0 such that

$$\beta(n) = d(i_0,\alpha,n), \quad \forall n \in \omega .$$

Then

$$\beta' \in \{\beta\} \Longleftrightarrow \exists\alpha' \in \Delta^1_{2n+1}(\beta') \; [\alpha' \in \{\alpha\} \wedge \forall n \; (\beta'(n) = d(i_0,\alpha',n))] .$$

So $\{\beta\} \in \Pi^1_{2n+1}$.

(ii) The set of Π^1_{2n+1}-singletons is clearly a subset of C_{2n+1} and $\leq_{2n+1}$ is a prewellordering on C_{2n+1}. ⊣

There is a simple but very useful fact concerning Q_{2n+1} and Π^1_{2n+1}-singletons, which we shall state after introducing the following convenient terminology.

5.2 <u>Definition</u>. The <u>trivial</u> Π^1_{2n+1}-<u>singletons</u> are the Δ^1_{2n+1} reals.

5.3 <u>Proposition</u> ($\underset{\sim}{\Delta}^1_{2n}$-DET). The set Q_{2n+1} contains no non-trivial Π^1_{2n+1}-singletons.

<u>Proof</u>. Say $\{\alpha_0\} \in \Pi^1_{2n+1}$ and $\alpha_0 \in Q_{2n+1}$. We shall prove that $\alpha_0 \in \Delta^1_{2n+1}$. For that it is enough to show that the graph of α_0 is Π^1_{2n+1}. But this is clear from the following equivalence

$$\alpha_0(n) = m \Longleftrightarrow \exists\alpha \in Q_{2n+1} \; [\alpha \in \{\alpha_0\} \wedge \alpha(n) = m] .$$ ⊣

The following corollary expresses a strong reflection property of Q_{2n+1}.

5.4 <u>Theorem</u> ($\underset{\sim}{\Delta}^1_{2n}$-DET). For each $P \subseteq \omega^\omega$ in Π^1_{2n+1}, we have

$$\exists\alpha \in Q_{2n+1} P(\alpha) \Longleftrightarrow \exists\alpha \in \Delta^1_{2n+1} P(\alpha) .$$

<u>Proof</u>. Assume $P \cap Q_{2n+1} \neq \emptyset$. Since $P \cap Q_{2n+1}$ is Π^1_{2n+1} it must contain, by the Basis Theorem for Π^1_{2n+1}, a Π^1_{2n+1}-singleton α, which by 5.3 must be trivial, i.e. Δ^1_{2n+1}. ⊣

Since the Δ^1_{2n+1}-degrees of nontrivial Π^1_{2n+1}-singletons are wellordered under their natural partial ordering, there is in particular a first nontrivial Δ^1_{2n+1}-degree of Π^1_{2n+1}-singletons. Let us denote any representative of this Δ^1_{2n+1}-degree by y^0_{2n+1}, and call it the <u>first non-trivial</u> Π^1_{2n+1}-<u>singleton</u>. Later on we shall choose an official y^0_{2n+1}, but for the time being the results we shall state depend only on the Δ^1_{2n+1}-degree of y^0_{2n+1}, so the particular choice of it is irrelevant.

For $n = 0$, since the Kleene $\mathcal{O}$ is a Π^1_1-singleton and clearly minimal in $\leq_1$ among nontrivial Π^1_1-singletons, we can take $y^0_1 = \mathcal{O}$. For higher $n > 0$, y^0_{2n+1} has Δ^1_{2n+1}-degree much higher than that of the complete Π^1_{2n+1} set of integers, as we shall see later on.

The relativized operation $\alpha \longmapsto y^\alpha_{2n+1}$ will provide the jump operator for the Q_{2n+1}-degrees. Before we discuss this however we need to develop first some basic results about y^0_{2n+1} and its relationship with the basis problem for Σ^1_{2n+1} sets. Let us start with a definition.

5.5 <u>Definition</u>. We call a real $y \in \omega^\omega$ a <u>basis</u> for Σ^1_{2n+1} if every nonempty Σ^1_{2n+1} set contains a real $\leq_{2n+1} y$.

Note that this depends only on the Δ^1_{2n+1}-degree of y. We have now the following basis theorem.

5.6 <u>Theorem</u> (Martin-Solovay; see [Mo3], 6C.10). Assume $\underset{\sim}{\Delta}^1_{2n}$-DET. Then the first nontrivial Π^1_{2n+1}-singleton y^0_{2n+1} is a basis for Σ^1_{2n+1}.

<u>Proof</u>. Let B be a nonempty Σ^1_{2n+1} set. Let A be a Π^1_{2n} set such that $\alpha \in B \Leftrightarrow \exists\beta\ (\langle\alpha,\beta\rangle \in A)$. Then A is nonempty and if we can show that A contains a real $\gamma \leq_{2n+1} y^0_{2n+1}$ then writing $\gamma = \langle\alpha,\beta\rangle$, we have also that $\alpha \in B$ and $\alpha \leq_{2n+1} y^0_{2n+1}$.

It is thus enough to show that every nonempty Δ^1_{2n+1} set A contains a real $\leq_{2n+1} y^0_{2n+1}$. Let $\{\psi_m\}$ be a Δ^1_{2n+1}-scale on A and define a new Δ^1_{2n+1}-scale $\{\varphi_m\}$ on A as follows:

$$\varphi_m(\alpha) = \langle\psi_0(\alpha),\alpha(0),\ldots,\psi_m(\alpha),\alpha(m)\rangle\ ,$$

where $\langle\xi_0,\ldots,\xi_i\rangle$ refers to the ordinal of $(\xi_0,\ldots,\xi_i)$ in the lexicographical ordering of tuples. Then $\{\varphi_m\}$ is an <u>excellent scale</u>, i.e. it satisfies the following properties:

$$(1)\quad \varphi_m(\alpha) \leq \varphi_m(\beta) \Rightarrow \forall i \leq m,\ \varphi_i(\alpha) \leq \varphi_i(\beta)\ ,$$

$$(2)\quad \varphi_m(\alpha) = \varphi_m(\beta) \Rightarrow \alpha \restriction m+1 = \beta \restriction m+1\ .$$

(This is a stronger version of the definition of <u>very good scale</u> in [Mo3], where instead of (2) it is required only that if $\alpha_0,\alpha_1,\ldots \in A$ and for each m and all large enough i, $\varphi_m(\alpha_i)$ is constant then $\lim_i \alpha_i = \alpha$ exists.)

Let now

$$A^* = \{\alpha \in A : y^0_{2n+1} \not\leq_{2n+1} \alpha\}\ .$$

By 5.3 $A^* \neq \emptyset$, since otherwise $y^0_{2n+1} \in \mathrm{Hull}_{2n+1}(A)$. So for each m, let β_m

be an element of A^* of least $\varphi_m(\beta)$. By property (1) of $\{\varphi_m\}$ we have that $\varphi_m(\beta_m) = \varphi_m(\beta_n)$ for all $n \geq m$, and by property (2) $\beta_m \upharpoonright m+1 = \beta_n \upharpoonright m+1$ for all $n \geq m$. So $\beta_m \to \overline{\beta} \in A$. We claim now that $\overline{\beta} \leq_{2n+1} y^0_{2n+1}$, which will finish our proof.

For that we compute

$$\overline{\beta}(n) = m \Longleftrightarrow \exists\beta\ [\beta \in A^* \wedge \forall\gamma(\gamma \in A^* \Rightarrow \varphi_n(\beta) \leq \varphi_n(\gamma)) \wedge \beta(n) = m]\ .$$

Now $A^* \in \Sigma^1_{2n+1}$, since

$$\alpha \notin A^* \Longleftrightarrow \alpha \notin A \vee [\alpha \in A \wedge \exists\beta \in \Delta^1_{2n+1}(\alpha)\ (\beta \in \{y^0_{2n+1}\})]\ .$$

So it will be enough to show that if

$$R(n,\beta) \Longleftrightarrow \forall\gamma[\gamma \in A^* \Rightarrow \varphi_n(\beta) \leq \varphi_n(\gamma)]$$

then $R \in \Sigma^1_{2n+1}(y^0_{2n+1})$, for $\beta \in A^*$. Clearly R is Π^1_{2n+1}, so if W is universal Π^1_{2n+1}, let n_0 be such that

$$R(n,\beta) \Longleftrightarrow (n_0,\langle n,\beta\rangle) \in W\ .$$

Let also m_0 be such that

$$\alpha \in \{y^0_{2n+1}\} \Longleftrightarrow (m_0,\alpha) \in W\ .$$

Finally let $\varphi : W \twoheadrightarrow \underset{\sim}{\delta}^1_{2n+1}$ be a Π^1_{2n+1}-norm on W. Then we claim that for $\beta \in A^*$ we have

$$(*) \qquad R(n,\beta) \Rightarrow \varphi(n_0,\langle n,\beta\rangle) < \varphi(m_0,y^0_{2n+1})\ ,$$

so that

$$R(n,\beta) \Longleftrightarrow \varphi(n_0,\langle n,\beta\rangle) < \varphi(m_0,y^0_{2n+1})$$

i.e. $R \in \Delta^1_{2n+1}(y^0_{2n+1})$, for $\beta \in A^*$, and we are done. To prove (*), note that if it fails, towards a contradiction, then for some n, β with $\beta \in A^*$ we have

$$R(n,\beta) \wedge \varphi(m_0,y^0_{2n+1}) \leq \varphi(n_0,\langle n,\beta\rangle)$$

so

$$\alpha \in \{y^0_{2n+1}\} \Longleftrightarrow \varphi(m_0,\alpha) \leq \varphi(n_0,\langle n,\beta\rangle)\ ,$$

therefore $\{y^0_{2n+1}\}$ is $\Delta^1_{2n+1}(\beta)$, i.e. $y^0_{2n+1} \leq_{2n+1} \beta$, which is absurd, since $\beta \in A^*$.

$\dashv$

The following corollary strengthening 5.6 will be needed below. We let

$$\alpha <_{2n+1} \beta \Longleftrightarrow \alpha \leq_{2n+1} \beta \wedge \beta \not\leq_{2n+1} \alpha .$$

5.7 Corollary ($\underset{\sim}{\Delta}^1_{2n}$-DET). If $\alpha <_{2n+1} y^0_{2n+1}$, then every nonempty $\Sigma^1_{2n+1}(\alpha)$ set A contains a real $<_{2n+1} y^0_{2n+1}$.

Proof. Clearly y^0_{2n+1} is a $\Pi^1_{2n+1}(\alpha)$-singleton, and is also nontrivial since $y^0_{2n+1} \notin \Delta^1_{2n+1}(\alpha)$. So by the relativized version of 5.6 every nonempty $\Sigma^1_{2n+1}(\alpha)$ set contains a real $\leq_{2n+1} \langle y^0_{2n+1},\alpha\rangle \equiv_{2n+1} y^0_{2n+1}$. Apply this fact now to the $\Sigma^1_{2n+1}(\alpha)$ set

$$A' = \{\alpha' \in A : y^0_{2n+1} \not\leq_{2n+1} \alpha'\} .$$

⊣

The next result will give us a concrete way of visualizing y^0_{2n+1} and will be useful in helping us to pick a canonical representative among the first nontrivial Π^1_{2n+1}-singletons. To state it in a compact way we introduce the following terminology:

Given a set $A \subseteq \omega^\omega$ and an excellent scale $\overline{\varphi} = \{\varphi_m\}$ on A let β_m be a real of least $\varphi_m(\beta)$ and let $y = \lim_{m\to\infty} \beta_m$. This y is clearly independent of the choice of $\{\beta_m\}$, depending only on A, $\overline{\varphi}$. We will denote it by

$$y_{A,\overline{\varphi}}$$

and call it the leftmost real associated with A, $\overline{\varphi}$. The procedure for defining this real is of course the standard one employed in the proof of the Uniformization Theorem. We have now

5.8 Theorem (Harrington). Assume $\underset{\sim}{\Delta}^1_{2n}$-DET. Let A be a nonempty Δ^1_{2n+1} set, let $\overline{\varphi} = \{\varphi_m\}$ be an excellent Δ^1_{2n+1}-scale on A and let $y_{A,\overline{\varphi}}$ be the leftmost real associated with A, $\overline{\varphi}$. Then

$$y_{A,\overline{\varphi}} \leq_{2n+1} y^0_{2n+1} .$$

Proof. In the notation of the proof of 5.6 it is sufficient to show that $\overline{\beta} = y_{A,\overline{\varphi}}$. For that it is again sufficient to show that for each m, there is $\beta <_{2n+1} y^0_{2n+1}$ such that $\forall\alpha \in A(\varphi_m(\beta) \leq \varphi_m(\alpha))$. If this fails, towards a contradiction, and we pick $\beta <_{2n+1} y^0_{2n+1}$ of least $\varphi_m(\beta)$, then $\{\alpha \in A : \varphi_m(\alpha) < \varphi_m(\beta)\}$ is a nonempty $\Delta^1_{2n+1}(\beta)$ set, which contains no real $<_{2n+1} y^0_{2n+1}$, contradicting 5.7.

⊣

5.9 <u>Corollary</u> ($\underset{\sim}{\Delta}^1_{2n}$-DET). Let A be a nonempty Δ^1_{2n+1} set, let $\overline{\varphi} = \{\varphi_m\}$ be an excellent Δ^1_{2n+1}-scale on A and let $y_{A,\overline{\varphi}}$ be the leftmost real associated with A, $\overline{\varphi}$. Then if A contains no Δ^1_{2n+1} real, we have

$$y_{A,\overline{\varphi}} \equiv_{2n+1} y^0_{2n+1} .$$

<u>Proof</u>. It is enough to check that $y_{A,\overline{\varphi}}$ is a Π^1_{2n+1}-singleton. Let β_m have minimal $\varphi_m(\beta)$. Then $\beta_m \to y_{A,\overline{\varphi}}$ and by the semicontinuity property of scales

$$\varphi_m(y_{A,\overline{\varphi}}) \leq \varphi_m(\beta_m) ,$$

so $\varphi_m(y_{A,\overline{\varphi}})$ is also minimal. So $y_{A,\overline{\varphi}}$ is the unique real α in A with the property that $\varphi_m(\alpha)$ is minimal for all $m \in \omega$. So

$$\alpha \in \{y_{A,\overline{\varphi}}\} \Longleftrightarrow \forall m\, \forall \beta \in A\, (\varphi_m(\alpha) \leq \varphi_m(\beta)) ,$$

i.e. $\{y_{A,\overline{\varphi}}\} \in \Pi^1_{2n+1}$. ⊣

5.10 <u>Corollary</u> ($\underset{\sim}{\Delta}^1_{2n}$-DET). If $A \in \Delta^1_{2n+1}$ is nonempty, and contains no Δ^1_{2n+1} member, then it contains a real of the same Δ^1_{2n+1}-degree as the first non-trivial Π^1_{2n+1} singleton y^0_{2n+1}.

§6. <u>The jump operator for</u> Q_{2n+1}-<u>degrees</u>. We assume in this section $\underset{\sim}{\Delta}^1_{2n}$-DET.

We will now define for each real α a canonical representative y^α_{2n+1} of the least non-trivial $\Pi^1_{2n+1}(\alpha)$-singletons. Recall first that we denote by $\{e\}^\alpha$ the eth partial recursive in α function from ω into ω. A real α is <u>recursive in</u> β, in symbols

$$\alpha \leq_T \beta$$

if $\alpha = \{e\}^\beta$ for some $e \in \omega$. Call α <u>Turing equivalent</u> to β, in symbols

$$\alpha \equiv_T \beta$$

if $\alpha \leq_T \beta, \wedge\ \beta \leq_T \alpha$. The <u>Turing degree of</u> α is

$$[\alpha]_T = \{\beta : \beta \equiv_T \alpha\} .$$

We have now

6.1 <u>Theorem</u> (Harrington). Assume $\underset{\sim}{\Delta}^1_{2n}$-DET. Then to each real α we can associate a real y^α_{2n+1} such that:

(i) For each α, y^α_{2n+1} is a (representative of the $\Delta^1_{2n+1}(\alpha)$-degree of the) first non-trivial $\Pi^1_{2n+1}(\alpha)$-singleton. The relation

$$A_{2n+1}(\alpha,\beta) \Longleftrightarrow \beta = y^{\alpha}_{2n+1}$$

is Π^1_{2n+1}.

(ii) For each α, $\alpha \leq_T y^{\alpha}_{2n+1}$ and $\alpha \leq_T \beta \Rightarrow y^{\alpha}_{2n+1} \leq_T y^{\beta}_{2n+1}$. In fact these reducibilities are uniform. For instance there is total recursive $p : \omega \to \omega$ such that

$$\alpha = \{e\}^{\beta} \Rightarrow y^{\alpha}_{2n+1} = \{p(e)\}^{y^{\beta}_{2n+1}} .$$

(iii) For all α, β

$$\alpha \leq^{Q}_{2n+1} \beta \Rightarrow y^{\alpha}_{2n+1} \leq_T y^{\beta}_{2n+1} .$$

(iv) The Turing degree of y^{α}_{2n+1} has the following intrinsic characterization. Let

$$\mathcal{L}_{2n+1}(\alpha) = \{y_{A,\overline{\varphi}} : A \in \Delta^1_{2n+1}(\alpha),\ A \neq \emptyset,\ \overline{\varphi} \text{ an excellent } \Delta^1_{2n+1}(\alpha)\text{-scale on } A\} .$$

Then $y^{\alpha}_{2n+1} \in \mathcal{L}_{2n+1}(\alpha)$ and every real in $\mathcal{L}_{2n+1}(\alpha)$ is recursive in y^{α}_{2n+1}.

In particular, y^{α}_{2n+1} is a recursive basis for $\Sigma^1_{2n+1}(\alpha)$, i.e. every nonempty $\Sigma^1_{2n+1}(\alpha)$ set contains a real recursive in y^{α}_{2n+1}.

Proof. We will define the notion of an α-chain. If y is an α-chain and $z \in \mathcal{L}_{2n+1}(\alpha)$ then $z \leq_T y$. Finally we show how to construct (uniformly on α) A, $\overline{\varphi}$ in $\Delta^1_{2n+1}(\alpha)$ such that $y_{A,\overline{\varphi}}$ is an α-chain. We take $y^{\alpha}_{2n+1} = y_{A,\overline{\varphi}}$.

For $\alpha \in \omega^{\omega}$, let $\alpha^* = \lambda t.\alpha(t+1)$. Let $\langle m,n\rangle = 2^m(2n+1) - 1$, so that $(m,n) \longmapsto \langle m,n\rangle$ is a recursive bijection of $\omega \times \omega$ onto ω. Let $\langle k,\ell,m\rangle = \langle k,\langle \ell,m\rangle\rangle$ and $(\alpha)_i = \lambda t.\alpha(\langle i,t\rangle)$, so that $\alpha \longmapsto \{(\alpha)_i : i \in \omega\}$ is a recursive bijection of ω^{ω} with $(\omega^{\omega})^{\omega}$. Finally let for $\alpha, \beta \in \omega^{\omega}$ $\alpha \oplus \beta = \gamma$, where $\gamma(2n) = \alpha(n)$, $\gamma(2n+1) = \beta(n)$.

An α-chain is a real y such that

(a) $(y)_0 = \alpha$;

(b) $(y)_{\langle 1,i,j\rangle} = y_i \oplus y_j$;

(c) $(y)_{\langle 2,i,j\rangle} = \begin{cases} \{i\}^{(y)_j}, & \text{if this is total,} \\ \lambda t.0, & \text{otherwise;} \end{cases}$

(d) Let $P(e,\beta,\gamma)$ be Π^1_{2n} such that for each $R \subseteq \omega^{\omega} \times \omega^{\omega}$ in Π^1_{2n} there is $e \in \omega$ with $R = P_e = \{(\beta,\gamma) : P(e,\beta,\gamma)\}$, i.e. P is universal Π^1_{2n}. Let $w = (y)_{\langle 3,i,j\rangle}$. Then

(d1) $w(0) = 0 \Longleftrightarrow \exists\beta\, P(i,\beta,(y)_j)$

(d2) $w(0) = 0 \Rightarrow P(i,w^*,(y)_j)$.

If y satisfies clauses (a), (b), (c) and (d2) only we call y an α-prechain. Clearly α-chains exist and

$$PC(y,\alpha) \Longleftrightarrow y \text{ is an } \alpha\text{-prechain}$$

is Π^1_{2n}.

Let y be an α-chain. Given integers $n_0,\dots,n_k$ and a Σ^1_{2n+1} code j for a $\Sigma^1_{2n+1}((y)_{n_0},\dots,(y)_{n_k})$ set R we can effectively from y determine if R is nonempty and if so determine a k such that $(y)_k \in R$.

Using these observations let us show first that every real in $\mathcal{L}_{2n+1}(\alpha)$ is recursive in every α-chain y. Let A be nonempty $\Delta^1_{2n+1}(\alpha)$, $\overline{\varphi}$ an excellent $\Delta^1_{2n+1}(\alpha)$-scale and let $w = y_{A,\overline{\varphi}}$ be the leftmost real of A, $\overline{\varphi}$. We show how to compute w from y. To find $w(n)$ it suffices to compute (effectively from n) a j such that $(y)_j \in A$ and $\varphi_n((y)_j)$ is minimal among $\{\varphi_n(x) : x \in A\}$, for then $w(n) = (y)_j(n)$. To find this define a sequence $n_0, n_1, \dots$ as follows: First n_0 is such that $(y)_{n_0} \in A$. Having found n_k such that $(y)_{n_k} \in A$ check if $\{x \in A : \varphi_n(x) < \varphi_n((y)_{n_k})\}$ is nonempty and if so find n_{k+1} such that $(y)_{n_{k+1}}$ belongs to it. This process must stop after finitely many steps, say m, and then $j = n_m$ is such that $\varphi_n((y)_j)$ is φ_n-minimal.

Let now

$$B = PC = \{(y,\alpha) : y \text{ is an } \alpha\text{-prechain}\} .$$

Given a norm $\psi : B \to \kappa$ we say that ψ has support r if $\psi(y,\alpha)$ depends only on $\{y_i : i < r\}$, α. First note that we can find a Δ^1_{2n+1}-scale $\{\psi_m\}$ on B such that ψ_m has support m. This is because $PC(y,\alpha)$ is effectively equivalent to a countable set of Π^1_{2n} conditions each involving only finitely many $(y)_i$'s. By padding the sequence $\{\psi_m\}$ by trivial norms we can then construct a new Δ^1_{2n+1}-scale $\{\psi'_m\}$ such that

$$m \leq \langle r,0\rangle \Rightarrow \psi'_m \text{ has support } \leq r .$$

Now define as usual an excellent Δ^1_{2n+1}-scale on B by

$$\chi_m(y,\alpha) = \langle \psi'_0(y,\alpha), y(0), \alpha(0), \dots, \psi'_m(y,\alpha), y(m), \alpha(m)\rangle .$$

For a fixed α let

$$A = B(\alpha) = \{y : y \text{ is an } \alpha\text{-prechain}\}$$

and let $\overline{\varphi}$ = the restriction of $\overline{\chi}$ to A, given by $\overline{\varphi}(y) = \overline{\chi}(y,\alpha)$. Clearly $\overline{\varphi}$ is an excellent $\Delta^1_{2n+1}(\alpha)$-scale on A and let

$$y^\alpha_{2n+1} = y_{A,\overline{\varphi}} .$$

We first show that $y = y_{A,\overline{\varphi}}$ is an α-chain. If not then (d1) fails for some $(y)_{\langle 3,i,j\rangle} = w$. Pick $r = \langle 3,i,j\rangle$ minimal. By (d2) we must have

$$\exists\beta\ P(i,\beta,(y)_j) \wedge w(0) > 0 .$$

There is an α-chain z with $(z)_i = (y)_i$ for all $i < r$ and $(z)_r(0) = 0$, by the minimality of r. But then we claim that $\varphi_{\langle r,0\rangle}(z) < \varphi_{\langle r,0\rangle}(y)$, which contradicts the fact that y is the leftmost real of A, φ. Indeed if $k = \langle r,0\rangle$ then

$$\varphi_k(z) = \langle \psi_0'(z,\alpha), z(0), \alpha(0), \ldots, \psi_k'(z,\alpha), z(k), \alpha(k)\rangle ,$$

and

$$\varphi_k(y) = \langle \psi_0'(y,\alpha), y(0), \alpha(0), \ldots, \psi_k'(y,\alpha), y(k), \alpha(k)\rangle ,$$

But $\psi_m'(z,\alpha) = \psi_m'(y,\alpha)$ for $m \leq k = \langle r,0\rangle$ since for such $m \in \psi_m'$ has support $\leq r$ and $(z)_i = (y)_i$ for $i < r$. Moreover if $m < k$ then $m = \langle i,t\rangle$ for some $i < r$ thus $z(m) = z_i(t) = y_i(t) = y(m)$ and finally $z(k) = z(\langle r,0\rangle) = z_r(0) = 0 < y_r(0) = y(k)$, so $\varphi_k(z) < \varphi_k(y)$, and we are done.

We complete the proof by verifying (i)-(iv) of the statement of the theorem. By condition (d), y^{α}_{2n+1} being an α-chain is not $\Delta^1_{2n+1}(\alpha)$, so by (the relativized) 5.8 it has the same $\Delta^1_{2n+1}(\alpha)$-degree as the first non-trivial Π^1_{2n+1}-singleton. That $A_{2n+1} \in \Pi^1_{2n+1}$ is obvious from the construction of y^{α}_{2n+1}. So are (ii) and (iv).

To prove (iii) note that as in our proof that every element of $\mathcal{L}_{2n+1}(\gamma)$ is recursive in any γ-chain, we can easily see that for any β-chain y every member of $\mathcal{L}_{2n+1}((y)_i)$ is recursive in y. Thus $y^{(y)_i}_{2n+1} \leq_T y$. Note now that if $\alpha \leq^Q_{2n+1} \beta$, i.e. $\alpha \in Q_{2n+1}(\beta)$, then for some nonempty $\Pi^1_{2n}(\beta)$ set C, α is recursive in every member of C and for some j, $(y)_j \in C$. Thus $\alpha = (y)_i$ for some i so $y^{\alpha}_{2n+1} \leq_T y$. Taking $y = y^{\beta}_{2n+1}$ we complete the proof of (iii). ⊣

For $n = 0$, if we let $\mathfrak{O}^{\alpha}$ be the relativized to α Kleene $\mathfrak{O}$, i.e. the complete $\Pi^1_1(\alpha)$ set of integers, it is not hard to check that

$$\mathfrak{O}^{\alpha} \equiv_T y^{\alpha}_1 ,$$

so $\mathfrak{O}^{\alpha}$ could be another choice for y^{α}_1 at this level. For $n > 0$ we will see later that the complete $\Pi^1_{2n+1}(\alpha)$ set of integers has Δ^1_{2n+1}-degree much lower than that of y^{α}_{2n+1}.

We can now define the jump operator.

6.2 <u>Definition</u>. For each Q_{2n+1}-degree $q = [\alpha]^Q_{2n+1}$ we define its jump q' by

$$q' = [y^{\alpha}_{2n+1}]^Q_{2n+1} .$$

This is well-defined by 6.1 (iii) (in fact $[\alpha]^Q_{2n+1} = [\beta]^Q_{2n+1} \Rightarrow [y^\alpha_{2n+2}]_T = [y^\beta_{2n+1}]_T$) and we also have the basic property

$$p \leq q \Rightarrow p' \leq q' .$$

It remains to offer some justification for the use of the first non-trivial $\Pi^1_{2n+1}(\alpha)$-singleton as the jump of the Q_{2n+1}-degree of α. In some sense the jump of the Q_{2n+1}-degree of α should be the Q_{2n+1}-degree of the "least natural" real not in $Q_{2n+1}(\alpha)$. One way to make this precise is in terms of the "jump operators" of Steel [Stl], whose definition we repeat below.

First we say that for $\alpha, \beta \in \omega^\omega$ $\alpha \equiv_T \beta$ <u>via</u> $e = \langle m,n \rangle$ if $\alpha = \{m\}^\beta$ and $\beta = \{n\}^\alpha$. Let now $F : \omega^\omega \to \omega^\omega$ and $\gamma \in \omega^\omega$. We say that F is <u>uniformly degree invariant above</u> γ if

$$\exists x,y \leq_T \gamma \, \forall \alpha,\beta \in \omega^\omega \, \forall e \, (\gamma \leq_T \alpha \wedge \alpha \equiv_T \beta \text{ via } e \Rightarrow \langle F(\alpha),y \rangle \equiv_T \langle F(\beta),y \rangle \text{ via } x(e)) .$$

Finally we call $f : \mathcal{D} \to \mathcal{D}$, where $\mathcal{D}$ is the set of Turing degrees, a <u>jump operator</u> if

(i) On a cone of degrees d, $d \leq f(d)$.

(ii) $\exists F : \omega^\omega \to \omega^\omega \, \exists \gamma \in \omega^\omega$ (F is uniformly degree invariant above γ $\wedge$ $\forall \alpha \geq_T \gamma \, ([F(\alpha)]_T = f([\alpha]_T))$).

Assuming AD, Steel [Stl] shows that the relation

$$f \leq g \Longleftrightarrow \text{on a cone of Turing degrees } d, \; f(d) \leq g(d) ,$$

is a prewellordering on jump operators. So there is a least possible jump operator $j^Q_{2n+1}(d)$ such that $j^Q_{2n+1}(d) \notin Q_{2n+1}(d)$, on a cone of Turing degrees d. We shall see in §11 that on a cone of Turing degrees d, the Q_{2n+1}-degree of $j^Q_{2n+1}(d)$ is that of the first non-trivial $\Pi^1_{2n+1}(d)$-singleton.

Another sense in which y^α_{2n+1} has "naturally" the least possible Δ^1_{2n+1}-degree among reals not in $Q_{2n+1}(\alpha)$ is expressed in the following result (stated for simplicity in its "lightface" form).

6.3 <u>Theorem</u> ($\underset{\sim}{\Delta}^1_{2n}$-DET). Consider the prewellordering $\leq_{2n+1}$ on the largest thin Π^1_{2n+1} set C_{2n+1}. Then Q_{2n+1} is a proper initial segment of $\leq_{2n+1}$ and y^0_{2n+1} has minimal Δ^1_{2n+1}-degree in $C_{2n+1} - Q_{2n+1}$ (i.e. comes immediately after Q_{2n+1} in $\leq_{2n+1} \restriction C_{2n+1}$).

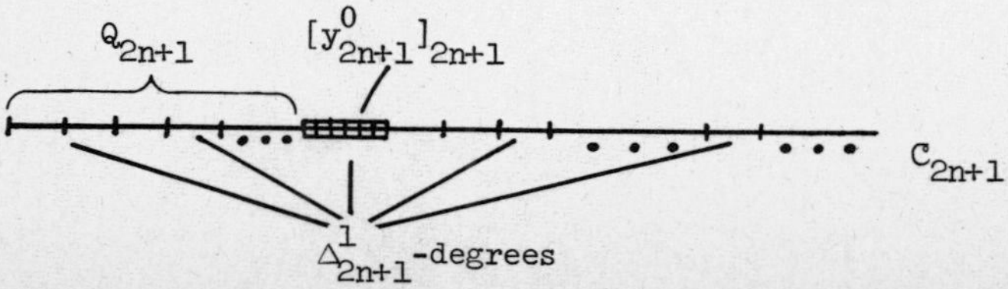

<u>Proof</u>. We know from §2 that Q_{2n+1} is downward closed under $\leq_{2n+1}$, so it is an initial segment of $\leq_{2n+1} \upharpoonright C_{2n+1}$.

Let now $<$ be a Δ^1_{2n+1}-good wellordering of C_{2n+1}. As for $\alpha, \beta \in C_{2n+1}$

$$\alpha \leq \beta \Rightarrow \alpha \leq_{2n+1} \beta$$

clearly Q_{2n+1} is also an initial segment of $<$ as well. Let z_0 be $<$-least in $C_{2n+1} - Q_{2n+1}$. Clearly $z_0 \leq y^0_{2n+1}$ so $z_0 \leq_{2n+1} y^0_{2n+1}$. So it is enough to show that $y^0_{2n+1} \leq_{2n+1} z_0$.

It will be convenient again to distinguish cases as $n = 0$, $n > 0$.

<u>Case $n = 0$</u>. Since z_0 belongs to the first non-zero Δ^1_1-degree in C_1, it is Δ^1_1-equivalent to the Kleene $\mathcal{O}$, which in turn has the same Δ^1_1-degree as y^0_1, so we are done.

<u>Case $n > 0$</u>. Let $\varphi : C_{2n+1} \twoheadrightarrow \rho_{2n+1}$ be the norm associated with $<$. Then if $\alpha \in C_{2n+1}$ and $\varphi(\alpha) = |w|$, where $w \in WO$, α is a $\Pi^1_{2n+1}(w)$-singleton. This is because

$$\beta \in \{\alpha\} \Longleftrightarrow \beta \in C_{2n+1} \wedge \varphi(\beta) = |w| ,$$

and as in the proof of 2.3 the relations

$$|w| \leq \varphi(\beta), \qquad \varphi(\beta) \leq |w|$$

are Δ^1_{2n+1} for $w \in WO$, $\beta \in C_{2n+1}$.

Let

$$S = \{w \in WO : \forall\alpha \in Q_{2n+1}(\varphi(\alpha) \leq |w|) .$$

Then $S \in \Sigma^1_{2n+1}$ and if $w \in S$ then $\varphi(z_0) \leq |w|$, therefore z_0 is a $\Pi^1_{2n+1}(w)$-singleton. Put

$$S' = \{w \in S : z_0 \not\leq_{2n+1} w\}$$

As $z_0 \notin Q_{2n+1} = \mathrm{Hull}_{2n+1}(S)$, we have $S' \neq \emptyset$ and clearly $S' \in \Sigma^1_{2n+1}(z_0)$. Moreover z_0 is a proper $\Pi^1_{2n+1}(w)$-singleton for all $w \in S'$. Thus

$$y^0_{2n+1} \leq_{2n+1} y^w_{2n+1} \leq_{2n+1} \langle z_0, w\rangle ,$$

for all $w \in S'$. So

$$y^0_{2n+1} \in \mathrm{Hull}^{z_0}_{2n+1}(S') \subseteq Q_{2n+1}(z_0)$$

and as y^0_{2n+1} is a $\Pi^1_{2n+1}(z_0)$-singleton we must have

$$y^0_{2n+1} \leq_{2n+1} z_0 . \qquad \dashv$$

As a corollary we obtain the following further characterization of Q_{2n+1}.

6.4 <u>Corollary</u> ($\underset{\sim}{\Delta}^1_{2n}$-DET). The set Q_{2n+1} is the largest set contained in a thin Π^1_{2n+1} set and closed downward under $\leq_{2n+1}$.

<u>Proof</u>. If C is a set contained in a thin Π^1_{2n+1} set and closed downward under $\leq_{2n+1}$, then $C \subseteq C_{2n+1}$ and $y^0_{2n+1} \notin C$, since $\omega^\omega - C_{2n+1}$ is a nonempty Σ^1_{2n+1} set, thus by 5.6 it contains a real $\leq_{2n+1} y^0_{2n+1}$. So by 6.3 $C \subseteq Q_{2n+1}$. $\dashv$

6.5 <u>Open problem</u>. Is Q_{2n+1} the largest (except for ω^ω) Π^1_{2n+1} set closed downward under $\leq_{2n+1}$, for $n > 0$?

For $n = 0$, $\mathfrak{D}^1_1 = Q_1$ has this property as proved in [K2]. The proof depends on the following result of Martin [Ma2]:

Every uncountable Δ^1_1 set contains reals of every Δ^1_1-degree above $\mathcal{O}$.

6.6 <u>Open problem</u>. Does every uncountable Δ^1_{2n+1} set contain reals of every Δ^1_{2n+1} (or even Q_{2n+1}-degree) above y^0_{2n+1}, for $n > 0$?

§7. <u>A generalization of the Gandy-Kreisel-Tait Theorem</u>. We assume in this section $\underset{\sim}{\Delta}^1_{2n}$-DET.

We have already seen a number of different characterizations of Q_{2n+1}. For convenience we summarize them here:

The set Q_{2n+1} is equal to:

(1) The largest Π^1_{2n+1}-bounded set,

(2) The largest Σ^1_{2n+1}-hull,

(3) The largest countable Π^1_{2n+1} set which is closed downwards under $\leq_{2n+1}$.

We will give now a totally new description of Q_{2n+1}, based on a generalization of the Gandy-Kreisel-Tait Theorem. Let us first recall the statement of that result.

An ω-model (for the language of analysis) is just a nonempty set M of reals. If $\varphi(x_1,\ldots,x_n)$ is a formula of analysis and $\alpha_1,\ldots,\alpha_n \in M$, then

$M \models \varphi[\alpha_1\ldots\alpha_n]$ iff the formula φ with the assignment $x_i \longmapsto \alpha_i$ is satisfied when the real variables in φ are restricted to M.

The Σ^1_n-comprehension axiom schema in symbols Σ^1_n-CA is the collection of all sentences

$$\forall\alpha_1\ldots\forall\alpha_n \exists\alpha \forall m\ [\alpha(m) = 0 \Longleftrightarrow \varphi(m,\alpha_1\ldots\alpha_n)]$$

for all Σ^1_n formulas φ.

The Σ^1_n-dependent choices schema in symbols Σ^1_n-DC is the collection of all sentences

$$\forall\alpha_1\ldots\forall\alpha_m\ (\forall\alpha\,\exists\beta\ \varphi(\alpha,\beta,\alpha_1\ldots\alpha_m) \Rightarrow \exists\alpha\,\forall n\ \varphi((\alpha)_n,(\alpha)_{n+1},\alpha_1\ldots\alpha_m))\ ,$$

for all Σ^1_n formulas φ.

We abbreviate:

$$\text{Analysis} = \bigcup_n (\Sigma^1_n - \text{CA})$$

$$\text{Dependent Choices} = \bigcup_n (\Sigma^1_n - \text{DC})\ .$$

Finally we call an ω-model M <u>Σ^1_n-correct</u> iff for all Σ^1_n formulas $\varphi(x_1\ldots x_n)$ and all $\alpha_1,\ldots,\alpha_n \in M$

$$M \models \varphi[\alpha_1,\ldots,\alpha_n] \ \text{ iff } \ \varphi[\alpha_1,\ldots,\alpha_n]\ .$$

For $n = 0$ this is automatically true as $\Sigma^1_0 = \Sigma^0_1$.

In the preceding terminology the Gandy-Kreisel-Tait Theorem says now: For every recursive set of axioms $T \supseteq$ Analysis having ω-models,

$$\Delta^1_1 = \bigcap\,\{M : M \text{ is an } \omega\text{-model of } T\}\ .$$

We have now the following generalization.

7.1 <u>Theorem</u> ($\underset{\sim}{\Delta}^1_{2n}$-DET). Let T be a recursive set of axioms in the language of analysis extending

$$\text{Analysis} + \text{Dependent Choices} + \underset{\sim}{\Delta}^1_{2n}\text{-DET}$$

which has Σ^1_{2n}-correct ω-models. Then

$$Q_{2n+1} = \bigcap\,\{M : M \text{ is a } \Sigma^1_{2n}\text{-correct } \omega\text{-model of } T\}\ .$$

<u>Proof</u>. We say that a real γ codes an ω-model M if $[\gamma] = \{(\gamma)_n : n \in \omega\} = M$. Let

$$S = \{\gamma : \gamma \text{ codes a } \Sigma^1_{2n}\text{-correct } \omega\text{-model of } T\}\ .$$

Then by our assumption and Skolem-Löwenheim $S \neq \emptyset$ and an easy computation shows that $S \in \Pi^1_{2n}$ (if $n > 0$; otherwise $S \in \Delta^1_1$ which is enough for the argument below), the main complexity coming from the requirement of Σ^1_{2n}-correctness. Since

$$\alpha \in \bigcap \{M : M \text{ is a } \Sigma^1_{2n}\text{-correct } \omega\text{-model of } T\} \Rightarrow$$

$$\forall \gamma \in S\ (\alpha \in [\gamma]) \Rightarrow$$

$$\forall \gamma \in S\ (\alpha \leq_T \gamma) \Rightarrow$$

$$\alpha \in \text{Hull}_{2n+1}(S) \Rightarrow$$

$$\alpha \in Q_{2n+1} ,$$

it is enough to show

$$\alpha \in Q_{2n+1} \Rightarrow \alpha \in \bigcap \{M : M \text{ is a } \Sigma^1_{2n}\text{-correct } \omega\text{-model of } T\} .$$

We shall need first the following lemma:

7.2 <u>Lemma</u> (Moschovakis). Assume $\underset{\sim}{\Delta}^1_{2n}$-DET. If M is a Σ^1_{2n}-correct ω-model of Analysis + Dependent Choices + $\underset{\sim}{\Delta}^1_{2n}$-DET, then M is downward closed under $\leq_{2n+1}$.

<u>Proof</u>. We shall prove that every Δ^1_{2n+1} subset of ω is in M. By relativizing the proof to any $\alpha \in M$ this will show the downward closure of M under $\leq_{2n+1}$.

Let $A \subseteq \omega$, $A \in \Delta^1_{2n+1}$. Then

$$m \in A \Leftrightarrow \exists\alpha\ R(m,\alpha)$$

$$m \notin A \Leftrightarrow \exists\beta\ P(m,\beta) ,$$

$R, P \in \Pi^1_{2n}$. Let

$$R'(m,\alpha,\beta) \Leftrightarrow R(m,\alpha)$$

$$P'(m,\alpha,\beta) \Leftrightarrow P(m,\beta) .$$

Then $R' \cap P' = \emptyset$, so by the Separation Theorem for Π^1_{2n} let $S \in \Delta^1_{2n}$ be such that S separates R', P', i.e. $R' \subseteq S$, $S \cap P' = \emptyset$. Let $\varphi(m,x,y)$, $\psi(m,x,y)$ be Σ^1_{2n} formulas such that φ, ψ define S and $\omega^\omega - S$ respectively. Then

$$\forall m\ \forall\alpha\ \forall\beta\ [\varphi(m,\alpha,\beta) \Leftrightarrow \neg\ \psi(m,\alpha,\beta)] ,$$

so by the Σ^1_{2n}-correctness of M,

$$M \models \forall m\ \forall\alpha\ \forall\beta\ [\varphi(m,\alpha,\beta) \Leftrightarrow \neg\ \psi(m,\alpha,\beta)] .$$

We also have

$$m \in A \Leftrightarrow \exists\alpha(0)\,\forall\beta(0)\,\exists\alpha(1)\,\forall\beta(1)\cdots S(m,\alpha,\beta)\ ,$$

where this infinite string has the usual game interpretation. But then

$$m \in A \Leftrightarrow M \models \exists\alpha(0)\,\forall\beta(0)\,\exists\alpha(1)\,\forall\beta(1)\cdots \varphi(m,\alpha,\beta)\ .$$

This is because $M \models \underset{\sim}{\Delta}^1_{2n}$-DET and, since S has the same Δ^1_{2n} definition in M and in V, the concept "is a winning strategy for I (resp. II) in the game S" is a Π^1_{2n} notion, and so is absolute between M and V. Since M is an ω-model of analysis this implies that $A \in M$. ⊣

At this stage it is convenient to distinguish again cases as $n = 0$ or $n > 0$.

Case n = 0. By the lemma $\bigcap\{M : M$ is a Σ^1_{2n}-correct ω-model of $T\} \supseteq \mathfrak{D}^1_1 = Q_1$, so we are done.

Case n > 0. For each real α let

$$\delta^1_{2n+1}(\alpha) = \sup\{\xi : \xi \text{ is the length of a } \Delta^1_{2n+1}(\alpha) \text{ wellordering of } \omega\}\ .$$

By the proof of 7.2 it follows that if M is a Σ^1_{2n}-correct ω-model of Analysis + Dependent Choices + $\underset{\sim}{\Delta}^1_{2n}$-DET, then

(*) $$\alpha \in M \wedge \beta \in \Delta^1_{2n+1}(\alpha) \Rightarrow \beta \in M \wedge M \models \beta \in \Delta^1_{2n+1}(\alpha)\ .$$

From this we have

$$\alpha \in M \Rightarrow \exists w\,[w \in M \wedge w \in WO \wedge |w| \geq \delta^1_{2n+1}(\alpha)]\ ,$$

since by (*) for $\alpha \in M$

$$\delta^1_{2n+1}(\alpha) \leq (\delta^1_{2n+1}(\alpha))^M\ ,$$

where the superscript M denotes the relativization of the concept to M.

Let now M be an ω-model of T. We have to show that $Q_{2n+1} \subseteq M$. First note that all our work on Q_{2n+1} so far, although officially done within the theory ZF + DC + $\underset{\sim}{\Delta}^1_{2n}$-DET, is purely analytical, so it is actually developed within "Analysis + Dependent Choices + $\underset{\sim}{\Delta}^1_{2n}$-DET".

In particular,

$$M \models \text{"}Q_{2n+1} \text{ is countable"}\ .$$

So let $\alpha_0 \in M$ be such that

$$\forall\beta \in Q^M_{2n+1}(\beta \leq_T \alpha_0)\ .$$

Then in particular we have

$$\beta \in Q^M_{2n+1} \Rightarrow \delta^1_{2n+1}(\beta) \leq \delta^1_{2n+1}(\alpha_0) .$$

But Q_{2n+1} being Π^1_{2n+1}, we have by the Σ^1_{2n}-correctness of M

$$Q_{2n+1} \cap M \subseteq Q^M_{2n+1} ,$$

so

$$\beta \in Q_{2n+1} \cap M \Rightarrow \delta^1_{2n+1}(\beta) \leq \delta^1_{2n+1}(\alpha_0) ,$$

and by our preceding remarks there is $w_0 \in M$, $w_0 \in WO$, such that

$$\beta \in Q_{2n+1} \cap M \Rightarrow \delta^1_{2n+1}(\beta) \leq |w_0| .$$

If now $Q_{2n+1} \not\subseteq M$ let β_0 be a $<$-least real in $Q_{2n+1} - M$, where $<$ is a Δ^1_{2n+1}-good wellordering on C_{2n+1}. If $\sigma : C_{2n+1} \twoheadrightarrow \rho_{2n+1}$ is the associated norm, then by the proof of 2.3 β_0 is Δ^1_{2n+1} is $\sigma(\beta_0)$. But for each $\beta \in C_{2n+1}$

$$\sigma(\beta) < \delta^1_{2n+1}(\beta) ,$$

since $\{\beta' : \beta' \in C_{2n+1} \wedge \beta' < \beta\}$ is a countable $\Delta^1_{2n+1}(\beta)$ set, so it can be enumerated by a $\Delta^1_{2n+1}(\beta)$ real. Thus

$$\begin{aligned} \sigma(\beta_0) &\leq \sup\{\delta^1_{2n+1}(\beta) : \beta < \beta_0\} \\ &\leq \sup\{\delta^1_{2n+1}(\beta) : \beta \in Q_{2n+1} \cap M\} \\ &\leq |w_0| . \end{aligned}$$

So $\beta_0 \in \Delta^1_{2n+1}(w_0)$ and since M is downward closed under $\leq_{2n+1}$, $\beta_0 \in M$, a contradiction.

$\dashv$

<u>Remark</u>. For $n = 0$ the proof shows that the result holds for $T \supseteq$ Analysis only, since Dependent Choices is not needed in this case, and Analysis $\Rightarrow$ $(\underset{\sim}{\Delta}^1_0 \equiv)$ $\underset{\sim}{\Delta}^0_1$-DET.

§8. <u>Divergence between Δ^1_{2n+1}- and Q_{2n+1}-degrees for</u> $n > 0$.

We assume $\underset{\sim}{\Delta}^1_{2n}$-DET in this section.

Consider the structure of Δ^1_{2n+1}-degrees under the usual partial ordering

$$[\alpha]_{2n+1} \leq [\beta]_{2n+1} \Leftrightarrow \alpha \leq_{2n+1} \beta .$$

The jump operator in this structure is defined as usual: For each real α let W^α_{2n+1} be a complete $\Pi^1_{2n+1}(\alpha)$ subset of ω. Then the Δ^1_{2n+1}-jump of the Δ^1_{2n+1}-degree of α is given by

$$[\alpha]'_{2n+1} = [W^{\alpha}_{2n+1}]_{2n+1} .$$

For $n = 1$ we have that $\leq_1 = \leq^Q_1$ and $W^{\alpha}_1 \equiv_1 \mathcal{O}^{\alpha} \equiv_1 y^{\alpha}_1$. So the Δ^1_1- and Q_1-degrees and their jump operators coincide. However for $n > 0$ the two notions differ drastically. This can be seen immediately from the following result.

8.1 <u>Theorem</u> ($\underset{\sim}{\Delta}^1_{2n}$-DET). Assume $n > 0$. Then Q_{2n+1} is closed under the Δ^1_{2n+1}-jump, i.e. for each $\alpha \in Q_{2n+1}$ the complete $\Pi^1_{2n+1}(\alpha)$ subset of ω is also in Q_{2n+1}.

<u>Proof</u>. We will need the following lemma.

8.2 <u>Lemma</u> ($\underset{\sim}{\Delta}^1_{2n}$-DET). Let $n > 0$ and let $A \subseteq \omega$ be Π^1_{2n+1}. If $w \in WO$ and $|w| \geq \delta^1_{2n+1}$ $(= \delta^1_{2n+1}$ $(\lambda t\ .0))$, then $A \in \Delta^1_{2n+1}(w)$, i.e. A is Δ^1_{2n+1} in δ^1_{2n+1}.

<u>Proof</u>. Let $\varphi : A \twoheadrightarrow \lambda$ be a Π^1_{2n+1}-norm. Then $\lambda \leq \delta^1_{2n+1}$. Let $v \in WO$ be such that $|v| = \lambda$. It is enough to show that $A \leq_{2n+1} v$.

Let $\leq_{\varphi}$ be the prewellordering on A associated with φ. Put

$E(\lesssim) \Longleftrightarrow \underset{\sim}{\leq}$ is a prewellordering of a subset of ω which is a (not necessarily proper) end extension of $\leq_{\varphi}$.

$\Longleftrightarrow \underset{\sim}{\leq}$ is a prewellordering $\wedge\ \forall m \forall n\, [m \leq_{\varphi} n \Rightarrow m \lesssim n] \wedge \forall m \in A\, \forall n\, [n \leq_{\varphi} m \Longleftrightarrow n \lesssim m]$.

So $E \in \Sigma^1_{2n+1}$. Then we have

$$m \in A \Longleftrightarrow \exists \lesssim [E(\lesssim) \wedge \text{length}(\lesssim) = |v| \wedge m \lesssim m] ,$$

so $A \in \Sigma^1_{2n+1}(v)$, thus since $A \in \Pi^1_{2n+1}$, $A \in \Delta^1_{2n+1}(v)$. ⊣

By the relativized version of this lemma, if $\alpha \in Q_{2n+1}$ and $A \subseteq \omega$ is $\Pi^1_{2n+1}(\alpha)$, then A is Δ^1_{2n+1} in α and a countable ordinal, so by the Case $n > 0$ in the proof of 2.3 $A \in Q_{2n+1}(\alpha) \subseteq Q_{2n+1}$, and we are done. ⊣

Thus for $n > 0$ the picture is this:

Let d^{2n+1}_0, d^{2n+1}_1, ..., d^{2n+1}_{ξ}, ... $(\xi < \rho_{2n+1})$ be the increasing enumeration of the Δ^1_{2n+1}-degrees in C_{2n+1}. Those that belong to Q_{2n+1} form a proper initial segment, say

$$Q_{2n+1} = \bigcup_{\eta < \pi_{2n+1}} d^{2n+1}_{\eta} ;$$

Thus $d^{2n+1}_{\pi_{2n+1}} = [y^0_{2n+1}]_{2n+1}$. Now by the preceding result and the fact that for each $\xi < \rho_{2n+1}$,

$$d^{2n+1}_{\xi+1} = (d^{2n+1}_{\xi})' ,$$

we have that π_{2n+1} is a limit ordinal. In fact, we will see later that it is quite large. From §12 it follows that

$$\omega^{\omega} \cap L(Q_{2n+1}) = Q_{2n+1}$$

and

$$\pi_{2n+1} = \omega_1^{L(Q_{2n+1})} ;$$

also $L(Q_{2n+1}) \models$ CH. So Q_{2n+1} consists of a lot of Δ^1_{2n+1}-degrees bunched together, in fact in an effective sense "as many" as there are elements of Q_{2n+1} itself.

An immediate corollary of 8.1 is that the direct analog of the Kleene Basis Theorem for Σ^1_1 fails for $n > 0$, even when restricted to very large sets.

8.3 Corollary ($\underset{\sim}{\Delta}^1_{2n}$-DET). For $n > 0$ there is a co-countable Σ^1_{2n+1} set which contains no real Δ^1_{2n+1} in the complete Π^1_{2n+1} set of integers.

Proof. The set Q_{2n+1} is downward closed under $\leq_{2n+1}$ and contains the complete Π^1_{2n+1} set of integers, so the set $\omega^{\omega} - Q_{2n+1}$ is the required set. ⊣

Of course 5.6 provides the correct analog of the Kleene Basis Theorem for $n > 0$.

§9. Summary of characterizations of Q_{2n+1} for $n > 0$. Assume in this section $\underset{\sim}{\Delta}^1_{2n}$-DET.

Beyond the general descriptions of Q_{2n+1} (collected in the beginning of §7) that hold for every n, we have seen a number of special characterizations of Q_{2n+1} for $n > 0$, which we summarize below.

9.1 Theorem. For $n > 0$ the following are equivalent for each real α:

(i) $\alpha \in Q_{2n+1}$;

(ii) α is Δ^1_{2n+1} in an ordinal $< \omega_1$;

(iii) α is Δ^1_{2n+1} in an ordinal $< \delta^1_{2n+1}(\alpha)$;

(iv) $\alpha \in C_{2n+1}$ and α is Δ^1_{2n+1} in $\varphi(\alpha)$, where $\varphi : C_{2n+1} \twoheadrightarrow \rho_{2n+1}$ is the norm associated with a Δ^1_{2n+1}-good wellordering $<$ on C_{2n+1};

(v) For all β,

$$\delta^1_{2n+1}(\alpha) \leq \delta^1_{2n+1}(\beta) \Longleftrightarrow \alpha \leq_{2n+1} \beta ;$$

(vi) α belongs to every Σ^1_{2n}-correct standard model of T, where T is a recursive set of axioms in the language of ZF extending $ZF_N + DC + \underset{\sim}{\Delta}^1_{2n}$-DET, provided that T has such models. (Here N is a fixed large enough integer.)

Proof. (ii)-(iv) come from the case $n > 0$ of the proof of 2.3. Since for $\alpha \in C_{2n+1}$, $\varphi(\alpha) < \delta^1_{2n+1}(\alpha)$ the direction (iv) $\Rightarrow$ (v) is clear. Conversely, if (v) holds for α then if $w \in WO \wedge |w| = \delta^1_{2n+1}(\alpha)$ we have $\delta^1_{2n+1}(w) > |w| = \delta^1_{2n+1}(\alpha)$, so $w \geq_{2n+1} \alpha$, i.e. α is $\Delta^1_{2n+1}(w)$, and (ii) holds. Finally (vi) is proved exactly as in §7. $\dashv$

For $n = 0$, (ii) characterizes $C_2 = L \cap \omega^\omega$, while (iii)-(iv) characterize C_1 = the largest thin Π^1_1 set. If we let for each real α,

$$\lambda_{2n+1}(\alpha) = \sup\{\xi : \xi \text{ is the length of a } \Delta^1_{2n+1}(\alpha) \text{ prewellordering of } \omega^\omega\},$$

then $\lambda_1(\alpha) = \delta^1_1(\alpha)$, but for $n > 0$ $\lambda_{2n+1}(\alpha)$ is uncountable. It has been shown in Guaspari-Harrington [GH] that (v) with δ^1_{2n+1} replaced by λ_{2n+1} characterizes again C_{2n+1}.

For even $2n > 0$, the set of reals which are Δ^1_{2n} in a countable ordinal is exactly C_{2n}, the largest countable Σ^1_{2n} set (see [K2]). In this and several other respects that will become clear in the sequel (see e.g. §10), the sets Q_{2n+1} appear to be the analogs of the sets C_{2n} at the odd levels of the analytical hierarchy. Correspondingly the Q_{2n+1}-degrees are the analogs of the C_{2n}-degrees, where the C_{2n}-degrees are the equivalence classes of the equivalence relation

$$\alpha \in C_{2n}(\beta) \wedge \beta \in C_{2n}(\alpha) .$$

Note that for $2n = 2$, C_2-degrees = constructibility- (or L-) degrees.

§10. Explicit formulas for the reals in Q_{2n+1} for $n > 0$.

We start by recalling the definition of the game quantifier ∂. If $P(\alpha,x)$ is a pointset then we let

$$\partial\alpha\, P(\alpha,x) \iff \exists\alpha(0)\, \forall\alpha(1)\, \exists\alpha(2)\, \forall\alpha(3) \cdots P(\alpha,x)$$

$\iff$ Player I has a winning strategy in the game:

I	II	
$\alpha(0)$		I wins iff $P(\alpha,x)$.
	$\alpha(1)$	
$\alpha(2)$		
	$\alpha(3)$	
⋮		

If Γ is a pointclass, we let

$$\Game\Gamma = \{\Game\alpha\, P(\alpha,x) : P \in \Gamma\} .$$

Assume now that Γ is just closed under recursive substitutions and disjoint unions, where if $A, B \subseteq \mathcal{X}$ their disjoint union $A \dot{\cup} B$ is defined to be $\{(n,x) \in \omega \times \mathcal{X} : (n = 0 \wedge x \in A) \vee (n \neq 0 \wedge x \in B)\}$. Then it is easy to verify that $\Game\Gamma$ is closed under $\wedge$, $\vee$, $\exists n$, $\forall n$ and recursive substitutions. Let also inductively

$$\Game^{n+1}\Gamma = \Game(\Game^n\Gamma) ,$$

where we agree that

$$\Game^0\Gamma = \Gamma .$$

For instance $\Game\Sigma^0_1 = \Pi^1_1$, $\Game\Pi^1_{2n+1} = \Sigma^1_{2n+2}$ and assuming $\mathbf{\Delta}^1_{2n}$-DET ($\Leftrightarrow \mathbf{\Sigma}^1_{2n}$-DET) we have $\Game\Sigma^1_{2n} = \Pi^1_{2n+1}$. (For other elementary properties of $\Game$, see [Mo3], 6D.) We shall look however here also at pointclasses lying properly between successive levels of the analytical hierarchy, and the result of applying the game quantifier to them.

If ξ is a recursive ordinal let $\xi - \Pi^1_1$ be the pointclass consisting of all pointsets A of the following form: For some recursive sequence $\{A_\eta\}_{\eta<\xi}$ of Π^1_1 sets we have, letting $A_\xi = \emptyset$,

$$x \in A \Leftrightarrow \text{the least } \eta \leq \xi \text{ such that } x \notin A_\eta, \text{ is even} .$$

The corresponding boldface pointclass will be denoted as usual by $\xi - \mathbf{\Pi}^1_1$.

Combining results of Martin [Ma1], [Ma4], and Harrington [H1] we have the following important equivalence: For any $\xi < \omega^2$:

$$\mathbf{\Pi}^1_1\text{-DET} \Leftrightarrow 0^{\#} \text{ exists} \Leftrightarrow (\xi - \Pi^1_1)\text{-DET} .$$

The pointclasses we are interested in here are the $\omega.n - \Pi^1_1$ for $n = 1,2,\ldots$. For convenience abbreviate

$$M_n = \omega.n - \Pi^1_1; \qquad n = 1,2,\ldots .$$

Note that the pointclasses M_n are closed under recursive substitutions and disjoint unions.

The following result underlines the significance of the M_n's in the structure theory of projective sets.

10.1 Theorem (Steel [St2]). Every Σ^1_1 set A admits a very good (see 5.6) scale $\{\varphi_k\}$ such that for each k the norm φ_k is a M_{k+1}-norm (uniformly on k).

From this and the proof of the Transfer Theorem for scales under the game quantifier (see [Mo3], 6E.15) it follows immediately that we have

10.2 Corollary. Let $n \geq 1$. Then

(i) If $\bigcup_k \Game^{2n-2} \underset{\sim}{M}_k$ - DET holds, every $\underset{\sim}{\Pi}^1_{2n}$ set A admits an excellent scale $\{\varphi_k\}$ such that for each k φ_k is a $\Game^{2n-1} M_{k+1}$-norm (uniformly on k).

(ii) If $\underset{\sim}{\Delta}^1_{2n}$-DET holds, every Σ^1_{2n+1} set A admits an excellent scale $\{\varphi_k\}$ such that for each k φ_k is a $\Game^{2n} M_{k+1}$-norm (uniformly on k).

It has been proved in [KW] that for each $n \geq 1$,

$$\underset{\sim}{\Delta}^1_{2n}\text{-DET} \Longleftrightarrow \bigcup_k \Game^{2n-1} \underset{\sim}{M}_k\text{-DET} .$$

By analogy with the result that

$$\underset{\sim}{\Pi}^1_1\text{-DET} \Longleftrightarrow \bigcup_k \underset{\sim}{M}_k\text{-DET} ,$$

we raise the following question.

10.3 Open problem. Let $n \geq 2$ be odd. Is it true that

$$\underset{\sim}{\Pi}^1_n\text{-DET} \Longleftrightarrow \bigcup_k \Game^{n-1} \underset{\sim}{M}_k\text{-DET} ?$$

Closer to our context Martin has shown that the pointclasses M_k are the basis of an explicit characterization of the reals in C_{2n}, the largest countable Σ^1_{2n} set. We say as usual that a real $\alpha \in \omega^\omega$ is in a pointclass Γ iff $\{(n,m) : \alpha(n) = m\} \in \Gamma$.

10.4 Theorem (Martin [Ma3]). Let $n \geq 1$ and assume $\bigcup_k \Game^{2n-2} \underset{\sim}{M}_k$-DET. Then the following are equivalent for $\alpha \in \omega^\omega$:

(i) $\alpha \in C_{2n}$,

(ii) For some k, $\alpha \in \Game^{2n-1} M_k$.

In particular, assuming sharps exist, this characterizes the constructible reals $(C_2 = L \cap \omega^\omega)$ as exactly those in $\Game M_k$, for some $k = 1,2,\ldots$.

As it turns out the analog of 10.4 for odd levels > 1 characterizes Q_{2n+1}. This reinforces the feeling that this set is the analog of C_{2n} at odd levels > 1.

10.5 Theorem. Let $n \geq 1$ and assume $\underset{\sim}{\Delta}^1_{2n}$-DET. Then the following are equivalent for $\alpha \in \omega^\omega$:

(i) $\alpha \in Q_{2n+1}$,

(ii) For some k, $\alpha \in \Game^{2n} M_k$.

Proof. (i) $\Rightarrow$ (ii). Let P be a Π^1_{2n} nonempty set such that $Q_{2n+1} = \{\alpha : \forall\beta \in P(\alpha \leq_T \beta)\}$, using 3.3. Consider the following game G:

I	II	
s_0		$s_i \in \{0,1\}^{<\omega}$; $a_i \in \{0,1\}$; $\beta(i) \in \omega$.
	$a_0, \beta(0)$	Let $x = s_0{}^\frown a_0{}^\frown s_1{}^\frown a_1{}^\frown \cdots \in 2^\omega$.
s_1		I wins iff $(\beta \in P \Rightarrow x \leq_T \beta)$.
	$a_1, \beta(1)$	
	$\vdots$	

This game has payoff Σ^1_{2n} for Player I, so it is determined by Martin's theorem that $ZF + DC + \underset{\sim}{\Delta}^1_{2n}\text{-DET} \Rightarrow \underset{\sim}{\Sigma}^1_{2n}\text{-DET}$. (Martin's proof is unpublished, but see a different proof in [KS].) If I has a winning strategy τ, let β_0 be any element in P. If we look at all x's coming from runs of this game, in which I follows τ and II plays arbitrary $a_0, a_1, \ldots$ and $\beta = \beta_0$, then we obtain a perfect set of reals, all recursive in β_0, a contradiction.

So II has a winning strategy, say σ. Fix $\alpha \in Q_{2n+1}$. As in the argument for the $*$-games, call a sequence $s_0, a_0, \beta(0), \ldots, s_i, a_i, \beta(i)$ <u>good</u> if it has been played according to σ and $s_0{}^\frown a_0{}^\frown \cdots {}^\frown s_i{}^\frown a_i$ is an initial segment of α. If every good sequence has a proper good extension, we obtain a run of the game G in which II follows σ and $x = \alpha$, β are produced. By the winning condition for II, we have $\beta \in P$ and $\alpha \not\leq_T \beta$, which is absurd. So there is a maximal good sequence $s_0, a_0, \beta(0), \ldots, s_\ell, a_\ell, \beta(\ell)$. Say $s_0{}^\frown a_0{}^\frown \cdots {}^\frown s_\ell{}^\frown a_\ell = \alpha \upharpoonright t$. Then for $j > t$

$$\alpha(j) = 1 - \sigma(s_0, \ldots, s_\ell, (\alpha(t), \ldots, \alpha(j-1))) ,$$

thus $\alpha \leq_T \sigma \upharpoonright (\ell + 1)$, where by $\sigma \upharpoonright m$ we understand the restriction of σ to the first $m + 1$ moves of I. (Recall that $\sigma : X^{<\omega} \to Y$, where $X = \{0,1\}^{<\omega}$, $Y = \{0,1\} \times \omega$, so $\sigma \upharpoonright m = \sigma \upharpoonright \bigcup_{i \leq m} X^i$.)

So we only need to show that there is a strategy σ for II in G, with the property that for each m, $\sigma \upharpoonright m$ is in $\mathfrak{D}^{2n} M_k$, for some k. (It is easy to verify that the $\mathfrak{D}^{2n} M_k$ reals are downward closed under $\leq_T$, since for any pointclass Γ closed under recursive substitutions, and disjoint unions, $\mathfrak{D}\Gamma$ is closed under $\wedge$, $\vee$, $\exists n$, $\forall n$ and recursive substitutions.) But since the payoff set for II is Π^1_{2n} this is exactly what the Third Periodicity Theorem will give us when applied to 10.2. For convenience let us formulate explicitly the appropriate version of the Third Periodicity Theorem that is needed here.

10.6 <u>Theorem</u> (Moschovakis; see the proof of [Mo3], 6E.1). Let $A \subseteq \omega^\omega$ be a set which admits a very good (see 5.6) scale $\{\varphi_k\}$ such that each φ_k is a Γ_k-norm, where $\{\Gamma_k\}$ is a sequence of pointclasses with the following properties:

(i) Each Γ_k is closed under recursive substitutions and disjoint unions,

(ii) $\Gamma_0 \subseteq \Gamma_1 \subseteq \Gamma_2 \subseteq \cdots$

If we assume $\bigcup_k \Gamma_k$-DET, and Player I has a winning strategy in the game with payoff A, then he has such a winning strategy τ such that for each m, $\tau \restriction m \in \Game\, \Gamma_m$.

The proof is immediate from the proof of 6E.1 in [Mo3].

So our proof of (i) $\Rightarrow$ (ii) is now complete.

(ii) $\Rightarrow$ (i). Exactly as in the corresponding part of the proof of 10.4 (see [Ma3]) we see that if $\alpha \in \Game^{2n} M_k$ then α belongs to a thin Π^1_{2n+1} set, thus to C_{2n+1}. So

$$A = \{\alpha : \exists k\, (\alpha \in \Game^{2n} M_k)\} \subseteq C_{2n+1} .$$

To conclude that actually $A \subseteq Q_{2n+1}$, it is enough by 6.4 to show that A is downward closed under $\leq_{2n+1}$. For the reader familiar with the basic theory of Spector pointclasses ([Mo3]; Chapter 4 and 6D.4) this is an immediate consequence of the following two facts:

(*) Each $\Game^{2n} M_k$ is a Spector pointclass containing Σ^1_{2n+1}.

(**) If Γ is a Spector pointclass, the relation $\alpha \in \Delta(\beta)$ is transitive.

Alternatively we can give the following direct argument (which is actually part of the proof of (*)).

Let $\alpha_0 \in \Game^{2n} M_k$ and let $\beta_0 \in 2^\omega$, $\beta_0 \leq_{2n+1} \alpha_0$. We will show that $\beta_0 \in \Game^{2n} M_k$. Let $R \in \Pi^1_{2n}$ be such that

$$\beta_0(m) = 0 \Longleftrightarrow \Game x\, R(m,x,\alpha_0) .$$

Let also S be in $\Game^{2n-1} M_k$ such that

$$\alpha_0(t) = \ell \Longleftrightarrow \Game y\, S(t,\ell,y) .$$

Consider now the following game G_m for each $m \in \omega$, where we let $\langle a,b\rangle'$ be a 1-1 recursive correspondence between $\omega \times \omega$ and $\omega - \{0\}$

I	II	
$x(0)$		I wins iff
	$x(1)$	1. $R(m,x,\alpha)$ and
$\alpha(0)$		2. Either γ is constantly 0 or else if
	$\gamma(0)$	j is least such that $\gamma(j) \neq 0$ and
$y(0)$		$\gamma(j) = \langle t,\ell\rangle'$, then either
	$y(1)$	
$x(2)$		$\alpha(t) \neq \ell$
	$x(3)$	or else
$\alpha(1)$		$S(t,\ell,\lambda i.y(2j+i))$.
	$\gamma(1)$	
$y(2)$		
	$y(3)$	
	$\vdots$	

Clearly the payoff set for this game G_m is $\partial^{2n-1}M_k$, uniformly on m. So it is enough to check that $\beta_0(m) = 0 \Longleftrightarrow$ I has a winning strategy in G_m.

($\Rightarrow$): Assume $\beta_0(m) = 0$. Then I plays as follows to win G_m. He lets $\alpha = \alpha_0$ and he plays x so that he wins $R(m,x,\alpha_0)$. As long as II plays $\gamma(j) = 0$ he plays for y arbitrarily. Otherwise if j_0 is least such that $\gamma(j_0) \neq 0$ and $\gamma(j_0) = \langle t_0,\ell_0\rangle'$, either $\alpha_0(t_0) \neq \ell_0$, in which case he again plays from then on y arbitrarily, or else $\alpha_0(t_0) = \ell_0$, in which case he plays from then on for y, i.e. $y(2j_0), y(2j_0+2), \ldots$ so that he wins $S(t,\ell_0,y')$, where $y' = \lambda i.y(2j_0+i)$.

($\Leftarrow$): Assume I has a winning strategy σ in G_m. Consider runs of the game G_m in which I follows σ and II plays for γ and y the constant 0. If we can show that for any such run $\alpha = \alpha_0$, then the x part of this game gives us a winning strategy for I in $R(m,x,\alpha_0)$, so $\beta_0(m) = 0$ and we are done.

If this assertion fails, then for some run, call it r_1, an α is produced such that $\alpha \neq \alpha_0$. Say m is least such that $\alpha(m) \neq \alpha_0(m)$. Then $\alpha(m) = p \neq p' = \alpha_0(m)$. Let $j > m$ and consider a run, call it r_2, of the game in which I follows σ and II plays as in the run r_1 up to the point where his turn is to play $\gamma(j)$. Instead of 0 then he plays $\gamma(j) = \langle m,p\rangle'$. Since $\alpha_0(m) \neq p$ II has a winning strategy in $S(m,p,y)$, so he follows in r_2 this run for his y part from then on, and for γ he plays anything from then on. Let α_1,γ_1,y_1 be the appropriate reals produced in r_2. Then as $j > m$, $\alpha_1(m) = \alpha(m) = p$ and $\gamma_1(j) = \langle m,p\rangle'$, where j is the least integer at which $\gamma_1(j) \neq 0$. Moreover $\neg S(m,p,\lambda i y_1(2j+i))$. Thus the second of the winning conditions of I fails, i.e. II won a run of the game in which I followed σ, a contradiction. $\dashv$

§11. Explicit formulas for y^0_{2n+1} for $n > 0$: An open problem. For each $k \geq 1$, $m \in \omega$ the pointclass $\mathfrak{D}^m M_k$ is clearly ω-parametrized (i.e. for each space $\mathfrak{X}$ there is $W \subseteq \omega \times \mathfrak{X}$ universal for the $\mathfrak{D}^m M_k$ subsets of $\mathfrak{X}$), and we can pick a canonical sequence U^m_k of subsets of ω which are universal for $\mathfrak{D}^m M_k$, i.e. for $A \subseteq \omega$,

$$A \in \mathfrak{D}^m M_k \Longleftrightarrow \exists e \, \forall t \, (t \in A \Longleftrightarrow \langle e,t \rangle \in U^m_k) \, .$$

Let for each $n \geq 2$

$$Y^0_n = \{\langle k,t \rangle : t \in U^{n-1}_k\}$$

be the subset of ω canonically coding the sequence $U^{n-1}_1, U^{n-1}_2, \ldots$. Under the appropriate hypotheses of 10.4 and 10.5 it is clear that for each $n \geq 1$

(1) $Y^0_{2n} \notin C_{2n}$,

(2) For each k, $(Y^0_{2n})_k = \{t : \langle k,t \rangle \in Y^0_{2n}\} \in C_{2n}$,

(3) $\alpha \in C_{2n} \Longleftrightarrow \exists k,e \, (\{\langle p,q \rangle : \alpha(p) = q\} = ((Y^0_{2n})_k)_e)$.

Thus Y^0_{2n} is a real coding a canonical decomposition of C_{2n} in an ω sequence. Similarly for odd levels we have (for each $n \geq 1$)

(1) $Y^0_{2n+1} \notin Q_{2n+1}$,

(2) For each k, $(Y^0_{2n+1})_k \in Q_{2n+1}$,

(3) $\alpha \in Q_{2n+1} \Longleftrightarrow \exists k \, \exists e \, (\{\langle p,q \rangle : \alpha(p) = q\} = ((Y^0_{2n+1})_k)_e)$.

Clearly Y^0_{2n} appears as the "least canonical real" not in C_{2n} and similarly for Y^0_{2n+1} and Q_{2n+1}.

One way to make this precise is using the jump operators of Steel [Stl]. First let for each real α, Y^α_n be the relativized version of Y^0_n to α (thus $Y^0_n = Y^{\lambda t.0}_n$). Let also $U^m_k(\alpha)$ be the relativized version of U^m_k. Let for each Turing degree $d = [\alpha]_T$,

$$u^m_k(d) = [U^m_k(\alpha)]_T$$

$$j^C_{2n}(d) = [Y^\alpha_{2n}]_T, \qquad j^Q_{2n+1}(d) = [Y^\alpha_{2n+1}]_T \, .$$

Then u^m_k, j^C_{2n}, j^Q_{2n+1} are jump operators and since Y^α_n is the code of the canonical sequence $U^n_1(\alpha), U^n_2(\alpha), \ldots$ we have by Lemma 4 in [Stl] that j^C_{2n} is the lub of $u^{2n}_1, u^{2n}_2, \ldots$ and j^Q_{2n+1} the lub of $u^{2n+1}_1, u^{2n+1}_2, \ldots$ in the prewellordering of jump operators, granting AD. From this it immediately follows that $j^C_{2n}(d)$ is the least jump operator f such that $f(d) \notin C_{2n}(d)$ on a cone of Turing degrees d, and similarly for $j^Q_{2n+1}(d)$ and $Q_{2n+1}(d)$.

By now it is clear that Y^0_{2n+1} ought to have the same Δ^1_{2n+1}-degree as y^0_{2n+1} and thus could serve as a most canonical representative of (the Δ^1_{2n+1}-degree of) the first non-trivial Π^1_{2n+1}-singleton. Although this is undoubtedly true, we can only prove a relativized version at this stage.

11.1 <u>Theorem</u>. Assume AD. For some real α_0 and all $\alpha \geq_T \alpha_0$,

$$y^{\alpha}_{2n+1} \equiv_T Y^{\alpha}_{2n+1} ,$$

for all $n \geq 1$.

<u>Proof</u>. Both $[\alpha]_T \longmapsto [y^{\alpha}_{2n+1}]_T$ and $j^Q_{2n+1}([\alpha]_T) = [Y^{\alpha}_{2n+1}]_T$ are jump operators and $y^{\alpha}_{2n+1} \notin Q_{2n+1}(\alpha)$, so by the preceding remarks $[y^{\alpha}_{2n+1}]_T \geq [Y^{\alpha}_{2n+1}]_T$ for all $\alpha \geq_T \alpha_0$, where α_0 is some fixed real.

We will see now that if in the definition of y^{α}_{2n+1} we choose the scale used in the definition appropriately then we can make sure that

$$y^{\alpha}_{2n+1} \leq_T Y^{\alpha}_{2n+1}, \quad \text{for all } \alpha .$$

Note that although $[y^{\alpha}_{2n+1}]_T$ is intrinsically determined, the actual y^{α}_{2n+1} we use depends on the choice of a Δ^1_{2n+1}-scale $\{\psi_m\}$ on $B = PC = \{(y,\alpha) : y$ is an α-prechain$\}$ as in the proof of 6.1, such that each ψ_m has support m. By 10.2 it is easy to see that $\{\psi_m\}$ can be chosen to be such that for each m ψ_m is a $\Game^{2n-1}M_{m+1}$-norm, uniformly on m. Then in the notation of the proof of 6.1 the scale $\{\varphi_m\}$ on $A = B(\alpha)$ is such that each φ_m is a $\Game^{2n-1}M_{m+1}(\alpha)$-norm uniformly on m, α. Now the leftmost real $y_{A,\overline{\varphi}} = y^{\alpha}_{2n+1}$ can be computed as follows:

To each finite sequence s from ω we will associate a $\Game^{2n-1}M_{m+1}(\alpha)$ game $G(s)$, with $m = \text{length}(s)$, such that

$$\text{I wins } G(s) \Longleftrightarrow s \subseteq y_{A,\overline{\varphi}} .$$

Moreover the $\Game^{2n-1}M_{m+1}(\alpha)$ code of $G(s)$ will be a recursive function of s. It will be then clear that $y^{\alpha}_{2n+1} = y_{A,\overline{\varphi}} \leq_T Y^{\alpha}_{2n+1}$.

The game $G(s)$ is as follows:

I	II
$\alpha(0)$	
	$\beta(0)$
$\alpha(1)$	
	$\beta(1)$
$\vdots$	

I wins iff

$$(1)\ \alpha \in A \wedge \ (2)\ \alpha \upharpoonright m = s \wedge$$
$$(3)\ (\beta \upharpoonright m \neq s \Rightarrow \alpha <^*_{\varphi_m} \beta) .$$

Since $\{\varphi_m\}$ is an excellent scale, all $\alpha \in A$ of minimal φ_m norm agree with $y_{A,\overline{\varphi}}$ on the first m values, so our claim is clear. ⊣

Let us state now explicitly the basic conjecture here.

11.2 Conjecture. Assume $n \geq 1$ and $\underset{\sim}{\Delta}^1_{2n}$-DET. Then $Y^\alpha_{2n+1} \equiv_{2n+1} y^\alpha_{2n+1}$, for every $\alpha \in \omega^\omega$.

There is also another way to characterize the real Y^0_2 given by the following result.

11.3 Theorem (Martin [Ma3]). Assume Σ^1_1-DET (i.e. $0^\#$ exists). Then

$$Y^0_2 \equiv_T 0^\# .$$

The preceding theorem suggests that for $n \geq 1$ and up to some degree equivalence Y^0_{2n} (Y^0_{2n+1}) should be the "sharp" of some inner model ("higher analog of L") whose reals are $C_{2n}(Q_{2n+1})$. Not much is known however about this very interesting (although a bit vague) open problem; see however the remark at the end of §13. One formal property of $0^\#$ however (that of being a Π^1_2 singleton) extends to all Y^0_n for $n \geq 2$.

11.4 Theorem. Let $n \geq 2$ and assume $\bigcup_k \Game^{n-2}\underset{\sim}{M}_k$-DET. Then Y^0_n is a Π^1_n-singleton.

Proof. We will need the following strengthening of 10.2.

11.5 Theorem (Steel [St2]). Assume $\bigcup_k \Game^{m-1}\underset{\sim}{M}_k$-DET. Then for each $k \geq 1$, every set in $\Game^m M_k$ admits a very good scale $\{\varphi_i\}$ such that each φ_i is a $\Game^m M_{k+i+1}$-norm (uniformly in m, k).

This fact and the version of the Third Periodicity Theorem given in 10.6 implies that every game in $\Game^{n-2}M_k$ which Player I wins has a strategy recursive in Y^0_n. Note also that for every set $A \in \Game^{n-2}M_k$, its complement $\omega^\omega - A$ is in $\Game^{n-2}M_{k+1}$. Thus every game in $\Game^{n-2}M_k$ which Player II wins has a strategy recursive in Y^0_n as well. Denoting then by $\sigma * \beta$ the run of a game in which Player I follows a strategy σ and II plays β, and by $\alpha * \tau$ the analogous run with the roles of I and II interchanged, we have, letting for each m $U^m_k = \{t : \Game x R^{m-1}_k(t,x)\}$, $R^{m-1}_k \in \Game^{m-1}M_k$:

$$y = Y_n^0 \Longleftrightarrow y \subseteq \omega \wedge (y)_0 = \emptyset \wedge \forall k \geq 1\ ((y)_k = U_k^{n-1})$$

$$\Longleftrightarrow y \subseteq \omega \wedge (y)_0 = \emptyset \wedge \forall k \geq 1$$

$$[\forall t\ (t \in (y)_k \Rightarrow \text{Э} x R_k^{n-2}(t,x)) \wedge$$

$$\forall t\ (t \notin (y)_k \Rightarrow \neg\, \text{Э} x R_k^{n-2}(t,x))]$$

$$\Longleftrightarrow y \subseteq \omega \wedge (y)_0 = \emptyset \wedge \forall k \geq 1$$

$$[\forall t (t \in (y)_k \Rightarrow \exists \sigma \leq_T y\ \forall \beta\ R_k^{n-2}(t, \sigma * \beta)) \wedge$$

$$\forall t\ (t \notin (y)_k \Rightarrow \exists \tau \leq_T y\ \forall \alpha \neg R_k^{n-2}(t, \alpha * \tau))] .$$

Since $\text{Э}^{n-2} M_k \subseteq \Pi_n^1$ it follows that $\{Y_n^0\} \in \Pi_n^1$, and the proof is complete. ⊣

§12. The model $L(Q_{2n+1})$ for $n > 0$. In this and the next section we shall study some interesting inner models associated with Q_{2n+1} for $n > 0$. We start with the smallest one. In this section we assume again $\underset{\sim}{\Delta}_{2n}^1$-DET.

12.1 Definition. Fix $n > 0$. The model $L(Q_{2n+1})$ is the smallest inner model of ZF containing Q_{2n+1} (as an element).

We summarize the basic properties of $L(Q_{2n+1})$ in the next theorem.

12.2 Theorem ($\underset{\sim}{\Delta}_{2n}^1$-DET). Let $n > 0$ and put $L^{2n+1} = L(Q_{2n+1})$. Then

(i) $\omega^\omega \cap L^{2n+1} = Q_{2n+1}$,

(ii) L^{2n+1} is downward closed under $\leq_{2n+1}$, so that L^{2n+1} is Σ_{2n}^1-correct, but L^{2n+1} is not Σ_{2n+1}^1-correct.

(iii) To each Π_{2n+1}^1 formula $\varphi(\alpha)$ we can effectively assign a Σ_{2n+1}^1 formula $\varphi^*(\alpha)$ such that for $\alpha \in \omega^\omega$:

$$\alpha \in L^{2n+1} \Rightarrow [\varphi(\alpha) \Longleftrightarrow L^{2n+1} \models \varphi^*(\alpha)] .$$

Similarly interchanging Π_{2n+1}^1 and Σ_{2n+1}^1. In particular, if $A \subseteq \omega^\omega$ is Π_{2n+1}^1 then $A \cap Q_{2n+1} \in L^{2n+1}$ and $L^{2n+1} \models$ "$A \cap Q_{2n+1}$ is Σ_{2n+1}^1". Dually for $A \in \Sigma_{2n+1}^1$.

(iv) $L^{2n+1} \models$ AC + GCH.

(v) $L^{2n+1} \models$ "There is a Δ_{2n+1}^1-good wellordering of ω^ω".

(vi) $L^{2n+1} \models$ "Every provable in ZFC $\underset{\sim}{\Delta}_{2n}^1$ game is determined, but there is an undetermined Δ_{2n}^1 game".

<u>Proof</u>. (i) Let α be a real such that $Q_{2n+1} \subseteq \{(\alpha)_k : k \in \omega\}$. Let $A_\alpha = \{k : (\alpha)_k \in Q_{2n+1}\}$. Then $A_\alpha \in \Pi^1_{2n+1}(\alpha)$. Clearly $Q_{2n+1} \in L[\alpha, A_\alpha]$, thus $L^{2n+1} \subseteq L[\alpha, A_\alpha]$, so

$$\gamma \in L^{2n+1} \Rightarrow \gamma \leq_3 \langle \alpha, A_\alpha \rangle .$$

Put now

$$S_0 = \{\alpha : \forall \beta \in Q_{2n+1} \exists k \, (\beta = (\alpha)_k)\}$$

and

$$S = \{\varepsilon : (\varepsilon)_0 \in S_0 \wedge (\varepsilon)_1 \in WO \wedge |(\varepsilon)_1| \geq \delta^1_{2n+1}((\varepsilon)_0)\} .$$

Clearly S_0, $S \in \Sigma^1_{2n+1}$. If $\varepsilon \in S$ then $\alpha = (\varepsilon)_0 \in S_0$ and by 8.2 $\langle (\varepsilon)_0, (\varepsilon)_1 \rangle \geq_{2n+1} A_\alpha$ (where A_α is as above). So for all $\varepsilon \in S$,

$$\gamma \in L^{2n+1} \Rightarrow \gamma \leq_{2n+1} \varepsilon$$

i.e. $L^{2n+1} \cap \omega^\omega \subseteq \mathrm{Hull}_{2n+1}(S) \subseteq Q_{2n+1}$ and we are done.

(ii) Obvious from (i) and the fact that Q_{2n+1} is downward closed under $\leq_{2n+1}$. To see that L^{2n+1} is not Σ^1_{2n+1}-correct notice that if the opposite were true every nonempty Σ^1_{2n+1} set would have a member in L^{2n+1}, contradicting the fact that $\omega^\omega - Q_{2n+1} = \omega^\omega - (\omega^\omega \cap L^{2n+1})$ is Σ^1_{2n+1}.

(iii) Let $\varphi(\alpha)$ be a Π^1_{2n+1} formula. By the Spector-Gandy Theorem for Π^1_{2n+1} we can find a Π^1_{2n} formula $\varphi'(\alpha,\beta)$ such that

$$\varphi(\alpha) \Longleftrightarrow \exists \beta \in \Delta^1_{2n+1}(\alpha) \, \varphi'(\alpha,\beta) .$$

But by the reflection property 5.4 (relativized) we have also

$$\varphi(\alpha) \Longleftrightarrow \exists \beta \in Q_{2n+1}(\alpha) \, \varphi'(\alpha,\beta) .$$

Put

$$\varphi^*(\alpha) \Longleftrightarrow \exists \beta \, \varphi'(\alpha,\beta) .$$

Clearly φ^* is a Σ^1_{2n+1} formula and for all $\alpha \in Q_{2n+1} = \omega^\omega \cap L^{2n+1}$, we have

$$\begin{aligned} \varphi(\alpha) &\Longleftrightarrow \exists \beta \in Q_{2n+1}(\alpha) \varphi'(\alpha,\beta) \\ &\Longleftrightarrow \exists \beta \in Q_{2n+1} \varphi'(\alpha,\beta) \\ &\Longleftrightarrow L^{2n+1} \models \varphi^*(\alpha) . \end{aligned}$$

by the Σ^1_{2n}-correctness of L^{2n+1}. By taking negations we obtain the corresponding result for Σ^1_{2n+1} formulas.

(iv), (v) Let $<$ be a Δ^1_{2n+1}-good wellordering on C_{2n+1}. Clearly $< \in \Pi^1_{2n+1}$, so $<^* = < \cap (Q_{2n+1} \times Q_{2n+1}) \in L^{2n+1}$ by (iii). Clearly Q_{2n+1} is an initial segment of $<$, so that $<^*$ is a Δ^1_{2n+1}-good wellordering on Q_{2n+1}, thus an easy checking shows that

$$L^{2n+1} \models \text{"}<^* \text{ is a } \Delta^1_{2n+1}\text{-good wellordering of } \omega^{\omega}\text{"} .$$

So (v) holds, and then by standard set theory (iv) holds as well.

(vi) Since L^{2n+1} satisfies that there exists a Δ^1_{2n+1}-good wellordering of the reals, it follows immediately that the prewellordering property holds for Σ^1_{2n+1} in L^{2n+1}. On the other hand the First Periodicity Theorem shows that from $ZF + DC + \underset{\sim}{\Delta}^1_{2n-2}\text{-DET} + \Delta^1_{2n}\text{-DET}$ we can deduce that the class of Π^1_{2n+1} sets <u>of numbers</u> has the prewellordering property. But $\underset{\sim}{\Delta}^1_{2n-2}$-DET is a Π^1_{2n} assertion that holds in V, so by the Σ^1_{2n}-correctness of L^{2n+1} it also holds in L^{2n+1}. Therefore

$$L^{2n+1} \models \neg\, \Delta^1_{2n}\text{-DET} .$$

Let us demonstrate now that L^{2n+1} satisfies provable in ZFC $\underset{\sim}{\Delta}^1_{2n}$-DET. Working in L^{2n+1}, we have that if $A \subseteq \omega^\omega$ is provable in ZFC $\underset{\sim}{\Delta}^1_{2n}$, then there are Σ^1_{2n} formulas φ, ψ such that $ZFC \vdash \forall\alpha\, \forall\beta\, (\varphi(\alpha,\beta) \Leftrightarrow \neg\, \psi(\alpha,\beta)]$, and there is a β_0 such that $\alpha = \{\alpha : \varphi(\alpha,\beta_0)\}$. Now we must also have $ZFC \vdash \forall\alpha\, \forall\beta\, (\varphi(\alpha,\beta) \Leftrightarrow \neg\, \psi(\alpha,\beta)]$ in V. Note now the following:

<u>Lemma</u>. Let $n > 0$ and assume $ZF + DC + \underset{\sim}{\Delta}^1_{2n}$-DET. If Φ is a Π^1_{2n+2} sentence then

$$(ZFC \vdash \Phi) \Rightarrow \Phi .$$

<u>Proof</u>. Say $ZFC \vdash \forall\alpha\, \vartheta(\alpha)$, where ϑ is Σ^1_{2n+1}, but for some α_0, $\neg\, \vartheta(\alpha_0)$. Let $M = L(C_{2n}(\alpha_0))$. Since $C_{2n}(\alpha_0)$ is countable, M has a sharp. So from the fact that $ZFC \vdash \forall\alpha\, \vartheta(\alpha)$, we can infer that $M \models \vartheta(\alpha_0)$, so by the Σ^1_{2n}-correctness of M, $\vartheta(\alpha_0)$ holds, a contradiction. ⊣

So we conclude that $\forall\alpha\, \forall\beta\, (\varphi(\alpha,\beta) \Leftrightarrow \neg\, \psi(\alpha,\beta))$ also holds in the real world. Working in the real world now, let $A' = \{\alpha : \varphi(\alpha,\beta_0)\}$. This is $\Delta^1_{2n}(\beta_0)$ and so by the Third Periodicity Theorem there is a $\Delta^1_{2n+1}(\beta_0)$ winning strategy σ in this game, so $\sigma \in L^{2n+1}$. But the statement that σ is a winning strategy (for the appropriate player) is Π^1_{2n+1} and so it holds in L^{2n+1} by the Σ^1_{2n}-correctness of L^{2n+1}. Thus $L^{2n+1} \models$ "A is determined". ⊣

From (iii) of the preceding theorem it follows that although we do not have (by (ii)) the analog of Shoenfield absoluteness for L^{2n+1}, we have a kind of "dual absoluteness", which allows us to know living within L^{2n+1}, whether Σ^1_{2n+1} and Π^1_{2n+1} statements are true in the universe. It also follows that for all Π^1_{2n+1} or Σ^1_{2n+1} $A \subseteq \omega^\omega$ we have

$$L[A] \subseteq L^{2n+1} ,$$

where $L[A]$ is the smallest inner model of ZFC M for which $M \cap A \in M$. This implies the failure of another possible generalization of the Kleene Basis Theorem to higher levels Σ^1_{2n+1}, for $n > 0$ (see 8.3). Namely, there is a cocountable Σ^1_{2n+1} set of reals which contains no real Kleene recursive in the complete Π^1_{2n+1} set of reals W_{2n+1} (viewed as a type 2 object). This is because all such reals belong to $L[W_{2n+1}]$.

From the proof of (vi) we see that we could replace "provable in ZFC" by "T-provable", where T is any recursively axiomatizable theory with the property that statements in analysis that are provable in T are true in the universe (e.g. $T = ZFC_N + \underset{\sim}{\Delta}^1_{2n}$-DET). From (vi) it also follows that provable (in these T's) $\underset{\sim}{\Delta}^1_{2n}$-determinacy does not imply Δ^1_{2n}-DET, unless $\underset{\sim}{\Delta}^1_{2n}$-DET is inconsistent.

Although in L^{2n+1} we do not have for instance measurable cardinals, we can remedy this easily by considering inner models like $L_\mu(Q_{2n+1})$ for which we can easily prove analogous theorems.

One drawback of the inner model $L(Q_{2n+1})$ is its behavior under relativization. For each inner model M of ZF and each real α let $M[\alpha]$ be the smallest inner model of ZF containing M and α. Let L^{2n+1} be the relativization of L^{2n+1} to α i.e.

$$L^{2n+1}_\alpha = L(Q_{2n+1}(\alpha)) .$$

Then in general we have $L^{2n+1}_\alpha \neq L^{2n+1}[\alpha]$. An example is $\alpha = y^0_{2n+1}$, because $L^{2n+1}[y^0_{2n+1}] \subseteq L[y^0_{2n+1}]$. On the other hand however for various types of generic α's, we do have $L^{2n+1}_\alpha = L^{2n+1}[\alpha]$.

In the next section we shall see another model whose set of reals is Q_{2n+1} and which properly relativizes as well, so it is much more satisfactory than the minimal model $L(Q_{2n+1})$.

§13. The models $HOD^{L[\alpha]}$ and generalizations. We will present first some new interesting models whose reals are exactly Q_3, and then discuss the generalizations to higher Q_{2n+1}.

For each real α consider the inner model $L[\alpha]$ of sets constructible from α and let

$$HOD^{L[\alpha]}$$

be the inner model of all sets which are hereditarily ordinal definable in $L[\alpha]$. Woodin was probably the first one to focus attention on these models and use their properties, for instance the fact that $L[\alpha]$ is a generic extension of $HOD^{L[\alpha]}$ (Vopenka's Theorem; see [J], p. 293), in unpublished work of his concerning the relationship between core models for large cardinals and various forms of projective determinacy. One of his early observations was that the set $\omega^\omega \cap HOD^{L[\alpha]}$ stabilizes on a cone of constructiblity degrees, and it was noticed by him and Kechris that this set is included in Q_3.

Later Martin independently "rediscovered" $HOD^{L[\alpha]}$ and proved the exact computation below.

13.1 Theorem (Martin). Assume $\underset{\sim}{\Delta}^1_2$-DET. Then there is a real z_0 such that

$$z_0 \in L[\alpha] \Rightarrow \omega^\omega \cap HOD^{L[\alpha]} = Q_3 .$$

Proof. Let $\beta \in Q_3$. Let $\varphi : C_3 \twoheadrightarrow \rho_3$ be the norm associated with a Δ^1_3-good wellordering $<$ on C_3. Then if $\varphi(\beta) = \xi$ we have for all $w \in WO$, $|w| = \xi$:

$$\beta(n) = m \Longleftrightarrow \forall \gamma \in Q_3 \ (\varphi(\gamma) = |w| \Rightarrow \gamma(n) = m)$$
$$\Longleftrightarrow \exists \delta \ P(n,m,\delta,w) ,$$

where $P \in \Pi^1_2$. Fix now some $w_0 \in WO$, $|w_0| = \xi$ and for each n, m with $\beta(n) = m$ let $\delta_{n,m}$ be a witness to $P(n,m,\delta,w)$. Let $z_1 = \langle w_0, (n,m) \longmapsto \delta_{n,m} \rangle$. Then if $z_1 \in L[\alpha]$ we have

$$\beta(n) = m \Longleftrightarrow L[\alpha] \models \exists w \exists \delta \ [w \in WO \wedge |w| = \xi \wedge P(n,m,\delta,w)] ,$$

so $\beta \in HOD^{L[\alpha]}$. Since Q_3 is countable, it follows that there is z_0' such that

$$z_0' \in L[\alpha] \Rightarrow Q_3 \subseteq HOD^{L[\alpha]} .$$

For each real α and each $\omega < \xi < \omega_1$ let $<_{\xi,\alpha}$ be a canonical wellordering of the reals which are OD in $L_\xi[\alpha]$. Note that $<_{\xi,\alpha}$ depends only on the Turing degree of α. Define the following wellordering $<_\alpha$ on the set of all reals H^α which are OD in some $L_\xi[\alpha]$ for $\omega < \xi < \omega_1$: If $x, y \in H^\alpha$ then

$$x <_\alpha y \iff [\text{least } \xi \text{ such that } x \text{ is OD in } L_\xi[\alpha] < \text{least } \xi \text{ such that } y \text{ is OD in } L_\xi[\alpha]] \vee [\text{the least such ordinal, say } \xi_0, \text{ is the same for both } x,y \text{ and } x <_{\xi_0,\alpha} y] .$$

Let θ^α be the order type of this wellordering. Clearly $\theta^\alpha \leq \omega_1^{L[\alpha]} < \omega_1$. Note also that $HOD^{L[\alpha]} \cap \omega^\omega \subseteq H^\alpha$. Moreover $<_\alpha$ depends only on the Turing degree of α.

For $\xi < \theta^\alpha$ let

$$x_\xi^\alpha = \text{the } \xi^{\underline{th}} \text{ real in } <_\alpha .$$

Again x_ξ^α depends only on the Turing degree of α. The following is an easy computation: If $\xi < \omega_1$, the set

$$P_\xi(\alpha) \iff \xi < \theta^\alpha$$

in $\underset{\sim}{\Sigma}_2^1$. Thus by $\underset{\sim}{\Sigma}_2^1$-DET ($\iff \underset{\sim}{\Delta}_2^1$-DET) for each ξ, either P_ξ or its complement contains all the reals in a cone of Turing degrees. Let

$$\begin{aligned} A &= \{\xi : \exists\alpha_0 \, \forall\alpha \geq_T \alpha_0 \; P_\xi(\alpha)\} \\ &= \{\xi : \exists\alpha_0 \, \forall\alpha \geq_T \alpha_0 \; (\xi < \theta^\alpha)\} . \end{aligned}$$

Clearly A is an initial segment of ω_1. Now if $\xi \in A$ we also claim that for all α's in a cone of Turing degrees we have that

$$x_\xi^\alpha = x_\xi \quad \text{is fixed ,}$$

where

$$x_\xi(n) = m \iff \exists\alpha_0 \, \forall\alpha \geq_T \alpha_0 \; (x_\xi^\alpha(n) = m) .$$

To see this notice that for each ξ the relation

$$R_\xi(\alpha,n,m) \iff \xi < \theta^\alpha \wedge x_\xi^\alpha(n) = m$$

in $\underset{\sim}{\Sigma}_2^1$ and so for each fixed ξ,n,m either $\{\alpha : R_\xi(\alpha,n,m)\}$ or its complement contain all reals in a cone of Turing degrees, and thus for some β_0 of sufficiently high Turing degree and all n,m, if $\beta_0 \leq_T \alpha$ we have

$$x_\xi^\alpha(n) = m \iff x_\xi^{\beta_0}(n) = m ,$$

and we are done.

Since the relation

$$w \in WO \wedge x_{|w|}^\alpha(n) = m$$

is Σ^1_2 it follows from

$$x_\xi(n) = m \iff \exists\alpha_0\ \forall\alpha \geq_T \alpha_0\ (x^\alpha_\xi(n) = m)$$
$$\iff \forall\beta\ \exists\alpha \geq_T \beta\ (x^\alpha_\xi(n) = m)\ ,$$

that each x_ξ is Δ^1_3 in a countable ordinal, thus

$$\{x_\xi : \xi \in A\} \subseteq Q_3 .$$

But the map $\xi \longmapsto x_\xi$ for $\xi \in A$ is 1-1, since if $\xi \neq \eta$ and α_0 is large enough so that $\xi, \eta < \theta^{\alpha_0}$ and $\alpha \geq_T \alpha_0 \Rightarrow x^\alpha_\xi = x_\xi$, $x^\alpha_\eta = x_\eta$ we clearly have $x^\alpha_\xi \neq x^\alpha_\eta$. So A is countable. Let $\xi_0 = \sup\{\xi + 1 : \xi \in A\}$. As $\xi_0 \notin A$ we have that $\forall\alpha\ \exists\beta \geq_T \alpha\ (\xi_0 \geq \theta^\beta)$, thus $\exists\alpha_0\ \forall\alpha \geq_T \alpha_0\ (\xi_0 \geq \theta^\alpha)$. So pick z''_0 such that for all $\alpha \geq_T z''_0$, $\theta^\alpha \leq \xi_0$ and moreover for $\xi < \xi_0$ thus in particular for $\xi < \theta^\alpha$, $x^\alpha_\xi = x_\xi$. Then for all $\alpha \geq_T z''_0$,

$$\mathrm{HOD}^{L[\alpha]} \cap \omega^\omega \subseteq H^\alpha = \{x^\alpha_\xi : \xi < \theta^\alpha\} \subseteq \{x_\xi : \xi \in A\} \subseteq Q_3 .$$

If finally $z_0 = \langle z'_0, z''_0\rangle$ we clearly have that $\alpha \geq_T z_0 \Rightarrow \mathrm{HOD}^{L[\alpha]} \cap \omega^\omega = Q_3$, thus $z_0 \in L[\alpha] \Rightarrow \mathrm{HOD}^{L[\alpha]} \cap \omega^\omega = \mathrm{HOD}^{L[\alpha, z_0]} \cap \omega^\omega = Q_3$, and we are done. ⊣

For each set x let HOD_x be the class of sets hereditarily ordinal definable from x. Then the relativized version of 13.1 asserts that for each real β, there is a real z (depending on β) such that

$$\beta,\ z \in L[\alpha] \Rightarrow \mathrm{HOD}_\beta^{L[\alpha]} \cap \omega^\omega = Q_3(\beta) .$$

In general $\mathrm{HOD}[x] \subsetneq \mathrm{HOD}_x$, even for reals, but the next result shows that in our particular case relativization amounts to the same thing as adjunction.

13.2 <u>Theorem</u> (Woodin). Assume $\underset{\sim}{\Delta}^1_2$-DET. Then for each real β, there is a real z such that

$$\beta,\ z \in L[\alpha] \Rightarrow \mathrm{HOD}[\beta]^{L[\alpha]} \cap \omega^\omega = \mathrm{HOD}_\beta^{L[\alpha]} \cap \omega^\omega = Q_3(\beta) .$$

<u>Proof</u>. We need to review first Vopenka's Theorem ([J], p. 293) and some of its consequences that we will need below.

Let $x \subseteq \omega$ and work in $L[x]$ below. Let B' be the collection of all ordinal definable subsets of $P(\omega)$ partially ordered under $\subseteq$. There is an ordinal definable isomorphism π between $(B', \subseteq)$ and a partial ordering $(B, \leq)$ in HOD. Clearly $(B, \leq)$ is a Boolean algebra which is complete in HOD. Let $f(n) = \pi\{y \subseteq \omega : n \in y\}$. Then $f : \omega \to B$ and $f \in \mathrm{HOD}$. If y is a real let $G_y = \pi\{b \in B' : y \in b\} \subseteq B$. Then G_y is a generic ultrafilter over HOD and since

$$n \in y \Longleftrightarrow f(n) \in G_y ,$$

we have $y \in \mathrm{HOD}[G_y]$. In particular $L[x] = \mathrm{HOD}[G_x]$, thus $L[x]$ is a generic extension of HOD. Note also that B has size $\leq \omega_2$ (in $L[x]$).

We claim now that actually

$$\mathrm{HOD}^{L[x]} = L[B,\leq,f] .$$

Proof of claim: Let $M_2 = \mathrm{HOD}^{L[x]}$ and let $M_1 = L[B,\leq,f]$. Clearly $M_1 \subseteq M_2$. Let $y \subseteq \omega$ be such that $L[y] = L[x]$. Let G_y be the M_2-generic filter on B determined by y. Clearly G_y is M_1-generic as well. Also $L[y] \subseteq M_1[G_y] \subseteq M_2[G_y] \subseteq L[x] = L[y]$. So $M_1[G_y] = L[y]$.

Work now in $L[x]$. Let

$$b_0' = \{y \subseteq \omega : L[y] = L[x]\} ,$$

$b_0 = \pi(b_0')$. Let $A \subseteq \mathrm{ORD}$ be in M_2. Then for some formula φ and some ordinal η_0

$$\xi \in A \Longleftrightarrow \varphi(\xi,\eta_0) .$$

We claim now that

$$\xi \in A \Longleftrightarrow b_0 \Vdash^{M_1}_{(B,\leq)} \varphi(\check{\xi},\check{\eta}_0) ,$$

thus $A \in M_1$, so $M_2 \subseteq M_1$ and we are done. If $b_0 \Vdash^{M_1}_{(B,\leq)} \varphi(\check{\xi},\check{\eta}_0)$, then as $b_0 \in G_x$, $M_1[G_x] = L[x] \models \varphi(\check{\xi},\check{\eta}_0)$, so $\xi \in A$. If $b_0 \not\Vdash^{M_1}_{(B,\leq)} \varphi(\check{\xi},\check{\eta}_0)$, then let $0 \neq b \leq b_0$ be such that $b \Vdash^{M_1}_{(B,\leq)} \neg\varphi(\check{\xi},\check{\eta}_0)$. Now $b = \pi(b')$, $b' \in B'$ and b' is a nonempty subset of b_0', so let $y \in b'$. Then $M_1[G_y] = L[y] = L[x]$, so $L[x] \models \neg\varphi(\xi,\eta_0)$, thus $\xi \notin A$.

This completes our review of Vopenka's Theorem.

Fix now a real β. By the relativized version of 13.1 find a real z such that

$$\beta,\ z \in L[\alpha] \Rightarrow \mathrm{HOD}_\beta^{L[\alpha]} \cap \omega^\omega = Q_3(\beta) .$$

We will show that if β, $z \in L[\alpha]$, then

$$Q_3(\beta) = \omega^\omega \cap \mathrm{HOD}_\beta^{L[\alpha]} \subseteq \mathrm{HOD}[\beta]^{L[\alpha]} .$$

By Vopenka's Theorem and standard results on intermediate forcing extensions (see [J], Lemma 25.3 and p. 237) there is a notion of forcing $C \in \mathrm{HOD}[\beta]^{L[\alpha]}$ and a $\mathrm{HOD}[\beta]^{L[\alpha]}$-generic set $H \subseteq C$ such that $\mathrm{HOD}[\beta]^{L[\alpha]}[H] = \mathrm{HOD}[\beta]^{L[\alpha]}[z]$. Moreover C has size $\leq \omega_2$ in $L[\alpha]$. Let $\kappa = \omega_2^{L[\alpha]}$ and let C_κ be the notion of forcing for collapsing κ to ω. Let F be a generic over $\mathrm{HOD}[\beta]^{L[\alpha]}[z]$ subset of C_κ (such an F exists since $\omega_3^{L[\alpha]} < \omega_1$). Then

$$\mathrm{HOD}[\beta]^{L[\alpha]}[z][F] = \mathrm{HOD}[\beta]^{L[\alpha]}[H][F]$$

is a generic extension of $\mathrm{HOD}[\beta]^{L[\alpha]}$ by a notion of forcing of size $\kappa > \aleph_0$ (in $\mathrm{HOD}[\beta]^{L[\alpha]}$), which always collapses κ to $\aleph_0$, thus (see [J], Lemma 25.11) it is isomorphic to C_κ itself. So $N = \mathrm{HOD}[\beta]^{L[\alpha]}[z][F]$ is a homogeneous generic extension of $\mathrm{HOD}[\beta]^{L[\alpha]}$. In particular, $\mathrm{HOD}_\beta^N \subseteq \mathrm{HOD}[\beta]^{L[\alpha]}$. If we can show that $N = L[\alpha']$ for some real α', then since $\beta, z \in L[\alpha']$ we have $Q_3(\beta) = \mathrm{HOD}_\beta^{L[\alpha']} \cap \omega^\omega$, thus $Q_3(\beta) \subseteq \mathrm{HOD}[\beta]^{L[\alpha]}$ and our proof is complete. But by our review of Vopenka's Theorem $\mathrm{HOD} = L[P]$, where $P \subseteq \kappa$, so $N = \mathrm{HOD}[\beta]^{L[\alpha]}[z][F] = L[P,\beta,z,F]$ and F collapses κ to ω, so clearly $N = L[\alpha']$, for some real α'.

$\dashv$

It follows from the results of [KS] that assuming Δ_2^1-DET there is a real x_0 such that

$$x_0 \in L[\alpha] \Rightarrow L[\alpha] \models \text{OD-DET} .$$

But if $L[\alpha] \models$ OD-DET, the standard proof that $AD \Rightarrow \omega_1$ is measurable, shows that $\omega_1^{L[\alpha]}$ is measurable in $\mathrm{HOD}^{L[\alpha]}$. Similarly it can be shown that there are higher order measurables in $\mathrm{HOD}^{L[\alpha]}$. It is open whether there exist even larger cardinals, like strongly compact or supercompact in $\mathrm{HOD}^{L[\alpha]}$, for sufficiently high constructibility degrees of α.

For each constructibility degree $d = [\alpha]_C$, let

$$L[d] = L[\alpha] ,$$

and assuming AD consider the Martin measure μ on constructibility degrees (i.e. the one generated by cones), and then the (transitive realization of the) ultraproduct

$$M_3 = \Pi_d \ \mathrm{HOD}^{L[d]}/\mu .$$

Then by 13.1

$$M_3 \cap \omega^\omega = Q_3 .$$

Since

$$M_3[\alpha] = \Pi_d \ \mathrm{HOD}[\alpha]^{L[d]}/\mu ,$$

we also have by 13.2

$$M_3[\alpha] \cap \omega^\omega = Q_3(\alpha) ,$$

thus for reals M_3 relativizes by adjunction. In general the relativized version of M_3 is

$$M_3(\alpha) = \prod_d \mathrm{HOD}_\alpha^{L[d]} /\mu$$

and clearly

$$M_3[\alpha] \cap \omega^\omega = M_3(\alpha) \cap \omega^\omega = Q_3(\alpha) ,$$

but at this stage it is not known whether $M_3[\alpha] = M_3(\alpha)$ as well.

From this fact about adjunctions and the argument used in 12.2 (iii) it follows easily that although M_3 is not Σ^1_3-correct, it satisfies the following "dual Shoenfield absoluteness" theorem at the third level:

To each Π^1_3 formula $\varphi(\alpha)$ we can effectively assign a Σ^1_3 formula $\varphi^*(\alpha)$ such that for $\alpha \in \omega^\omega$:

$$\varphi(\alpha) \Longleftrightarrow M_3[\alpha] \models \varphi^*(\alpha) .$$

Similarly interchanging the roles of Π^1_3 and Σ^1_3.

There are a number of very interesting open problems concerning the internal structure of M_3. We mentioned before the question of the existence of very large cardinals (beyond measurable or higher order measurable) in M_3. Another unresolved problem is whether GCH holds in it.

The preceding ideas can be also used to give a stronger version of 9.1 (vi). The result is due to Woodin.

13.3 Theorem (Woodin). Assume $\utilde{\Delta}^1_2$-DET. Then the following are equivalent:

(i) $\alpha \in Q_3$;

(ii) α belongs to every Σ^1_2-correct standard model of T, where T is a recursive set of axioms in the language of ZF extending $ZF_N + DC + \Delta^1_2$-DET (lightface!), provided T has such models. (Here N is a large enough integer.)

Proof. It is enough to show that if M is a countable Σ^1_2-correct standard model of such a T, then $Q_3 \subseteq M$. Let $\xi = o(M)$ be the least ordinal not in M. Since $M \models \Delta^1_2$-DET, and by [KS], $\Delta^1_2\text{-DET} \Rightarrow$ For a cone of x's, $L[x] \models$ OD-DET, we can find $x \in M$ with $L_\xi[x] \models ZF_n$ + OD-DET, for some large enough n, such that $L_\xi[x]$ is also Σ^1_2-correct. So it is enough to show that for such $L_\xi[x]$, $Q_3 \subseteq L_\xi[x]$. We will do this by establishing the following:

There is a Π^1_2 formula $\psi(\alpha)$, such that

(i) $\forall\alpha\, [\psi(\alpha) \Rightarrow \forall\beta\, [\beta \in Q \Rightarrow \beta \leq_T \alpha]]$

(ii) ZF_n + OD-DET $\vdash \exists\alpha\, \psi(\alpha)$

The predicate $\psi(\alpha)$ will say the following:

$(\alpha)_1$ codes a well-ordered sequence of reals $\langle z_\xi : \xi < \lambda \rangle$ such that for any real y with $(\alpha)_0 \leq_T y$, the wellordering of H^y (see the second paragraph of the proof of 13.1 for this notation) is an initial segment of $\langle z_\xi : \xi < \lambda \rangle$.

Thus ψ is easily seen to be Π^1_2 and (i) follows from the proof of Theorem 13.1. It remains to prove (ii). We work in ZF_n + OD-DET.

The proof is a variant of the argument in the proof of 13.1, whose notation we use below.

Since the set

$$P_\xi(\alpha) \Longleftrightarrow \xi < \theta^\alpha$$

is ordinal definable for each fixed $\xi < \omega_1$, either P_ξ or its complement contains all the reals in a cone of Turing degrees. Let $A = \{\xi : \exists \alpha_0 \, \forall \alpha \geq_T \alpha_0 \, (\xi < \theta^\alpha)\}$. Then A is an initial segment of ω_1. If $\xi \in A$ then for all α's in a cone of Turing degrees, $x^\alpha_\xi = x_\xi$ is fixed by using again OD-DET (the set R_ξ in the proof of 13.1 is ordinal definable). Again the sequence $\xi \longmapsto x_\xi$ for $\xi \in A$ is 1-1 and it lies in HOD. Hence the cardinality of A is at most the cardinality of $(2^{\aleph_0})^{HOD}$. But from OD-DET we can conclude that ω_1 is measurable in HOD (use for instance the degree theoretic proof from AD that ω_1 is measurable). Thus $(2^{\aleph_0})^{HOD}$ is countable in the real world, and so A is countable. But this immediately implies that $\exists \alpha \, \psi(\alpha)$, and our proof is complete. $\dashv$

We discuss now the generalizations to higher Q_{2n+3}. For each real α, let, assuming $\underset{\sim}{\Delta}^1_{2n}$-DET,

$$L(C_{2n+2}(\alpha))$$

be the smallest inner model of ZF containing $C_{2n+2}(\alpha)$. Then as shown in [Kl], Ch. 3 the reals in $L(C_{2n+2}(\alpha))$ are exactly $C_{2n+2}(\alpha)$ and we can think of these models as higher level analogs of $L[\alpha]$, since $L(C_2(\alpha)) = L[\alpha]$. Then it is easy to prove using the ideas in [Kl], Ch. 3 and the method of proof of 13.1 the following:

13.4 <u>Theorem</u>. Assume $\underset{\sim}{\Delta}^1_{2n}$-DET and $n > 0$. Then there is a real z_0 such that

$$z_0 \in C_{2n}(\alpha) \Rightarrow \omega^\omega \cap HOD^{L(C_{2n}(\alpha))} = Q_{2n+1} \, .$$

Also we can easily obtain by generalizing the proof of 13.2 and using the $L(C_{2n+2}(\alpha))$'s that the following also holds:

13.5 Theorem. Assume $\underset{\sim}{\Delta}^1_{2n}$-DET and $n > 0$. Then the following are equivalent:

(1) $\alpha \in Q_{2n+1}$;

(2) α belongs to every Σ^1_{2n}-correct model of T, where T is a recursive set of axioms in the language of ZF extending $ZF_N + DC + \underset{\sim}{\Delta}^1_{2n-2}\text{-DET} + \Delta^1_{2n}\text{-DET}$, provided T has such models. (Here N is a large enough integer.)

More satisfactory generalizations to higher levels can be obtained by working with the models $L[T_{2n+1}]$ instead of $L(C_{2n+2})$. Here T_{2n+1} is the tree coming from a Π^1_{2n+1}-scale on a complete Π^1_{2n+1} set (see [Mo3], p. 547). Again $L[T_1] = L$ and $L[T_{2n+1}] \cap \omega^\omega = C_{2n+2}$. The proofs here require at present in general more than $\underset{\sim}{\Delta}^1_{2n}$-DET, so for convenience we will work with PD and we will not bother to specify the exact level of determinacy needed. We have now

13.6 Theorem (PD). There is a real z_0 such that

$$z_0 \in L[T_{2n+1},\alpha] \Rightarrow \omega^\omega \cap \mathrm{HOD}_{T_{2n+1}}^{L[T_{2n+1},\alpha]} = Q_{2n+3} \,.$$

The proof is exactly as in 13.1 using the results in [HK] (especially §4,7,8) to compute the required definability estimates. (Actually the proof also shows that $\mathrm{HOD}_{T_{2n+1}}^{L[T_{2n+1},\alpha]}$ can be replaced by $\mathrm{HOD}^{L[T_{2n+1},\alpha]}$ as well.)

We also have the analog of 13.2.

13.7 Theorem (PD). For each real β, there is a real z such that

$$\beta, z \in L[T_{2n+1},\alpha] \Rightarrow \mathrm{HOD}_{T_{2n+1}}[\beta]^{L[T_{2n+1},\alpha]} \cap \omega^\omega = \mathrm{HOD}^{L[\alpha]}_{T_{2n+1},\beta} \cap \omega^\omega = Q_{2n+3}(\beta) \,.$$

Consider now for each C_{2n+2}-degree

$$d = [\alpha]_{C_{2n+2}} = \{\beta : \alpha \in C_{2n+2}(\beta) \wedge \beta \in C_{2n+2}(\alpha)\}$$

$$= \{\beta : \alpha \in L[T_{2n+1},\beta] \wedge \beta \in L[T_{2n+1},\alpha]\} \,,$$

the inner model

$$L[T_{2n+1},d] = L[T_{2n+1},\alpha]$$

and assuming AD let μ be the Martin measure on these degrees (i.e. the one generated by cones). Then define

$$M_{2n+3} = \prod_d \mathrm{HOD}_{T_{2n+1}}^{L[T_{2n+1},d]} \Big/ \mu \,.$$

Again

$$M_{2n+3} \cap \omega^\omega = Q_{2n+3}$$

and for each real α

$$M_{2n+3}[\alpha] \cap \omega^\omega = Q_{2n+3}(\alpha) .$$

Moreover by similar reasoning high order measurable cardinals exist in M_{2n+3}.

Remark. Assume $\underset{\sim}{\Delta}^1_2$-DET. Let $\infty^\#$ be the "eventual sharp" of a real, i.e. let

$$\ulcorner\varphi(v_1\ldots v_n)\urcorner \in \infty^\# \Longleftrightarrow \text{For a cone of Turing degrees } \alpha, \ \ulcorner\varphi(v_1\ldots v_n)\urcorner \in \alpha^\# ,$$

where φ is a formula in the language of set theory (in particular φ does not contain a constant for α). Using the results in [Ma3] one can easily see that for every ∂M_n set A there is a formula $\varphi(v_1\ldots v_n,v)$ of the language of set theory such that

$$x \in A \Longleftrightarrow L[x] \models \varphi(\aleph_1,\ldots,\aleph_n,x) .$$

Conversely, if $\varphi(v_1\ldots v_n,v)$ is a formula of the language of set theory, then

$$\{x : L[x] \models \varphi(\aleph_1\ldots\aleph_n,x)\}$$

is a ∂M_{n+1} set. Let now $R \subseteq \omega$ be $\partial^2 M_n$. Then there is $A \subseteq \omega \times \omega^\omega$ in ∂M_n with

$$\begin{aligned} R(m) &\Longleftrightarrow \partial\alpha\, A(m,\alpha) \\ &\Longleftrightarrow \partial\alpha\, (L[\alpha] \models \psi(m,\aleph_1\ldots\aleph_n,\alpha)) , \end{aligned}$$

for some ψ. Thus

$$\begin{aligned} R(m) \Longleftrightarrow & \text{ For a cone of Turing degrees } x, \\ & L[x] \models \text{"}\partial\alpha\, L[\alpha] \models \psi(m,\aleph_1\ldots\aleph_n,\alpha)\text{"} . \end{aligned}$$

(We are using here $\bigcup_n \partial\underset{\sim}{M}_n$-DET, which by [KW] is equivalent to $\underset{\sim}{\Delta}^1_2$-DET.)

It follows immediately that

$$Y^0_3 \equiv_T \infty^\# .$$

Recall now from 13.2 that $HOD^{L[x]} = L[P_x]$, where $P_x \subseteq \aleph_2^{L[x]}$ and P_x depends only on $L[x]$, $x \in \omega^\omega$. Assuming AD, let $M_3 = \prod_d HOD^{L[d]}/\mu$. Then $M_3 = L[P]$, P a set of ordinals. Since by Vopenka's Theorem each $L[x]$, $x \in \omega^\omega$, is a generic extension of $HOD^{L[x]}$ by a notion of forcing coded in P_x it follows easily that also

$$\infty^\# \equiv_T \{\ulcorner\varphi(v,v_1\ldots v_n)\urcorner : M_3 \models \varphi(P,\kappa_1,\ldots,\kappa_n)\} ,$$

where $\{\kappa_i\}$ are large enough cardinals (therefore indiscernibles for $L[P] = M_3$). Thus (up to Turing equivalence) $\infty^\#$, and thus Y_3^0, is the part of $P^\#$ which does not mention ordinals beyond the indiscernibles, i.e. roughly speaking "$P^\# \cap \omega$".

§14. The ordinal of the Q_{2n+1}-degrees. The ordinal assignment

$$\alpha \longmapsto \omega_1^\alpha$$

where

$$\omega_1^\alpha = \text{first non-recursive in } \alpha \text{ ordinal}$$
$$= \delta_1^1(\alpha) = \lambda_1(\alpha) \ ,$$

(where $\lambda_{2n+1}(\alpha)$ is defined in §9) plays a crucial role in the theory of Π_1^1 sets and Δ_1^1-degrees. An important relationship between Δ_1^1-degrees and this ordinal assignment is the so-called Spector Criterion

$$d \leq f \Rightarrow [d' \leq f \Longleftrightarrow \lambda_1(d) < \lambda_1(f)] \ .$$

We assume again in this section $\underset{\sim}{\Delta}{}_{2n}^1$-DET.

It is not hard to verify, using standard prewellordering theory, that the Spector Criterion goes through for the Δ_{2n+1}^1-degrees and the ordinal assignment $\alpha \longmapsto \lambda_{2n+1}(\alpha)$. We shall define an appropriate ordinal assignment that behaves analogously for the Q_{2n+1}-degrees.

If the Spector Criterion is to be satisfied then we must have, if $\kappa'_{2n+1}(\alpha)$ is the proposed ordinal assignment,

$$\alpha \leq_{2n+1}^Q \beta \Rightarrow [y_{2n+1}^\alpha \leq_{2n+1}^Q \beta \Longleftrightarrow \kappa'_{2n+1}(\alpha) < \kappa'_{2n+1}(\beta)] \ ,$$

and $\kappa'_{2n+1}(\alpha)$ should be invariant under $\equiv_{2n+1}^Q$, and monotone i.e.

$$\alpha \leq_{2n+1}^Q \beta \Rightarrow \kappa'_{2n+1}(\alpha) \leq \kappa'_{2n+1}(\beta) \ .$$

This requirement imposes the following lower bound on $\kappa'_{2n+1}(\alpha)$, assuming at least the reasonable requirement that $\kappa'_{2n+1}(\alpha) \geq \lambda_{2n+1}(\alpha)$: For all α,

$$(*) \qquad \kappa'_{2n+1}(\alpha) \geq \sup\{\lambda_{2n+1}(\langle\alpha,\beta\rangle) : y_{2n+1}^\alpha \not\leq_{2n+1} \langle\alpha,\beta\rangle\} \ .$$

Indeed, if $y_{2n+1}^\alpha \not\leq_{2n+1} \langle\alpha,\beta\rangle$ but $\kappa'_{2n+1}(\alpha) < \lambda_{2n+1}(\langle\alpha,\beta\rangle)$, we have $\kappa'_{2n+1}(\alpha) < \lambda_{2n+1}(\langle\alpha,\beta\rangle) \leq \kappa'_{2n+1}(\langle\alpha,\beta\rangle)$. Therefore by the Spector Criterion $y_{2n+1}^\alpha \leq_{2n+1}^Q \langle\alpha,\beta\rangle$ so $y_{2n+1}^\alpha \leq_{2n+1} \langle\alpha,\beta\rangle$, a contradiction. (Granting AD, we can easily see by a Solovay-type game, that for any ordinal assignment $\alpha \longmapsto \kappa(\alpha)$ unbounded in $\underset{\sim}{\delta}{}_{2n+1}^1$, we have $\kappa(\alpha) \geq \lambda_{2n+1}(\alpha)$ on a cone of α's.)

We shall now define our ordinal assignment to be the lower bound in (*) and show that it works.

14.1 Definition. For each real α, let

$$\kappa_{2n+1}(\alpha) = \sup\{\lambda_{2n+1}(\langle\alpha,\beta\rangle) : y^{\alpha}_{2n+1} \not\leq_{2n+1} \langle\alpha,\beta\rangle\} .$$

In particular

$$\kappa_{2n+1} = \kappa_{2n+1}(\lambda t.0) = \sup\{\lambda_{2n+1}(\beta) : y^{0}_{2n+1} \not\leq_{2n+1} \beta\} .$$

We shall now derive a number of basic properties of this ordinal assignment in a series of lemmas.

14.2 Lemma ($\underset{\sim}{\Delta}^1_{2n}$-DET). For all α, β

$$y^{\alpha}_{2n+1} \leq_{2n+1} \langle\alpha,\beta\rangle \iff \lambda_{2n+1}(y^{\alpha}_{2n+1}) \leq \lambda_{2n+1}(\langle\alpha,\beta\rangle) ,$$

thus

$$\kappa_{2n+1}(\alpha) = \sup\{\lambda_{2n+1}(\langle\alpha,\beta\rangle) : \lambda_{2n+1}(\langle\alpha,\beta\rangle) < \lambda_{2n+1}(y^{\alpha}_{2n+1})\} .$$

Moreover for all α,

$$\kappa_{2n+1}(\alpha) < \lambda_{2n+1}(y^{\alpha}_{2n+1}) .$$

Proof. Let W be universal Π^1_{2n+1} and let n_0 be such that

$$\delta = y^{\alpha}_{2n+1} \iff (n_0,\langle\delta,\alpha\rangle) \in W .$$

Let $\varphi : W \twoheadrightarrow \underset{\sim}{\delta}^1_{2n+1}$ be a Π^1_{2n+1}-norm on W. Then for each real γ we have

$$\begin{aligned}\lambda_{2n+1}(\gamma) &= \sup\{\varphi(x) : x \in W \wedge x \leq_{2n+1} \gamma\} \\ &= \sup\{\varphi(x) : x \in W \wedge x \leq_T \gamma\} .\end{aligned}$$

This follows from 4C.14 in [Mo3]. Since $\varphi(n_0,\langle y^{\alpha}_{2n+1},\alpha\rangle) < \lambda_{2n+1}(y^{\alpha}_{2n+1})$, we have then that, if

$$\lambda_{2n+1}(y^{\alpha}_{2n+1}) \leq \lambda_{2n+1}(\langle\alpha,\beta\rangle)$$

then

$$\varphi(n_0,\langle y^{\alpha}_{2n+1},\alpha\rangle) \leq \varphi(x) ,$$

for some $x \in W$, $x \leq_{2n+1} \langle\alpha,\beta\rangle$, thus y^{α}_{2n+1} is a $\Delta^1_{2n+1}(\alpha,\beta)$-singleton i.e. $y^{\alpha}_{2n+1} \leq_{2n+1} \langle\alpha,\beta\rangle$.

From this argument it also follows that

$$\kappa_{2n+1}(\alpha) \le \varphi(n_0, \langle y^{\alpha}_{2n+1},\alpha\rangle) < \lambda_{2n+1}(y^{\alpha}_{2n+1}) \ . \qquad \dashv$$

14.3 <u>Lemma</u> ($\underset{\sim}{\Delta}^1_{2n}$-DET). (i) For $n = 0$, $\kappa_{2n+1}(\alpha) = \kappa_1(\alpha) = \lambda_1(\alpha)$.

(ii) For $n > 0$, $\lambda_{2n+1}(\alpha) < \kappa_{2n+1}(\alpha)$ and the sup defining $\kappa_{2n+1}(\alpha)$ is <u>not</u> attained, i.e. for all α,β we have $\kappa_{2n+1}(\alpha) \ne \lambda_{2n+1}(\alpha,\beta)$.

<u>Picture for</u> $\alpha = \lambda t.0$

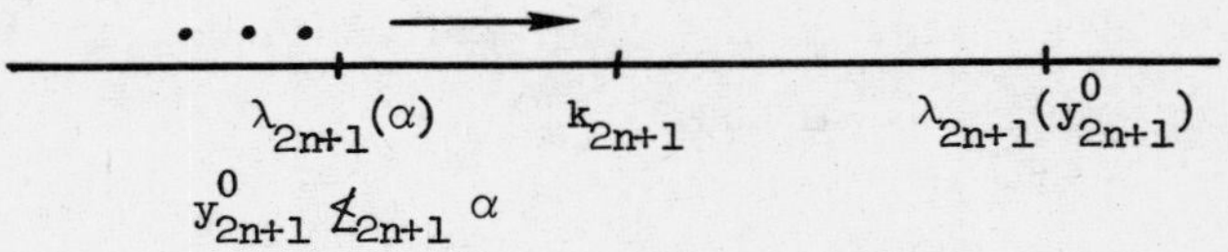

<u>Proof</u>. (i) By the classical Spector Criterion

$$y^{\alpha}_1 \not\le_1 \langle\alpha,\beta\rangle \Leftrightarrow \mathcal{O}^{\alpha} \not\le_1 \langle\alpha,\beta\rangle \Leftrightarrow \lambda_1(\alpha) = \lambda_1(\langle\alpha,\beta\rangle) \ .$$

Thus $\kappa_1(\alpha) = \lambda_1(\alpha)$.

(ii) Let W^{α}_{2n+1} be a complete $\Pi^1_{2n+1}(\alpha)$ set of numbers. Then for $n > 0$ $W^{\alpha}_{2n+1} \in Q_{2n+1}(\alpha)$ (by 8.1), thus $y^{\alpha}_{2n+1} \not\le_{2n+1} W^{\alpha}_{2n+1}$, so $\kappa_{2n+1}(\alpha) \ge \lambda_{2n+1}(W^{\alpha}_{2n+1}) > \lambda_{2n+1}(\alpha)$.

Assume now $\kappa_{2n+1}(\alpha) = \lambda_{2n+1}(\gamma)$, for some $\alpha \le_{2n+1} \gamma$, towards a contradiction. Let A be a complete $\Pi^1_{2n+1}(\gamma)$ subset of ω. Then by Spector's Criterion for Δ^1_{2n+1}-degrees, $\kappa_{2n+1}(\alpha) = \lambda_{2n+1}(\gamma) < \lambda_{2n+1}(A)$ and $\alpha \le_T A$, so by the definition of $\kappa_{2n+1}(\alpha)$ we have $y^{\alpha}_{2n+1} \le_{2n+1} A$. Since $A \in Q_{2n+1}(\gamma)$ we have $y^{\alpha}_{2n+1} \in Q_{2n+1}(\gamma)$ and so since y^{α}_{2n+1} is a $\Pi^1_{2n+1}(\gamma)$-singleton, $y^{\alpha}_{2n+1} \le_{2n+1} \gamma$. Thus $\lambda_{2n+1}(y^{\alpha}_{2n+1}) \le \lambda_{2n+1}(\gamma) = \kappa_{2n+1}(\alpha)$, a contradiction. $\dashv$

14.4 <u>Lemma</u> ($\underset{\sim}{\Delta}^1_{2n}$-DET). For all $\alpha,\beta \in \omega^{\omega}$,

$$\alpha \le^{Q}_{2n+1} \beta \Leftrightarrow [\kappa_{2n+1}(\alpha) < \kappa_{2n+1}(\beta) \Leftrightarrow y^{\alpha}_{2n+1} \le^{Q}_{2n+1} \beta] \ .$$

<u>Proof</u>. We distinguish cases as $n = 0$ or $n > 0$. For $n = 0$ this is just the standard Spector Criterion, so assume $n > 0$.

Assume $\alpha \le^{Q}_{2n+1} \beta$. We will show that

$$\kappa_{2n+1}(\alpha) < \kappa_{2n+1}(\beta) \Leftrightarrow y^{\alpha}_{2n+1} \le^{Q}_{2n+1} \beta \ .$$

($\Rightarrow$): Assume $\kappa_{2n+1}(\alpha) < \kappa_{2n+1}(\beta)$. Since $\alpha \leq^Q_{2n+1} \beta$ we have also $\alpha \in C_{2n+1}(\beta)$ and thus $y^\alpha_{2n+1} \in C_{2n+1}(\beta)$. (To see this consider the set $C' = \{\gamma : \exists\alpha \leq_T \gamma\ (\alpha \in C_{2n+1}(\beta) \wedge \gamma = y^\alpha_{2n+1})\}$ and check easily that C' is a thin $\Pi^1_{2n+1}(\beta)$ set, so $C' \subseteq C_{2n+1}(\beta)$.) So if $y^\alpha_{2n+1} \leq^Q_{2n+1} \beta$, towards a contradiction, we must have $y^\beta_{2n+1} \leq_{2n+1} \langle y^\alpha_{2n+1},\beta\rangle$, since the relation $x \leq_{2n+1} \langle y,\beta\rangle$ is a prewellordering on $C_{2n+1}(\beta)$. Since $\kappa_{2n+1}(\alpha) < \kappa_{2n+1}(\beta)$, find γ such that $\kappa_{2n+1}(\alpha) < \lambda_{2n+1}(\langle\beta,\gamma\rangle) < \kappa_{2n+1}(\beta)$ and $y^\beta_{2n+1} \not\leq_{2n+1} \langle\beta,\gamma\rangle$. Thus $\kappa_{2n+1}(\alpha) < \lambda_{2n+1}(\langle\alpha,\beta,\gamma\rangle)$, so $y^\alpha_{2n+1} \leq_{2n+1} \langle\alpha,\beta,\gamma\rangle$ and since $\alpha \leq^Q_{2n+1} \beta$ we have $y^\alpha_{2n+1} \leq^Q_{2n+1} \langle\beta,\gamma\rangle$. So $y^\beta_{2n+1} \leq^Q_{2n+1} \langle\beta,\gamma\rangle$, thus $y^\beta_{2n+1} \leq_{2n+1} \langle\beta,\gamma\rangle$, a contradiction.

($\Leftarrow$): If $y^\alpha_{2n+1} \leq^Q_{2n+1} \beta$, then we have $\kappa_{2n+1}(\alpha) < \lambda_{2n+1}(y^\alpha_{2n+1}) \leq \lambda_{2n+1}(\langle y^\alpha_{2n+1},\beta\rangle) < \kappa_{2n+1}(\beta)$ (as $\langle y^\alpha_{2n+1},\beta\rangle \in Q_{2n+1}(\beta)$, so that $y^\beta_{2n+1} \not\leq_{2n+1} \langle y^\alpha_{2n+1},\beta\rangle$). ⊣

14.5 <u>Lemma</u> ($\underset{\sim}{\Delta}^1_{2n}$-DET). For all α,β

$$\alpha \equiv^Q_{2n+1} \beta \Rightarrow \kappa_{2n+1}(\alpha) = \kappa_{2n+1}(\beta) .$$

<u>Proof</u>. This is obvious for $n = 0$. If $n > 0$, assume $\alpha \equiv^Q_{2n+1} \beta$ and assume by symmetry $\kappa_{2n+1}(\alpha) < \kappa_{2n+1}(\beta)$, towards a contradiction. Then by the preceding lemma we have $y^\alpha_{2n+1} \leq^Q_{2n+1} \beta \leq^Q_{2n+1} \alpha$, which is absurd. ⊣

Thus for each Q_{2n+1}-degree $p = [\alpha]^Q_{2n+1}$, we can define without ambiguity

$$\kappa_{2n+1}(p) = \kappa_{2n+1}(\alpha) .$$

Then 14.4 expresses the Spector Criterion for Q_{2n+1}-degrees

$$p \leq q \Rightarrow [p' \leq q \Leftrightarrow \kappa_{2n+1}(p) < \kappa_{2n+1}(q)] .$$

14.6 <u>Lemma</u> ($\underset{\sim}{\Delta}^1_{2n}$-DET). For all α,

$$\kappa_{2n+1}(\alpha) = \sup\{\lambda_{2n+1}(\langle\alpha,\beta\rangle) : \langle\alpha,\beta\rangle <_{2n+1} y^\alpha_{2n+1}\}$$

$$= \sup\{\lambda_{2n+1}(\beta) : \alpha \leq_{2n+1} \beta <_{2n+1} y^\alpha_{2n+1}\} .$$

In particular, for all α

$$\mathrm{cof}(\kappa_{2n+1}(\alpha)) = \omega .$$

<u>Proof</u>. Again for $n = 0$ the result is obvious, so assume $n > 0$. We prove the result for κ_{2n+1}, since the proof relativizes immediately.

Let A be a nonempty Δ^1_{2n+1} real, and let $\overline{\varphi} = \{\varphi_m\}$ be an excellent Δ^1_{2n+1}-scale on A. Let $\alpha_0 = y_{A,\overline{\varphi}}$ be the leftmost real associated with $A,\overline{\varphi}$; see 5.8.

By 5.9 $\alpha_0 \equiv_{2n+1} y^0_{2n+1}$. We shall prove that

$$(*) \qquad \sup\{\lambda_{2n+1}(\gamma) : \gamma <_{2n+1} y^0_{2n+1}\} \leq \lambda_{2n+1}(\beta) \Rightarrow \alpha_0 \leq_{2n+1} \beta ,$$

from which it follows immediately that $\kappa_{2n+1} = \sup\{\lambda_{2n+1}(\beta) : \beta <_{2n+1} y^0_{2n+1}\}$.

So assume the hypothesis of (*). Then note that we can actually assume without loss of generality that

$$\sup\{\lambda_{2n+1}(\gamma) : \gamma <_{2n+1} y^0_{2n+1}\} < \lambda_{2n+1}(\beta) ,$$

since otherwise we can replace β by the complete $\Pi^1_{2n+1}(\beta)$ set of integers W^{β}_{2n+1}. Let W be a universal Π^1_{2n+1} set and $\sigma : W \twoheadrightarrow \underset{\sim}{\delta}^1_{2n+1}$ a Π^1_{2n+1}-norm on W. Then there is $x \in W$, $x \leq_T \beta$ such that

$$\sup\{\lambda_{2n+1}(\gamma) : \gamma <_{2n+1} y^0_{2n+1}\} \leq \sigma(x) .$$

Now recall that in the proof of 5.8 we showed that for each m there is a $\gamma <_{2n+1} y^0_{2n+1}$ of minimal $\varphi_m(\gamma)$. Thus

$$\alpha_0(m) = k \Leftrightarrow \exists\gamma\ [\gamma \in A \wedge \forall\gamma'\ (\gamma' \in A \Rightarrow \varphi_m(\gamma) \leq \varphi_m(\gamma')) \wedge \gamma(m) = k]$$

$$\Leftrightarrow \exists\gamma\ [y^0_{2n+1} \not\leq_{2n+1} \gamma \wedge (\ell_m,\gamma) \in W \wedge \gamma(m) = k] ,$$

(where $m \longmapsto \ell_m$ is a recursive function such that

$$\gamma \in A \wedge \forall\gamma'\ [\gamma' \in A \Rightarrow \varphi_m(\gamma) \leq \varphi_m(\gamma')] \Leftrightarrow (\ell_m,\gamma) \in W) ,$$

$$\Leftrightarrow \exists\gamma\ (\gamma <_{2n+1} y^0_{2n+1} \wedge (\ell_m,\gamma) \in W \wedge \gamma(m) = k)$$

$$\Leftrightarrow \exists\gamma\ [y^0_{2n+1} \not\leq_{2n+1} \gamma \wedge \sigma(\ell_m,\gamma) \leq \sigma(x) \wedge \gamma(m) = k] .$$

The last equivalence shows that $\alpha_0 \in \Delta^1_{2n+1}(x)$, thus $\alpha_0 \leq_{2n+1} \beta$ and we are done. ⊣

14.7 <u>Lemma</u> (Harrington). Assume $\underset{\sim}{\Delta}^1_{2n}$-DET. Then for all Q_{2n+1}-degrees p,q we have

$$p \leq q \Rightarrow \kappa_{2n+1}(p) \leq \kappa_{2n+1}(q) .$$

<u>Proof</u>. For $n = 0$ this is again obvious, so assume $n > 0$.

Let $\alpha \leq^Q_{2n+1} \beta$, but towards a contradiction $\kappa_{2n+1}(\beta) < \kappa_{2n+1}(\alpha)$. Pick γ with $\alpha \leq_{2n+1} \gamma$, $y^{\alpha}_{2n+1} \not\leq_{2n+1} \gamma$ and

$$\kappa_{2n+1}(\beta) < \lambda_{2n+1}(\gamma) < \kappa_{2n+1}(\alpha) .$$

Let A be a nonempty $\Delta^1_{2n+1}(\alpha)$ set containing no $\Delta^1_{2n+1}(\alpha)$ real and let $\{\varphi_m\}$ be an excellent $\Delta^1_{2n+1}(\alpha)$-scale on A and let $u_0 = y_{A,\overline{\varphi}}$ be the leftmost real associated with $A,\overline{\varphi}$. By the relativized 5.9, if we can show that $u_0 \leq_{2n+1} \gamma$, we conclude that $y^{\alpha}_{2n+1} \leq_{2n+1} \langle\gamma,\alpha\rangle \leq_{2n+1} \gamma$, a contradiction.

So it is enough to prove that $u_0 \leq_{2n+1} \gamma$.

Claim. For each m, there is $\alpha^* \in A$ with $\forall\delta \in A(\varphi_m(\alpha^*) \leq \varphi_m(\delta))$, such that $\langle\alpha,\beta,\alpha^*\rangle <_{2n+1} y^{\beta}_{2n+1}$.

Proof of Claim. Fix m. First notice that there is $\alpha' \in A$ with $y^{\beta}_{2n+1} \not\leq_{2n+1} \langle\alpha,\beta,\alpha'\rangle$, since otherwise $y^{\beta}_{2n+1} \in \mathrm{Hull}^{\langle\alpha,\beta\rangle}_{2n+1}(A)$ i.e. $y^{\beta}_{2n+1} \in Q_{2n+1}(\langle\alpha,\beta\rangle)$, so $y^{\beta}_{2n+1} \in Q_{2n+1}(\beta)$, a contradiction. So let

$$A' = \{\alpha' \in A : y^{\beta}_{2n+1} \not\leq_{2n+1} \langle\alpha,\beta,\alpha'\rangle\} .$$

Then $A' \in \Sigma^1_{2n+1}(\alpha,\beta)$ and A' is nonempty, so it contains by the relativized 5.7 a real α'' such that

$$\langle\alpha,\beta,\alpha''\rangle <_{2n+1} y^{\langle\alpha,\beta\rangle}_{2n+1} \leq_{2n+1} y^{\beta}_{2n+1} .$$

Pick then $\alpha^* \in A$ such that $\langle\alpha,\beta,\alpha^*\rangle < y^{\beta}_{2n+1}$ and $\varphi_m(\alpha^*)$ is minimal with that property. We shall show that $\forall\delta \in A(\varphi_m(\alpha^*) \leq \varphi_m(\delta))$. Otherwise $\{\delta \in A : \varphi_m(\delta) < \varphi_m(\alpha^*)\}$ is a nonempty $\Delta^1_{2n+1}(\alpha,\beta,\alpha^*)$ set, so it must contain a real δ^* with $\langle\alpha,\beta,\alpha^*,\delta^*\rangle <_{2n+1} y^{\langle\alpha,\beta,\alpha^*\rangle}_{2n+1}$. Now $\langle\alpha,\beta,\alpha^*\rangle <_{2n+1} y^{\beta}_{2n+1}$ and y^{β}_{2n+1} is a $\Pi^1_{2n+1}(\alpha,\beta,\alpha^*)$-singleton, so it is a non-trivial $\Pi^1_{2n+1}(\alpha,\beta,\alpha^*)$-singleton, thus $y^{\langle\alpha,\beta,\alpha^*\rangle}_{2n+1} \leq_{2n+1} \langle y^{\beta}_{2n+1},\langle\alpha,\beta,\alpha^*\rangle\rangle \leq_{2n+1} y^{\beta}_{2n+1}$. So $\langle\alpha,\beta,\delta^*\rangle < y^{\beta}_{2n+1}$, contradicting the minimality of $\varphi_m(\alpha^*)$. This concludes the proof of the claim.

Let now W be universal Π^1_{2n+1}, $\sigma : W \twoheadrightarrow \underset{\sim}{\delta}^1_{2n+1}$ a Π^1_{2n+1}-norm on W and $m \longmapsto \ell_m$ a recursive sequence such that $\alpha^* \in A \wedge \forall\delta \in A(\varphi_m(\alpha^*) \leq \varphi_m(\delta)) \Longleftrightarrow (\ell_m,\langle\alpha,\alpha^*\rangle) \in W$. Let $x \leq_T \gamma$ be such that $x \in W$ and $\kappa_{2n+1}(\beta) \leq \sigma(x)$. Then we have

$$u_0(m) = k \Longleftrightarrow \exists\alpha' \ (\alpha' \in A \wedge \forall\delta \in A \ (\varphi_m(\alpha') \leq \varphi_m(\delta)) \wedge \alpha'(m) = k)$$

$$\Longleftrightarrow \exists\alpha^* \ (\langle\alpha,\beta,\alpha^*\rangle <_{2n+1} y^{\beta}_{2n+1} \wedge \alpha^* \in A \wedge \forall\delta \in A \ (\varphi_m(\alpha^*) \leq \varphi_m(\delta)) \wedge \alpha^*(m) = k)$$

$$\Longleftrightarrow \exists\alpha^* \ (\alpha^* \in A \wedge \sigma(\ell_m,\langle\alpha,\alpha^*\rangle) \leq \sigma(x) \wedge \alpha^*(m) = k) ,$$

where the bound $\sigma(\ell_m,\langle\alpha,\alpha^*\rangle) \leq \sigma(x)$ comes from the fact that if $\langle\alpha,\beta,\alpha^*\rangle <_{2n+1} y^{\beta}_{2n+1}$, then $\lambda_{2n+1}(\langle\alpha,\alpha^*\rangle) \leq \lambda_{2n+1}(\langle\alpha,\beta,\alpha^*\rangle) < \kappa_{2n+1}(\beta) \leq \sigma(x)$. Thus $u_0 \in \Delta^1_{2n+1}(\langle\alpha,\sigma\rangle)$, so $u_0 \leq_{2n+1} \gamma$ and we are done. ⊣

14.8 Lemma. Assume $n > 0$ and $\underset{\sim}{\Delta}^1_{2n}$-DET. For all α,β

$$\lambda_{2n+1}(\langle\alpha,\beta\rangle) < \kappa_{2n+1}(\alpha) \Longleftrightarrow \kappa_{2n+1}(\langle\alpha,\beta\rangle) = \kappa_{2n+1}(\alpha)$$

$$\Longleftrightarrow y^{\alpha}_{2n+1} \not\leq_{2n+1} \langle\alpha,\beta\rangle .$$

Proof. If $\lambda_{2n+1}(\langle\alpha,\beta\rangle) < \kappa_{2n+1}(\alpha)$, then $y^{\alpha}_{2n+1} \not\leq_{2n+1} \langle\alpha,\beta\rangle$ by 14.2. If $y^{\alpha}_{2n+1} \not\leq_{2n+1} \langle\alpha,\beta\rangle$, then $\kappa_{2n+1}(\langle\alpha,\beta\rangle) = \kappa_{2n+1}(\alpha)$ by 14.4 and 14.7. Finally, if $\kappa_{2n+1}(\langle\alpha,\beta\rangle) = \kappa_{2n+1}(\alpha)$ then $\lambda_{2n+1}(\langle\alpha,\beta\rangle) < \kappa_{2n+1}(\langle\alpha,\beta\rangle) = \kappa_{2n+1}(\alpha)$ by 14.3 (ii). ⊣

The Gandy Basis Theorem asserts that each nonempty Σ^1_1 set A contains a real α with $\omega^{\alpha}_1 = \omega^{ck}_1 = \omega^{\lambda t.0}_1$. The first part of the next result generalizes this fact.

14.9 Lemma ($\underset{\sim}{\Delta}^1_{2n}$-DET). (i) If A is a nonempty Σ^1_{2n+1} set, then it contains a real α with $\kappa_{2n+1}(\alpha) = \kappa_{2n+1}$.

(ii) If A is a nonempty Π^1_{2n+1} set, then it contains a real $\alpha <_{2n+1} y^0_{2n+1}$ iff it contains a real α with $\kappa_{2n+1}(\alpha) = \kappa_{2n+1}$.

Proof. (i) If $\forall\alpha \in A\ (\kappa_{2n+1}(\alpha) > \kappa_{2n+1})$, then $\forall\alpha \in A\ (y^0_{2n+1} \leq_{2n+1} \alpha)$, so $y^0_{2n+1} \in \mathrm{Hull}_{2n+1}(A)$, a contradiction.

(ii) If $n = 0$ and there is $\alpha_0 \in A$ with $\lambda_1(\alpha_0) = \lambda_1$, let W be universal Π^1_1 and $\varphi : W \twoheadrightarrow \underset{\sim}{\delta}^1_1$ a Π^1_1-norm and choose $n_0 \in \omega$ such that

$$\alpha \in A \Longleftrightarrow (n_0,\alpha) \in W .$$

As $\sigma(n_0,\alpha) < \lambda_1$, choose x recursive, $x \in W$ such that $\varphi(n_0,\alpha_0) \leq \varphi(x)$. Then $\{\alpha : \varphi(n_0,\alpha) \leq \varphi(x)\} \subseteq A$ is a nonempty Δ^1_1 set so it contains a real $\alpha <_{2n+1} y^0_1$.

Assume now $n > 0$. Let again W be universal Π^1_{2n+1}, $\varphi : W \twoheadrightarrow \underset{\sim}{\delta}^1_{2n+1}$ a Π^1_{2n+1}-norm and pick $n_0 \in \omega$ such that

$$\alpha \in A \Longleftrightarrow (n_0,\alpha) \in W .$$

Pick $\alpha_0 \in A$ such that $\kappa_{2n+1}(\alpha_0) = \kappa_{2n+1}$. Then let $\alpha_1 <_{2n+1} y^0_{2n+1}$ be such that

$$\lambda_{2n+1}(\alpha_0) < \lambda_{2n+1}(\alpha_1) .$$

Find $x \in W$, $x \leq_T \alpha_1$ with $\lambda_{2n+1}(\alpha_0) \leq \varphi(x)$. Then $A' = \{\alpha : \varphi(n_0,\alpha) \leq \varphi(x)\}$ is a nonempty $\Delta^1_{2n+1}(x)$ set, so it contains a real $\alpha <_{2n+1} y^0_{2n+1}$, by 5.7. Since $A' \subseteq A$ we are done. ⊣

Our last result is the following definability estimate generalizing the standard fact that the relation

$$\omega^{\alpha}_1 \leq \omega^{\beta}_1$$

is Σ^1_1. The proof is motivated by the proof of an analogous result of Harrington in recursion in higher types.

14.10 Lemma ($\underset{\sim}{\Delta}^1_{2n}$-DET). The relation

$$\kappa_{2n+1}(\alpha) \leq \kappa_{2n+1}(\beta)$$

is Σ^1_{2n+1}.

Proof. For $n = 0$ this is the above mentioned fact, so assume $n > 0$.

Sublemma 1. Let A be a Σ^1_{2n+1} set. If $A^* = \{\alpha \in A : \kappa_{2n+1}(\alpha) = \kappa_{2n+1}\}$, then $A^* \in \Delta^1_{2n+1}(y^0_{2n+1})$.

Proof. Put

$$S(\alpha,\beta) \Longleftrightarrow \kappa_{2n+1}(\langle\alpha,\beta\rangle) = \kappa_{2n+1}(\alpha) \quad (\Longleftrightarrow y^\alpha_{2n+1} \not\leq_{2n+1} \langle\alpha,\beta\rangle) .$$

Then $S \in \Sigma^1_{2n+1}$. Let W be a universal Π^1_{2n+1} set and $\sigma : W \twoheadrightarrow \underset{\sim}{\delta}^1_{2n+1}$ a Π^1_{2n+1}-norm. Let n_0 be such that

$$\alpha \notin A \Longleftrightarrow (n_0,\alpha) \in W ,$$

and k_0 be such that

$$\beta = y^0_{2n+1} \Longleftrightarrow (k_0,\beta) \in W .$$

Then we claim that

$$\alpha \in A^* \Longleftrightarrow \neg\, S(\alpha,y^0_{2n+1}) \wedge \neg\, (\sigma(n_0,\alpha) < \sigma(k_0,y^0_{2n+1})) ,$$

which implies that $A^* \in \Pi^1_{2n+1}(y^0_{2n+1})$ and since also $A^* = \{\alpha \in A : y^0_{2n+1} \not\leq_{2n+1} \alpha\} \in \Sigma^1_{2n+1}$, we have $A^* \in \Delta^1_{2n+1}(y^0_{2n+1})$. To verify our claim, assume first that $\alpha \in A^*$. Then $\alpha \in A$, so $(n_0,\alpha) \notin W$. Also $\kappa_{2n+1}(\alpha) = \kappa_{2n+1}$, so that $\kappa_{2n+1}(\alpha) = \kappa_{2n+1} < \kappa_{2n+1}(\langle\alpha,y^0_{2n+1}\rangle)$ i.e. $\neg\, S(\alpha,y^0_{2n+1})$. Also $(k_0,y^0_{2n+1}) \in W$ thus $\neg\, (\sigma(n_0,\alpha) < \sigma(k_0,y^0_{2n+1}))$. Conversely, assume the right hand side of the above equivalence holds. If $\alpha \notin A$, then $(n_0,\alpha) \in W$ so that $\sigma(k_0,y^0_{2n+1}) \leq \sigma(n_0,\alpha)$, thus $\beta = y^0_{2n+1} \Longleftrightarrow \sigma(k_0,\beta) \leq \sigma(n_0,\alpha)$ and therefore $y^0_{2n+1} \leq_{2n+1} \alpha$. Then $\kappa_{2n+1}(\langle\alpha,y^0_{2n+1}\rangle) = \kappa_{2n+1}(\alpha)$, i.e. $S(\alpha,y^0_{2n+1})$, a contradiction. So $\alpha \in A$. To see that $\kappa_{2n+1}(\alpha) = \kappa_{2n+1}$, notice that otherwise $y^0_{2n+1} \leq_{2n+1} \alpha$, thus again $S(\alpha,y^0_{2n+1})$, a contradiction.

Sublemma 2. Let W be a universal Π^1_{2n+1} set, $\sigma : W \twoheadrightarrow \underset{\sim}{\delta}^1_{2n+1}$ a Π^1_{2n+1}-norm on W. Then there is a recursive function $i \longmapsto n_i$ such that

(a) $\forall\alpha\, \forall i\, \exists\beta\, [\langle\alpha,\beta\rangle <_{2n+1} y^\alpha_{2n+1} \wedge (n_i,\langle\alpha,\beta\rangle) \in W]$

(b) $\forall\alpha\, \forall\{\beta_i\}$ [If for all i, $(n_i,\langle\alpha,\beta_i\rangle) \in W$ and $\lambda_{2n+1}(\langle\alpha,\beta_i\rangle) < \kappa_{2n+1}(\alpha)$, then $\sup_i \varphi(n_i,\langle\alpha,\beta_i\rangle) = \kappa_{2n+1}(\alpha)$].

Proof. Let $P(\alpha,\beta)$ be a Π^1_{2n} set such that $\forall\alpha\, \exists\beta\, P(\alpha,\beta)$ but $\forall\alpha\, \neg\, \exists\beta \leq_{2n+1} \alpha P(\alpha,\beta)$. Let $\{\varphi_m\}$ be an excellent Δ^1_{2n+1}-scale on P. Put

$$(n_i,\langle\alpha,\beta\rangle) \in W \Longleftrightarrow P(\alpha,\beta) \wedge \forall\gamma\, [P(\alpha,\gamma) \Rightarrow \varphi_i(\alpha,\beta) \leq \varphi_i(\alpha,\gamma)] .$$

Then (a) follows from the proof of 5.8. If (b) fails fix $\alpha,\{\beta_i\}$ which are counterexamples. Then find γ and $x \leq_T \langle\alpha,\gamma\rangle$ such that $x \in W$ and

$$\sup_i \sigma(n_i,\langle\alpha,\beta_i\rangle) < \sigma(x) < \lambda_{2n+1}(\langle\alpha,\gamma\rangle) < \kappa_{2n+1}(\alpha) .$$

Let u_0 be the leftmost real associated with P_α and $\overline{\varphi}$ restricted to it. Then as usual

$$u_0(i) = k \Longleftrightarrow \exists\beta\ [P(\alpha,\beta) \wedge \sigma(n_i,\langle\alpha,\beta\rangle) < \sigma(x) \wedge \beta(i) = k] ,$$

so $u_0 \leq_{2n+1} \langle\alpha,\gamma\rangle$, so $y^\alpha_{2n+1} \leq_{2n+1} \langle\alpha,\gamma\rangle$, a contradiction as $\lambda_{2n+1}(\langle\alpha,\gamma\rangle) < \kappa_{2n+1}(\alpha)$.

Sublemma 3. The relations (in the notation of Sublemma 2) below are all Π^1_{2n+1}

(a) $z \in W \wedge \sigma(z) < \kappa_{2n+1}(\alpha)$,

(b) $\alpha \leq_{2n+1} z \wedge z \in W \wedge \kappa_{2n+1}(\alpha) \leq \sigma(z)$,

(c) $\kappa_{2n+1}(\alpha) < \kappa_{2n+1}(\beta) < \lambda_{2n+1}(\langle\alpha,\beta\rangle)$.

Proof. (a) We have in the notation of Sublemma 2:

$$z \in W \wedge \sigma(z) < \kappa_{2n+1}(\alpha) \Longleftrightarrow z \in W \wedge \exists i \neg \exists\beta\ [\sigma(n_i,\langle\alpha,\beta\rangle) < \sigma(z) \wedge \lambda_{2n+1}(\langle\alpha,\beta\rangle) < \kappa_{2n+1}(\alpha)] .$$

(b) Note that

$$\begin{aligned} &\alpha \leq_{2n+1} z \wedge z \in W \wedge \kappa_{2n+1}(\alpha) \leq \sigma(z) \\ &\Longleftrightarrow \alpha \leq_{2n+1} z \wedge z \in W \wedge \end{aligned}$$

$$(*)\quad \begin{cases} \exists y \leq_{2n+1} z\ [y = y^\alpha_{2n+1} \wedge \forall i\, \exists\beta \leq_{2n+1} y\ (\lambda_{2n+1}(\langle\alpha,\beta\rangle) < \kappa_{2n+1}(\alpha) \\ \qquad \wedge (n_i,\langle\alpha,\beta\rangle) \in W \wedge \sigma(n_i,\langle\alpha,\beta\rangle) < \sigma(z))] . \end{cases}$$

This is because $\kappa_{2n+1}(\alpha) \leq \sigma(z) \Rightarrow \kappa_{2n+1}(\alpha) < \lambda_{2n+1}(z) \Rightarrow y^\alpha_{2n+1} \leq_{2n+1} \langle z,\alpha\rangle$.

That (*) is Π^1_{2n+1} follows from the fact that by Sublemma 1 relativized,

$$\{\beta : \lambda_{2n+1}(\langle\alpha,\beta\rangle) < \kappa_{2n+1}(\alpha)\} = \{\beta : \kappa_{2n+1}(\langle\alpha,\beta\rangle) = \kappa_{2n+1}(\alpha)\}$$

is $\Delta^1_{2n+1}(y^\alpha_{2n+1})$, uniformly on α.

(c) Note that

$$\begin{aligned} &\kappa_{2n+1}(\alpha) < \kappa_{2n+1}(\beta) < \lambda_{2n+1}(\langle\alpha,\beta\rangle) \\ &\Longleftrightarrow \kappa_{2n+1}(\alpha) < \lambda_{2n+1}(\langle\alpha,\beta\rangle) \wedge \kappa_{2n+1}(\beta) < \lambda_{2n+1}(\langle\alpha,\beta\rangle) \\ &\quad \wedge \exists y \leq_{2n+1} \langle\alpha,\beta\rangle\ \exists z \leq_{2n+1} \langle\alpha,\beta\rangle \\ &\quad \{y = y^\alpha_{2n+1} \wedge z = y^\beta_{2n+1} \wedge \exists x \leq_{2n+1} z\ [x \in W \wedge \sigma(x) < \kappa_{2n+1}(\beta) \wedge \\ &\quad \forall i\, \exists\gamma \leq_{2n+1} y\ (\lambda_{2n+1}(\langle\alpha,\gamma\rangle < \kappa_{2n+1}(\alpha)) \wedge \\ &\quad (n_i,\langle\alpha,\gamma\rangle) \in W \wedge \sigma(n_i,\langle\alpha,\gamma\rangle) < \sigma(x))]\} , \end{aligned}$$

which completes the proof of (c).

We put now together these sublemmas to complete the proof of 14.10.

We compute that the relation

$$\kappa_{2n+1}(\alpha) < \kappa_{2n+1}(\beta)$$

is Π^1_{2n+1}. Observe that $\kappa_{2n+1}(\alpha) < \kappa_{2n+1}(\beta) \Rightarrow \kappa_{2n+1}(\alpha) < \kappa_{2n+1}(\langle\alpha,\beta\rangle)$ $y^\alpha_{2n+1} \leq_{2n+1} \langle\alpha,\beta\rangle \Rightarrow \kappa_{2n+1}(\alpha) < \lambda_{2n+1}(\langle\alpha,\beta\rangle)$. Thus

$$\begin{aligned}\kappa_{2n+1}(\alpha) < \kappa_{2n+1}(\beta) \Longleftrightarrow {} & [\kappa_{2n+1}(\alpha) < \kappa_{2n+1}(\beta) < \lambda_{2n+1}(\langle\alpha,\beta\rangle)] \\ & \vee [\kappa_{2n+1}(\alpha) < \lambda_{2n+1}(\langle\alpha,\beta\rangle) \wedge \\ & \quad \exists n\, ((n,\langle\alpha,\beta\rangle) \in W \wedge \kappa_{2n+1}(\alpha) \leq \sigma(n,\langle\alpha,\beta\rangle) \\ & \quad \wedge \sigma(n,\langle\alpha,\beta\rangle) < \kappa_{2n+1}(\beta))] ,\end{aligned}$$

so by Sublemma 3 we are done. ⊣

We collect now the results of the preceding lemmas in the following theorem.

14.11 <u>Theorem</u> ($\underset{\sim}{\Delta}^1_{2n}$-DET). For each real α, let

$$\kappa_{2n+1}(\alpha) = \sup\{\lambda_{2n+1}(\langle\alpha,\beta\rangle) : y^\alpha_{2n+1} \not\leq_{2n+1} \langle\alpha,\beta\rangle\} .$$

Then the following hold:

(1) $\lambda_{2n+1}(\alpha) \leq \kappa_{2n+1}(\alpha) < \lambda_{2n+1}(y^\alpha_{2n+1})$

(2) For $n = 0$, $\kappa_1(\alpha) = \lambda_1(\alpha) = \omega^\alpha_1$. For $n > 0$, $\lambda_{2n+1}(\alpha) < \kappa_{2n+1}(\alpha)$ and for each β, $\kappa_{2n+1}(\alpha) \neq \lambda_{2n+1}(\langle\alpha,\beta\rangle)$, so that the sup above is not attained.

(3) $\kappa_{2n+1}(\alpha) = \sup\{\lambda_{2n+1}(\langle\alpha,\beta\rangle) : \langle\alpha,\beta\rangle <_{2n+1} y^\alpha_{2n+1}\}$, so that $\mathrm{cof}(\kappa_{2n+1}(\alpha)) = \omega$.

(4) For $n > 0$, $\lambda_{2n+1}(\langle\alpha,\beta\rangle) < \kappa_{2n+1}(\alpha) \Longleftrightarrow \kappa_{2n+1}(\langle\alpha,\beta\rangle) = \kappa_{2n+1}(\alpha)$ $\Longleftrightarrow y^\alpha_{2n+1} \not\leq_{2n+1} \langle\alpha,\beta\rangle$.

(5) A Σ^1_{2n+1} set A is nonempty iff it contains a real α with $\kappa_{2n+1}(\alpha) = \kappa_{2n+1}$ $(= \kappa_{2n+1}(\lambda t.0))$.

(6) A Π^1_{2n+1} set A contains a real $\alpha <_{2n+1} y^0_{2n+1}$ iff it contains a real α with $\kappa_{2n+1}(\alpha) = \kappa_{2n+1}$.

(7) The relation

$$\kappa_{2n+1}(\alpha) \leq \kappa_{2n+1}(\beta)$$

is Σ^1_{2n+1}.

Moreover this ordinal assignment is invariant under $\equiv^Q_{2n+1}$, and if we let for each Q_{2n+1}-degree $p = [\alpha]^Q_{2n+1}$, $\kappa_{2n+1}(p) = \kappa_{2n+1}(\alpha)$ we have

(8) $p \leq q \Rightarrow \kappa_{2n+1}(p) \leq \kappa_{2n+1}(q)$

(9) $p \leq q \Rightarrow [p' \leq q \Longleftrightarrow \kappa_{2n+1}(p) < \kappa_{2n+1}(q)]$.

§15. Q-theory for the 2-envelope of ^3E. In this and the next section we shall examine versions of Q-theory applicable to other interesting pointclasses beyond the analytical hierarchy. We concentrate here mainly on the pointclass consisting of those pointsets which are Kleene semirecursive in ^3E, the so-called 2-envelope of ^3E, in symbols

$$_2\mathrm{ENV}(^3E) .$$

In the next section we will take up the case of the pointclass of (absolutely) inductive sets, in symbols

$$\mathrm{IND} .$$

It will be convenient to present first an abstract version of Q-theory applicable to certain pointclasses satisfying some general conditions. Examples will be the Π^1_{2n+1} for $n > 0$, $_2\mathrm{ENV}(^3E)$ and IND, under appropriate determinacy hypotheses.

15.1 Definition. Let Γ be a pointclass. We call Γ Q-adequate if Γ is a scaled Spector pointclass closed under universal quantification over ω^ω and such that the following conditions are also met:

(i) $WO \in \Delta$, where $\Delta = \Gamma \cap \check{\Gamma}$ ($\check{\Gamma}$ is the dual class of Γ).

(ii) Every thin set in $\underset{\sim}{\Gamma}$ is countable and every thin set in $\Delta(\alpha)$ ($= \Gamma(\alpha) \cap \check{\Gamma}(\alpha)$) contains only $\Delta(\alpha)$ reals.

(iii) The ideal of thin sets if $\underset{\sim}{\Gamma}$-additive, i.e. if $\{A_\xi\}_{\xi<\vartheta}$ is a transfinite sequence of thin sets for which the associated prewellordering

$$x \leq y \Longleftrightarrow x,y \in \bigcup_{\xi<\vartheta} A_\xi \wedge \text{least } \xi\ [x \in A_\xi] \leq \text{least } \xi\ [y \in A_\xi] ,$$

is in $\underset{\sim}{\Gamma}$, then $\bigcup_{<\vartheta} A_\xi$ is thin.

Note that if Γ is Q-adequate, so are all its relativizations $\Gamma(\beta)$.

Clearly if $\underset{\sim}{\Pi}^1_{2n+1}$-DET holds and $n > 0$, Π^1_{2n+1} is Q-adequate (for (iii) see [K2], 1A-1). If $_2\underset{\sim}{\mathrm{ENV}}(^3E)$-DET holds then $_2\mathrm{ENV}(^3E)$ is Q-adequate. This follows from [Mo2], p. 192. Finally if $\underset{\sim}{\mathrm{IND}}$-DET holds then IND is Q-adequate. Again this follows from [Mo2].

Now exactly as in [K2] we can prove

15.2 Theorem. Let Γ be a Q-adequate pointclass. Then there exists a largest countable Γ set, which will be denoted by C_Γ.

Thus $C_{\Pi^1_{2n+1}} = C_{2n+1}$.

The relativized version of C_Γ for any $\beta \in \omega^\omega$ will be denoted by $C_\Gamma(\beta)$.

The concepts of Δ-reducibility $\leq_\Delta$, Δ-degrees and Δ-jump are defined in the usual way. Let also for each real α,

$$[\alpha] = \{(\alpha)_n : n \in \omega\}$$

and

$$(\alpha)_{n,k} = ((\alpha)_n)_k .$$

We say that $A \subseteq \omega^\omega$ is <u>closed under pairings</u> if $\alpha,\beta \in A \Rightarrow \langle\alpha,\beta\rangle \in A$. Define now

$$(\alpha,\beta) \in S_\Gamma \Longleftrightarrow \text{(i)}\ \beta \in [\alpha] \wedge [\alpha] \text{ is closed under pairings,}$$

$$\text{(ii)}\ [\alpha] \text{ is downward closed under } \leq_\Delta \text{ and } \Delta\text{-jumps,}$$

$$\text{(iii)}\ C_\Gamma(\beta) \cap [\alpha] \text{ is countable in } [\alpha], \text{ i.e. there is } n \text{ such that}$$

$$C_\Gamma(\beta) \cap [\alpha] = \{(\alpha)_{n,k} : k \in \omega\} .$$

15.3 <u>Lemma</u>. Let Γ be Q-adequate. Then S_Γ is in $\check{\Gamma}$.

<u>Proof</u>. From standard properties of Spector pointclasses (see [Mo3], Ch. 4) it follows that (i) and downward closure under $\leq_\Delta$ are properties in $\check{\Gamma}$. For closure under the Δ-jump notice that as in 8.2, if $A \subseteq \omega$ is in $\Gamma(\beta)$, $w \in WO$ and $|w| \geq \delta_\Gamma(\beta)$, where

$$\delta_\Gamma(\beta) = \sup\{|w| : w \in WO \wedge w \in \Delta(\beta)\} ,$$

then A is Δ in w. So closure under the Δ-jump translates (in the presence of downward closure under $\leq_\Delta$) to

$$\forall m\, \exists k\, ((\alpha)_k \in WO \wedge \delta_\Gamma((\alpha)_m) \leq |(\alpha)_k|) ,$$

which is again in $\check{\Gamma}$.

Finally (iii) translates to

$$\exists n\, \forall m\, [(\alpha)_m \in C_\Gamma(\beta) \Rightarrow \exists k\, ((\alpha)_m = (\alpha)_{n,k})] ,$$

because if $C_\Gamma(\beta) \cap [\alpha] \subseteq \{(\alpha)_{n,k} : k \in \omega\}$ then since $\{k : (\alpha)_{n,k} \in C_\Gamma(\beta)\} = A \in \Gamma((\alpha)_n,\beta)$ we have, by closure under the Δ-jump and $\leq_\Delta$, that $A \in [\alpha]$ and thus there is some n' with $C_\Gamma(\beta) \cap [\alpha] = \{(\alpha)_{n',k} : k \in \omega\}$. $\dashv$

15.4 <u>Definition</u>. For each β let

$$\alpha \in Q_\Gamma(\beta) \Longleftrightarrow \forall\gamma\, ((\gamma,\beta) \in S_\Gamma \Rightarrow \alpha \in [\gamma]) .$$

Again with the obvious definition of $\check{\Gamma}(\beta)$-hull and $\Gamma(\beta)$-bounded set we have

15.5 Theorem. Assume Γ is a Q-adequate pointclass. Then $Q_{\check{\Gamma}}(\beta)$ is the largest countable $\Gamma(\beta)$-bounded set of reals and also the largest $\check{\Gamma}(\beta)$-hull.

Proof. For notational simplicity we drop the β, since the proof below relativizes immediately. First, exactly as in 3.2 (ii), every countable Γ-bounded set is contained in a $\check{\Gamma}$-hull. Also Q_Γ $(= Q_\Gamma(\lambda t.0))$ is Γ-bounded (as in 3.2 (i)), and countable as $\forall\beta\,\exists\alpha\, S_\Gamma(\alpha,\beta)$, thus it is enough to show that every $\check{\Gamma}$-hull A is contained in Q_Γ.

As A is countable and in Γ it is contained in C_Γ. Let $(\gamma,\lambda t.0) \in S_\Gamma$. We have to show that $A \subseteq [\gamma]$. Since Γ is Q-adequate, the Δ-degrees in C_Γ are wellordered under their natural partial ordering, say in a sequence $\{d_0, d_1, \ldots, d_\xi, \ldots\}_{\xi<\rho_\Gamma}$. Also $d_{\xi+1} = d_\xi{}'$. Let ξ be the least ordinal such that $d_\xi \subseteq A - [\gamma]$, towards a contradiction. (Clearly $A = \bigcup_{\eta<\xi_0} d_\eta$ for some $\xi_0 \le \rho_\Gamma$.) Let $\varphi : C_\Gamma \twoheadrightarrow \kappa$ be a Γ-norm. Then let β be in d_ξ and have least $\varphi(\beta)$. Since $\varphi(x) \le \varphi(y) \Rightarrow x \le_\Delta y$ we have $\varphi(\beta) \le \sup\{\delta_\Gamma(\beta') : \beta' \in \bigcup_{\eta<\xi} d_\eta\}$. But $\bigcup_{\eta<\xi} d_\eta \subseteq C_\Gamma \cap [\gamma] = \{(\gamma)_{n,k} : k \in \omega\}$ for some n. So there is m with $(\gamma)_m \in WO$ and $|(\gamma)_m| \ge \sup\{\delta_\Gamma(\beta') : \beta' \in \bigcup_{\eta<\xi} d_\eta\} = \varphi(\beta)$. Then we claim that $\beta \le_\Delta (\gamma)_m$, thus $\beta \in [\gamma]$, a contradiction, and the proof is complete. Our claim follows from the following general fact, and the observation that $\psi = \varphi \restriction A$ is a Γ-norm on A.

Fact. Let Γ be Q-adequate, A a countable Γ-bounded set, and let $\psi : A \twoheadrightarrow \lambda$ a Γ-norm on A. Then

$$\forall\alpha \in A\, \forall w \in WO\, (\psi(\alpha) \le |w| \Rightarrow \alpha \le_\Delta w)\ .$$

Proof. Let

$$\begin{aligned} T(\varepsilon,\lesssim) \Leftrightarrow\ & \lesssim \text{ is a prewellordering on } \omega \\ & \wedge \forall m\, \forall n\, (m \ne n \Rightarrow (\varepsilon)_m \ne (\varepsilon)_n) \\ & \wedge \forall\alpha \in A\, \exists m\, (\alpha = (\varepsilon)_m) \\ & \wedge \forall m\, \forall n\, ((\varepsilon)_m, (\varepsilon)_n \in A \Rightarrow [\psi((\varepsilon)_m) \le \psi((\varepsilon)_n) \Leftrightarrow m \lesssim n]) \\ & \wedge \forall m\, \forall n\, (m \lesssim n \wedge (\varepsilon)_n \in A \Rightarrow (\varepsilon)_m \in A)\ . \end{aligned}$$

Notice that the last condition, in view of the preceding one, can be replaced by

$$\forall m\, \forall n\, (m \lesssim n \wedge (\varepsilon)_n \in A \Rightarrow \psi((\varepsilon)_m) \le \psi((\varepsilon)_n))\ .$$

Thus $T \in \check{\Gamma}$. Now notice that if $w \in WO$ and $|w| < \lambda$ we have

$$\alpha \in A \wedge \psi(\alpha) < |w| \Leftrightarrow \exists \varepsilon \exists \precsim [T(\varepsilon,\precsim) \wedge \exists n (\{(m,m') : m \precsim m' \wedge m' \precsim n\} \text{ has order type } < |w| \wedge \psi(\alpha) = \psi((\varepsilon)_n))] ,$$

$$\Leftrightarrow \forall \varepsilon \forall \precsim [T(\varepsilon,\precsim) \Rightarrow \exists n (\{(m,m') : m \precsim m' \wedge m' \precsim n\} \text{ has order type } < |w| \wedge \psi(\alpha) = \psi((\varepsilon)_n))] ,$$

so $\{\alpha \in A : \psi(\alpha) < |w|\}$ is a countable $\Delta(w)$ set, thus it contains only $\Delta(w)$ reals.

⊣

In general we cannot assert that Q_Γ is the largest Γ-bounded set, because if Γ is closed under existential quantification over ω^ω, then ω^ω is itself Γ-bounded.

15.6 <u>Definition</u>. For $\alpha,\beta \in \omega^\omega$ let

$$\alpha \leq_\Gamma^Q \beta \Leftrightarrow \alpha \in Q_\Gamma(\beta) .$$

Again we have as in 4.5

15.7 <u>Proposition</u>. Let Γ be a Q-adequate pointclass. Then $\leq_\Gamma^Q$ is transitive.

We define then $\alpha \equiv_\Gamma^Q \beta$, and $[\alpha]_\Gamma^Q$ as usual. The analog of Proposition 4.7 also goes through with the same proof, i.e. C_Γ is closed under $\equiv_\Gamma^Q$.

In order to define now a jump operator for Q_Γ-degrees we have to impose a further condition on Γ. If Γ is closed under existential quantification over ω^ω, then actually $Q_\Gamma = C_\Gamma$, (because C_Γ is countable and Γ-bounded, so $C_\Gamma \subseteq Q_\Gamma$) and there are no nontrivial Γ-singletons. This is a limiting case of Q-theory, which we will take up in the last section. Its least example is IND. So in the rest of this section we shall postulate the (uniform) non-closure of Γ under existential quantification over ω^ω.

15.8 <u>Definition</u>. Let Γ be a pointclass. Then we say that Γ is <u>uniformly not closed under</u> $\exists^{\mathbb{R}}$ if $\check{\Gamma} \subseteq \exists^{\mathbb{R}}\Gamma$, where $\exists^{\mathbb{R}}$ = existential quantification over ω^ω.

Notice that if Γ is uniformly not closed under $\exists^{\mathbb{R}}$ then so are its relativizations $\Gamma(\beta)$. So everything we do below relativizes immediately to all $\beta \in \omega^\omega$.

If Γ is uniformly not closed under $\exists^{\mathbb{R}}$ and Q-adequate there exist nontrivial Γ-singletons, as otherwise by uniformization, every nonempty Γ set would contain a Δ real thus every $\exists^{\mathbb{R}}\Gamma$ set of integers would be in Γ, a contradiction. Also as in 5.1 the Γ-singletons are closed under $\equiv_\Delta$ and $\leq_\Delta$ is a prewellordering on Γ-singletons. Also no nontrivial Γ-singleton belongs to Q_Γ, and the analog of 5.4 goes through.

At this stage it is not clear that one can prove any further interesting properties of the first nontrivial singleton for the general Γ's we have been considering. So from now on we shall restrict ourselves to the class $\Gamma = {}_2\mathrm{ENV}({}^3E)$ and other pointclasses very closely resembling it. Their basic theory is worked out in [Mo1], [H2], [K3]. These last two references contain all the results about these classes that we will need below, and we assume that the reader is familiar with them. We summarize in the following definition the extra properties that hold for ${}_2\mathrm{ENV}({}^3E)$.

15.9 <u>Definition</u>. A Spector pointclass Γ closed under $\forall^{\mathbb{R}}$ is <u>${}_2\mathrm{ENV}({}^3E)$-like</u> if it satisfies the following:

(i) Γ is closed under ${}^3E$, i.e. for any two disjoint relations $P(x,\alpha)$, $S(x,\alpha)$ in Γ, the relation

$$T(x) \Longleftrightarrow \forall\alpha\, [P(x,\alpha) \vee S(x,\alpha)] \wedge \exists\alpha\, P(x,\alpha)$$

is also in Γ.

(ii) Γ satisfies the <u>Moschovakis representation</u>, i.e. there is a universal Γ set $W \subseteq \omega \times \omega^\omega$ such that for some $P \subseteq \omega^\omega \times \omega^\omega$ with the property that $P \in \Gamma$ and for each $\alpha \in \omega^\omega$, $P_\alpha \in \Delta(\alpha)$ we have

$$(n,\alpha) \notin W \Longleftrightarrow \exists\beta\, ((\beta)_0 = \langle n,\alpha\rangle \wedge \forall i\, P((\beta)_i,(\beta)_{i+1})) .$$

That ${}_2\mathrm{ENV}({}^3E)$ has these properties is proved in [Mo1]; see also [KM2]. Note that by condition (ii), if Γ is ${}_2\mathrm{ENV}({}^3E)$-like then it is uniformly not closed under $\exists^{\mathbb{R}}$. So let us denote by y^0_Γ any representative of the first nontrivial Γ-singleton, and similarly for its relativizations y^α_Γ. We first have

15.10 <u>Theorem</u>. Let Γ be a ${}_2\mathrm{ENV}({}^3E)$-like pointclass. If Γ is Q-adequate then y^0_Γ is a basis for Γ, i.e. every nonempty Γ set contains a member in $\Delta(y^0_\Gamma)$.

<u>Proof</u>. We review first some aspects of the reflection theory of a ${}_2\mathrm{ENV}({}^3E)$-like Γ (see [H2] and [K3]). Let W be universal in Γ and let $\varphi : W \twoheadrightarrow \kappa$ be a Γ-norm. We call an ordinal ξ Γ,α-<u>reflecting</u> if

$$\forall n\ [\exists\beta\ (\varphi(n,\langle\alpha,\beta\rangle) < \xi) \Rightarrow \exists\beta\ (\varphi(n,\langle\alpha,\beta\rangle) < \lambda_\Gamma(\alpha))]\ ,$$

where

$$\lambda_\Gamma(\alpha) = \lambda(\alpha) = \sup\{\xi : \xi \text{ is the length of a } \Delta(\alpha) \text{ prewellordering of } \omega^\omega\}$$

(<u>Caution</u>: In [K3], $\lambda_\Gamma(\alpha)$ is denoted $\kappa^{\Gamma,\alpha}$ and $\lambda^{\Gamma,\alpha}$ is reserved for another ordinal assignment.) The ordinal

$$\kappa_r^\Gamma(\alpha) = \kappa_r(\alpha) = \sup\{\lambda(\alpha,\beta) : \lambda(\alpha,\beta) \text{ is } \alpha\text{-reflecting}\}$$

is called the <u>largest</u> Γ,α-<u>reflecting ordinal</u> and plays an important role in the theory of Γ. We will need here the following facts (see [K3], 5.7 (ii), 5.12 and 5.7 (iv)):

(a) A set $A \in \Gamma(\alpha)$ contains a nonempty $\Delta(\alpha)$ subset iff it contains a real β with $\lambda(\alpha,\beta) < \kappa_r(\alpha)$.

(b) Every nonempty $A \in \check{\Gamma}(\alpha)$ contains a nonempty $B \subseteq A$ in $\Delta(\alpha,\beta)$, for any β with $\lambda(\alpha,\beta) > \kappa_r(\alpha)$.

Let now $A \neq \emptyset$ be in Γ. Since $\{y_\Gamma^0\} \in \Gamma$ contains no nonempty Δ set (since $y_\Gamma^0 \notin \Delta$), we have by (a) that $\lambda(y_\Gamma^0) > \kappa_r(\lambda t.0)$ (since it can be shown, [K3] 5.7 (vi), that $\kappa_r(\alpha) \neq \lambda(\alpha,\beta)$, for all β). So by (b) there is $B \subseteq A$, $B \neq \emptyset$ in $\Delta(y_\Gamma^0)$. Since Γ, therefore $\Delta(\alpha)$ for all α, is scaled and each $\Delta(\alpha)$ is closed under real quantification, it follows by the standard uniformization procedure that B contains a real in $\Delta(y_\Gamma^0)$, and we are done. ⊣

Before we choose a canonical representative of the first nontrivial $\Gamma(\alpha)$ singleton we shall need the following lemma. Our assumptions on Γ are as in 15.10.

15.11 <u>Lemma</u>. For any reals α,β

$$\alpha \leq_T \beta \Rightarrow y_\Gamma^\alpha \leq_\Delta y_\Gamma^\beta\ ,$$

where for each real α, y_Γ^α is any nontrivial $\Gamma(\alpha)$-singleton of least $\Delta(\alpha)$-degree, such that $\alpha \leq_\Delta y_\Gamma^\alpha$.

<u>Proof</u>. Assume $\alpha \leq_T \beta$. It is enough to show that $\lambda(y_\Gamma^\alpha) \leq \lambda(y_\Gamma^\beta)$. Because then if W is universal for Γ and $\varphi : W \twoheadrightarrow \kappa$ is a Γ-norm, find n such that

$$y = y_\Gamma^\alpha \Leftrightarrow (n,\langle\alpha,y\rangle) \in W\ .$$

Then $\varphi(n,\langle\alpha,y\rangle) < \lambda(y_\Gamma^\alpha) \leq \lambda(y_\Gamma^\beta) = \sup\{\varphi(m,x) : (m,x) \in W \wedge x \leq_\Delta y_\Gamma^\beta\}$. Then let $x \leq_\Delta y_\Gamma^\beta$ be such that $\varphi(n,\langle\alpha,y\rangle) \leq \varphi(m,x)$, for some m. Then

$$y = y_\Gamma^\alpha \Leftrightarrow \varphi(n,\langle\alpha,y\rangle) \leq \varphi(m,x)\ ,$$

so $y_\Gamma^\alpha \leq_\Delta \langle x,\alpha\rangle \leq y_\Gamma^\beta$ and we are done.

Let now for each real γ,

$$\bar{\kappa}_r(\gamma) = \min\{\lambda(\gamma,\delta) : \lambda(\gamma,\delta) > \kappa_r(\gamma)\} .$$

Then

$$\kappa_r(\gamma) = \sup\{\lambda(\gamma,\delta) : \lambda(\gamma,\delta) < \kappa_r(\gamma)\}$$

$$= \sup\{\lambda(\gamma,\delta) : \lambda(\gamma,\delta) < \bar{\kappa}_r(\gamma)\} ,$$

since $\lambda(\gamma,\delta) \neq \kappa_r(\gamma)$, for all δ, by Theorem 5.7 (vi) of [K3]. We claim now that

(*) $$\lambda(y_\Gamma^\alpha) = \bar{\kappa}_r(\alpha) .$$

Granting this we have, by Harrington's Theorem (see 5.10 of [K3]) which asserts that

$$\alpha \leq_T \beta \Rightarrow \kappa_r(\alpha) \leq \kappa_r(\beta) ,$$

that

$$\kappa_r(\alpha) \leq \kappa_r(\beta) < \bar{\kappa}_r(\beta) = \lambda(y_\Gamma^\beta) .$$

So $\kappa_r(\alpha) < \lambda(y_\Gamma^\beta) = \lambda(\alpha,y_\Gamma^\beta)$. Thus $\bar{\kappa}_r(\alpha) \leq \lambda(y_\Gamma^\beta)$, therefore $\lambda(y_\Gamma^\alpha) = \bar{\kappa}_r(\alpha) \leq \lambda(y_\Gamma^\beta)$ and we are done.

We prove now (*): For notational simplicity take the case $\alpha = \lambda t.0$, as the proof relativizes immediately. Let $\bar{\kappa}_r = \bar{\kappa}_r(\lambda t.0)$.

Put

$$\alpha \in S \Leftrightarrow \lambda(\alpha) < \kappa_r .$$

Then by Theorem 5.7 (iii) of [K3], $S \in \check{\Gamma}$, so if $W \subseteq \omega \times \omega^\omega$ is universal in Γ let $n_0 \in \omega$ be such that

$$\alpha \notin S \Leftrightarrow (n_0,\alpha) \in W .$$

Let $\{\varphi_m\}$ be an excellent Γ-scale on W and let $\{\psi_m\}$ be the corresponding scale it induces on $S' = \omega^\omega - S$. Let α_0 be the leftmost real associated with $S',\bar{\psi}$. Clearly α_0 is a Γ-singleton and $\lambda(\alpha_0) > \kappa_r$. We first show that $\lambda(\alpha_0) = \bar{\kappa}_r$. For that it is enough to prove that $\alpha_0 \leq_\Delta \alpha'$ for all $\alpha' \in S'$. But if $\alpha' \in S'$ then for each m, the initial segment of $\leq^{\psi_m}$ (where $x \leq^{\psi_m} y \Leftrightarrow x,y \in S' \wedge \psi_m(x) \leq \psi_m(y)$) determined by α' is in $\Delta(\alpha')$, uniformly on m, so $\alpha_0 \in \Delta(\alpha')$, since

$$\gamma = \alpha_0 \Leftrightarrow \forall m\, \forall \beta\ (\beta \leq^{\psi_m} \alpha' \Rightarrow \psi_m(\gamma) \leq \psi_m(\beta)) .$$

Finally we clain that $\alpha_0 \equiv_\Delta y^0_\Gamma$, which completes our proof. Clearly, since $\alpha_0 \notin \Delta$, $y^0_\Gamma \leq_\Delta \alpha_0$. But also $\lambda(y^0_\Gamma) > \kappa_r$ (as in the proof of 15.10), so $y^0_\Gamma \in S'$, thus by the above $\alpha_0 \leq_\Delta y^0_\Gamma$, and we are done. ⊣

Still under the assumptions of the preceding theorem let

$$P(\beta,\alpha) \Longleftrightarrow \beta \text{ is a } \Gamma(\alpha)\text{-singleton} \wedge \lambda(\alpha,\beta) > \kappa_r(\alpha) .$$

Then $P \in \Gamma$, because if $R(e,\beta,\alpha)$ is universal in Γ and $R^* \subseteq R$ in Γ uniformizes R on β, then

$$\beta \text{ is a } \Gamma(\alpha)\text{-singleton} \Longleftrightarrow \exists e\, R^*(e,\beta,\alpha) ,$$

while the relation

$$\lambda(\alpha,\beta) > \kappa_r(\alpha)$$

is in Γ, by Theorem 5.9 (due to Harrington) of [K3]. Let $P^*(\beta,\alpha)$ uniformize P on β and let $\{z_\alpha\} = \{\beta : P^*(\beta,\alpha)\}$.

We assume of course here that P^* is constructed from P by the canonical leftmost branch uniformization procedure from an excellent Γ-scale on P.

15.12 Definition. For each $\alpha \in \omega^\omega$ let y^α_Γ be the unique real y such that (assuming $x \longmapsto \{(x)_n : n \in \omega\}$ is a recursive 1-1 correspondence between ω^ω and $(\omega^\omega)^\omega$):

$$\begin{aligned}(y)_e &= \langle z_\beta,\beta\rangle, \text{ if } \{e\}^\alpha = \beta \\ &= \lambda t.0, \text{ if } \{e\}^\alpha \text{ is not total.}\end{aligned}$$

Again we have

15.13 Proposition. Assume Γ is a ${}_2\mathrm{ENV}({}^3E)$-like pointclass and Γ is Q-adequate. Then

(i) For each α, y^α_Γ is a (representative of the $\Delta(\alpha)$-degree of the) first nontrivial $\Gamma(\alpha)$-singleton. The relation

$$R(\alpha,\beta) \Longleftrightarrow \beta = y^\alpha_\Gamma$$

is in Γ.

(ii) For each α, $\alpha \leq_T y^\alpha_\Gamma$ and $\alpha \leq_T \beta \Rightarrow y^\alpha_\Gamma \leq_T y^\beta_\Gamma$ (uniformly).

Proof. (i) It is clear that $R(\alpha,\beta)$ is in Γ. Now let u_α be any real of the same $\Delta(\alpha)$-degree as the first nontrivial $\Gamma(\alpha)$-singleton. We will show that y^α_Γ has the same $\Delta(\alpha)$-degree as u_α. Clearly $u_\alpha \leq_\Delta y^\alpha_\Gamma$. By the proof of

Lemma 15.11 each z_α is a representative of the $\Delta(\alpha)$-degree of the least non-trivial $\Gamma(\alpha)$-singleton. So if $\beta \leq_T \alpha$ then $z_\beta \leq_\Delta z_\alpha \leq_\Delta u_\alpha$, by 15.11. Now we have

$$(y_\Gamma^\alpha)_e(n) = m \Longleftrightarrow (\{e\}^\alpha \text{ is not total} \wedge m = 0)$$
$$\vee\ [\{e\}^\alpha \text{ is total} \wedge$$
$$\exists z \in \Delta(u_\alpha)(z = z_{\{e\}^\alpha} \wedge \langle z, \{e\}^\alpha\rangle(n) = m)] \ .$$

So $y_\Gamma^\alpha \leq_\Delta u_\alpha$.

(ii) is obvious. ⊣

Now under the same assumptions the analog of 6.1 (iii) goes through, i.e.

$$\alpha \leq_\Gamma^Q \beta \Rightarrow y_\Gamma^\alpha \leq_\Delta y_\Gamma^\beta \ .$$

(*Proof*. Assume $\alpha \leq_\Gamma^Q \beta$. Since by 15.11 we have $y_\Gamma^\alpha \leq_\Delta y_\Gamma^{\langle\alpha,\beta\rangle}$ it is enough to show that $y_\Gamma^{\langle\alpha,\beta\rangle} \leq_\Delta y_\Gamma^\beta$. Since both $y_\Gamma^{\langle\alpha,\beta\rangle}$, y_Γ^β are $\Gamma(\langle\alpha,\beta\rangle)$-singletons, we have that if

$$y_\Gamma^{\langle\alpha,\beta\rangle} \not\leq_\Delta \langle y_\Gamma^\beta, \langle\alpha,\beta\rangle\rangle \ ,$$

then

$$y_\Gamma^\beta <_\Delta \langle y_\Gamma^{\langle\alpha,\beta\rangle}, \langle\alpha,\beta\rangle\rangle \equiv_T y_\Gamma^{\langle\alpha,\beta\rangle} \ ,$$

so y_Γ^β is a trivial $\Gamma(\langle\alpha,\beta\rangle)$-singleton i.e.

$$y_\Gamma^\beta \leq_\Delta \langle\alpha,\beta\rangle \ .$$

But $\langle\alpha,\beta\rangle \in Q_\Gamma(\beta)$, so $y_\Gamma^\beta \in Q_\Gamma(\beta)$, a contradiction. So we must have that

$$y_\Gamma^{\langle\alpha,\beta\rangle} \leq_\Delta \langle y_\Gamma^\beta, \langle\alpha,\beta\rangle\rangle \ .$$

But clearly

$$\langle\alpha,\beta\rangle \in Q_\Gamma(\beta) \Rightarrow \langle\alpha,\beta\rangle \leq_\Delta y_\Gamma^\beta \ ,$$

so

$$y_\Gamma^{\langle\alpha,\beta\rangle} \leq_\Delta y_\Gamma^\beta$$

and our proof is complete.)

15.14 Definition. For each Q_Γ-degree $q = [\alpha]_\Gamma^Q$, its jump q' is defined by

$$q' = [y_\Gamma^\alpha]_\Gamma^Q \ .$$

Finally we can prove the analog of Theorem 6.3.

15.15 Theorem. Assume Γ is $_2\mathrm{ENV}(^3E)$-like and Q-adequate. Then if we consider the prewellordering $\leq_\Delta$ on C_Γ, Q_Γ is a proper initial segment of it and y^0_Γ has minimal Δ-degree in $C_\Gamma - Q_\Gamma$.

Proof. Let $<$ be a Δ-good wellordering of C_Γ as in [K2]. Let α_0 be the $<$-least real in $C_\Gamma - Q_\Gamma$. (Clearly Q_Γ is a proper $\leq_\Delta$-initial segment of C_Γ, thus also a proper $<$-initial segment.) Put $Q^* = Q_\Gamma \cup \{\alpha_0\}$. Then $Q^* \in \Gamma$, since

$$\alpha \in Q^* \Longleftrightarrow \alpha \in C_\Gamma \wedge \forall\beta < \alpha \ (\beta \in Q_\Gamma) .$$

For any $B \in \check{\Gamma}$, $B \neq \emptyset$ put

$$\overline{B} = \{w \in WO : \exists\beta \in B\ (\beta \leq_T w \wedge |w| \geq \delta_\Gamma(\beta))\} .$$

Again $\overline{B} \in \check{\Gamma}$ and $\overline{B} \neq \emptyset$. If now $A \neq \emptyset$, $A \in \check{\Gamma}$ and $\mathrm{Hull}_\Delta(A) = Q_\Gamma$, we put

$$A' = \{\alpha : \alpha_0 \notin \Delta(\alpha) \wedge \alpha \in \overline{A}\} .$$

Since $\alpha_0 \notin \mathrm{Hull}_\Delta(\overline{A})$, we have $A' \neq \emptyset$. Also $A' \in \check{\Gamma}(\alpha_0)$.

Another basic property of $\kappa_r(\alpha)$ is the analog of the Gandy Basis Theorem (see [K3], 5.7 (iv)).

If $A \in \check{\Gamma}(\alpha)$ is nonempty, it contains a β with $\lambda(\alpha,\beta) < \kappa_r(\alpha)$ (which is equivalent to $\kappa_r(\alpha,\beta) = \kappa_r(\alpha)$).

So pick $w_0 \in A'$ with $\lambda(\alpha_0,w_0) < \kappa_r(\alpha_0)$.

For all $\alpha \in Q_\Gamma$ and all $\beta \in A$ we have $\alpha \leq_\Delta \beta$, but also by the above basis theorem A contains a β with $\kappa_r(\beta) = \kappa_r$ $(= \kappa_r(\lambda t.0))$. By a previously cited result of Harrington, if $\alpha \leq_\Delta \beta$, $\kappa_r(\alpha) \leq \kappa_r(\beta)$. So if $\alpha \in Q_\Gamma$ then $\kappa_r(\alpha) = \kappa_r$. Thus if $\kappa_r(\alpha_0) > \kappa_r$ we have

$$\alpha \in \{\alpha_0\} \Longleftrightarrow \alpha \in Q^* \wedge \kappa_r(\alpha) > \kappa_r \Longleftrightarrow \alpha \in Q^* \wedge \lambda(\alpha) > \kappa_r ,$$

i.e. α_0 is a Γ-singleton and we are done.

So we can assume that $\kappa_r(\alpha_0) = \kappa_r$, towards a contradiction. Since $w_0 \in \overline{A}$ let $\gamma_0 \in A$ be such that $\gamma_0 \leq_T w_0$ and $|w_0| \geq \delta_\Gamma(\gamma_0)$. Then if P^{γ_0} is a complete $\Gamma(\gamma_0)$ subset of ω, we have $P^{\gamma_0} \leq_\Delta w_0$. Let now $W \subseteq \omega \times \omega^\omega$ be universal in Γ and let $\varphi : W \twoheadrightarrow \kappa$ be a Γ-norm. Let $g(n,m)$ be a total recursive function such that

$$\alpha \in Q^* \wedge \alpha(n) = m \wedge \alpha \not\leq_T x \Longleftrightarrow (g(n,m),\langle x,\alpha\rangle) \in W .$$

Then we have the following

Claim. For all n,m

$$\alpha_0(n) = m \Longleftrightarrow \exists k\ [(k,P^{\gamma_0}) \in W \wedge \exists\alpha\ (\varphi(g(n,m),\langle P^{\gamma_0},\alpha\rangle) \leq \varphi(k,P^{\gamma_0}))].$$

Granting the claim it is clear that $\alpha_0 \leq_\Delta P^{\gamma_0}$ i.e. $\alpha_0 \leq_\Delta w_0$, which contradicts the fact that $w_0 \in A'$, thus $\alpha_0 \not\leq_\Delta w_0$, and completes the proof.

Proof of the claim: ($\Leftarrow$) Let $(k,P^{\gamma_0}) \in W$ be such that for some α $(g(n,m),\langle P^{\gamma_0},\alpha\rangle) \in W$, thus $\alpha \in Q^*$ and $\alpha(n) = m$ and $\alpha \not\leq_T P^{\gamma_0}$. If $\alpha \in Q_\Gamma$ then $\alpha \leq_\Delta \gamma_0$, so $\alpha \leq_T P^{\gamma_0}$. So $\alpha \notin Q_\Gamma$, i.e. $\alpha = \alpha_0$ and so $\alpha_0(n) = m$.

($\Rightarrow$) Assume $\alpha_0(n) = m$. Then $(g(n,m),\langle P^{\gamma_0},\alpha_0\rangle) \in W$, since $\alpha_0 \in Q^*$, $\alpha_0(n) = m$ and $\alpha_0 \not\leq_T P^{\gamma_0}$ (otherwise $\alpha_0 \leq_\Delta w_0$).

But also

$$\lambda(\alpha_0,P^{\gamma_0}) \leq \lambda(\alpha_0,w_0) < \kappa_r(\alpha_0) = \kappa_r \leq \kappa_r(P^{\gamma_0}) ,$$

since by [K3], 5.10, $\kappa_r(\alpha) \leq \kappa_r(\alpha,\beta)$, for all α,β. Thus, since $\varphi(g(n,m),\langle P^{\gamma_0},\alpha_0\rangle) < \lambda(\alpha_0,P^{\gamma_0})$, we have that $\varphi(g(n,m),\langle P^{\gamma_0},\alpha_0\rangle) < \kappa_r(P^{\gamma_0})$. Then by the definition of $\kappa_r(P^{\gamma_0})$, i.e. using reflection, we have

$$\exists\alpha\ (\varphi(g(n,m),\langle P^{\gamma_0},\alpha\rangle) < \kappa_r(P^{\gamma_0})) \Rightarrow$$

$$\exists\alpha\ (\varphi(g(n,m),\langle P^{\gamma_0},\alpha\rangle) < \lambda(P^{\gamma_0})) \Rightarrow$$

$$\exists\alpha\, \exists k\ ((k,P^{\gamma_0}) \in W \wedge \varphi(g(n,m),\langle P^{\gamma_0},\alpha\rangle) < \varphi(k,P^{\gamma_0})) ,$$

since $\lambda(\alpha) = \sup\{\varphi(k,\alpha) : (k,\alpha) \in W\}$, and the proof of the claim is complete. ⊣

Concerning the ordinal assignment for Q_Γ-degrees we have again the following

15.16 Definition. For each real α, let

$$\kappa_\Gamma(\alpha) = \sup\{\lambda(\langle\alpha,\beta\rangle) : y^\alpha \not\leq_\Delta \langle\alpha,\beta\rangle\} .$$

Then for Γ as above one can easily prove the analogs of 14.2, 14.3 (ii), 14.4, 14.5 so that for each Q_{2n+1}-degree $p = [\alpha]^Q$ we can define

$$\kappa_\Gamma(p) = \kappa_\Gamma(\alpha) ,$$

and the Spector Criterion goes through.

The reader has probably noticed at this stage some similarities between the properties of the ordinal assignments $\kappa_r(\alpha)$ and $\kappa_\Gamma(\alpha)$. We will conclude this section by proving that $\kappa_\Gamma(\alpha)$ equals $\kappa_r(\alpha)$, thereby establishing the coincidence of these two different notions from Q-theory and recursion in higher types respectively. From this one can quote known results in recursion in higher types (see [H1], [K3]) to establish the analogs of 14.6, 14.7 and 14.10. (The analog of 14.9 holds as well, as we pointed out in the proof of 15.15.)

15.17 Theorem. Assume Γ is ${}_2\mathrm{ENV}({}^3E)$-like and Q-adequate. Let $\kappa_r(\alpha)$ be the largest Γ,α-reflecting ordinal. Then

$$\kappa_\Gamma(\alpha) = \kappa_r(\alpha) \ .$$

Proof. We have seen in the proof of 15.11 that $\bar{\kappa}_r(\alpha) = \lambda(y_\Gamma^\alpha)$. From this we have

$$\begin{aligned}\kappa_\Gamma(\alpha) &= \sup\{\lambda(\alpha,\beta) : y_\Gamma^\alpha \not\leq_\Delta \langle\alpha,\beta\rangle\}\\ &= \sup\{\lambda(\alpha,\beta) : \lambda(\alpha,\beta) < \lambda(y_\Gamma^\alpha)\}\\ &= \sup\{\lambda(\alpha,\beta) : \lambda(\alpha,\beta) < \bar{\kappa}_r(\alpha)\}\\ &= \kappa_r(\alpha) \ ,\end{aligned}$$

and we are done. ⊣

§16. Q-theory for inductive sets. In this final section we study a limiting case of Q-theory, that of a Q-adequate pointclass Γ closed under $\exists^{\mathbb{R}}$. In this case as we pointed out in §15 we have $Q_\Gamma = C_\Gamma$ and there are no nontrivial Γ-singletons. The strong closure property of Γ however implies some very interesting new phenomena, which we shall explain below, after establishing some notation.

Let Γ be an ω-parametrized pointclass closed under $\wedge$, $\vee$, $\exists^{\mathbb{R}}$ and $\forall^{\mathbb{R}}$, and recursive substitution, like for example Γ = IND. We define the pointclasses $\Sigma_n^*(\Gamma)$, $\Pi_n^*(\Gamma)$, $\Delta_n^*(\Gamma)$ for $n \geq 1$ as follows:

$$\Sigma_1^*(\Gamma) = \exists^{\mathbb{R}}(\Gamma \wedge \check{\Gamma})$$

= The closure of boolean combinations of Γ pointsets under $\exists^{\mathbb{R}}$,

$$\Sigma_{n+1}^*(\Gamma) = \exists^{\mathbb{R}}\Pi_n^*(\Gamma), \quad \text{if } n \geq 1, \text{ where}$$

$$\Pi_n^*(\Gamma) = \check{\Sigma}_n^*(\Gamma),$$

$$\Delta_n^*(\Gamma) = \Sigma_n^*(\Gamma) \cap \Pi_n^*(\Gamma) \ .$$

Then it is easy to check that each $\Sigma_n^*(\Gamma)$ is ω-parametrized and closed under $\wedge$, $\vee$, $\exists^{\mathbb{R}}$, $\exists^\omega$, $\forall^\omega$ and recursive substitutions. Analogously for $\Pi_n^*(\Gamma)$. For

$\Gamma = \mathrm{IND}$ we abbreviate

$$\Sigma_n^* = \Sigma_n^*(\mathrm{IND}), \qquad \Pi_n^* = \Pi_n^*(\mathrm{IND}), \qquad \Delta_n^* = \Delta_n^*(\mathrm{IND}) .$$

We have now the following result.

16.1 <u>Theorem</u> (Martin [Ma3]). Assume Γ is Q-adequate and closed under $\exists^{\mathbb{R}}$. Then if $\bigcup_n \underset{\sim}{\Sigma}_n^*(\Gamma)$-DET holds, every $\Sigma_n^*(\Gamma)$ real belongs to C_Γ $(= Q_\Gamma)$.

In general C_Γ can contain much more complicated reals than those in $\Sigma_n^*(\Gamma)$. For instance assuming $AD^{L[\mathbb{R}]}$ and letting $\Gamma = (\Sigma_1^2)^{L[\mathbb{R}]}$, it follows from the results of [MMS], that

$$C_\Gamma = \{\alpha : \alpha \text{ is ordinal definable in } L[\mathbb{R}]\} .$$

However for the smallest Γ satisfying the hypotheses of the preceding result i.e. $\Gamma = \mathrm{IND}$, we have the following exact converse.

16.2 <u>Theorem</u> (Martin [Ma3]). Assume $\bigcup_n \underset{\sim}{\Sigma}_n^*$-DET. Then

$$C_{\mathrm{IND}} = \{\alpha : \alpha \text{ is } \Sigma_n^*, \text{ for some } n\} .$$

This result implies for instance that there is a cocountable $\check{\mathrm{IND}}$ set which contains no $\underset{\sim}{\Sigma}_n^*$ real, thus in general $\check{\mathrm{IND}}$ sets cannot be uniformized by $\underset{\sim}{\Sigma}_n^*$ sets (for any n).

References

[B] H. Becker, Partially playful universes, Cabal Seminar 76-77, A. S. Kechris and Y. N. Moschovakis (eds.), Lecture Notes in Mathematics, 689, Springer-Verlag, 1978, 55-90.

[DJ] A. Dodd and R. Jensen, The core model, Ann. Math. Logic, 20 (1981), 43-75.

[GH] D. Guaspari and L. A. Harrington, Characterizing C_3 (the largest countable Π^1_3 set), Proc. Amer. Math. Soc, 57 (1976), 127-129.

[H1] L. A. Harrington, Analytic determinacy and $0^{\#}$, J. Symb. Logic, 43 (1978), 685-693.

[H2] L. A. Harrington, Contributions to recursion theory on higher types, Ph.D. Thesis, M. I. T., 1973.

[HK] L. A. Harrington and A. S. Kechris, On the determinacy of games on ordinals, Ann. Math. Logic, 20 (1981), 109-154.

[J] T. Jech, Set Theory, Academic Press, 1978.

[K1] A. S. Kechris, Projective ordinals and countable analytical sets, Ph.D. Thesis, UCLA, 1972.

[K2] A. S. Kechris, The theory of countable analytical sets, Trans. Amer. Math. Soc., 202 (1975), 259-297.

[K3] A. S. Kechris, The structure of envelopes: A survey of recursion theory in higher types, M. I. T. Logic Seminar Notes, 1973.

[K4] A. S. Kechris, Countable ordinals and the analytical hierarchy I, Pacific J. Math., 60 (1975), 223-227.

[K5] A. S. Kechris, Recent advances in the theory of higher level projective sets, The Kleene Symposium, J. Barwise, H. J. Keisler and K. Kunen (eds.), North Holland, 1980, 149-166.

[KM1] A. S. Kechris and Y. N. Moschovakis, Notes on the theory of scales, Cabal Seminar 76-77, ibid., 1-54.

[KM2] A. S. Kechris and Y. N. Moschovakis, Recursion in higher types, Handbook of Mathematical Logic, J. Barwise (ed.), North Holland, 1977, 681-737.

[KS] A. S. Kechris and R. M. Solovay, On the relative consistency strength of determinacy hypothesis, to appear.

[KW] A. S. Kechris and W. H. Woodin, The equivalence of partition properties and determinacy, mimeographed notes, March 1982.

[Ku] K. Kunen, Some applications of iterated ultrapowers in set theory, Ann. Math. Logic, 1 (1970), 179-227.

[Ma1] D. A. Martin, Measurable cardinals and analytic games, Fund. Math., 66 (1970), 287-291.

[Ma2] D. A. Martin, Proof of a conjecture of Friedman, Proc. Amer. Math. Soc., 55 (1976), 129.

[Ma3] D. A. Martin, The largest countable this, that, and the other, this volume.

[Ma4] D. A. Martin, Borel and projective games, to appear.

[Ma5] D. A. Martin, Infinite games, Proc. Intern. Congress of Mathematicians, Helsinki 1978, 269-273.

[MMS] D. A. Martin, Y. N. Moschovakis and J. R. Steel, The extend of scales, Bull. Amer. Math. Soc., 6 (1982), 435-440.

[Mi1] W. Mitchell, Sets constructible from sequences of ultrafilters, J. Symb. Logic, 39 (1974), 57-66.

[Mi2] W. Mitchell, Hypermeasurable cardinals, Logic Colloquium '78, M. Boffa et al. (eds.), North-Holland, 1979, 303-316.

[Mi3] W. Mitchell, The core model for sequences of ultrafilters, to appear.

[Mo1] Y. N. Moschovakis, Hyperanalytic predicates, Trans. Amer. Math. Soc., 129 (1967), 249-282.

[Mo2] Y. N. Moschovakis, Inductive scales on inductive sets, Cabal Seminar 76-77, ibid., 185-192.

[Mo3] Y. N. Moschovakis, Descriptive Set Theory, North-Holland, 1980.

[So] R. M. Solovay, On the cardinality of Σ^1_2 sets of reals, Foundations of Mathematics, Bullof et al. (eds.), Springer-Verlag, 1969, 58-73.

[St1] J. R. Steel, A classification of jump operators, J. Symb. Logic, 47 (1982), 347-358.

[St2] J. R. Steel, Scales on Σ^1_1 sets, this volume.

ADDENDUM TO "INTRODUCTION TO Q-THEORY"

The open problem referred to in §11 has now been solved by M. Crawshaw, who proved Conjecture 11.2 (i.e. for all $n \geq 1$, $\underset{\sim}{\Delta}^1_{2n}$ - DET implies that $Y^{\alpha}_{2n+1} \equiv_{2n+2} y^{\alpha}_{2n+1}$ for every $\alpha \in \omega^{\omega}$). The proof will appear in his Caltech Ph.D. thesis.

APPENDIX: PROGRESS REPORT ON THE VICTORIA DELFINO PROBLEMS

The five Delfino problems were posed in [1]. Three of the five have now been solved. (See [5] for the solution of the second problem.) Partial results have been obtained on one of the remaining two problems.

Concerning the first problem, it was announced in [4] that Martin had shown $\underset{\sim}{\delta}^1_5 \geq \aleph_{\omega^3+1}$ and that the ultrapowers of $\underset{\sim}{\delta}^1_3$ with respect to the three normal measures on $\underset{\sim}{\delta}^1_3$ are $\aleph_{\omega+2}$, $\aleph_{\omega\cdot 2+1}$, and $\aleph_{\omega^2+1}$. The proof of part of the last assertion, that the ultrapower by the ω_2-cofinal measure is $\leq \aleph_{\omega^2+1}$, was incorrect. Actually this ultrapower is larger $(\aleph_{\omega^\omega+1})$.

Steve Jackson has completely solved the first problem. He first proved that $\underset{\sim}{\delta}^1_5 \leq \aleph_{\omega^{(\omega^\omega)}+1}$. This result will appear in his UCLA Ph.D. Thesis. He next used the machinery for getting this upper bound to analyze all measures on $\underset{\sim}{\delta}^1_3$ and to get a good representation of functions with respect to these measures. Martin observed that this representation and ideas of Kunen allow one to show $\underset{\sim}{\delta}^1_3 \rightarrow (\underset{\sim}{\delta}^1_3)^{\underset{\sim}{\delta}^1_3}$. From this it follows by a result of Martin that the ultrapowers of $\underset{\sim}{\delta}^1_3$ with respect to any of its measures is a cardinal. Jackson's analysis then gives $\underset{\sim}{\delta}^1_5 \geq \aleph_{\omega^{(\omega^\omega)}+1}$ so $\underset{\sim}{\delta}^1_5 = \aleph_{\omega^{(\omega^\omega)}+1}$.

The third problem was solved by Becker and Kechris who showed that $L[T^3]$ is independent of the choice of T^3. This is a consequence of the following fact, which is a theorem of ZF + DC.

THEOREM. Let Γ be an ω-parametrized pointclass closed under & and recursive substitution and containing all recursive sets. Let $P \subset \mathcal{N}$ be a complete Γ set, $\bar{\varphi} = \{\varphi_i\}_{i\in\omega}$ be an $\exists^{\mathcal{N}}\Gamma$-scale on P such that all norms φ_i are regular, and $\kappa = \sup\{\varphi_i(x) : i \in \omega, x \in P\}$. Let $T(\bar{\varphi})$ be the tree on $\omega \times \kappa$ associated with $\bar{\varphi}$. For any set $A \subset \kappa$, if A is $\exists^{\mathcal{N}}\Gamma$-in-the-codes with respect to $\bar{\varphi}$ (that is, if the set $\{(i,x) \in \omega \times \mathcal{N} : x \in P \,\&\, \varphi_i(x) \in A\}$ is $\exists^{\mathcal{N}}\Gamma$), then $A \in L[T(\bar{\varphi})]$.

In general, given two such scales $\bar{\varphi},\bar{\psi}$, it is not known that $T(\bar{\psi})$ is $\exists^{\mathcal{N}}\Gamma$-in-the-codes with respect to $\bar{\varphi}$, so the invariance of $L[T(\bar{\varphi})]$ has not been shown in this generality. However there are special cases where invariance can be proved. Henceforth, assume AD.

In Moschovakis [3], page 562, a model H_Γ is defined for every pointclass Γ which resembles Π^1_1; this includes the pointclasses Π^1_n for odd n. It follows from the theorem, together with known results about the H_Γ's ([3] 8G), that for any $\Gamma, P, \bar{\varphi}$ such that Γ resembles Π^1_1 and $\Gamma, P, \bar{\varphi}$ satisfy the theorem, $L[T(\bar{\varphi})] = H_\Gamma$, and hence $L[T(\bar{\varphi})]$ is independent of the choice of P and $\bar{\varphi}$. For $\Gamma = \Pi^1_3$ this solves the third problem.

While the invariance problem for $L[T^n]$ is thus solved for odd n, for even n the situation is still unclear. Call a Σ^1_3-scale on a Π^1_2 set <u>good</u> if it satisfies the ordinal quantification property of Kechris-Martin [2]. It follows from the above theorem that $L[T^2]$ is independent of the choice of a complete Π^1_2 set and of the choice of a good scale. Whether or not it is independent of the choice of an arbitrary scale is unknown. For even $n > 2$, it is not known whether there exist any good scales.

T. Slaman and J. Steel have proved two theorems relevant to Problem 5. The first verifies a special case of conjecture (a):

THEOREM. (ZF + AD + DC). Let $f : \mathcal{D} \to \mathcal{D}$ be such that $f(d) < d$ a.e.; then for some c, $f(d) = c$ a.e.

The second verifies a special case of conjecture (b). Call $f : \mathcal{D} \to \mathcal{D}$ order-pre preserving a.e. iff $\exists c\, \forall a,b \geq c\ (a \leq b \Rightarrow f(a) \leq f(b))$.

THEOREM (ZF + AD + DC) Let $f : \mathcal{D} \to \mathcal{D}$ be order-preserving a.e. and such that $d < f(d)$ a.e. Then either

(i) $\exists \alpha < \omega_1\ (f(d) = d^\alpha$ a.e.),

or

(ii) for a.e. d, $\forall \alpha < \omega_1^d\ (f(d) > d^\alpha)$.

(Here ω_1 is the least uncountable ordinal, and ω_1^d is the least d-admissible ordinal greater than ω.)

REFERENCES

[1] APPENDIX: The Victoria Delfino problems, in Cabal Seminar 76-77, Lecture Notes in Mathematics #689, Springer 1978.

[2] A.S. Kechris and D.A. Martin, On the theory of Π^1_3 sets of reals, Bull. Amer. Math. Soc. (1978), 149-151.

[3] Y.N. Moschovakis, Descriptive Set Theory, North-Holland 1980.

[4] APPENDIX: Progress report on the Victoria Delfino Problems, in Cabal Seminar 77-79, Lecture notes in Mathematics #839, Springer 1981, 273-274.

[5] Y.N. Moschovakis, Scales on coinductive sets, this volume.

Vol. 873: Constructive Mathematics, Proceedings, 1980. Edited by F. Richman. VII, 347 pages. 1981.

Vol. 874: Abelian Group Theory. Proceedings, 1981. Edited by R. Göbel and E. Walker. XXI, 447 pages. 1981.

Vol. 875: H. Zieschang, Finite Groups of Mapping Classes of Surfaces. VIII, 340 pages. 1981.

Vol. 876: J. P. Bickel, N. El Karoui and M. Yor. Ecole d'Eté de Probabilités de Saint-Flour IX – 1979. Edited by P. L. Hennequin. XI, 280 pages. 1981.

Vol. 877: J. Erven, B.-J. Falkowski, Low Order Cohomology and Applications. VI, 126 pages. 1981.

Vol. 878: Numerical Solution of Nonlinear Equations. Proceedings, 1980. Edited by E. L. Allgower, K. Glashoff, and H.-O. Peitgen. XIV, 440 pages. 1981.

Vol. 879: V. V. Sazonov, Normal Approximation – Some Recent Advances. VII, 105 pages. 1981.

Vol. 880: Non Commutative Harmonic Analysis and Lie Groups. Proceedings, 1980. Edited by J. Carmona and M. Vergne. IV, 553 pages. 1981.

Vol. 881: R. Lutz, M. Goze, Nonstandard Analysis. XIV, 261 pages. 1981.

Vol. 882: Integral Representations and Applications. Proceedings, 1980. Edited by K. Roggenkamp. XII, 479 pages. 1981.

Vol. 883: Cylindric Set Algebras. By L. Henkin, J. D. Monk, A. Tarski, H. Andréka, and I. Németi. VII, 323 pages. 1981.

Vol. 884: Combinatorial Mathematics VIII. Proceedings, 1980. Edited by K. L. McAvaney. XIII, 359 pages. 1981.

Vol. 885: Combinatorics and Graph Theory. Edited by S. B. Rao. Proceedings, 1980. VII, 500 pages. 1981.

Vol. 886: Fixed Point Theory. Proceedings, 1980. Edited by E. Fadell and G. Fournier. XII, 511 pages. 1981.

Vol. 887: F. van Oystaeyen, A. Verschoren, Non-commutative Algebraic Geometry, VI, 404 pages. 1981.

Vol. 888: Padé Approximation and its Applications. Proceedings, 1980. Edited by M. G. de Bruin and H. van Rossum. VI, 383 pages. 1981.

Vol. 889: J. Bourgain, New Classes of $\mathcal{L}^p$-Spaces. V, 143 pages. 1981.

Vol. 890: Model Theory and Arithmetic. Proceedings, 1979/80. Edited by C. Berline, K. McAloon, and J.-P. Ressayre. VI, 306 pages. 1981.

Vol. 891: Logic Symposia, Hakone, 1979, 1980. Proceedings, 1979, 1980. Edited by G. H. Müller, G. Takeuti, and T. Tugué. XI, 394 pages. 1981.

Vol. 892: H. Cajar, Billingsley Dimension in Probability Spaces. III, 106 pages. 1981.

Vol. 893: Geometries and Groups. Proceedings. Edited by M. Aigner and D. Jungnickel. X, 250 pages. 1981.

Vol. 894: Geometry Symposium. Utrecht 1980, Proceedings. Edited by E. Looijenga, D. Siersma, and F. Takens. V, 153 pages. 1981.

Vol. 895: J.A. Hillman, Alexander Ideals of Links. V, 178 pages. 1981.

Vol. 896: B. Angéniol, Familles de Cycles Algébriques – Schéma de Chow. VI, 140 pages. 1981.

Vol. 897: W. Buchholz, S. Feferman, W. Pohlers, W. Sieg, Iterated Inductive Definitions and Subsystems of Analysis: Recent Proof-Theoretical Studies. V, 383 pages. 1981.

Vol. 898: Dynamical Systems and Turbulence, Warwick, 1980. Proceedings. Edited by D. Rand and L.-S. Young. VI, 390 pages. 1981.

Vol. 899: Analytic Number Theory. Proceedings, 1980. Edited by M.I. Knopp. X, 478 pages. 1981.

Vol. 900: P. Deligne, J. S. Milne, A. Ogus, and K.-Y. Shih, Hodge Cycles, Motives, and Shimura Varieties. V, 414 pages. 1982.

Vol. 901: Séminaire Bourbaki vol. 1980/81 Exposés 561–578. III, 299 pages. 1981.

Vol. 902: F. Dumortier, P.R. Rodrigues, and R. Roussarie, Germs of Diffeomorphisms in the Plane. IV, 197 pages. 1981.

Vol. 903: Representations of Algebras. Proceedings, 1980. Edited by M. Auslander and E. Lluis. XV, 371 pages. 1981.

Vol. 904: K. Donner, Extension of Positive Operators and Korovkin Theorems. XII, 182 pages. 1982.

Vol. 905: Differential Geometric Methods in Mathematical Physics. Proceedings, 1980. Edited by H.-D. Doebner, S.J. Andersson, and H.R. Petry. VI, 309 pages. 1982.

Vol. 906: Séminaire de Théorie du Potentiel, Paris, No. 6. Proceedings. Edité par F. Hirsch et G. Mokobodzki. IV, 328 pages. 1982.

Vol. 907: P. Schenzel, Dualisierende Komplexe in der lokalen Algebra und Buchsbaum-Ringe. VII, 161 Seiten. 1982.

Vol. 908: Harmonic Analysis. Proceedings, 1981. Edited by F. Ricci and G. Weiss. V, 325 pages. 1982.

Vol. 909: Numerical Analysis. Proceedings, 1981. Edited by J.P. Hennart. VII, 247 pages. 1982.

Vol. 910: S.S. Abhyankar, Weighted Expansions for Canonical Desingularization. VII, 236 pages. 1982.

Vol. 911: O.G. Jørsboe, L. Mejlbro, The Carleson-Hunt Theorem on Fourier Series. IV, 123 pages. 1982.

Vol. 912: Numerical Analysis. Proceedings, 1981. Edited by G. A. Watson. XIII, 245 pages. 1982.

Vol. 913: O. Tammi, Extremum Problems for Bounded Univalent Functions II. VI, 168 pages. 1982.

Vol. 914: M. L. Warshauer, The Witt Group of Degree k Maps and Asymmetric Inner Product Spaces. IV, 269 pages. 1982.

Vol. 915: Categorical Aspects of Topology and Analysis. Proceedings, 1981. Edited by B. Banaschewski. XI, 385 pages. 1982.

Vol. 916: K.-U. Grusa, Zweidimensionale, interpolierende Lg-Splines und ihre Anwendungen. VIII, 238 Seiten. 1982.

Vol. 917: Brauer Groups in Ring Theory and Algebraic Geometry. Proceedings, 1981. Edited by F. van Oystaeyen and A. Verschoren. VIII, 300 pages. 1982.

Vol. 918: Z. Semadeni, Schauder Bases in Banach Spaces of Continuous Functions. V, 136 pages. 1982.

Vol. 919: Séminaire Pierre Lelong – Henri Skoda (Analyse) Années 1980/81 et Colloque de Wimereux, Mai 1981. Proceedings. Edité par P. Lelong et H. Skoda. VII, 383 pages. 1982.

Vol. 920: Séminaire de Probabilités XVI, 1980/81. Proceedings. Edité par J. Azéma et M. Yor. V, 622 pages. 1982.

Vol. 921: Séminaire de Probabilités XVI, 1980/81. Supplément: Géométrie Différentielle Stochastique. Proceedings. Edité par J. Azéma et M. Yor. III, 285 pages. 1982.

Vol. 922: B. Dacorogna, Weak Continuity and Weak Lower Semicontinuity of Non-Linear Functionals. V, 120 pages. 1982.

Vol. 923: Functional Analysis in Markov Processes. Proceedings, 1981. Edited by M. Fukushima. V, 307 pages. 1982.

Vol. 924: Séminaire d'Algèbre Paul Dubreil et Marie-Paule Malliavin. Proceedings, 1981. Edité par M.-P. Malliavin. V, 461 pages. 1982.

Vol. 925: The Riemann Problem, Complete Integrability and Arithmetic Applications. Proceedings, 1979-1980. Edited by D. Chudnovsky and G. Chudnovsky. VI, 373 pages. 1982.

Vol. 926: Geometric Techniques in Gauge Theories. Proceedings, 1981. Edited by R. Martini and E.M.de Jager. IX, 219 pages. 1982.

Vol. 927: Y. Z. Flicker, The Trace Formula and Base Change for GL (3). XII, 204 pages. 1982.

Vol. 928: Probability Measures on Groups. Proceedings 1981. Edited by H. Heyer. X, 477 pages. 1982.

Vol. 929: Ecole d'Eté de Probabilités de Saint-Flour X – 1980. Proceedings, 1980. Edited by P.L. Hennequin. X, 313 pages. 1982.

Vol. 930: P. Berthelot, L. Breen, et W. Messing, Théorie de Dieudonné Cristalline II. XI, 261 pages. 1982.

Vol. 931: D.M. Arnold, Finite Rank Torsion Free Abelian Groups and Rings. VII, 191 pages. 1982.

Vol. 932: Analytic Theory of Continued Fractions. Proceedings, 1981. Edited by W.B. Jones, W.J. Thron, and H. Waadeland. VI, 240 pages. 1982.

Vol. 933: Lie Algebras and Related Topics. Proceedings, 1981. Edited by D. Winter. VI, 236 pages. 1982.

Vol. 934: M. Sakai, Quadrature Domains. IV, 133 pages. 1982.

Vol. 935: R. Sot, Simple Morphisms in Algebraic Geometry. IV, 146 pages. 1982.

Vol. 936: S.M. Khaleelulla, Counterexamples in Topological Vector Spaces. XXI, 179 pages. 1982.

Vol. 937: E. Combet, Intégrales Exponentielles. VIII, 114 pages. 1982.

Vol. 938: Number Theory. Proceedings, 1981. Edited by K. Alladi. IX, 177 pages. 1982.

Vol. 939: Martingale Theory in Harmonic Analysis and Banach Spaces. Proceedings, 1981. Edited by J.-A. Chao and W.A. Woyczyński. VIII, 225 pages. 1982.

Vol. 940: S. Shelah, Proper Forcing. XXIX, 496 pages. 1982.

Vol. 941: A. Legrand, Homotopie des Espaces de Sections. VII, 132 pages. 1982.

Vol. 942: Theory and Applications of Singular Perturbations. Proceedings, 1981. Edited by W. Eckhaus and E.M. de Jager. V, 363 pages. 1982.

Vol. 943: V. Ancona, G. Tomassini, Modifications Analytiques. IV, 120 pages. 1982.

Vol. 944: Representations of Algebras. Workshop Proceedings, 1980. Edited by M. Auslander and E. Lluis. V, 258 pages. 1982.

Vol. 945: Measure Theory. Oberwolfach 1981, Proceedings. Edited by D. Kölzow and D. Maharam-Stone. XV, 431 pages. 1982.

Vol. 946: N. Spaltenstein, Classes Unipotentes et Sous-groupes de Borel. IX, 259 pages. 1982.

Vol. 947: Algebraic Threefolds. Proceedings, 1981. Edited by A. Conte. VII, 315 pages. 1982.

Vol. 948: Functional Analysis. Proceedings, 1981. Edited by D. Butković, H. Kraljević, and S. Kurepa. X, 239 pages. 1982.

Vol. 949: Harmonic Maps. Proceedings, 1980. Edited by R.J. Knill, M. Kalka and H.C.J. Sealey. V, 158 pages. 1982.

Vol. 950: Complex Analysis. Proceedings, 1980. Edited by J. Eells. IV, 428 pages. 1982.

Vol. 951: Advances in Non-Commutative Ring Theory. Proceedings, 1981. Edited by P.J. Fleury. V, 142 pages. 1982.

Vol. 952: Combinatorial Mathematics IX. Proceedings, 1981. Edited by E. Billington, S. Oates-Williams, and A.P. Street. XI, 443 pages. 1982.

Vol. 953: Iterative Solution of Nonlinear Systems of Equations. Proceedings, 1982. Edited by R. Ansorge, Th. Meis, and W. Törnig. VII, 202 pages. 1982.

Vol. 954: S.G. Pandit, S.G. Deo, Differential Systems Involving Impulses. VII, 102 pages. 1982.

Vol. 955: G. Gierz, Bundles of Topological Vector Spaces and Their Duality. IV, 296 pages. 1982.

Vol. 956: Group Actions and Vector Fields. Proceedings, 1981. Edited by J.B. Carrell. V, 144 pages. 1982.

Vol. 957: Differential Equations. Proceedings, 1981. Edited by D.G. de Figueiredo. VIII, 301 pages. 1982.

Vol. 958: F.R. Beyl, J. Tappe, Group Extensions, Representations, and the Schur Multiplicator. IV, 278 pages. 1982.

Vol. 959: Géométrie Algébrique Réelle et Formes Quadratiques, Proceedings, 1981. Edité par J.-L. Colliot-Thélène, M. Coste, L. Mahé, et M.-F. Roy. X, 458 pages. 1982.

Vol. 960: Multigrid Methods. Proceedings, 1981. Edited by W. Hackbusch and U. Trottenberg. VII, 652 pages. 1982.

Vol. 961: Algebraic Geometry. Proceedings, 1981. Edited by J.M. Aroca, R. Buchweitz, M. Giusti, and M. Merle. X, 500 pages. 1982.

Vol. 962: Category Theory. Proceedings, 1981. Edited by K.H. Kamps, D. Pumplün, and W. Tholen, XV, 322 pages. 1982.

Vol. 963: R. Nottrot, Optimal Processes on Manifolds. VI, 124 pages. 1982.

Vol. 964: Ordinary and Partial Differential Equations. Proceedings, 1982. Edited by W.N. Everitt and B.D. Sleeman. XVIII, 726 pages. 1982.

Vol. 965: Topics in Numerical Analysis. Proceedings, 1981. Edited by P.R. Turner. IX, 202 pages. 1982.

Vol. 966: Algebraic K-Theory. Proceedings, 1980, Part I. Edited by R.K. Dennis. VIII, 407 pages. 1982.

Vol. 967: Algebraic K-Theory. Proceedings, 1980. Part II. VIII, 409 pages. 1982.

Vol. 968: Numerical Integration of Differential Equations and Large Linear Systems. Proceedings, 1980. Edited by J. Hinze. VI, 412 pages. 1982.

Vol. 969: Combinatorial Theory. Proceedings, 1982. Edited by D. Jungnickel and K. Vedder. V, 326 pages. 1982.

Vol. 970: Twistor Geometry and Non-Linear Systems. Proceedings, 1980. Edited by H.-D. Doebner and T.D. Palev. V, 216 pages. 1982.

Vol. 971: Kleinian Groups and Related Topics. Proceedings, 1981. Edited by D.M. Gallo and R.M. Porter. V, 117 pages. 1983.

Vol. 972: Nonlinear Filtering and Stochastic Control. Proceedings, 1981. Edited by S.K. Mitter and A. Moro. VIII, 297 pages. 1983.

Vol. 973: Matrix Pencils. Proceedings, 1982. Edited by B. Kågström and A. Ruhe. XI, 293 pages. 1983.

Vol. 974: A. Draux, Polynômes Orthogonaux Formels – Applications. VI, 625 pages. 1983.

Vol. 975: Radical Banach Algebras and Automatic Continuity. Proceedings, 1981. Edited by J.M. Bachar, W.G. Bade, P.C. Curtis Jr., H.G. Dales and M.P. Thomas. VIII, 470 pages. 1983.

Vol. 976: X. Fernique, P.W. Millar, D.W. Stroock, M. Weber, Ecole d'Eté de Probabilités de Saint-Flour XI – 1981. Edited by P.L. Hennequin. XI, 465 pages. 1983.

Vol. 977: T. Parthasarathy, On Global Univalence Theorems. VIII, 106 pages. 1983.

Vol. 978: J. Ławrynowicz, J. Krzyż, Quasiconformal Mappings in the Plane. VI, 177 pages. 1983.

Vol. 979: Mathematical Theories of Optimization. Proceedings, 1981. Edited by J.P. Cecconi and T. Zolezzi. V, 268 pages. 1983.

Vol. 980: L. Breen. Fonctions thêta et théorème du cube. XIII, 115 pages. 1983.

Vol. 981: Value Distribution Theory. Proceedings, 1981. Edited by I. Laine and S. Rickman. VIII, 245 pages. 1983.

Vol. 982: Stability Problems for Stochastic Models. Proceedings, 1982. Edited by V.V. Kalashnikov and V.M. Zolotarev. XVII, 295 pages. 1983.

DATE DUE

GAYLORD			PRINTED IN U.S.A.